Advanced Techniques in Logic Synthesis, Optimizations and Applications

Sunil P. Khatri · Kanupriya Gulati
Editors

Advanced Techniques in Logic Synthesis, Optimizations and Applications

Springer

Editors
Sunil P. Khatri
Department of ECE
333F WERC, MS 3259
Texas A&M University
College Station, TX 77843-3259,
USA
sunilkhatri@tamu.edu

Kanupriya Gulati
Intel Corporation
2501 NW 229th Ave
Hillsboro, OR 97124,
USA
kanupriya.gulati@intel.com

ISBN 978-1-4419-7517-1 e-ISBN 978-1-4419-7518-8
DOI 10.1007/978-1-4419-7518-8
Springer New York Dordrecht Heidelberg London

Printed on acid-free paper

Springer is part of Springer Science+Business Media (www.springer.com)

Preface

The last few decades have seen a stupendous growth in the speed and complexity of VLSI integrated circuits. This growth has been enabled by a powerful set of electronic design automation (EDA) tools. The earliest EDA tools were two-level logic minimization and PLA folding tools. Subsequently, EDA tools were developed to address other aspects of the VLSI design flow (in addition to logic optimization) such as technology mapping, layout optimization, formal verification. However, research in logic synthesis and optimization continued to progress rapidly. Some of the research in logic synthesis tools saw broader application, to areas far removed from traditional EDA, and routinely continue to do so. While observing the recent developments and publications in logic synthesis and optimization, we felt that there was a need for a single resource which presents some recent significant developments in this area. This is how the idea of this edited monograph came about. We decided to cover some key papers in logic synthesis, optimization, and its applications, in an effort to provide an advanced practitioner a single reference source that covers the important papers in these areas over the last few years.

This monograph is organized into five sections, dealing with logic decomposition, Boolean satisfiability, Boolean matching, logic optimization, and applications of logic techniques to special design scenarios. Each of the chapters in any section is an expanded, archival version of the original paper by the chapter authors, with additional examples, results, and/or implementation details.

We dedicate this book to the area of logic synthesis and hope that it can stimulate new and exciting ideas which expand the contribution of logic synthesis to areas far beyond its traditional stronghold of VLSI integrated circuit design.

College Station, Texas — Sunil P. Khatri
Hillsboro, Oregon — Kanupriya Gulati

Contents

Contributors

Kia Bazargan University of Minnesota, Minneapolis, MN USA, kia@umn.edu

Anna Bernasconi Department of Computer Science, Universit‘a di Pisa, Pisa, Italy, annab@di.unipi.it

Kalyana C. Bollapalli NVIDIA Corporation, San Jose, CA, USA, kbollapalli@nvidia.com

Robert Brayton University of California, Berkeley, CA, USA, brayton@eecs.berkeley.edu

Chung-Ming Chan Fu Jen Catholic University, Taipei County, Taiwan, vicax95@csie.fju.edu.tw

Satrajit Chatterjee Strategic CAD Labs, Intel Corporation, Hillsboro, OR, USA, satrajit.chatterjee@intel.com

Valentina Ciriani Department of Information Technologies, Universit‘a degli Studi di Milano, Milano, Italy, valentina.ciriani@unimi.it

Jeff L. Cobb Texas Instruments, Sugar Land, TX USA, jcobb@ti.com

Xin Fan Innovations for High Performance Microelectronics, Frankfurt (Oder), Brandenburg, Germany, fan@ihp-microelectronics.com

Lei Fang Microsoft Corporation, Redmond, WA, USA, lei.fang@microsoft.com

Eugene Goldberg Northeastern University, Boston, MA, USA, eigold@ccs.neu.edu

Sivaram Gopalakrishnan Synopsys Inc., Hillsboro, OR, USA, sivaram.gopalakrishnan@synopsys.com

Eckhard Grass Innovations for High Performance Microelectronics, Frankfurt (Oder), Brandenburg, Germany, grass@ihp-microelectronics.com

Kanupriya Gulati Intel Corporation, Hillsboro, OR, USA, kanupriya.gulati@intel.com

Michael S. Hsiao Virginia Tech, Blacksburg, VA, USA, mhsiao@vt.edu

Wei-Lun Hung National Taiwan University, Taipei, Taiwan, b91076@csie.ntu.edu.tw

Jie-Hong Roland Jiang National Taiwan University, Taipei, Taiwan, jhjiang@cc.ee.ntu.edu.tw

Priyank Kalla University of Utah, Salt Lake City, UT, USA, kalla@ece.utah.edu

Hadi Katebi University of Michigan, Ann Arbor, MI, USA, hadik@eecs.umich.edu

Sunil P. Khatri Department of ECE, Texas A&M University, College Station, TX, USA, sunilkhatri@tamu.edu

Laszlo B. Kish Department of ECE, Texas A&M University, College Station, TX, USA, Laszlo.Kish@ece.tamu.edu

Nathan B. Kitchen University of Berkeley, Berkeley, CA, USA, nbk@eecs.berkeley.edu

Victor N. Kravets IBM TJ Watson Research Center, Yorktown Heights, NY, USA, kravets@us.ibm.com

Smita Krishnaswamy IBM TJ Watson Research Center, Yorktown Heights, NY, USA, skrishn@us.ibm.com

Miloš Krstić Innovations for High Performance Microelectronics, Frankfurt (Oder), Brandenburg, Germany, krstic@ihp-microelectronics.com

Andreas Kuehlmann Cadence Design Systems, Inc., San Jose, CA, USA, kuehl@cadence.com

Ruei-Rung Lee National Taiwan University, Taipei, Taiwan, r95943156@ntu.edu.tw

David J. Lilja University of Minnesota, Minneapolis, MN USA, lilja@umn.edu

Hsuan-Po Lin National Taiwan University, Taipei, Taiwan, centau-ricog@hotmail.com

Jung-Chang Liu Fu Jen Catholic University, Taipei County, Taiwan, binize97@csie.fju.edu.tw

Panagiotis Manolios Northeastern University, Boston, MA, USA, bjchamb@ccs.neu.edu

Igor Markov University of Michigan, Ann Arbor, MI, USA, imarkov@eecs.umich.edu

Joao Marques-Sila University College Dublin, Dublin, Ireland, jpms@ucd.ie

Alan Mishchenko Department of EECS, University of California, Berkeley, CA, USA, alanmi@eecs.berkeley.edu

Nilesh Modi IBM TJ Watson Research Center, Yorktown Heights, NY, USA, nilesh@ece.ucsb.edu

Nasir Mohyuddin Department of Electrical Engineering – Systems, University of Southern California, Los Angeles, CA, USA, mohyuddi@usc.edu

Ehsan Pakbaznia Department of Electrical Engineering – Systems, University of Southern California, Los Angeles, CA, USA, pakbazni@usc.edu

Massoud Pedram Department of Electrical Engineering – Systems, University of Southern California, Los Angeles, CA, USA, pedram@usc.edu

Jordi Planes University de Lleida, Lleida, Spain, jplanes@diei.udl.cat

Ruchir Puri IBM TJ Watson Research Center, Yorktown Heights, NY, USA, ruchir@us.ibm.com

Weikang Qian University of Minnesota, Minneapolis, MN, USA, qianx030@umn.edu

Haoxing Ren IBM TJ Watson Research Center, Yorktown Heights, NY, USA, haoxing@us.ibm.com

Marc D. Riedel University of Minnesota, Minneapolis, MN, USA, mriedel@umn.edu

Alberto Sangiovanni-Vincentelli University of Berkeley, Berkeley, CA, USA, alberto@eecs.berkeley.edu

Subarna Sinha Synopsys Inc., Mountain View, CA, USA, subarna.sinha@synopsys.com

Duncan Smith Vennsa Technologies, Inc., Toronto, ON, Canada, duncan.smith@vennsa.com

Gabriella Trucco Department of Information Technologies, Universit'a degli Studi di Milano, Milano, Italy, gabriella.trucco@unimi.it

Andreas Veneris University of Toronto, Toronto, ON, Canada, veneris@eecg.utoronto.ca

Tiziano Villa Department of Computer Science, Universit'a degli Studi di Verona, Verona, Italy, tiziano.villa@univr.it

Kuo-Hua Wang Fu Jen Catholic University, Taipei County, Taiwan, khwang@csie.fju.edu.tw

Yu-Shen Yang University of Toronto, Ontario, Canada, terry.yang@utoronto.ca

Qi Zhu Intel Corporation, Hillsboro, OR, USA, qi.dts.zhu@intel.com

Chapter 1
Introduction

Sunil P. Khatri and Kanupriya Gulati

With the advances in VLSI technology, enhanced optimization techniques are required to enable the design of faster electronic circuits that consume less power and occupy a smaller area. In the VLSI design cycle, significant optimization opportunities exist in the logic design stage. In recent times, several research works have been proposed in the area of logic synthesis, which can prove to be very valuable for VLSI/CAD engineers. A solid understanding and sound implementation of these advanced techniques would enable higher levels of optimization, and thus enable better electronic design. This text is a systematic collection of important recent work in the field of logic design and optimization.

Conventional logic synthesis consists of the following phases. Given an initial netlist, technology-independent optimizations [1, 3, 4] are first carried out in order to optimize various design criteria such as gate, literal, and net count. Both Boolean and algebraic techniques are used, such as kernel and cube extraction, factorization, node substitution, don't care-based optimizations [2]. During the logic decomposition phase, large gates are decomposed into smaller gates, which allows for efficient technology mapping and technology-dependent optimizations. Finally, technology mapping is applied on the decomposed netlist which is followed by technology-dependent optimizations. This book presents recent research in some of the above areas.

In order to enhance the scalability and performance of logic synthesis approaches, newer optimization styles are continually investigated. Boolean satisfiability (SAT) plays a big role in some of the recent logic optimization methodologies. Therefore, this edited volume also includes some of the latest research in SAT techniques. Further, several non-CAD systems can be viewed as an instance of logic optimization, and thus these too can take advantage of the rich body of recent research in logic synthesis and optimization. Such non-traditional applications of logic synthesis to specialized design scenarios are also included in this volume.

S.P. Khatri (✉)
Department of ECE, Texas A&M University, College Station, TX, USA
e-mail: sunilkhatri@tamu.edu

S.P. Khatri, K. Gulati (eds.), *Advanced Techniques in Logic Synthesis, Optimizations and Applications*, DOI 10.1007/978-1-4419-7518-8_1,

The approaches described in this text are enhanced and archival versions of the corresponding original conference publications. The modifications include enhancements to the original approach, more experimental data, additional background and implementation details, along with as yet unpublished graphs and figures.

The different sections of this volume are described next.

1.1 Logic Decomposition

This section discusses the latest research in logic decomposition. The first chapter investigates restructuring techniques based on decomposition and factorization. In this chapter the authors describe new types of factorization that extend Shannon cofactoring, using projection functions that change the Hamming distance of the original minterms, to favor logic minimization of the component blocks.

The next chapter uses reachable state analysis and symbolic decomposition to improve upon the synthesis of sequential designs. The approach described uses under-approximation of unreachable states of a design to derive incomplete specification of the combinational logic. The resulting incompletely specified functions are decomposed to optimize technology-dependent synthesis. The decomposition choices are implicitly computed by using recursive symbolic bi-decomposition.

The third chapter in the topic of Boolean decomposition employs fast Boolean techniques to restructure logic networks. The techniques used are a cut-based view of a logic network, and heuristic disjoint-support decompositions. Local transformations to functions with a small number of inputs allow fast manipulations of truth tables. The use of Boolean methods reduces the structural bias associated with algebraic methods, while still allowing for high-speed.

The fourth chapter investigates Ashenhurst decomposition wherein both single and multiple output decomposition can be formulated with satisfiability solving, Craig interpolation, and functional dependency. In comparison to existing BDD-based approaches for functional decomposition, Ashenhurst decomposition does not suffer from memory explosion and scalability issues. A key feature of this approach is that variable partitioning can be automated and integrated into the decomposition process without the bound-set size restriction. Further, the approach naturally extends to nondisjoint decomposition.

The last chapter in the Boolean decomposition section focuses on scalability and quality of Boolean function bi-decomposition. The quality of a bi-decomposition is mainly determined by its variable partition. Disjoint and balanced decompositions reduce communication and circuit complexity and yield simple physical design solutions. Furthermore, finding a good or feasible partition may require costly enumeration, requiring separate decomposability checks. This chapter uses interpolation and incremental SAT solving to address these problems.

1.2 Boolean Satisfiability

In the area of Boolean satisfiability, this book presents some key ideas to make SAT more effective. The first chapter studies resolution proofs using boundary points elimination. Given a CNF formula F, boundary points are complete assignments that falsify only certain clauses of the formula. Since any resolution proof has to eventually eliminate all boundary points of F, this approach focuses on resolution proofs from the viewpoint of boundary point elimination. The authors use equivalence checking formulas to compare unsatisfiability proofs built by a conflict-driven SAT-solver. They show how every resolution of a specialized proof eliminates a boundary point, and how this enables building resolution SAT-solvers that are driven by elimination of cut boundary points.

The next chapter presents a methodology called SAT sweeping for simplifying And-Inverter Graphs (AIGs) by systematically merging graph vertices in a topological fashion starting from the inputs, using a combination of structural hashing, simulation, and SAT queries. This chapter presents the details of a SAT-sweeping approach that exploits local observability don't cares (ODCs) to increase the number of vertices merged. In order to enhance the efficiency and scalability of the approach, the authors bound the ODCs and thus the computational effort to generate them. They demonstrate that the use of ODCs in SAT sweeping results in significant graph simplification, with great benefits for Boolean reasoning in functional verification and logic synthesis techniques.

SAT-solvers find a satisfiable assignment for a propositional formula, but finding the "optimal" solution for a given function is very expensive. The next chapter discusses MIN-ONE SAT, an optimization problem which requires the satisfying assignment with the minimal number of ones, which can be easily applied to minimize an arbitrary linear objective function. The chapter proposes an approximation algorithm for MIN-ONE SAT that is efficient and achieves a tight bound on the solution quality. RelaxSAT generates a set of constraints from the objective function to guide the search, and then these constraints are gradually relaxed to eliminate the conflicts with the original Boolean SAT formula until a solution is found. The experiments demonstrate that RelaxSAT is able to handle very large instances which cannot be solved by existing MIN-ONE algorithms. The authors further show that RelaxSAT is able to obtain a very tight bound on the solution with one to two orders of magnitude speedup.

The last chapter in the Boolean satisfiability category presents an algorithm for MaxSAT that improves existing state-of-the-art solvers by orders of magnitude on industrial benchmarks. The proposed algorithm is based on efficient identification of unsatisfiable subformulas. Moreover, the new algorithm draws a connection between unsatisfiable subformulas and the maximum satisfiability problem.

1.3 Boolean Matching

Three research works are presented under the Boolean matching category. The first work proposes a methodology for Boolean matching under permutations of inputs and outputs that enables incremental logic design by identifying sections of netlist that are unaffected by incremental changes in design specifications. Identifying and reusing the equivalent subcircuits accelerates design closure. By integrating graph-based, simulation-driven, and SAT-based techniques, this methodology makes Boolean matching feasible for large designs.

The second approach in the Boolean matching category is DeltaSyn, a tool and methodology for generating the logic difference between a modified high-level specification and an implemented design. By using fast functional and structural analysis techniques, the approach first identifies equivalent signals between the original and the modified circuits. Then, by using a topologically guided dynamic matching algorithm, reusable portions of logic close to the primary outputs are identified. Finally, functional hash functions are employed to locate similar chunks of logic throughout the remainder of the circuit. Experiments on industrial designs show that together, these techniques successfully implement incremental changes while preserving an average of 97% of the pre-existing logic.

The last approach discussed in the Boolean matching section proposes an incremental learning-based algorithm, along with a Boolean satisfiability search, for solving Boolean matching. The proposed algorithm utilizes functional properties like unateness and symmetry to reduce the search space. This is followed by the simulation phase in which three types of input vector generation and checking methods are used to match the inputs of two target functions. Experimental results on large benchmark circuits demonstrate that the matching algorithm can efficiently solve the Boolean matching for large Boolean networks.

1.4 Logic Optimization

The first advanced logic optimization approach presents algebraic techniques to enhance common sub-expression extraction to allow circuit optimization. Common sub-expression elimination (CSE) is a useful optimization technique in the synthesis of arithmetic datapaths described at the RTL level.

The next chapter investigates a comprehensive methodology to automate logic restructuring in combinational and sequential circuits. This technique algorithmically constructs the required transformation by utilizing Set of Pairs of Function to be Distinguished (SPFDs). SPFDs can express more functional flexibility than traditional don't cares and have been shown to provide additional degrees of flexibility during logic synthesis. In practice, however, computing SPFDs may suffer from memory or runtime problems. This approach presents Approximate SPFDs (ASPFDs) that approximate the information contained in SPFDs by using the results of test-vector simulation, thereby yielding an efficient and robust optimization platform.

The third chapter presents an approach to enhance the determinization of a Boolean relation by using interpolation. Boolean relations encapsulate the flexibility of a design and are therefore an important tool in system synthesis, optimization, and verification to characterize solutions to a set of Boolean constraints. For physical realization, a deterministic function often has to be extracted from a relation. Existing methods are limited in their handling of large problem instances. Experimental results for the interpolation-based relation determinization approach show that Boolean relations with thousands of variables can be effectively determinized and the extracted functions are of reasonable quality.

In this section, the fourth approach presented is a scalable approach for dual-node technology-independent optimization. This technique scales well and can minimize large designs typical of industrial circuits. The methodology presented first selects the node pairs to be minimized that are likely to give gains. For each node pair, a window or a subnetwork is created around the nodes. This windowing is done in order to allow the approach to scale to larger designs. Once the subnetwork is created, the Boolean relation, which represents the flexibility of the nodes, is computed. During this process, early quantification is performed which further extends the scalability of the approach. The Boolean relation is minimized, and the new nodes replace the original nodes in the original circuit. These steps are repeated for all selected node pairs. The authors experimentally demonstrate that this technique produces minimized technology-independent networks that are on average 12% smaller than networks produced by state-of-the-art single-node minimization techniques.

1.5 Applications to Specialized Design Scenarios

This volume presents applications of logic synthesis in non-traditional CAD areas. The first approach investigates techniques for synthesizing logic that generates new arbitrary probabilities from a given small set of probabilities. These ideas can be used in probabilistic algorithms. Instead of using different voltage levels to generate different probability values, which can be very expensive, the technique presented in the chapter alleviates this issue by generating probabilities using combinational logic.

The next chapter presents a gate-level probabilistic error propagation model which takes as input the Boolean function of the gate, the signal and error probabilities of the gate inputs, and the gate error probability and produces the error probability at the output of the gate. The presented model uses Boolean difference calculus and can be applied to the problem of calculating the error probability at the primary outputs of a multilevel Boolean circuit. The time complexity of the approach is linear in the number of gates in the circuit, and the results demonstrate the accuracy and efficiency of the approach compared to the other known methods for error calculation in VLSI circuits.

In the third chapter in the applications category, a novel realization of combinational logic circuit is presented. In this approach, logic values 0 and 1 are implemented as sinusoidal signals of the same frequency that are phase shifted by π. The properties of such sinusoids can be used to identify a logic value without ambiguity, and hence a realizable system of logic is created. The chapter further presents a family of logic gates that can operate using such sinusoidal signals. In addition, due to orthogonality of sinusoid signals with different frequencies, multiple sinusoids could be transmitted on a single wire simultaneously, thereby naturally allowing the approach to implement multilevel logic. One advantage of such a logic family is its immunity from external additive noise and an improvement in switching (dynamic) power.

The last chapter focuses on asynchronous circuit design issues. In Systems-on-a-Chip (SOCs) and Networks-on-a-Chip (NoCs), using globally asynchronous locally synchronous (GALS) system design for pausible clocking is widely popular. This chapter investigates throughput reduction and synchronization failures introduced by existing GALS pausible clocking schemes and proposes an optimized scheme for more reliable GALS system design with higher performance. The approach minimizes the acknowledge latency and maximizes the safe timing region for inserting the clock tree.

References

1. Brayton, R.K., Hachtel, G.D., Sangiovanni-Vincentelli, A.L.: Multilevel logic synthesis. In: Proceedings of IEEE, 78(2):264–270 (1990)
2. Hassoun, S. (ed.): Logic Synthesis and Verification. San Jose, CA, USA (2001)
3. Sinha, S., Brayton, R.K.: Implementation and use of SPFDs in optimizing Boolean networks. In: Proceedings of International Conference on Computer-Aided Design, pp. 103–110. Paris, France (1998)
4. Wurth, B., Wehn, N.: Efficient calculation of Boolean relations for multi-level logic optimization. In: Proceedings of European Design and Test Conference, pp. 630–634. (1994)

Part I
Logic Decomposition

Under logic decomposition this book presents five research works. The first chapter proposes hypergraph partitioning and Shannon decomposition-based techniques for logic decomposition. The second chapter uses reachable state analysis and symbolic decomposition to improve upon the synthesis of sequential designs. Fast Boolean decomposition techniques employing a cut-based view of a logic network and heuristic disjoint-support decompositions are presented in the third work. The fourth approach performs Ashenhurst decomposition formulated using satisfiability, Craigs interpolation, and functional dependency. This last chapter in this category uses interpolation and incremental SAT solving to improve the quality of Boolean function decomposition.

Chapter 2
Logic Synthesis by Signal-Driven Decomposition

Anna Bernasconi, Valentina Ciriani, Gabriella Trucco, and Tiziano Villa

Abstract This chapter investigates some restructuring techniques based on decomposition and factorization, with the objective to move critical signals toward the output while minimizing area. A specific application is synthesis for minimum switching activity (or high performance), with minimum area penalty, where decompositions with respect to specific critical variables are needed (the ones of highest switching activity, for example). In order to reduce the power consumption of the circuit, the number of gates that are affected by the switching activity of critical signals is maintained constant. This chapter describes new types of factorization that extend Shannon cofactoring and are based on projection functions that change the Hamming distance among the original minterms to favor logic minimization of the component blocks. Moreover, the proposed algorithms generate and exploit don't care conditions in order to further minimize the final circuit. The related implementations, called P-circuits, show experimentally promising results in area with respect to classical Shannon cofactoring.

2.1 Introduction

In recent years, power has become an important factor during the design phase. This trend is primarily due to the remarkable growth of personal computing devices, embedded systems, and wireless communications systems that demand high-speed computation and complex functionality with low power consumption. In these applications, average power consumption is a critical design concern.

Low-power design methodologies must consider power at all stages of the design process. At the logic synthesis level, logic transformations proved to be an effective technique to reduce power consumption by restructuring a mapped circuit through permissible signal substitution or perturbation [1]. A fundamental step in VLSI design is logic synthesis of high-quality circuits matching a given specification. The

A. Bernasconi (✉)
Department of Computer Science, Università di Pisa, Pisa, Italy
e-mail: annab@di.unipi.it

Based on [5], pp.1464–1469, 20–24 April 2009 © [2009] IEEE.

S.P. Khatri, K. Gulati (eds.), *Advanced Techniques in Logic Synthesis, Optimizations and Applications*, DOI 10.1007/978-1-4419-7518-8_2,

performance of the circuit can be expressed in terms of several factors, such as area, delay, power consumption, and testability properties. Unfortunately, these factors often contradict each other, in the sense that it is very difficult to design circuits that guarantee very good performances with respect to all of them. In fact, power consumption is often studied as a single minimization objective without taking into account important factors such as area and delay.

In CMOS technology, power consumption is characterized by three components: dynamic, short-circuit, and leakage power dissipation, of which dynamic power dissipation is the predominant one. Dynamic power dissipation is due to the charge and discharge of load capacitances, when the logic value of a gate output toggles; switching a gate may trigger a sequence of signal changes in the gates of its output cone, increasing dynamic power dissipation. So, reducing switching activity reduces dynamic power consumption. Previous work proposed various transformations to decrease power consumption and delay (for instance [11, 14, 16] for performance and [1, 13, 15] for low power), whereby the circuit is restructured in various ways, e.g., redeploying signals to avoid critical areas, bypassing large portions of a circuit. For instance, if we know the switching frequency of the input signals, a viable strategy to reduce dynamic power is to move the signals with the highest switching frequency closer to the outputs, in order to reduce the part of the circuit affected by the switching activity of these signals. Similarly for performance, late-arriving signals are moved closer to the outputs to decrease the worst-case delay.

The aim of our research is a systematic investigation of restructuring techniques based on decomposition/factorization, with the objective to move critical signals toward the output and avoid losses in area. A specific application is synthesis for minimum switching activity (or high performance), with minimum area penalty. Differently from factorization algorithms developed only for area minimization, we look for decompositions with respect to specific critical variables (the ones of highest switching activity, for example). This is exactly obtained by Shannon cofactoring, which decomposes a Boolean function with respect to a chosen splitting variable; however, when applying Shannon cofactoring, the drawback is that too much area redundancy might be introduced because large cubes are split between two disjoint subspaces, whereas no new cube merging will take place as the Hamming distance among the projected minterms do not change.

In this chapter we investigate thoroughly the more general factorization introduced in [5], a decomposition that extends straightforward Shannon cofactoring; instead of cofactoring a function f only with respect to single variables as Shannon does, we cofactor with respect to more complex functions, expanding f with respect to the orthogonal basis $\overline{x}_i \oplus p$ (i.e., $x_i = p$) and $x_i \oplus p$ (i.e., $x_i \neq p$), where $p(x)$ is a function defined over all variables except x_i. We study different functions $p(x)$ trading-off quality vs. computation time. Our factorizations modify the Hamming distance among the on-set minterms, so that more logic minimization may be performed on the projection of f onto the two disjoint subspaces $x_i = p$ and $x_i \neq p$, while signals are moved in the circuit closer to the output. We then introduce and study another form of decomposition, called *decomposition with intersection*, where a function f is projected onto three overlapping subspaces of the Boolean space $\{0, 1\}^n$ in order to favor area minimization avoiding cube fragmentation (e.g., cube splitting for the cubes intersecting both subspaces $x_i = p$ and $x_i \neq p$). More precisely, we partition the on-set minterms of f into three sets: $f|_{x_i=p}$ and $f|_{x_i \neq p}$, representing the projections of f onto the two disjoint subspaces $x_i = p$ and $x_i \neq p$, and a third set $I = f|_{x_i=p} \cap f|_{x_i \neq p}$, which contains all minterms of f whose projections onto $x_i = p$ and $x_i \neq p$ are identical. Observe that each point in I corresponds to two different points of f that could be merged in a cube, but are split into the two spaces $x_i = p$ and $x_i \neq p$. Thus, we can avoid cube fragmentation keeping the points in I unprojected. Moreover, given that the points in the intersection I must be covered, we can project them as *don't cares* in the two spaces $f|_{x_i=p}$ and $f|_{x_i \neq p}$ to ease the minimization of $f|_{x_i=p} \setminus I$ and $f|_{x_i \neq p} \setminus I$. Observe that, while classical don't care sets are specified by the user or are derived from the surrounding environment, our don't cares are dynamically constructed during the synthesis phase.

The circuits synthesized according to these decompositions are called *Projected Circuits*, or *P-circuits*, without and with intersection. We provide minimization algorithms to compute optimal P-circuits and argue how augmenting P-circuits with at most a pair of multiplexers guarantees full testability under the single stuck-at-fault model. We also show that the proposed decomposition technique can be extended and applied to move all critical signals, and not just one, toward the output, still avoiding losses in area.

The chapter is organized as follows. Section 2.2 describes the new theory of decomposition based on generalized cofactoring, which is applied in Section 2.3 to the synthesis of Boolean functions as *P-circuits*. Section 2.4 extends the decomposition from single to multiple variables. Experiments and conclusions are reported in Sections 2.5 and 2.6, respectively.

2.2 Decomposition Methods

How to decompose Boolean functions is an ongoing research area to explore alternative logic implementations. A technique to decompose Boolean functions is based on expanding them according to an orthogonal basis (see, for example [8], section 3.15), as in the following definition, where a function f is decomposed according to the basis $(g, \overline{g})$.

Definition 2.1 Let $f = (f_{\text{on}}, f_{\text{dc}}, f_{\text{off}})$ be an incompletely specified function and g be a completely specified function, the *generalized cofactor* of f with respect to g is the incompletely specified function $\text{co}(f, g) = (f_{\text{on}} \cdot g, f_{\text{dc}} + \overline{g}, f_{\text{off}} \cdot g)$.

This definition highlights that in expanding a Boolean function we have two degrees of freedom: choosing the basis (in this case, the function g) and choosing one completely specified function included in the incompletely specified function $\text{co}(f, g)$. This flexibility can be exploited according to the purpose of the expansion. For instance, when $g = x_i$, we have $\text{co}(f, x_i) = (f_{\text{on}} \cdot x_i, f_{\text{dc}} + \overline{x}_i, f_{\text{off}} \cdot x_i)$. Notice that the well-known Shannon cofactor $f_{x_i} = f(x_1, \ldots, (x_i = 1), \ldots, x_n)$ is a completely specified function contained in $\mathrm{co}(f, x_i) = (f_{\mathrm{on}} \cdot x_i, f_{\mathrm{dc}} + \overline{x}_i, f_{\mathrm{off}} \cdot x_i)$ (since $f_{\mathrm{on}} \cdot x_i \subseteq f_{x_i} \subseteq f_{\mathrm{on}} \cdot x_i + f_{\mathrm{dc}} + \overline{x}_i = f_{\mathrm{on}} + f_{\mathrm{dc}} + \overline{x}_i$); moreover, f_{x_i} is the unique cover of $\mathrm{co}(f, x_i)$ independent from the variable x_i.

We introduce now two types of expansion of a Boolean function that yield decompositions with respect to a chosen variable (as in Shannon cofactoring), but are also area-efficient because they favor minimization of the logic blocks so obtained. Let $f(X) = (f_{\mathrm{on}}(X), f_{\mathrm{dc}}(X), f_{\mathrm{off}}(X))$ be an incompletely specified function depending on the set $X = \{x_1, x_2, \ldots, x_n\}$ of n binary variables. Let $X^{(i)}$ be the subset of X containing all variables but x_i, i.e., $X^{(i)} = X \setminus \{x_i\}$, where $x_i \in X$. Consider now a completely specified Boolean function $p(X^{(i)})$ depending only on the variables in $X^{(i)}$. We introduce two decomposition techniques based on the projections of the function f onto two complementary subsets of the Boolean space $\{0, 1\}^n$ defined by the function p. More precisely, we note that the space $\{0, 1\}^n$ can be partitioned into two sets: one containing the points for which $x_i = p(X^{(i)})$ and the other containing the points for which $x_i \neq p(X^{(i)})$. Observe that the characteristic functions of these two subsets are $(\overline{x}_i \oplus p)$ and $(x_i \oplus p)$, respectively, and that these two sets have equal cardinality. We denote by $f|_{x_i=p}$ and $f|_{x_i \neq p}$ the projections of the points of $f(X)$ onto the two subsets where $x_i = p(X^{(i)})$ and $x_i \neq p(X^{(i)})$, respectively. Note that these two functions only depend on the variables in $X^{(i)}$. The first decomposition technique, already described in [12] and [6], is defined as follows.

Definition 2.2 Let $f(X)$ be an incompletely specified function, $x_i \in X$, and $p(X^{(i)})$ be a completely specified function. The (x_i, p)-*decomposition* of f is the algebraic expression

$$f = (\overline{x}_i \oplus p) f|_{x_i=p} + (x_i \oplus p) f|_{x_i \neq p}.$$

First of all we observe that each minterm of f is projected onto one and only one subset. Indeed, let $m = m_1 m_2 \cdots m_n$ be a minterm of f; if $m_i = p(m_1, \ldots, m_{i-1}, m_{i+1}, \ldots, m_n)$, then m is projected onto the set where $x_i = p(X^{(i)})$, otherwise m is projected onto the complementary set where $x_i \neq p(X^{(i)})$. The projection simply consists in eliminating m_i from m. For example, consider the function f shown on the left side of Fig. 2.1 with $f_{\mathrm{on}} = \{0000, 0001, 0010, 0101, 1001, 1010, 1100, 1101\}$ and $f_{\mathrm{dc}} = \{0111\}$. Let p be the simple Boolean function x_2, and x_i be x_1. The Boolean space $\{0, 1\}^4$ can be partitioned into the two sets $x_1 = x_2$ and $x_1 \neq x_2$ each containing 2^3 points. The projections of f onto these two sets are $f_{\mathrm{on}}|_{x_1=x_2} = \{000, 001, 010, 100, 101\}$, $f_{\mathrm{dc}}|_{x_1=x_2} = \emptyset$, and $f_{\mathrm{on}}|_{x_1 \neq x_2} = \{101, 001, 010\}$, $f_{\mathrm{dc}}|_{x_1 \neq x_2} = \{111\}$.

Second, observe that these projections do not preserve the Hamming distance among minterms, since we eliminate the variable x_i from each minterm, and two minterms projected onto the same subset could have different values for x_i. The Hamming distance is preserved only if the function $p(X^{(i)})$ is a constant, that is when the (x_i, p)-decomposition corresponds to the classical Shannon decomposition. The fact that the Hamming distance may change could be useful when f is

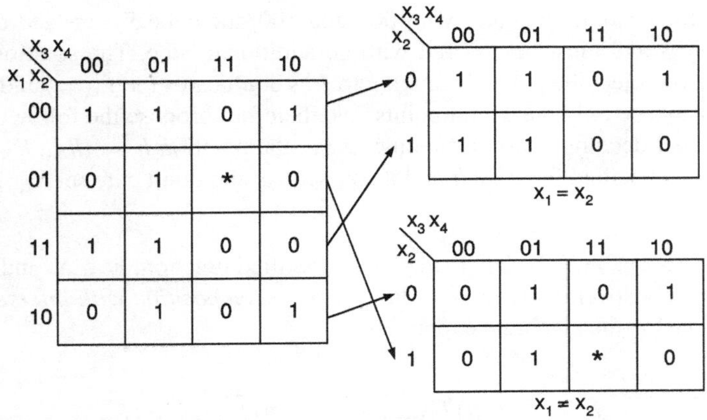

Fig. 2.1 An example of projection of the incompletely specified function f onto the spaces $x_1 = x_2$ and $x_1 \neq x_2$

represented in SOP form, as bigger cubes could be built in the projection sets. For example, consider again the function f shown on the left side of Fig. 2.1. The points 0000 and 1100 contained in f_{on} have Hamming distance equal to 2, and thus cannot be merged in a cube, while their projections onto the space $f_{\text{on}}|_{x_1=x_2}$ (i.e., 000 and 100, respectively) have Hamming distance equal to 1, and they form the cube $\overline{x}_3\overline{x}_4$.

On the other hand, the cubes intersecting both subsets $x_i = p(X^{(i)})$ and $x_i \neq p(X^{(i)})$ are divided into two smaller subcubes. For instance, in our running example, the cube $\overline{x}_3 x_4$ of function f_{on} is split into the two sets $x_1 = x_2$ and $x_1 \neq x_2$ forming a cube in $f_{\text{on}}|_{x_1=x_2}$ and one in $f_{\text{on}}|_{x_1 \neq x_2}$, as shown on the right side of Fig. 2.1.

Observe that the cubes that end up to be split may contain pairs of minterms, whose projections onto the two sets are identical. In our example, $\overline{x}_3 x_4$ is the cube corresponding to the points $\{0001, 0101, 1001, 1101\}$, where 0001 and 1101 are projected onto $f_{\text{on}}|_{x_1=x_2}$ and become 001 and 101, respectively, and 0101 and 1001 are projected onto $f_{\text{on}}|_{x_1 \neq x_2}$ and again become 101 and 001, respectively. Therefore, we can characterize the set of these minterms as $I = f|_{x_i=p} \cap f|_{x_i \neq p}$. Note that the points in I do not depend on x_i. In our example $I_{\text{on}} = f_{\text{on}}|_{x_1=x_2} \cap f_{\text{on}}|_{x_1 \neq x_2} = \{001, 010, 101\}$, and $I_{\text{dc}} = \emptyset$.

In order to overcome the splitting of some cubes, we could keep I unprojected and project only the points in $f|_{x_i=p} \setminus I$ and $f|_{x_i \neq p} \setminus I$, obtaining the expression $f = (\overline{x}_i \oplus p)(f|_{x_i=p} \setminus I) + (x_i \oplus p)(f|_{x_i \neq p} \setminus I) + I$.

However, we are left with another possible drawback: some points of I could also belong to cubes covering points of $f|_{x_i=p}$ and/or $f|_{x_i \neq p}$, and their elimination could cause the fragmentation of these cubes. Thus, eliminating these points from the projected subfunctions would not be always convenient. On the other hand, some points of I are covered only by cubes entirely contained in I. Therefore keeping them both in I and in the projected subfunctions would be useless and expensive. In our example, since $I_{\text{on}} = \{001, 010, 101\}$, in $f_{\text{on}}|_{x_1=x_2}$ the points 001 and 101 are useful for forming, together with 000 and 100, the cube $\overline{x}_3$; instead the point 010 is useless and must be covered with an additional cube. The solution to this problem is to project the points belonging to I as don't cares for $f|_{x_i=p}$ and $f|_{x_i\neq p}$, in order to choose only the useful points. We therefore propose the following more refined second decomposition technique, using the notation $h = (h_{\text{on}}, h_{\text{dc}})$ for an incompletely specified function h and its on-set h_{on} and don't care set h_{dc}.

Definition 2.3 Let $f(X)$ be an incompletely specified function, $x_i \in X$, and $p(X^{(i)})$ be a completely specified function. The (x_i, p)*-decomposition with intersection* of $f = (f_{\text{on}}, f_{\text{dc}})$ is the algebraic expression

$$f = (\overline{x}_i \oplus p)\tilde{f}|_{x_i=p} + (x_i \oplus p)\tilde{f}|_{x_i\neq p} + I,$$

where

$$\tilde{f}|_{x_i=p} = (f_{\text{on}}|_{x_i=p} \setminus I_{\text{on}}, f_{\text{dc}}|_{x_i=p} \cup I_{\text{on}}),$$
$$\tilde{f}|_{x_i\neq p} = (f_{\text{on}}|_{x_i\neq p} \setminus I_{\text{on}}, f_{\text{dc}}|_{x_i\neq p} \cup I_{\text{on}}),$$
$$I = (I_{\text{on}}, I_{\text{dc}}),$$

with $I_{\text{on}} = f_{\text{on}}|_{x_i=p} \cap f_{\text{on}}|_{x_i\neq p}$ and $I_{\text{dc}} = f_{\text{dc}}|_{x_i=p} \cap f_{\text{dc}}|_{x_i\neq p}$.

For our example, the projections of f become $\tilde{f}|_{x_1=x_2} = (f_{\text{on}}|_{x_1=x_2} \setminus I_{\text{on}}, f_{\text{dc}}|_{x_1=x_2} \cup I_{\text{on}}) = (\{000, 100\}, \{001, 010, 101\})$ and $\tilde{f}|_{x_1\neq x_2} = (f_{\text{on}}|_{x_1\neq x_2} \setminus I_{\text{on}}, f_{\text{dc}}|_{x_1\neq x_2} \cup I_{\text{on}}) = (\emptyset, \{111\} \cup \{001, 010, 101\})$. The Karnaugh maps of this decomposition are shown in Fig. 2.2.

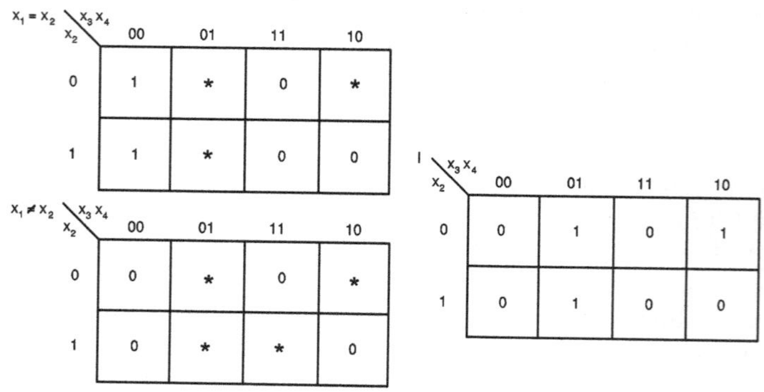

Fig. 2.2 An example of projection *with intersection* of the function f of Fig. 2.1 onto the spaces $x_1 = x_2$, $x_1 \neq x_2$, and I

Observe that, fixing the function p and a variable x, these decompositions are canonical. We now study these decomposition methods for some choices of the function p.

Case $p = 0$.

As we have already observed, if p is a constant function, then the (x_i, p)-decomposition is indeed the classical Shannon decomposition: $f = \overline{x}_i f|_{x_i=0} + x_i f|_{x_i=1}$. Recall that $(\overline{x}_i \oplus 0)$ is equivalent to $\overline{x}_i$, while $(x_i \oplus 0)$ is equivalent to x_i. Also observe that choosing $p = 1$ we would get exactly the same form. For the (x_i, p)-decomposition with intersection we have the following particular form:

$$f = \overline{x}_i \tilde{f}|_{x_i=0} + x_i \tilde{f}|_{x_i=1} + I.$$

Observe that in this particular case, the set I is

$$I = f(x_1, \ldots, x_{i-1}, 0, x_{i+1}, \ldots, x_n) \cap f(x_1, \ldots, x_{i-1}, 1, x_{i+1}, \ldots, x_n).$$

This implies the following property.

Proposition 2.1 *The characteristic function χ_I of I is the biggest subfunction of f that does not depend on x_i.*

Proof Let $\chi_1, \ldots, \chi_k$ be the subfunctions of f that do not depend on x_i, and let χ be their union, i.e., $\chi = \chi_1 + \chi_2 + \cdots + \chi_k$. Observe that χ is still a subfunction of f and it does not depend on x_i. Therefore χ is the biggest subfunction that does not depend on x_i. We must show that $\chi = \chi_I$. First note that χ_I is one of the functions $\chi_1, \ldots, \chi_k$. Suppose $\chi_I = \chi_j$, with $1 \leq j \leq k$. By construction, χ_j is a subfunction of χ. On the other hand, if $\chi(X) = 1$, then there exists an index h such that $\chi_h(X) = 1$. Since χ_h does not depend on x_i, we have

$$\chi_h(x_1, \ldots, x_{i-1}, 1, x_{i+1}, \ldots, x_n) = \chi_h(x_1, \ldots, x_{i-1}, 0, x_{i+1}, \ldots, x_n) = 1.$$

Moreover, since χ_h is a subfunction of f, on the same input X we have that

$$f(x_1, \ldots, x_{i-1}, 1, x_{i+1}, \ldots, x_n) = f(x_1, \ldots, x_{i-1}, 0, x_{i+1}, \ldots, x_n) = 1.$$

This implies that

$$\chi_j = f(x_1, \ldots, x_{i-1}, 1, x_{i+1}, \ldots, x_n) \cap f(x_1, \ldots, x_{i-1}, 0, x_{i+1}, \ldots, x_n) = 1,$$

which means that χ is a subfunction of χ_j. As $\chi_j = \chi_I$, we finally have that $\chi = \chi_I$.

Note that if χ_I is equal to f, then f does not depend on x_i. We conclude the analysis of this special case observing how the $(x_i, 0)$-decomposition, i.e., the classical Shannon decomposition, and the $(x_i, 0)$-decomposition with intersection show a different behavior when the subfunctions $f|_{x_i=0}$, $f|_{x_i=1}$, $\tilde{f}|_{x_i=0}$, $\tilde{f}|_{x_i=1}$ and the intersection I are represented as sum of products. Consider a minimal sum of products *SOP*(f) for the function f. The number of products in *SOP*(f) is always less than or equal to the overall number of products in the minimal SOP representations for $f|_{x_i=0}$ and $f|_{x_i=1}$. This easily follows from the fact that each product in *SOP*(f) that does not depend on x_i is split into two products, one belonging to a minimal SOP for $f|_{x_i=0}$ and the other belonging to a minimal SOP for $f|_{x_i=1}$. On the other hand, the $(x_i, 0)$-decomposition with intersection contains the same number of products as *SOP*(f), and its overall number of literals is less or equal to the number of literals in *SOP*(f).

Theorem 2.1 *An $(x_i, 0)$-decomposition with intersection for a Boolean function f, where $\tilde{f}|_{x_i=0}$, $\tilde{f}|_{x_i=1}$, and I are represented as minimal sums of products, contains an overall number of products equal to the number of products in a minimal SOP for f and an overall number of literals less or equal to the number of literals in a minimal SOP for f.*

Proof First observe how we can build minimal SOP representations for $\tilde{f}|_{x_i=0}$, $\tilde{f}|_{x_i=1}$, and I starting from a minimal SOP, *SOP*(f), for f. Indeed, the sum of the projections of all products in *SOP*(f) containing the literal x_i gives a minimal SOP for $\tilde{f}|_{x_i=1}$, the sum of the projections of all products in $SOP(f)$ containing the literal $\overline{x}_i$ gives a minimal SOP for $\tilde{f}|_{x_i=0}$, while all remaining products, that do not depend on x_i or $\overline{x}_i$, give a minimal SOP covering exactly the points in the intersection I. The minimality of these SOPs follows from the fact that the $(x_i, 0)$-decomposition with intersection does not change the Hamming distance among the minterms, so that no bigger cubes can be built in the projection sets.

Let us now analyze the overall number of literals in the $(x_i, 0)$-decomposition with intersection built from *SOP*(f). Let ℓ_{SOP} denote the number of literals in *SOP*(f). The products in the SOP for I are left unchanged, so that their overall number of literals ℓ_I is preserved. Suppose that r products in *SOP*(f) contain x_i, and let ℓ_{x_i} denote their overall number of literals. The projection of these r products forms a SOP for $\tilde{f}|_{x_i=1}$, whose number of literals is equal to $\ell_{x_i} - r$, as projecting a product simply consists in eliminating x_i from it. Analogously, if s products in *SOP*(f) contain $\overline{x}_i$, and $\ell_{\overline{x}_i}$ is their overall number of literals, the SOP for $\tilde{f}|_{x_i=0}$ contains $\ell_{\overline{x}_i} - s$ literals. Thus, the $(x_i, 0)$-decomposition with intersection contains exactly $\ell_I + \ell_{x_i} - r + \ell_{\overline{x}_i} - s + 2 = \ell_{SOP} - r - s + 2$ literals, where the two additional literals represent the characteristic functions of the projection sets.

Case $p = x_j$.

For $p = x_j$, with $j \neq i$, the two decomposition techniques are based on the projection of f onto the two complementary subspaces of $\{0, 1\}^n$ where $x_i = x_j$ and $x_i \neq x_j$. For the (x_i, x_j)-decomposition we get the expression $f = (\overline{x}_i \oplus x_j) f|_{x_i=x_j} + (x_i \oplus x_j) f|_{x_i \neq x_j}$, while the (x_i, x_j)-decomposition with intersection is given by $f = (\overline{x}_i \oplus x_j)\tilde{f}|_{x_i=x_j} + (x_i \oplus x_j)\tilde{f}|_{x_i \neq x_j} + I$, where

$$\tilde{f}|_{x_i=x_j} = (f_{\text{on}}|_{x_i=x_j} \setminus I_{\text{on}},\ f_{\text{dc}}|_{x_i=x_j} \cup I_{\text{on}}),$$
$$\tilde{f}|_{x_i \neq x_j} = (f_{\text{on}}|_{x_i \neq x_j} \setminus I_{\text{on}},\ f_{\text{dc}}|_{x_i \neq x_j} \cup I_{\text{on}}),$$

with $I_{\text{on}} = f_{\text{on}}|_{x_i=x_j} \cap f_{\text{on}}|_{x_i \neq x_j}$ and $I_{\text{dc}} = f_{\text{dc}}|_{x_i=x_j} \cap f_{\text{dc}}|_{x_i \neq x_j}$. These expressions share some similarities with the *EXOR Projected Sum of Products* studied in [3]. In particular, if we represent the subfunctions as sums of products, the (x_i, x_j)-decomposition corresponds to an *EP-SOP form*, while the (x_i, x_j)-decomposition with intersection is only partially similar to an *EP-SOP with remainder form* [3]. The differences between the two expressions are due to the presence of don't cares in $\tilde{f}|_{x_i=x_j}$ and $\tilde{f}|_{x_i \neq x_j}$ and to the fact that the intersection I does not depend on the variable x_i, while the remainder in an EP-SOP may depend on all the n input variables. Also observe that, thanks to the presence of don't cares, the (x_i, x_j)-decomposition with intersection has a cost less or equal to the cost of an EP-SOP with remainder.

Cases $p = x_j \oplus x_k$ and $p = x_j x_k$.

In general the function p used to split the Boolean space $\{0, 1\}^n$ may depend on all input variables, but x_i. In this chapter we consider only two special cases, based on the use of two simple functions: an EXOR and an AND of two literals. The partition of $\{0, 1\}^n$ induced by the EXOR function does not depend on the choice of the variable complementations. Indeed, since $x_j \oplus x_k = \overline{x}_j \oplus \overline{x}_k$, and $\overline{(x_j \oplus x_k)} = \overline{x}_j \oplus x_k = x_j \oplus \overline{x}_k$, the choices $p = x_j \oplus x_k$ and $p = \overline{x}_j \oplus x_k$ give the same partition of the Boolean space. On the contrary, the partition of $\{0, 1\}^n$ induced by the AND function changes depending on the choice of the variable complementations, so that four different cases must be considered:

1. $p = x_j x_k$, corresponding to the partition into the sets where $x_i = x_j x_k$ and $x_i \neq x_j x_k$, i.e., $x_i = \overline{x}_j + \overline{x}_k$;
2. $p = x_j \overline{x}_k$, corresponding to the partition into the sets where $x_i = x_j \overline{x}_k$ and $x_i \neq x_j \overline{x}_k$, i.e., $x_i = \overline{x}_j + x_k$;
3. $p = \overline{x}_j x_k$, corresponding to the partition into the sets where $x_i = \overline{x}_j x_k$ and $x_i \neq \overline{x}_j x_k$, i.e., $x_i = x_j + \overline{x}_k$;
4. $p = \overline{x}_j \overline{x}_k$, corresponding to the partition into the sets where $x_i = \overline{x}_j \overline{x}_k$ and $x_i \neq \overline{x}_j \overline{x}_k$, i.e., $x_i = x_j + x_k$.

When the subfunctions are represented as SOPs, the resulting decomposition forms share some similarities with the *Projected Sum of Products (P-SOP)* introduced in [2]. Again, the two forms are different thanks to the presence of don't cares in the subfunctions and to the fact that the intersection I does not depend on x_i.

2.3 P-Circuits

We now show how the decomposition methods described in Section 2.2 can be applied to the logic synthesis of Boolean functions. The idea for synthesis is simply to construct a network for f using as building blocks networks for the projection function p, for the subfunctions $f|_{x_i=p}$, $f|_{x_i\neq p}$, $\tilde{f}|_{x_i=p}$, and $\tilde{f}|_{x_i\neq p}$, and a network for the intersection I. Observe that the overall network for f will require an EXOR gate for computing the characteristic functions of the projection subsets, two AND gates for the projections, and a final OR gate.

The function p, the projected subfunctions, and the intersection can be synthesized in any framework of logic minimization. In our experiments we focused on the standard *Sum of Products* synthesis, i.e., we represented p, $f|_{x_i=p}$, $f|_{x_i\neq p}$, $\tilde{f}|_{x_i=p}$, $\tilde{f}|_{x_i\neq p}$, and I as sums of products. In this way we derived networks for f which we called *Projected Circuit* and *Projected Circuit with Intersection*, in short *P-circuits*, see Fig. 2.3. If the SOPs representing p, $f|_{x_i=p}$, $f|_{x_i\neq p}$, $\tilde{f}|_{x_i=p}$, $\tilde{f}|_{x_i\neq p}$, and I are minimal, the corresponding circuits are called *Optimal P-circuits*. For instance, the function in Figs. 2.1 and 2.2 has minimal SOP form $\overline{x}_1\overline{x}_2\overline{x}_3 + x_1x_2\overline{x}_3 + \overline{x}_3x_4 + \overline{x}_2x_3\overline{x}_4$, while its corresponding optimal P-circuit is $(\overline{x}_1 \oplus x_2)\overline{x}_3 + \overline{x}_3x_4 + \overline{x}_2x_3\overline{x}_4$.

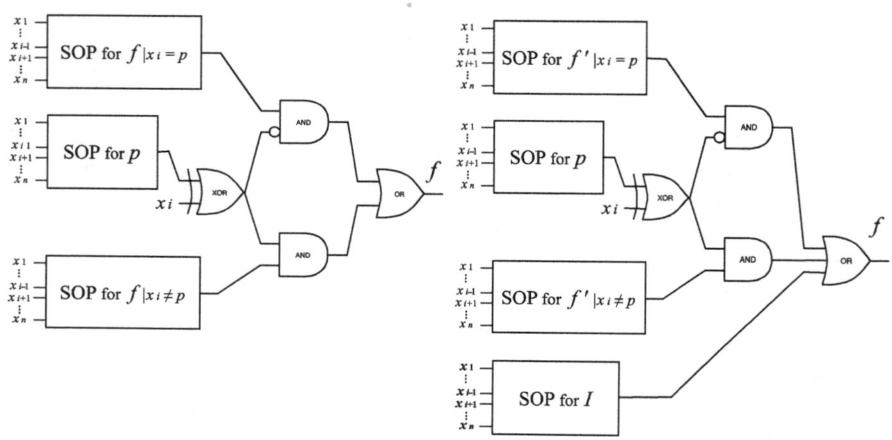

Fig. 2.3 P-circuit (*left*) and P-circuit with intersection (*right*)

The number of logic levels in a P-circuit varies from four to five: it is equal to four when the SOP for p consists in just one product and it is equal to five otherwise.

If we consider now the power consumption, we can observe in Fig. 2.3 that x_i, i.e., the variable with the highest switching frequency, is connected near the output of the overall logic network, thus triggering a sequence of switching events only for the last four gates. In this way, the contribution of x_i to the total power consumption is limited. Finally, we observe that it is possible to apply this decomposition when more than one variable switches with high frequency as shown in Section 2.4.

2.3.1 Synthesis Algorithms

We now describe two algorithms for computing optimal P-circuits, with and without intersection. Both algorithms can be implemented using OBDD data structures [9] for Boolean function manipulation and a classical SOP minimization procedure (e.g., ESPRESSO [7]).

The heuristic that finds a P-circuit with intersection (in Fig. 2.4) first computes the projections of the on-set and dc-set of f onto $x_i \neq p$ and $x_i = p$ and their intersections I_{on} and I_{dc}. The on-set of the intersection, I_{on}, is subtracted from the two on-sets ($f_{\text{on}}|_{x_i \neq p}$ and $f_{\text{on}}|_{x_i = p}$), and it is inserted in the two dc-sets ($f_{\text{dc}}|_{x_i \neq p}$ and $f_{\text{dc}}|_{x_i = p}$). This step guarantees that only the useful points of the intersection are covered in the SOP form of $f|_{x_i \neq p}$ and $f|_{x_i = p}$. Finally, the algorithm synthesizes the projected functions and the intersection with a SOP minimizer, and a P-circuit is then returned. The algorithm that computes a P-circuit without intersection is similar to the former but does not take into account the intersection, as shown in Fig. 2.5.

Synthesis of *P*-Circuits with intersection
INPUT: Functions f and p, and a variable x_i
OUTPUT: An optimal *P*-circuit for the (x_i,p)-decomposition with intersection of f
NOTATION: let $f = (f_{on}, f_{dc})$, i.e., f_{on} is the on-set of f, and f_{dc} is the don't care-set of f,

$I_{on} = f_{on}|_{x_i=p} \cap f_{on}|_{x_i \neq p}$;
$Idc = fdc|_{x_i=p} \cap f_{dc}|_{x_i \neq p}$;
$f_{on}^{(=)} = fon|_{x_i=p} \setminus I_{on}$;
$f_{on}^{(\neq)} = fon|_{x_i \neq p} \setminus I_{on}$;
$f_{dc}^{(=)} = f_{dc}|_{x_i=p} \cup I_{on}$;
$f_{dc}^{(\neq)} = f_{dc}|_{x_i \neq p} \cup I_{on}$;
$MinSOP^{(=)} = OptSOP(f_{on}^{(=)}, f_{dc}^{(=)})$; // optimal SOP for $f^{(=)}$
$MinSOP^{(\neq)} = OptSOP(f_{on}^{(\neq)}, f_{dc}^{(\neq)})$; // optimal SOP for $f^{(\neq)}$
$MinSOP^{I} = OptSOP(I_{on}, I_{dc})$; // optimal SOP for $I = (I_{on}, I_{dc})$
$MinSOP^{p} = OptSOP(p, \emptyset)$; // optimal SOP for p
P-circuit = $(\overline{x}_i \oplus MinSOP^{p})MinSOP^{(=)} + (x_i \oplus MinSOP^{p})MinSOP^{(\neq)} + MinSOP^{I}$
return *P*-circuit

Fig. 2.4 Algorithm for the optimization of P-circuits with intersection

The complexity of the algorithms depends on two factors: the complexity of OBDD operations, which is polynomial in the size of the OBDDs for the operands f and p, and the complexity of SOP minimization. Exact SOP minimization is superexponential in time, but efficient heuristics are available (i.e., ESPRESSO in the heuristic mode).

The algorithms compute correct covers as proved in the following theorem.

Theorem 2.2 (Correctness) *Algorithms in Figs. 2.4 and 2.5 compute a P-circuit C that covers the input function f.*

Proof Overloading the notation, let us denote with C the Boolean function that corresponds to the circuit C. In both cases we have to show that $f_{\text{on}} \subseteq C \subseteq f_{\text{on}} \cup$

Synthesis of *P*-Circuits

INPUT: Functions f and p, and a variable x_i
OUTPUT: An optimal P-circuit for the (x_i, p)-decomposition of f
NOTATION: let $f = (f_{on}, f_{dc})$, i.e., f_{on} is the on-set of f and f_{dc} is the don't care-set of f,
$f_{on}^{(=)} = f_{on}|_{x_i=p}$;
$f_{on}^{(\neq)} = f_{on}|_{x_i\neq p}$;
$f_{dc}^{(=)} = f_{dc}|_{x_i=p}$;
$f_{dc}^{(\neq)} = f_{dc}|_{x_i\neq p}$;
$MinSOP^{(=)} = OptSOP(f_{on}^{(=)}, f_{dc}^{(=)})$; // optimal SOP for $f^{(=)}$
$MinSOP^{(\neq)} = OptSOP(f_{on}^{(\neq)}, f_{dc}^{(\neq)})$; // optimal SOP for $f^{(\neq)}$
$MinSOP^{p} = OptSOP(p, \emptyset)$; // optimal SOP for p
P-circuit = $(\overline{x}_i \oplus MinSOP^p)MinSOP^{(=)} + (x_i \oplus MinSOP^p)MinSOP^{(\neq)}$
return P-circuit

Fig. 2.5 Algorithm for the optimization of P-circuits without intersection

f_{dc}. We first consider the algorithm in Fig. 2.4 for the (x_i, p)-decomposition with intersection of f that outputs the circuit C.

Let $y \in f_{\mathrm{on}}$ be the minterm $y = y_1, y_2, \ldots, y_n$, we show that $y \in C$. We have two cases: (1) if $y_i = p(y_1, \ldots, y_{i-1}, y_{i+1}, \ldots, y_n)$ we have that, for the synthesis algorithm, y is covered by $(\overline{x}_i \oplus MinSOP^p)MinSOP^{(=)}$ or by $MinSOP^I$; (2) if $y_i \neq p(y_1, \ldots, y_{i-1}, y_{i+1}, \ldots, y_n)$ we have that, for the synthesis algorithm, y is covered by $(x_i \oplus MinSOP^p)MinSOP^{(\neq)}$ or by $MinSOP^I$. Thus y is in C.

From the other side, let y be a point of C, we have to show that y is also in $f_{\mathrm{on}} \cup f_{\mathrm{dc}}$. We have two cases: (1) if y is covered by $MinSOP^I$, then y is in both $f|_{x_i=p}$ and $f|_{x_i\neq p}$, and – given that $MinSOP^I$ is synthesized with ESPRESSO – y is in $f_{\mathrm{on}} \cup f_{\mathrm{dc}}$; (2) if y is not covered by $MinSOP^I$, then it is covered by $(\overline{x}_i \oplus MinSOP^p)MinSOP^{(=)}$ or $(x_i \oplus MinSOP^p)MinSOP^{(\neq)}$. In both cases y must be in a projected space that is synthesized with ESPRESSO.

Consider now the algorithm in Fig. 2.5 for the computation of a (x_i, p)-decomposition without intersection. In this case the intersection is not computed thus each point of the function is simply projected onto one of the projecting spaces. The thesis immediately follows.

Considering the Stuck-At Fault Model (SAFM), we now briefly discuss the testability of P-circuits in the case where p is a constant function (i.e., $p = 0$). A fault in the Stuck-At Fault Model fixes exactly one input or one output pin of a node in a combinatorial logic circuit C to constant value (0 or 1) independently of the values applied to the primary inputs of the circuit. A node v in C is called *fully testable*, if there does not exist a redundant fault with fault location v. If all nodes in C are fully testable, then C is *fully testable*.

Theorem 2.3 *From a given P-circuit we can obtain a circuit that is fully testable in the SAFM by adding at most two more inputs and two multiplexers.*

Proof The proof of this theorem follows directly from the testability proof in [4] where the decomposed functions are synthesized in 2SPP form [10] instead of SOP forms. 2SPP expressions are direct generalizations of SOP forms where we can use EXORs of two literals instead of simple literals as inputs to the AND gates. We note that the testability theorem in [4] still holds for any form that is prime and irredundant. Since the SOP forms that we use for the synthesis of P-circuits have this property, the thesis immediately follows. In the case of P-circuits with intersection, the testable circuit that we obtain contains two MUXs before the inputs of the final OR gate. One is between the outputs of the decomposed parts and the second is after the output of the intersection. The MUXs are used to test the three single blocks of the circuit separately. In the case of P-circuits without intersection, just one MUX (between the outputs of the decomposed parts before the OR gate) is needed. In this case the proof still holds since we can consider a P-circuit without intersection as a special case of a P-circuit with intersection where the intersection is empty.

2.4 Multivariable Decomposition

In this section we show how our new decomposition technique can be extended from one to more variables, so that it could be applied to move all critical signals, and not just one, toward the output, still avoiding losses in area. A first naive solution for extending our technique could be to apply recursively the decompositions, i.e.,

- compute a decomposition of the function under study with respect to the variable with highest switching frequency among the variables in the set $X = \{x_1, \ldots, x_n\}$, say x_i;
- apply the same procedure to the functions $f|_{x_i=p}$ and $f|_{x_i \neq p}$ or to the functions $\tilde{f}|_{x_i=p}$, $\tilde{f}|_{x_i \neq p}$ and I (in case of decomposition with intersection), with respect to the variable with highest switching frequency in the set $X \setminus \{x_i\}$;
- if needed, recursively repeat the same procedure on the subfunctions derived in the previous decomposition step.

Observe that with this naive approach, the number of levels increases by three at each decomposition step. Moreover, the critical signals have different distances from the final output gate and their switching activity affects different portions of the circuit. In particular, the first variable selected is the one closest to the output, affecting only the last four gates.

In order to keep the number of levels constant and independent from the number of decomposition steps and to move all critical signals equally close to the output, so that the number of gates affected by their switching activity can be maintained constant, a different solution should be adopted. This solution is based on a "parallel decomposition" in which the points of the function are simultaneously partitioned and projected onto the 2^k subspaces defined by the k critical variables. For ease of exposition, we explain in detail only the case $k = 2$. The general case $k > 2$ can be easily derived from it, but at the expense of a quite heavy notation.

Definition 2.4 Let $f(X)$ be an incompletely specified Boolean function, $x_i, x_j \in X$, and p_i and p_j be two completely specified Boolean functions depending on all variables in $X \setminus \{x_i, x_j\}$. The $[(x_i, p_i), (x_j, p_j)]$-*decomposition* of f is the algebraic expression

$$f = (\overline{x}_i \oplus p_i)(\overline{x}_j \oplus p_j) f|_{\substack{x_i=p_i \\ x_j=p_j}} + (\overline{x}_i \oplus p_i)(x_j \oplus p_j) f|_{\substack{x_i=p_i \\ x_j\neq p_j}}$$
$$+(x_i \oplus p_i)(\overline{x}_j \oplus p_j) f|_{\substack{x_i\neq p_i \\ x_j=p_j}} + (x_i \oplus p_i)(x_j \oplus p_j) f|_{\substack{x_i\neq p_i \\ x_j\neq p_j}} .$$

The extension of the notion of decomposition with intersection will require the introduction of five new subfunctions representing the overall intersection among the four projections of f, and the four intersections between the projections of $f|_{x_i=p_i}$ and $f|_{x_i\neq p_i}$ w.r.t. x_j, and between the projections of $f|_{x_j=p_j}$ and $f|_{x_j\neq p_j}$ w.r.t. x_i, respectively. As for the decomposition w.r.t. one variable, the intersection sets will be added as don't cares to the projected subfunctions, in order to possibly improve their minimal SOP forms.

Definition 2.5 Let $f(X)$ be an incompletely specified Boolean function, $x_i, x_j \in X$, and p_i and p_j be two completely specified Boolean functions depending on all variables in $X \setminus \{x_i, x_j\}$. The $[(x_i, p_i), (x_j, p_j)]$-*decomposition with intersection* of $f = (f_{\text{on}}, f_{\text{dc}})$ is the algebraic expression

$$f = (\overline{x}_i \oplus p_i)(\overline{x}_j \oplus p_j) \tilde{f}|_{\substack{x_i=p_i \\ x_j=p_j}} + (\overline{x}_i \oplus p_i)(x_j \oplus p_j) \tilde{f}|_{\substack{x_i=p_i \\ x_j\neq p_j}} +$$
$$(x_i \oplus p_i)(\overline{x}_j \oplus p_j) \tilde{f}|_{\substack{x_i\neq p_i \\ x_j=p_j}} + (x_i \oplus p_i)(x_j \oplus p_j) \tilde{f}|_{\substack{x_i\neq p_i \\ x_j\neq p_j}} +$$
$$(\overline{x}_i \oplus p_i) I^{(i,=)} + (x_i \oplus p_i) I^{(i,\neq)} + (\overline{x}_j \oplus p_j) I^{(j,=)} + (x_j \oplus p_j) I^{(j,\neq)} + I,$$

where

$$\tilde{f}|_{\substack{x_i=p_i \\ x_j=p_j}} = (f_{\text{on}}|_{\substack{x_i=p_i \\ x_j=p_j}} \setminus (I^{(i,=)}_{\text{on}} \cup I^{(j,=)}_{\text{on}} \cup I_{\text{on}}), f_{\text{dc}}|_{\substack{x_i=p_i \\ x_j=p_j}} \cup I^{(i,=)}_{\text{on}} \cup I^{(j,=)}_{\text{on}} \cup I_{\text{on}}),$$
$$\tilde{f}|_{\substack{x_i=p_i \\ x_j\neq p_j}} = (f_{\text{on}}|_{\substack{x_i=p_i \\ x_j\neq p_j}} \setminus (I^{(i,=)}_{\text{on}} \cup I^{(j,\neq)}_{\text{on}} \cup I_{\text{on}}), f_{\text{dc}}|_{\substack{x_i=p_i \\ x_j\neq p_j}} \cup I^{(i,=)}_{\text{on}} \cup I^{(j,\neq)}_{\text{on}} \cup I_{\text{on}}),$$
$$\tilde{f}|_{\substack{x_i\neq p_i \\ x_j=p_j}} = (f_{\text{on}}|_{\substack{x_i\neq p_i \\ x_j=p_j}} \setminus (I^{(i,\neq)}_{\text{on}} \cup I^{(j,=)}_{\text{on}} \cup I_{\text{on}}), f_{\text{dc}}|_{\substack{x_i\neq p_i \\ x_j=p_j}} \cup I^{(i,\neq)}_{\text{on}} \cup I^{(j,=)}_{\text{on}} \cup I_{\text{on}}),$$
$$\tilde{f}|_{\substack{x_i\neq p_i \\ x_j\neq p_j}} = (f_{\text{on}}|_{\substack{x_i\neq p_i \\ x_j\neq p_j}} \setminus (I^{(i,\neq)}_{\text{on}} \cup I^{(j,\neq)}_{\text{on}} \cup I_{\text{on}}), f_{\text{dc}}|_{\substack{x_i\neq p_i \\ x_j\neq p_j}} \cup I^{(i,\neq)}_{\text{on}} \cup I^{(j,\neq)}_{\text{on}} \cup I_{\text{on}}),$$

with

$$I_{\text{on}} = f_{\text{on}}|_{\substack{x_i=p_i \\ x_j=p_j}} \cap f_{\text{on}}|_{\substack{x_i=p_i \\ x_j\neq p_j}} \cap f_{\text{on}}|_{\substack{x_i\neq p_i \\ x_j=p_j}} \cap f_{\text{on}}|_{\substack{x_i\neq p_i \\ x_j\neq p_j}} ,$$
$$I_{\text{dc}} = f_{\text{dc}}|_{\substack{x_i=p_i \\ x_j=p_j}} \cap f_{\text{dc}}|_{\substack{x_i=p_i \\ x_j\neq p_j}} \cap f_{\text{dc}}|_{\substack{x_i\neq p_i \\ x_j=p_j}} \cap f_{\text{dc}}|_{\substack{x_i\neq p_i \\ x_j\neq p_j}} ,$$
$$I^{(i,=)}_{\text{on}} = (f_{\text{on}}|_{\substack{x_i=p_i \\ x_j=p_j}} \cap f_{\text{on}}|_{\substack{x_i=p_i \\ x_j\neq p_j}}) \setminus I_{\text{on}} , \quad I^{(i,=)}_{\text{dc}} = (f_{\text{dc}}|_{\substack{x_i=p_i \\ x_j=p_j}} \cap f_{\text{dc}}|_{\substack{x_i=p_i \\ x_j\neq p_j}}) \cup I_{\text{on}},$$
$$I^{(i,\neq)}_{\text{on}} = (f_{\text{on}}|_{\substack{x_i\neq p_i \\ x_j=p_j}} \cap f_{\text{on}}|_{\substack{x_i\neq p_i \\ x_j\neq p_j}}) \setminus I_{\text{on}} , \quad I^{(i,\neq)}_{\text{dc}} = (f_{\text{dc}}|_{\substack{x_i\neq p_i \\ x_j=p_j}} \cap f_{\text{dc}}|_{\substack{x_i\neq p_i \\ x_j\neq p_j}}) \cup I_{\text{on}},$$

$$I_{\text{on}}^{(j,=)} = (f_{\text{on}}|_{\substack{x_i=p_i\\x_j=p_j}} \cap f_{\text{on}}|_{\substack{x_i\neq p_i\\x_j=p_j}}) \setminus I_{\text{on}}, \quad I_{\text{dc}}^{(j,=)} = (f_{\text{dc}}|_{\substack{x_i=p_i\\x_j=p_j}} \cap f_{\text{dc}}|_{\substack{x_i\neq p_i\\x_j=p_j}}) \cup I_{\text{on}},$$

$$I_{\text{on}}^{(j,\neq)} = (f_{\text{on}}|_{\substack{x_i=p_i\\x_j\neq p_j}} \cap f_{\text{on}}|_{\substack{x_i\neq p_i\\x_j\neq p_j}}) \setminus I_{\text{on}}, \quad I_{\text{dc}}^{(j,\neq)} = (f_{\text{dc}}|_{\substack{x_i=p_i\\x_j\neq p_j}} \cap f_{\text{dc}}|_{\substack{x_i\neq p_i\\x_j\neq p_j}}) \cup I_{\text{on}}.$$

Observe that we do not consider the intersections between $f|_{\substack{x_i=p_i\\x_j=p_j}}$ and $f|_{\substack{x_i\neq p_i\\x_j\neq p_j}}$ and between $f|_{\substack{x_i=p_i\\x_j\neq p_j}}$ and $f|_{\substack{x_i\neq p_i\\x_j=p_j}}$ as these two pairs of projections belong to non-adjacent subspaces and therefore there are no cubes split between them.

When the functions p_i and p_j, the four projected subfunctions, and the intersection sets are represented as minimal SOP forms, these two algebraic expressions give rise to P-circuits without and with intersection, both of depth 5, exactly as in the case of the decomposition w.r.t. a single variable. Moreover, the two critical signals x_i and x_j are equally close to the output and their switching activity affects only a constant number of gates, as p_i, p_j and the intersection sets do not depend on them.

The two circuits can be synthesized generalizing the algorithms shown in Figs. 2.4 and 2.5 in a straightforward way.

For example, consider the function f shown on the left side of Fig. 2.1. Let $p_i = 0$, $p_j = 0$, and x_i and x_j be x_1 and x_2, respectively. The Boolean space $\{0,1\}^4$ can be partitioned into the four sets: $(x_1 = 0, x_2 = 0)$, $(x_1 = 0, x_2 = 1)$, $(x_1 = 1, x_2 = 0)$, and $(x_1 = 1, x_2 = 1)$, each containing 2^2 points. The projections of f onto these four sets are

$$\begin{aligned} f_{\text{on}}|_{\substack{x_1=0\\x_2=0}} &= \{00, 01, 10\} & f_{\text{dc}}|_{\substack{x_1=0\\x_2=0}} &= \emptyset \\ f_{\text{on}}|_{\substack{x_1=0\\x_2\neq 0}} &= \{01\} & f_{\text{dc}}|_{\substack{x_1=0\\x_2\neq 0}} &= \{11\} \\ f_{\text{on}}|_{\substack{x_1\neq 0\\x_2=0}} &= \{01, 10\} & f_{\text{dc}}|_{\substack{x_1\neq 0\\x_2=0}} &= \emptyset \\ f_{\text{on}}|_{\substack{x_1\neq 0\\x_2\neq 0}} &= \{00, 01\} & f_{\text{dc}}|_{\substack{x_1\neq 0\\x_2\neq 0}} &= \emptyset \end{aligned}$$

The $[(x_1, 0), (x_2, 0)]$-decomposition of f thus determines the optimal P-circuit $\overline{x}_1\overline{x}_2(\overline{x}_3 + \overline{x}_4) + \overline{x}_1 x_2 x_4 + x_1\overline{x}_2(\overline{x}_3 x_4 + x_3\overline{x}_4) + x_1 x_2 \overline{x}_3$, containing 16 literals.

Let us now consider the $[(x_1, 0), (x_2, 0)]$-decomposition with intersection. The intersection sets are $I_{\text{on}} = \{01\}$, $I_{\text{dc}} = \emptyset$, $I_{\text{on}}^{(i,=)} = I_{\text{on}}^{(i,\neq)} = I_{\text{on}}^{(j,\neq)} = \emptyset$, $I_{\text{on}}^{(j,=)} = \{10\}$, $I_{\text{dc}}^{(i,=)} = I_{\text{dc}}^{(i,\neq)} = I_{\text{dc}}^{(j,=)} = I_{\text{dc}}^{(j,\neq)} = \{01\}$, and the projections become

$$\begin{aligned} \tilde{f}_{\text{on}}|_{\substack{x_1=0\\x_2=0}} &= \{00\} & \tilde{f}_{\text{dc}}|_{\substack{x_1=0\\x_2=0}} &= \{01, 10\} \\ \tilde{f}_{\text{on}}|_{\substack{x_1=0\\x_2\neq 0}} &= \emptyset & \tilde{f}_{\text{dc}}|_{\substack{x_1=0\\x_2\neq 0}} &= \{01, 11\} \\ \tilde{f}_{\text{on}}|_{\substack{x_1\neq 0\\x_2=0}} &= \emptyset & \tilde{f}_{\text{dc}}|_{\substack{x_1\neq 0\\x_2=0}} &= \{01, 10\} \\ \tilde{f}_{\text{on}}|_{\substack{x_1\neq 0\\x_2\neq 0}} &= \{00\} & \tilde{f}_{\text{dc}}|_{\substack{x_1\neq 0\\x_2\neq 0}} &= \{01\} \end{aligned}$$

The corresponding P-circuit with intersection is now $\overline{x}_1\overline{x}_2\overline{x}_3 + x_1 x_2 \overline{x}_3 + \overline{x}_2 x_3 \overline{x}_4 + \overline{x}_3 x_4$, with 11 literals.

2.5 Experimental Results

In this section we report experimental results for the two decomposition methods described in the previous sections. The methods have been implemented in C, using the CUDD library for OBDDs to represent Boolean functions. The experiments have been run on a Pentium 1.6 GHz CPU with 1 GB of main memory. The benchmarks are taken from LGSynth93 [17]. We report in the following a significant subset of the functions as representative indicators of our experiments.

In order to evaluate the performances of these new synthesis methods, we compare area and delay of different versions of P-circuits with P-circuits based on the classical Shannon decomposition, i.e., P-circuits representing $(x_i, 0)$-decomposition without intersection (referred as **Shannon** in Table 2.1). In particular we have considered P-circuits for the following choices of the projection function p:

- $p = 0$, decomposition with intersection (referred as **Constant** in Table 2.2);
- $p = x_j$, decomposition without and with intersection (referred as **VAR** in Tables 2.1 and 2.2);
- $p = x_j \oplus x_k$, decomposition without and with intersection (referred as **XOR** in Tables 2.1 and 2.2);
- $p = x_j x_k$, decomposition without and with intersection, choosing the complementations of variables giving the best area (referred as **AND** in Tables 2.1 and 2.2).

After the projection, all SOP components of the P-circuits have been synthesized with multioutput synthesis using ESPRESSO in the heuristic mode. Finally, to evaluate the obtained circuits, we ran our benchmarks using the SIS system with the MCNC library for technology mapping and the SIS command `map -W -f 3 -s`.

In Tables 2.1 and 2.2 we compare the mapped area and the synthesis time (in seconds) of P-circuits representing decomposition forms without intersection (Table 2.1) and with intersection (Table 2.2) for a subset of the benchmarks. Due to space limitation, the results shown refer only to decompositions with respect to the first input variable, x_0, of each benchmark. In the overall set of experiments we have considered decompositions with respect to each input variable of each benchmark.

The results, reported in Fig. 2.6, are quite interesting: about 79% of P-circuits based on the $(x_i, 0)$-decomposition with intersection have an area smaller than the P-circuits based on the classical Shannon decomposition, i.e., on the $(x_i, 0)$-decomposition without intersection; analogously, 32 and 59% of the P-circuits benefit from the (x_i, x_j)-decomposition without and with intersection, respectively; 22 and 50% of the circuits benefit from the $(x_i, x_j \oplus x_k)$-decomposition without and with intersection, respectively; and 28 and 58% of the circuits benefit from the $(x_i, x_j x_k)$-decomposition without and with intersection, respectively. These results support the conclusion that decompositions with intersection provide better results, and that the best choice for the projection function p is the simplest: $p = 0$.

Moreover synthesis for $p = 0$ with intersection is very efficient in computational time, as shown in Fig. 2.7; in fact, about 80% of P-circuits based on the $(x_i, 0)$-decomposition with intersection have a synthesis time smaller than the synthesis

Table 2.1 Comparison of area and synthesis time of P-circuits representing (x_0, p)-decomposition forms for different choices of the projection function p without intersection

	(x_0, p)-**Decomposition without intersection**							
	Shannon		**VAR**		**XOR**		**AND**	
Benchmark (in/out)	**Area**	**Time**	**Area**	**Time**	**Area**	**Time**	**Area**	**Time**
add6 (12/7)	908	0.65	507	5.19	669	24.58	524	90.84
adr4 (8/5)	284	0.05	172	0.14	223	0.45	237	1.76
alu2 (10/8)	355	0.45	382	0.79	416	3.60	356	12.93
alu3 (10/8)	256	0.34	330	0.67	402	2.54	354	9.22
amd (14/24)	162	0.17	1694	1.24	1800	8.65	1747	30.31
apla (10/1)	379	0.12	371	0.58	467	3.19	398	9.11
b9 (16/5)	436	0.15	463	1.08	492	8.30	472	29.36
b12 (15/9)	227	0.11	306	0.55	401	4.27	340	15.90
br1 (12/8)	347	0.05	381	0.19	435	0.88	418	3.53
br2 (12/8)	281	0.03	314	0.18	377	0.97	337	3.30
dc2 (8/7)	249	0.05	279	0.13	337	0.40	276	1.45
dist (8/7)	891	0.11	1266	0.34	1202	0.95	946	3.77
dk17 (10/11)	263	0.10	250	0.38	291	1.82	230	6.85
dk48 (15/17)	263	0.23	284	1.65	288	14.73	276	46.21
ex7 (16/5)	436	0.12	463	1.04	492	8.30	472	29.07
exp (8/18)	824	0.09	873	0.30	947	0.96	1011	3.15
f51m (8/8)	497	0.09	706	0.21	640	0.64	528	2.24
inc (7/9)	237	0.05	287	0.11	364	0.25	316	1.17
l8err (8/8)	301	0.08	328	0.30	356	0.70	311	2.43
life (9/1)	267	0.06	252	0.21	298	0.60	267	2.56
m181 (15/9)	227	0.42	308	0.58	404	4.44	341	16.39
m2 (8/16)	808	0.08	919	0.21	1282	0.55	1043	1.93
m3 (8/16)	1042	0.08	1392	0.24	1638	0.71	1184	2.92
m4 (8/16)	2766	0.19	3286	0.96	2846	2.10	2271	6.93
max1024 (10/6)	2534	0.34	2511	1.97	2973	8.74	2642	30.72
max128 (7/24)	2373	0.08	2711	0.35	3219	0.91	2391	3.14
max512 (9/6)	1470	0.15	1607	0.64	1116	2.27	1227	8.09
mlp4 (8/8)	1113	0.15	1031	0.36	1292	1.13	997	4.12
mp2d (14/14)	355	0.09	435	0.61	508	4.47	455	16.49
p1 (8/18)	724	0.18	781	0.96	821	3.07	842	10.77
p3 (8/14)	587	0.22	524	0.52	559	1.64	548	5.90
p82 (5/14)	244	0.02	321	0.06	394	0.11	370	0.33
rd73 (7/3)	312	0.05	437	0.60	355	9.12	388	35.82
root (8/5)	416	0.05	594	0.14	393	0.50	385	1.91
spla (8/5)	2239	0.79	2570	7.88	3142	74.99	2886	273.75
sqr6 (6/12)	443	0.05	656	6.05	561	43.76	532	170.62
sym10 (10/1)	559	0.30	414	0.64	309	2.92	416	14.31
t1 (21/23)	905	0.83	951	3.52	1186	41.02	982	155.28
t2 (17/16)	501	0.06	589	0.65	686	6.37	618	22.95
t3 (12/8)	156	0.14	212	0.74	275	5.21	236	20.77
tial (14/8)	3430	5.33	3337	23.68	4062	159.84	3823	557.19
tms (8/16)	670	0.03	787	23.00	904	161.92	737	548.21
vtx1 (27/6)	430	0.09	445	1.89	501	32.57	585	107.74
x9dn (27/7)	530	0.22	528	2.23	595	30.62	548	116.64
Z5xp1 (7/10)	479	0.08	593	0.12	743	0.33	547	1.24
Z9sym (9/1)	464	0.17	288	0.33	267	1.15	371	6.07

Table 2.2 Comparison of area and synthesis time of P-circuits representing (x_0, p)-decomposition forms for different choices of the projection function p with intersection

	(x_0, p)-**Decomposition with intersection**							
	Constant		**VAR**		**XOR**		**AND**	
Benchmark (in/out)	**Area**	**Time**	**Area**	**Time**	**Area**	**Time**	**Area**	**Time**
add6 (12/7)	672	0.51	814	4.44	759	23.70	651	80.93
adr4 (8/5)	203	0.03	125	0.18	161	0.40	175	1.58
alu2 (10/8)	283	0.18	308	1.03	310	4.72	298	16.79
alu3 (10/8)	263	0.16	276	0.42	295	1.67	283	5.91
amd (14/24)	1012	0.12	1085	1.55	1202	10.88	1180	37.65
apla (10/1)	379	0.08	371	0.38	470	1.42	398	6.11
b9 (16/5)	327	0.20	360	0.81	393	5.17	364	19.74
b12 (15/9)	199	0.18	248	0.65	367	5.25	292	18.13
br1 (12/8)	347	0.02	381	0.18	435	0.87	418	3.47
br2 (12/8)	281	0.01	314	0.18	377	0.86	337	3.16
dc2 (8/7)	238	0.02	281	0.14	355	0.28	268	1.46
dist (8/7)	1036	0.09	1507	0.26	1373	0.69	1048	3.24
dk17 (10/11)	263	0.06	250	0.46	291	1.99	230	7.21
dk48 (15/17)	263	0.17	284	0.69	288	4.34	276	17.89
ex7 (16/5)	327	0.09	360	1.56	393	10.39	364	38.51
exp (8/18)	838	0.05	877	0.22	930	0.66	1035	3.01
f51m (8/8)	277	0.09	290	0.28	314	0.85	323	4.11
inc (7/9)	270	0.02	134	0.10	372	0.22	348	0.85
l8err (8/8)	355	0.03	337	0.18	450	0.68	354	2.48
life (9/1)	197	0.05	227	0.12	216	0.43	224	2.03
m181 (15/9)	199	0.08	252	0.68	341	6.65	288	29.20
m2 (8/16)	808	0.05	919	0.24	1282	0.48	1043	2.23
m3 (8/16)	1042	0.05	1392	0.26	1638	0.76	1184	3.53
m4 (8/16)	2163	0.14	2981	0.38	3683	1.22	2496	4.77
max1024 (10/6)	2980	0.25	3043	2.12	2977	10.13	2829	34.28
max128 (7/24)	2155	0.06	2259	0.23	2704	0.58	1975	1.97
max512 (9/6)	1346	0.12	1533	0.39	1351	1.45	1265	5.02
mlp4 (8/8)	908	0.08	917	0.30	1081	0.98	938	3.34
mp2d (14/14)	276	0.16	357	0.75	411	6.82	359	22.56
p1 (8/18)	711	0.20	777	1.18	847	3.74	818	13.66
p3 (8/14)	520	0.12	552	0.37	554	0.79	504	2.68
p82 (5/14)	229	0.02	313	0.06	372	0.10	343	0.31
rd73 (7/3)	332	0.02	577	0.69	496	8.78	464	33.63
root (8/5)	417	0.02	536	0.17	602	0.55	446	1.94
spla (8/5)	2428	0.73	2761	8.82	3249	84.11	3107	336.30
sqr6 (6/12)	333	1.59	429	4.49	437	40.35	370	124.48
sym10 (10/1)	568	0.27	529	0.96	551	3.90	554	16.81
t1 (21/23)	463	0.61	510	6.06	655	78.07	585	277.38
t2 (17/16)	358	0.05	406	0.88	469	9.80	416	22.33
t3 (12/8)	218	0.08	270	0.74	336	3.77	295	14.03
tial (14/8)	3368	3.29	3319	31.12	3952	215.08	3827	741.85
tms (8/16)	670	3.00	787	14.06	904	103.07	737	317.65
vtx1 (27/6)	390	0.14	499	3.03	486	50.57	524	171.45
x9dn (27/7)	412	0.19	401	4.26	457	57.18	418	217.77
Z5xp1 (7/10)	324	0.03	369	0.19	441	0.41	302	1.29
Z9sym (9/1)	379	0.17	391	0.64	395	1.68	393	9.28

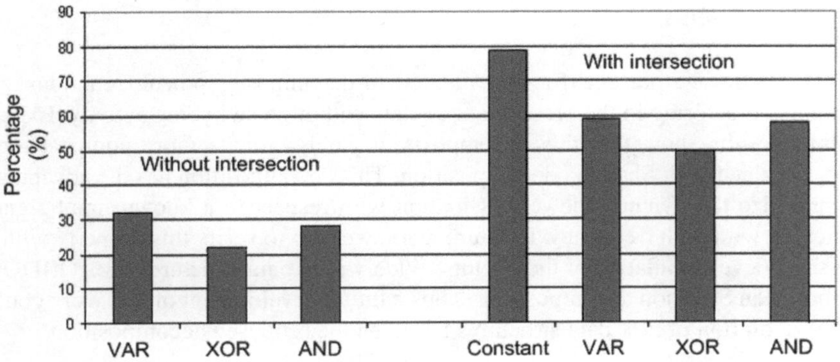

Fig. 2.6 Percentage of P-circuits, over all the benchmarks, having smaller area than the P-circuits based on Shannon decomposition

time of P-circuits based on the classical Shannon decomposition. When p is not constant, synthesis is time consuming, since the algorithm must choose the best combination of variables for p. In particular, 3 and 5% of the P-circuits benefit from the (x_i, x_j)-decomposition without and with intersection, respectively; 2% of the circuits benefit from the $(x_i, x_j \oplus x_k)$-decomposition both without and with intersection; and only 1% of the circuits benefit from the $(x_i, x_j x_k)$-decomposition both without and with intersection. Altogether, only 14% of the P-circuits achieve the smallest area when implemented according to the classical Shannon decomposition. The subset of results shown in Tables 2.1 and 2.2 reflects these percentages.

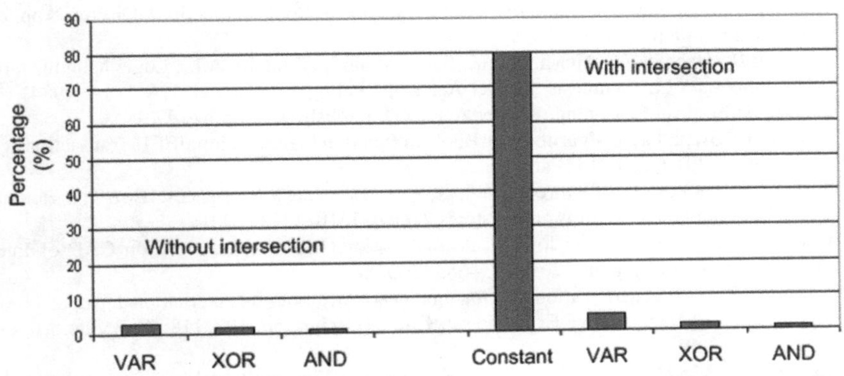

Fig. 2.7 Percentage of P-circuits, over all the benchmarks, having smaller synthesis time than the P-circuits based on Shannon decomposition

2.6 Conclusion

In conclusion, we presented a new method to decompose Boolean functions via complex cofactoring in the presence of signals with high switching activity. Experimental results show that this decomposition yields circuits more compact than those obtained with Shannon decomposition. This decomposition has the advantage to minimize the dynamic power dissipation with respect to a known input signal switching with high frequency. In future work, we plan to verify this property with a transistor-level simulation of the circuits. Widely used data structures (i.e., OBDDs) are based on Shannon decomposition. Thus a future development of this work could be the definition of new data structures based on the proposed decomposition.

Acknowledgments Tiziano Villa gratefully acknowledges partial support from the COCONUT EU project FP7-2007-IST-1-217069, and the CON4COORD EU Project FP7-ICT-2007.3.7.(c) grant agreement nr. INFSO-ICT-223844.

References

1. Benini, L., Micheli, G.D.: Logic synthesis for low power. In: S. Hassoun, T. Sasao (eds.) Logic Synthesis and Verification, pp. 197–223. Kluwer Academic Publishers Norwell, MA, USA (2002)
2. Bernasconi, A., Ciriani, V., Cordone, R.: On projecting sums of products. In: 11th Euromicro Conference on Digital Systems Design: Architectures, Methods and Tools. Parma, Italy (2008)
3. Bernasconi, A., Ciriani, V., Cordone, R.: The optimization of kEP-SOPs: Computational complexity, approximability and experiments. ACM Transactions on Design Automation of Electronic Systems **13**(2), 1–31 (2008)
4. Bernasconi, A., Ciriani, V., Trucco, G., Villa, T.: Logic Minimization and Testability of 2SPPP-Circuits. In: Euromicro Conference on Digital Systems Design (DSD). Patras, Greece (2009)
5. Bernasconi, A., Ciriani, V., Trucco, G., Villa, T.: On decomposing Boolean functions via extended cofactoring. In: Design Automation and Test in Europe. Nice, France (2009)
6. Bioch, J.C.: The complexity of modular decomposition of Boolean functions. Discrete Applied Mathematics **149**(1–3), 1–13 (2005)
7. Brayton, R., Hachtel, G., McMullen, C., Sangiovanni-Vincentelli, A.L.: Logic Minimization Algorithms for VLSI Synthesis. Kluwer Academic Publishers Norwell, MA, USA (1984)
8. Brown, F.: Boolean Reasoning. Kluwer Academic Publishers, Boston (1990)
9. Bryant, R.: Graph based algorithm for Boolean function manipulation. IEEE Transactions on Computers **35**(9), 667–691 (1986)
10. Ciriani, V.: Synthesis of SPP three-level logic networks using affine spaces. IEEE Transactions on CAD of Integrated Circuits and Systems **22**(10), 1310–1323 (2003)
11. Cortadella, J.: Timing-driven logic bi-decomposition. IEEE Transactions on CAD of Integrated Circuits and Systems **22**(6), 675–685 (2003)
12. Kerntopf, P.: New generalizations of Shannon decomposition. In: International Workshop on Applications of Reed-Muller Expansion in Circuit Design, pp. 109–118. Starkville, Mississippi, USA (2001)
13. Lavagno, L., McGeer, P.C., Saldanha, A., Sangiovanni-Vincentelli, A.L.: Timed Shannon circuits: A power-efficient design style and synthesis tool. In: 32nd ACM/IEEE Conference on Design automation, pp. 254–260. (1995)

14. McGeer, P.C., Brayton, R.K., Sangiovanni-Vincentelli, A.L., Sahni, S.: Performance enhancement through the generalized bypass transform. In: ICCAD, pp. 184–187. Santa Clara, CA, USA (1991)
15. Pedram, M.: Power estimation and optimization at the logic level. International Journal of High Speed Electronics and Systems **5**(2), 179–202 (1994)
16. Soviani, C., Tardieu, O., Edwards, S.A.: Optimizing sequential cycles through Shannon decomposition and retiming. In: DATE '06: Proceedings of the conference on Design, Automation and Test in Europe, pp. 1085–1090. European Design and Automation Association, 3001 Leuven, Belgium, Belgium (2006)
17. Yang, S.: Logic synthesis and optimization benchmarks user guide version 3.0. User Guide, Microelectronics Center of North Carolina (1991)

Chapter 3
Sequential Logic Synthesis Using Symbolic Bi-decomposition

Victor N. Kravets and Alan Mishchenko

Abstract In this chapter we use under-approximation of unreachable states of a design to derive incomplete specification of combinational logic. The resulting incompletely specified functions are decomposed to enhance the quality of technology-dependent synthesis. The decomposition choices are computed implicitly using novel formulation of symbolic bi-decomposition that is applied recursively to decompose logic in terms of simple primitives. The ability of binary decision diagrams to represent compactly certain exponentially large combinatorial sets helps us to implicitly enumerate and explore variety of decomposition choices improving quality of synthesized circuits. Benefits of the symbolic technique are demonstrated in sequential synthesis of publicly available benchmarks as well as on the realistic industrial designs.

3.1 Introduction and Motivation

Due to recent advances in verification technology [2] circuit synthesis of semiconductor designs no longer has to be limited to logic optimization of combinational blocks. Nowadays logic transformations may involve memory elements which change design's state encodings or its reachable state space and still be verified against its original description. In this chapter we focus on a more conservative synthesis approach that changes sequential behavior of a design only in unreachable states, leaving its intended "reachable" behavior unchanged. Unreachable states are used to extract incomplete specification of combinational blocks and are applied as *don't cares* during functional decomposition to improve circuit quality.

To implement combinational logic of a design we rely on a very simple, yet complete, form of functional decomposition commonly referred to as bi-decomposition.

V.N. Kravets (✉)
IBM TJ Watson Research Center, Yorktown, NY
e-mail: kravets@us.ibm.com

Based on Kravets, V.N.; Mishchenko, A.; "Sequential logic synthesis using symbolic bi-decomposition," Design, Automation & Test in Europe Conference & Exhibition, 2009. DATE '09, pp.1458–1463, 20–24 April 2009 © [2009] IEEE.

S.P. Khatri, K. Gulati (eds.), *Advanced Techniques in Logic Synthesis, Optimizations and Applications*, DOI 10.1007/978-1-4419-7518-8_3,

In general, for a given completely specified Boolean function its bi-decomposition has form

$$f(\mathbf{x}) = h(g_1(\mathbf{x}_1), g_2(\mathbf{x}_2))$$

where h is an arbitrary 2-input Boolean function. This decomposition is not unique and its quality varies depending on selected subsets $\mathbf{x}_1$ and $\mathbf{x}_2$ that form possibly overlapping (i.e., non-disjoint) partition of $\mathbf{x}$. The problem of finding good bi-decomposition has been studied in [1, 10, 18, 19, 21]. The main contribution of the material in this chapter is symbolic formulation of bi-decomposition for incompletely specified functions. The bi-decomposition is used as main computational step in the prototype sequential synthesis tool and is applied recursively to implement logic of combinational blocks whose incomplete specification is extracted from unreachable states of a design. Our symbolic formulation of bi-decomposition finds all feasible solutions and picks the best ones, without explicit enumeration.

Computation of variable partitions in our symbolic formulation of bi-decomposition favors implicit enumeration of decomposition subsets. They are represented compactly with a binary decision diagram (BDD) [4] and are selected based on optimization objective. Unlike previous approaches (e.g., [1, 23]) that rely on BDDs, the decomposition is not checked explicitly for a variable partition and is solved implicitly for all feasible partitions simultaneously utilizing fundamental property of BDDs to share partial computations across subproblems. Thus, no costly enumeration that requires separate and independent decomposability checks is needed. The technique was also used to tune greedy bi-decomposition when handling larger functions.

To overcome limitations of explicit techniques authors in [14] proposed solution that uses a satisfiability solver [11]. Their approach is based on proving that a problem instance is unsatisfied. The unsatisfiable core is then used to greedily select partition of variables that induces bi-decomposition. Authors demonstrate the approach to be efficient in runtime, when determining existence of non-trivial decomposition. The experimental results on a selected benchmark set, however, are primarily focused on the existence of decomposition and do not offer a qualitative synthesis data.

The problem of using unreachable states of a design to improve synthesis and verification quality has been studied before in various contexts. In general, these algorithms either avoid explicit computation of unreachable states or first compute them in pre-optimization stage. Approaches that do not explicitly compute unreachable states are mostly limited to incremental structural changes of a circuit and rely on ATPG environment or induction [5, 8, 12] to justify a change. In contrast, approaches that pre-compute subsets of unreachable states treat them as external don't cares [20] for re-synthesis of combinational logic blocks [6, 15]. In this chapter we adopt the later approach as it offers more flexibility in logic re-implementation through functional decomposition.

This chapter has the following structure. After brief introduction and motivation preliminary constructs are given in Section 3.2. Section 3.3 describes bi-decomposition existence requirements. They are used in Section 3.4 to formulate implicit computation of decomposition. Implementation details are described in Section 3.5. Experimental results are given in Section 3.6. Section 3.7 gives concluding remarks and possible directions for future work.

3.2 Preliminary Constructs

Basic constructs used by synthesis algorithms of the chapter are introduced in this section.

3.2.1 "Less-Than-or-Equal" Relation

Computational forms constructed in this chapter rely on the partial order relation between Boolean functions. Given functions $f(\mathbf{x})$ and $g(\mathbf{x})$, $f(\mathbf{x}) \leq g(\mathbf{x})$ indicates that $f(\mathbf{x})$ precedes $g(\mathbf{x})$ in the order. This "less-than-or-equal" relation ($\leq$) between the two functions can be expressed by one of the following three equivalent forms:

$$[f(\mathbf{x}) \Rightarrow g(\mathbf{x})] \equiv [f(\mathbf{x}) \leq g(\mathbf{x})] \equiv [\overline{f(\mathbf{x})} + g(\mathbf{x}) = 1]$$

The relation imposes *consistency* constraint on constructed computational forms. It allows us to represent incompletely specified Boolean functions in terms of *intervals* [3], defined as

$$[l(\mathbf{x}), u(\mathbf{x})] = \{f(\mathbf{x}) | l(\mathbf{x}) \leq f(\mathbf{x}) \leq u(\mathbf{x})\}$$

Here interval represents a set of completely specified functions using its two distinguished members $l(\mathbf{x})$ and $u(\mathbf{x})$, known as upper and lower bounds, respectively. It is non-empty (or consistent) if and only if $l(\mathbf{x}) \leq u(\mathbf{x})$ is satisfied.

Example 3.1 Consider interval $[\overline{x}y, x + y]$ which represents an incompletely specified function. It is composed of four completely specified functions: $\overline{x}y$, y, $x \oplus y$, and $x + y$. Each of them has a don't care set represented by function x. □

Application of existential quantification $\exists$ and universal quantification $\forall$ to lower and upper bounds of the interval enables convenient selection of its member functions that are vacuous, i.e., independent in certain variables.

Example 3.2 Consider abstraction of x from the interval in Example 3.1:

$$[\exists x(\overline{x}y), \forall x(x + y)]$$

The abstraction yields non-empty interval that is composed of a unique function that is vacuous in x: $[y, y]$. Abstraction of y, however, results in empty interval since the relation between its lower and upper bounds is not satisfied: $[\overline{x}, x]$ is empty. □

We will use notation $\nabla x[l(\mathbf{x}), u(\mathbf{x})]$ to represent abstraction $[\exists x l(\mathbf{x}), \forall x u(\mathbf{x})]$.

3.2.2 Parameterized Abstraction

To determine subsets of variables whose abstraction preserves consistency of a symbolic statement (or a formula) we use *parameterized abstraction* construct. It parameterizes computational form with a set of auxiliary decision variables c that are used to guide variable abstraction decisions. An assignment to c effects consistency of a computational form and thereby determines feasibility of abstracting a corresponding variable subset.

We use the "if-then-else" operator $ITE(c, x, y)$ to encode effect of quantifying variable subsets from a formula. Defined as $cx + \overline{c}y$, the operator selects between variables x and y depending on value of c. As stated, it provides a mechanism to parameterize signal dependencies in a Boolean function. It can be also generalized to the selection between functions. In particular, $ITE(c, f(\mathbf{x}), \exists x f(\mathbf{x}))$ encodes a decision of existential quantification of x from $f(\mathbf{x})$, similarly for the universal quantification.

Example 3.3 We can parameterize abstraction of variable x from interval $[\overline{x}y, x+y]$ using *ITE* operator and auxiliary variable c_x as

$$[ITE(c_x, \exists x(\overline{x}y), \overline{x}y), ITE(c_x, \forall x(x+y), x+y)]$$

or equivalently $[c_x y + \overline{c_x}(\overline{x}y), c_x y + \overline{c_x}(x+y)]$. Subsequent parameterization of y transforms lower bound $L_x = ITE(c_x, \exists x(\overline{x}y), \overline{x}y)$ into $L_{xy} = ITE(c_y, \exists y L_x, L_x)$. The effect of decisions on c variables then has a form depicted in the tree below:

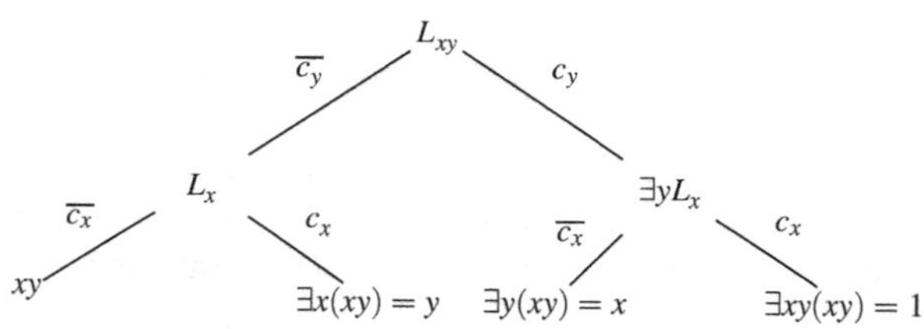

Parameterization of universal quantification has analogous application to an upper bound of the interval, producing U_{xy}. □

We rely on consistency of the "less-than-or-equal" relation to find decomposition of a Boolean function. In the example below it is illustrated determining feasible abstraction of the interval variable subsets, and in finding its member functions with smallest support (i.e., with fewest variables).

Example 3.4 For each assignment to c variables the consistency of relation $\leq$ between interval bounds in Example 3.3 determines existence of a function that is independent of the corresponding variables. The tree expansion over decision variables shows that there are only two feasible abstractions, marked with ✓:

Of the two, there is only one non-trivial abstraction $\nabla x[\overline{x}y, x+y]$ and it yields single member interval $[y, y]$. □

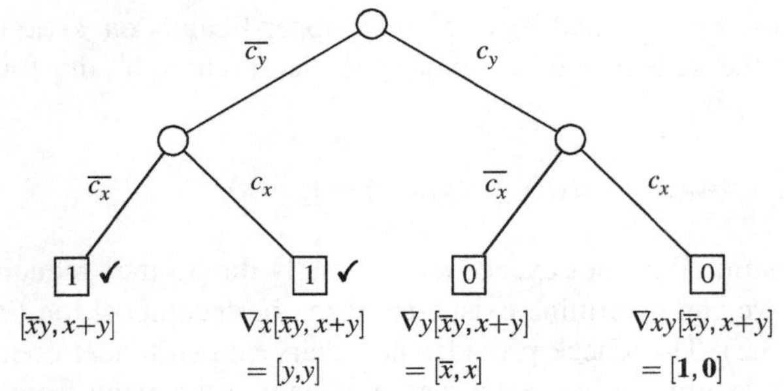

In a typical computational form, consistent assignment to the decision variables must hold universally, independent of non-decision variables. We can therefore compute *characteristic function* of consistent assignments implicitly, universally abstracting non-decision variables.

Example 3.5 For the parameterized bounds L_{xy} and U_{xy} from Example 3.3 the characteristic function of consistent assignments in $[L_{xy}, U_{xy}]$ is computed implicitly as

$$\forall x, y[L_{xy}, U_{xy}] \equiv \forall x, y[\overline{L_{xy}} + U_{xy} = \mathbf{1}] \equiv \overline{c_x}\,\overline{c_y} + c_x\,\overline{c_y} \equiv \overline{c_y}$$

The computed function states that abstraction of variable y yields an empty interval. Therefore, the interval contains no function that is independent of y. □

3.3 Bi-decomposition of Incompletely Specified Functions

This section gives formal statement of bi-decomposition over 2-input decomposition primitives, namely OR and XOR.

3.3.1 OR Decomposition

For a completely specified function $f(\mathbf{x})$, the decomposition of this type is described in terms of equation below:

$$f(\mathbf{x}) = g_1(\mathbf{x}_1) + g_2(\mathbf{x}_2) \tag{3.1}$$

When function is incompletely specified with interval $[l(\mathbf{x}), u(\mathbf{x})]$ we need to make sure that *OR* composition $g_1 + g_2$ is a member function of the interval.

Let $\underline{\mathbf{x}}_1$ and $\underline{\mathbf{x}}_2$ be signal subsets in which decomposition functions g_1 and g_2 are, respectively, vacuous, i.e., are functionally independent. (The underline in $\underline{\mathbf{x}}_i$ indicates that the computed g_i is independent in these variables.) Vacuous in $\underline{\mathbf{x}}_1$ function g_1 must not exceed largest member $u(\mathbf{x})$ in all its minterm points, independent of $\underline{\mathbf{x}}_1$, i.e., relation $g_1(\mathbf{x}_1) \leq \forall \underline{\mathbf{x}}_1 u(\mathbf{x})$ must be satisfied. Otherwise g_1 is either not contained in the interval or not independent of $\underline{\mathbf{x}}_1$. Similarly, $g_2(\mathbf{x}_2) \leq \forall \underline{\mathbf{x}}_2 u(\mathbf{x})$

must hold. Thus, $\forall \underline{\mathbf{x}}_1 u(\mathbf{x})$ and $\forall \underline{\mathbf{x}}_2 u(\mathbf{x})$ give upper bounds on $g_1(\mathbf{x}_1)$ and $g_2(\mathbf{x}_2)$. To ensure that the selection of g_1 and g_2 is "large enough" the following must hold:

$$l(u) \leq \forall \underline{\mathbf{x}}_1 u(\mathbf{x}) + \underline{\mathbf{x}}_2 u(\mathbf{x}) \qquad (3.2)$$

The OR composition does not exceed u universally due to the "reducing" effect of $\forall$ on u. Thus, we can determine existence of the bi-decomposition limiting check to relation in (3.2). This check provides necessary and sufficient condition for the existence of OR decomposition and is a re-statement of the result from [17].

AND Decomposition. As indicated in [17], AND decomposition of f can be obtained from OR decomposition utilizing dual property of the two gates. For an incompletely specified function $[l, u]$ we can find complemented g_1 and g_2 by establishing OR decomposability of the interval complement, derived as $\overline{[l, u]} = [\overline{u}, \overline{l}]$.

3.3.2 XOR Decomposition

We first describe XOR decomposability condition for a completely specified function $f(\mathbf{x})$. To derive an existence condition for the XOR decomposition

$$f(\mathbf{x}) = g_1(\mathbf{x}_1) \oplus g_2(\mathbf{x}_2) \qquad (3.3)$$

requires partitioning of $\mathbf{x}_1$ and $\mathbf{x}_2$ into finer subsets. Let $\underline{\mathbf{x}}_1$ and $\underline{\mathbf{x}}_2$ be subsets of variables in which g_1 and g_2 are, respectively, vacuous, and let $\mathbf{x}_3$ be a set of variables on which both decomposition functions depend. We can then state necessary and sufficient condition for the existence of XOR decomposition as follows:

Proposition 3.1 *XOR bi-decomposition*

$$f(\mathbf{x}) = g_1(\underline{\mathbf{x}}_2, \mathbf{x}_3) \oplus g_2(\underline{\mathbf{x}}_1, \mathbf{x}_3) \qquad (3.4)$$

exists if and only if

$$f(\underline{\mathbf{x}}_1, \underline{\mathbf{x}}_2, \mathbf{x}_3) \neq f(\underline{\mathbf{y}}_1, \underline{\mathbf{x}}_2, \mathbf{x}_3) \qquad (3.5)$$

$$\Downarrow$$

$$\forall \underline{\mathbf{y}}_2 [f(\underline{\mathbf{x}}_1, \underline{\mathbf{y}}_2, \mathbf{x}_3) \neq f(\underline{\mathbf{y}}_1, \underline{\mathbf{y}}_2, \mathbf{x}_3) \qquad (3.6)$$

We derived conditions in the above proposition when analyzing library requirements for an advanced technology [13]. In [14] authors recently and independently stated analogous proposition in terms of the unsatisfiability problem. We therefore show correctness of the above proposition giving only an information-theoretical argument: For (3.4) to hold, it must be that all onset/offset minterms in f that cannot be distinguished by g_1 (3.5) must be distinguished by g_2 (3.6).

For an incompletely specified function $[l(\mathbf{x}), u(\mathbf{x})]$ the consistency constrain (3.5) $\Rightarrow$ (3.6) of Proposition 3.1 changes to

$$[l(\underline{\mathbf{x}}_1, \underline{\mathbf{x}}_2, \mathbf{x}_3) \neq l(\underline{\mathbf{y}}_1, \underline{\mathbf{x}}_2, \mathbf{x}_3)] \wedge [h(\underline{\mathbf{x}}_1, \underline{\mathbf{x}}_2, \mathbf{x}_3) \neq h(\underline{\mathbf{y}}_1, \underline{\mathbf{x}}_2, \mathbf{x}_3)]$$
$$\Downarrow$$
$$\forall \underline{\mathbf{y}}_2 [[l(\underline{\mathbf{x}}_1, \underline{\mathbf{y}}_2, \mathbf{x}_3) \neq h(\underline{\mathbf{y}}_1, \underline{\mathbf{y}}_2, \mathbf{x}_3)] \vee [h(\underline{\mathbf{x}}_1, \underline{\mathbf{y}}_2, \mathbf{x}_3) \neq l(\underline{\mathbf{y}}_1, \underline{\mathbf{y}}_2, \mathbf{x}_3)]]$$

The above statement extends containment relation (3.5) $\Rightarrow$ (3.6) by reducing lower bound (3.5) and increasing upper bound (3.6) as much as possible. The relation provides the condition for `XOR` bi-decomposition of incompletely specified functions, previously unsolved problem [14].

3.4 Parameterized Decomposition

Section 3.3 decomposition checks assume that the $\mathbf{x}_1$ and $\mathbf{x}_2$ subsets are predetermined. Finding such feasible subsets, however, may not be straightforward and depending on the objectives could potentially require an exponential search if performed explicitly. Our solution to the problem is to perform the search implicitly, formulating the problem symbolically and solving it by leveraging the capability of binary decision diagrams to compactly represent certain combinatorial subsets.

3.4.1 OR Parameterization

We use (3.2) to find feasible `OR` decompositions implicitly. It is used to construct a computational form that parameterizes the $\forall$ operation applied to variables $\mathbf{x}$:

$U \leftarrow u$;
for each $x \in \mathbf{x}$ **do**
  $U \leftarrow ITE(c_x, U, \forall x U)$;
end for

Such iterative parameterization gives function $U(\mathbf{c}, \mathbf{x})$ that encodes the effect of abstracting all variable subsets from u, where variable x is abstracted *iff* $c_x = 0$.

The parameterized function $U(\mathbf{c}, \mathbf{x})$ can be used in (3.2) to encode possible supports to g_1 and g_2 in terms of the decision variables $\mathbf{c}_1$ and $\mathbf{c}_2$:

$$l(\mathbf{x}) \leq U_1(\mathbf{x}, \mathbf{c}_1) + U_2(\mathbf{x}, \mathbf{c}_2) \quad (3.7)$$

For any feasible assignment to $\mathbf{c}_1$ and $\mathbf{c}_2$, the above relation must hold universally, irrespective of values on $\mathbf{x}$. Thus, computational form

$$\mathrm{Bi}(\mathbf{c}_1, \mathbf{c}_2) \equiv \forall \mathbf{x}[\overline{l(\mathbf{x})} + U_1(\mathbf{x}, \mathbf{c}_1) + U_2(\mathbf{x}, \mathbf{c}_2)] \quad (3.8)$$

yields a characteristic function of all feasible supports for g_1 and g_2: it evaluates to truth *iff* assignments to $\mathbf{c}_1$ and $\mathbf{c}_2$ induce feasible supports for g_1 and g_2.

We illustrate potentially scalable nature of BDDs to handle computation in (3.8) decomposing multiplexer function for its various support sizes:

Max width		Bi computation		Best partition	
Control	Data	BDD size	Time(s)	$(\|x_1\|, \|x_2\|)$	No. of Choices
2	4	23	0.00	(4, 4)	6
3	8	43	0.01	(7, 7)	70
4	16	79	0.09	(12, 12)	12870
5	32	147	1.35	(21, 21)	6E8.0
6	64	279	20.56	(38, 38)	1.8E18

The above table gives results of the computation in terms of multiplexer widths, BDD size and time required to compute Bi, and the best support sizes of g_1 and g_2. As the table suggests, the amount of resources required in computation grows moderately for smaller problem instances and is tolerable even for a larger function.

We point out that the exhaustive computational form (3.8) could be relaxed to produce solution subsets to (3.7), instead of producing a complete solution. For example, in place of (3.8) a specialized satisfiability procedure could be used to produce solutions with additional optimization constraints. Specialized BDD-based abstraction techniques that monitor resource consumption could be also deployed to produce solution subsets. Another possibility is to rely on a greedy assignment selection to $\mathbf{c}_1$ and $\mathbf{c}_2$ targeting disjoint subsets $\mathbf{x}_1$ and $\mathbf{x}_2$. More detailed discussion on selecting best $\mathbf{x}_1$ and $\mathbf{x}_2$ is given later, in Section 3.5.

3.4.2 XOR Parameterization

To simplify presentation we compute characteristic function of all feasible support partitions for a completely specified function. As before, encoding of possible supports for g_1 and g_2 is performed using two sets of auxiliary variables $\mathbf{c}_1$ and $\mathbf{c}_2$. Using $\mathbf{c}_1$, (3.5) is transformed into $f(\mathbf{x}) \neq F_1(\mathbf{x}, y, \mathbf{c}_1)$, where F is derived from $f(\mathbf{x})$ replacing each of its variables x_i with $ITE(c_{1i}, x_i, y_i)$. Similarly, part $f(\underline{\mathbf{x}}_1, \underline{\mathbf{y}}_2, \mathbf{x}_3)$ from (3.6) is parameterized with $\mathbf{c}_2$ to construct $F_2'(\mathbf{x}, \mathbf{y}, \mathbf{c}_2)$. It encodes selection of vacuous variables for g_2. The last component $f(\underline{\mathbf{y}}_1, \underline{\mathbf{y}}_2, \mathbf{x}_3)$ is transformed into $F_2''(\mathbf{x}, \mathbf{y}, \mathbf{c}_1, \mathbf{c}_2)$, replacing each variable in $f(\mathbf{x})$ with $\overline{ITE}(c_{1i} \cdot c_{2i}, x_i, y_i)$. Universally abstracting $\mathbf{x}$ and $\mathbf{y}$ variables gives representation of all feasible supports for g_1 and g_2:

$$Bi(\mathbf{c}_1, \mathbf{c}_2) \equiv \forall \mathbf{x}, \mathbf{y}[(f \neq F_1) \Rightarrow (F_2' \neq F_2'')] \tag{3.9}$$

We compare implicit computation of decomposition choices to a greedy algorithm for the XOR decomposition, used by authors in [17, 22]. Starting from a seed partition, the algorithm greedily extends support subsets calling XOR decomposability check in its inner loop. Although efficient in general, the check has potentially formidable runtime. The profile of its behavior on a 16-bit adder is given in the table below; it is compared against our implicit computation:

Output			Time(s)	
Sum bit	No. of Inputs	Best part.	Implicit	[17] Check
s2	7	(2,5)	0.01	0.00
s4	11	(2,9)	0.06	0.13
s6	15	(2,13)	0.12	4.44
s8	19	(2,17)	0.13	71.05
⋮			⋮	
s16	33	(2,31)	0.42	Time out

For a subset of sum-bit functions the table lists runtime for both techniques. (The *best part.* column gives data generated by our implicit enumeration of feasible partitions.) Although not typical, it is interesting that where a rather efficient greedy check times out after an hour, an implicit exhaustive computation takes only 0.42 s.

In general, we can use best partition produced by the exhaustive implicit computation to evaluate and tune greedy algorithm or to improve some other approximate technique.

3.5 Implementation Details of Sequential Synthesis

This section describes a sequential synthesis flow that first extracts incompletely specified logic accounting for unreachable states in a design and then uses bi-decomposition to synthesize technology-independent circuit.

3.5.1 Extraction of Incompletely Specified Logic

Unreachable states of a design form don't cares for the combinational logic. Due to the complexity of computing unreachable states even in designs of modest size, incompletely specified combinational logic is extracted with respect to an approximation of unreachable states. Unlike other partitioning approaches that try to produce a good approximation of unreachable states in reasonable time [10, 16], our objective is to compute a good approximation with respect to support of individual functions. A similar approach to approximate unreachable states using induction was proposed in [7].

We perform state-space exploration with forward reachability analysis for overlapping subsets of registers. These subsets are selected using structural dependence of next-state and primary outputs on the design latches. The selection tries to create partitions maximizing accuracy of reachability analysis for present-state signals $supp\_ps(f)$ output function f. In particular, the partitioning tries to meet the following goals:

- For each function f, present-state inputs $supp\_ps(f)$ are represented in at least one partition.
- Each partition selects additional logic to maximize accuracy of reachability analysis.

After completing reachability analysis for a partition, an incomplete specification of signals that depend on the partition latches becomes available in the form of a interval notation. For each signal, its interval pre-processed with the ∇ operation eliminates vacuous variables, selecting a dependence on the least number of variables. The interval is then used for performing bi-decomposition. Figure 3.1 exemplifies OR bi-decomposition applied to function $f = \overline{a}\overline{b} + \overline{a}c + bc$ of its output signal. The bi-decomposition of $[f \cdot \overline{\overline{a}bc},\ f + \overline{a}bc]$ finds OR decomposition of f in $g_1 = \overline{a}\overline{b} + bc$ simplifying the circuit.

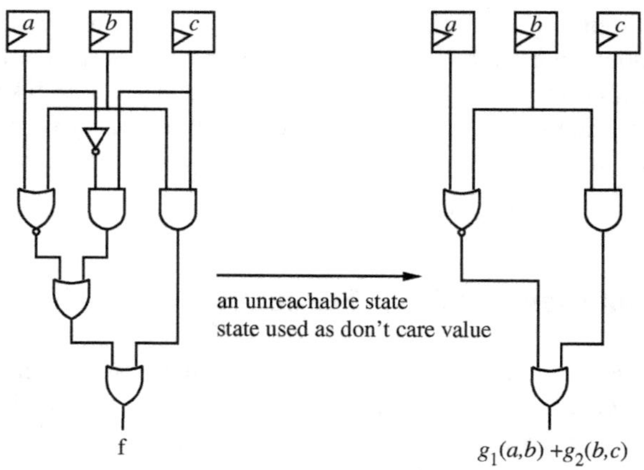

Fig. 3.1 Bi-decomposition with unreachable states. State $\overline{a}bc$ is used as a don't care condition to find OR decomposition that simplifies circuit

3.5.2 Exploring Decomposition Choices

The characteristic function Bi gives all feasible supports for decomposition functions. Since the provided variety of choices could be very large, we restrict them to a subset of desired solutions. The restriction targets minimization and balanced selection of supports in decomposition functions. It is achieved symbolically, as described below.

Let $w_i(\mathbf{c})$ be characteristic function of assignments to $\mathbf{c}$ that have weight i (i.e., have exactly i decision variables set to 1). For a given $n = |\mathbf{c}|$ it represents combinatorial subsets $\binom{n}{i}$. This function has compact representation in terms of BDDs. Given a desired support size $k_1 = |\mathbf{x}_1|$ of g_1, and of $k_2 = |\mathbf{x}_2|$ of g_2, existence of the decomposition is determined constraining Bi with its corresponding solution space:

$$Bi(\mathbf{c}_1, \mathbf{c}_2) \cdot w_{k_1}(\mathbf{c}_1) \cdot w_{k_2}(\mathbf{c}_2)$$

If the resulting function is not empty, then the desired decomposition exists. To target balanced decomposition of a function we seek feasible k_1 and k_2 minimizing $\max(k_1, k_2)$. Such simultaneous minimization of k_1 and k_2 balances supports $\mathbf{x}_1$ and $\mathbf{x}_2$, favoring their disjoint selection. (This is in contrast to [14], where different cost measures are used for each of the objectives.)

To avoid "trial-and-error" search for feasible support sizes we select desired k_1 and k_2 from a computed set of feasible pairs for g_1 and g_2. Suppose function $\kappa_i(\mathbf{e})$ encodes integer k_i in terms of the $\mathbf{e}$ variables. Then function $K(\mathbf{c}, \mathbf{e}) = \sum_{i=0}^{n} w_i(\mathbf{c}) \cdot \kappa_i(\mathbf{e})$ relates a decision variables assignment to integer encoding of the support size it induces. This function is used within the computational form

$$Bi_{\kappa}(\mathbf{e}_1, \mathbf{e}_2) \equiv \exists \mathbf{c}_1 \mathbf{c}_2 [Bi(\mathbf{c}_1, \mathbf{c}_2) \cdot K(\mathbf{c}_1, \mathbf{e}_1) \cdot K(\mathbf{c}_2, \mathbf{e}_2)]$$

to generate all feasible size pairs (k_1, k_2).

The Bi_{κ} function should be post-processed to purge pairs that are dominated by other, better solutions. For example, pair (3, 5) is dominated by pair (3, 4) since it produces smaller distribution of supports between g_1 and g_2. Let $\mathtt{gte}(\mathbf{e}, \mathbf{e}')$ describe "greater-than-or-equal" relation between a pair of integers encoded with $\mathbf{e}$ and $\mathbf{e}'$; similarly, let equ describe the equality relation. We then define the dominance relation between bi-tuples $\varepsilon \equiv (\mathbf{e}_1, \mathbf{e}_2)$ and $\varepsilon' \equiv (\mathbf{e}_1', \mathbf{e}_2')$ as

$$\mathtt{dom}(\varepsilon, \varepsilon') = \mathtt{gte}(\mathbf{e}_1, \mathbf{e}_1') \cdot \mathtt{gte}(\mathbf{e}_2, \mathbf{e}_2') \cdot \overline{\mathtt{equ}(\mathbf{e}_1, \mathbf{e}_1') \cdot \mathtt{equ}(\mathbf{e}_2, \mathbf{e}_2')}$$

Using this relation subtraction of the dominated solutions from Bi_{κ} is performed as

$$\forall \varepsilon' [Bi_{\kappa}(\varepsilon') \leq Bi_{\kappa}(\varepsilon) \cdot \overline{\mathtt{dom}(\varepsilon, \varepsilon')}]$$

It states that if an assignment to ε' is in Bi_{κ} (left-hand side of the relation), then its dominated assignments to ε should be subtracted from Bi_{κ} (right-hand side of the assignment).

To complete decomposition of a function we need to find functions g_1 and g_2. For the OR decomposition, possible functions g_1 and g_2 can be deduced directly from the corresponding existence condition (3.2), universally quantifying out variables in which g_1 and g_2 are vacuous. To construct XOR decomposition functions we use algorithm from [17].

3.5.3 Synthesis Algorithm

Our logic optimization algorithm selectively re-implements functions of circuit signals relying on bi-decomposition of extracted incompletely specified logic. The pseudocode code in Algorithm 1 captures general flow of the optimization.

Algorithm 1 Logic optimization loop

create latch partitions of a design;
selectively collapse logic;
while (more logic to decompose) **do**
  select a signal and its function $f(\mathbf{x})$;
  retrieve unreachable states $u(\mathbf{x})$;
  abstract vars from interval $[f \cdot \overline{u},\ f+u]$;
  apply bi-decomposition to interval;
end while

The algorithm first creates overlapping partitions of a design. These partitions are formed according to Section 3.5.2 and are typically limited to 100 latches. Additional connectivity cost measures are used to control size of a partition. For each partition computation of unreachable states is delayed until being requested by a function that depends on its present-state signals. BDDs for computed reachable states are then stored in a separate node space for each partition. When retrieving unreachable states for a given support, their conjunctive approximation is brought together to a common node space.

To re-decompose logic of a design the algorithm first creates functional representation for selected signals in terms of their cone inputs or in terms of other intermediate signals. The decision on whether to select a signal is driven by an assessed impact of bi-decomposition on circuit quality: if it has potential to improve variable partition, logic sharing, or timing over existing circuit structure, then signal is added to a list of re-decomposition candidates.

The logic of candidate signals is processed in topological order until it is fully implemented with simple primitives. This processing constitutes main loop of the algorithm. After a signal and its function $f(\mathbf{x})$ in the loop are selected, a set of unreachable states $u(\mathbf{x})$ are retrieved. This set is derived from reachability information of partitions that $f(\mathbf{x})$ depends on.

Before applying bi-decomposition to the incompletely specified function

$$[f(\mathbf{x}) \cdot \overline{u(\mathbf{x})},\ f(\mathbf{x}) + u(\mathbf{x})]$$

the algorithm tries to abstract some of the interval variables while keeping it consistent; this eliminates redundant inputs. The bi-decomposition is then applied targeting potential logic sharing and balanced partition of $\mathbf{x}$, as described in Section 3.5.3. From a generated set of choices, partition that best improves timing and logic sharing is selected. Figure 3.2 illustrates bi-decomposition that benefits from logic sharing. The transformation re-uses logic of g_1, which was present in the network but was not in the fanin of f.

3.6 Experimental Evaluation

From a suite of publicly available benchmarks we selected a subset of sequential circuits and assessed effect of unreachable states on bi-decomposition. Three types

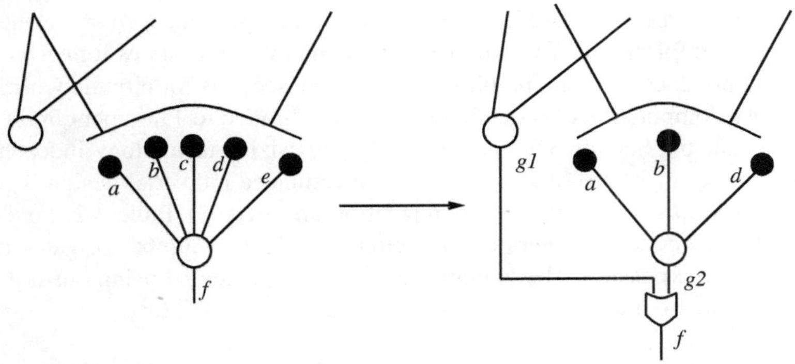

Fig. 3.2 Bi-decomposition which benefits from re-using existing logic: node g_1 is shared in the $f = g_1 + g_2$ decomposition

of bi-decomposition were applied to functions of their output and next-state logic: OR, AND, and XOR. They are evaluated in terms of their ability to reduce maximum support of functions g_1 and g_2. Experiments with and without reachable state-space analysis were performed.

The experimental results are given in Table 3.1. The table first lists circuit name, along with its corresponding number of inputs/outputs and latches. Each circuit was structurally pre-processed to remove cloned, dead, and constant latches. The *#dec.* column gives number of functions for which non-trivial decomposition was identified. The average ratio between maximum support sizes of g_1 and g_2 and support size of the function being decomposed is given in *avg. reduct.* column. Note that the reduction of less than 0.5 (as in s713 and s838) indicates that both g_1 and g_2 tend to be vacuous in some of the variables.

Table 3.1 Application of bi-decomposition to functions of next-state and output logic (without and with state analysis)

Name	Original circuit		No states		With states		
	Input/output	Latches	#Dec.	Avg. reduct.	$\log_2$ states	#Dec.	Avg. reduct
s344	10/11	15	18	0.781	12	18	0.634
s526	3/6	21	21	0.775	14	21	0.556
s713	36/23	19	40	0.652	11	40	0.453
s838	36/2	32	33	0.540	5	33	0.088
s953	17/23	29	29	0.607	13	29	0.565
s1269	18/10	37	39	0.672	31	39	0.671
s5378	36/49	163	145	0.609	125	145	0.603
s9234	36/39	145	97	0.754	141	97	0.774
Average reduction:				0.673			0.54

The results are collected for two experiments: with and without state-space information. The $\log_2$ of computed reachable states is also listed in the table. Computed average reduction ratios suggest that decomposability of a function improves as the number of unreachable states gets larger. The unreachable states did not contribute much to s5378 largely because its logic is highly decomposable even in the absence of state-space information. The runtime to compute reachable states for each of the circuits did not exceed 1 min, requiring at most few seconds for circuits with 32 or less latches. Computation of bi-decomposition was limited to 1 min per circuit.

We evaluate our Section 3.5.3 Algorithm 1 synthesizing technology-independent netlists for a set of macro-blocks of a high-performance industrial design. Results of the netlists optimized with bi-decomposition are given in Table 3.2. First four columns list general parameters of each circuit, including number of gates it has in its `and/inv` expansion. The circuits were first pre-processed using our in-house tool, by optimizing it against publicly available mcnc.genlib library.

Table 3.2 Results of applying bi-decomposition in synthesis of industrial circuits

Name	Original circuit			Pre-processed		Algor. 1	
	Input/Output	Latches	AND	Area	Delay	Area	Delay
seq4	108/202	253	1845	3638	44.8	2921	41.9
seq5	66/12	93	925	1951	47.2	1807	41.6
seq6	183/74	142	811	1578	34.9	1487	36.0
seq7	173/116	423	3173	6435	52.4	5348	48.3
seq8	140/23	201	2922	6183	50.1	5427	48.8
seq9	212/124	353	3896	8250	56.0	6938	45.2
Average reduction:						0.88	0.94

An implementation of the algorithm was then applied to improve each of the circuits. Columns *Pre-processed* and *Algor. 1* compare area (which corresponds to the number of literals) and delay (estimated with a load-dependent model) of mapped netlists before and after running our algorithm. The additional area and timing savings are due to the algorithm, with the average area and delay reductions of 0.88 and 0.94, respectively. We attribute these gains to the algorithm's ability to implicitly explore reach arsenal of decomposition choices during bi-decomposition. Optimization of each circuit was completed within 4 min of runtime.

3.7 Conclusions and Future Work

Extraction of incompletely specified logic using under-approximation of unreachable states in sequential designs offers valuable opportunity for reducing the circuit complexity. We developed a novel formulation of symbolic bi-decomposition and showed that the extracted logic has better implementation, with substantial area and delay improvements. The introduced symbolic bi-decomposition computes decomposition choices implicitly and enables their efficient subsetting using BDDs. Selecting best decomposition patterns during synthesis, we improved circuit quality of publicly available and realistic industrial design. We are currently working on ways to further maximize logic sharing through bi-decomposition and to apply it in a re-synthesis loop of well-optimized designs.

References

1. Ashenhurst, R.L.: The decomposition of switching functions. Annals of Computation Laboratory, Harvard University **29**, 74–116 (1959)
2. Baumgartner, J., Mony, H., Paruthi, V., Kanzelman, R., Janssen, G.: Scalable sequential equivalence checking across arbitrary design transformations. In: Proceedings of ICCD, San Jose, CA, pp. 259–266. (2006)
3. Brown, F.M.: Boolean Reasoning. Kluwer, Boston, MA (1990)
4. Bryant, R.E.: Graph-based algorithms for Boolean function manipulation. IEEE Transactions on Computers C-**35**(6), 677–691 (1986)
5. Case, M.L., Kravets, V.N., Mishchenko, A., Brayton, R.K.: Merging nodes under sequential observability. In: Proceedings of DAC, Anaheim, CA, pp. 540–545. (2008)
6. Case, M.L., Mishchenko, A., Brayton, R.K.: Inductive finding a reachable state-space overapproximation. In: IWLS, Vail, CO, pp. 172–179. (2006)
7. Case, M.L., Mishchenko, A., Brayton, R.K.: Cut-based inductive invariant computation. In: IWLS, Lake Tahoe, CA, pp. 172–179. (2008)
8. Cheng, K.T., Entrena, L.A.: Sequential logic optimization by redundancy addition and removal. In: Proceedings of ICCAD, San Jose, CA, pp. 310–315. (1993)
9. Cho, J., Hachtel, G., Macii, E., Poncino, M., Somenzi, F.: Automatic state decomposition for approximate FSM traversal based on circuit analysis. IEEE Transactions on CAD **15**(12), 1451–1464 (1996)
10. Cortadella, J.: Timing-driven logic bi-decomposition. IEEE Transactions on CAD **22**(6), 675–685 (2003)
11. Een, N., Sorensson, N.: An extensible SAT-solver. In: Proceedings of SAT, Santa Margherita Ligure, Italy, pp. 502–518. (2003)
12. van Eijk, C.: Sequential equivalence checking based on structural similarities. IEEE Transactions on CAD **19**(7), 814–819 (2000)
13. Kravets, V.N. et al.: Automated synthesis of limited-switch dynamic logic (LSDL) circuits. Prior Art Database (ip.com) (March 2008)
14. Lee, R.-R., Jiang, J.-H., Hung, W.-L.: Bi-decomposing large Boolean functions via interpolation and satisfiability solving. In: Proceedings of DAC, Anaheim, CA, pp. 636–641. (2008)
15. Lin, B., Touati, H., Newton, R.: Don't care minimization of multi-level sequential networks. In: Proceedings of ICCAD, San Jose, CA, pp. 414–417. (1990)
16. Mishchenko, A., Case, M.L., Brayton, R.K., Jang, S.: Scalable and scalable-verifiable sequential synthesis. In: Proceedings of ICCAD, San Jose, CA, pp. 234–241. (2008)
17. Mishchenko, A., Steinbach, B., Perkowski, M.: An algorithm for bi-decomposition of logic functions. In: Proceedings of DAC, Las Vegas, NV, pp. 103–108. (2001)
18. Roth, J.P., Karp, R.: Minimization over Boolean graphs. IBM Journal of Research and Development **6**(2), 227–238 (1962)
19. Sasao, T., Butler, J.: On bi-decomposition of logic functions. In: IWLS, Tahoe City, CA, (1997)
20. Savoj, H., Brayton, R.K.: The use of observability and external don't cares for the simplification of multi-level networks. In: Proceedings of DAC, Orlando, FL, pp. 297–301. (1990)
21. Stanion, T., Sechen, C.: Quasi-algebraic decomposition of switching functions. In: Proceedings of the 16th Conference on Advance Research in VLSI, Ann Arbor, MI, pp. 358–367. (1998)
22. Steinbach, B., Wereszczynski, A.: Synthesis of multi-level circuits using EXOR-gates. In: Proceedings of IFIP WG 10.5 – Workshop on Application of the Reed-Muller Expansion in Circuit Design, Chiba City, Japan, pp. 161–168. (1995)
23. Yang, C., Cieselski, M., Singhal, V.: BDS: A BDD-based logic optimization system. In: Proceedings of DAC, Los Angeles, CA, pp. 92–97. (2000)

References

Chapter 4
Boolean Factoring and Decomposition of Logic Networks

Robert Brayton, Alan Mishchenko, and Satrajit Chatterjee

Abstract This chapter presents new methods for restructuring logic networks based on fast Boolean techniques. The bases for these are (1) a cut-based view of a logic network, (2) exploiting the uniqueness and speed of disjoint-support decompositions, (3) a new heuristic for speeding up computations, (4) extension for more general decompositions, and (5) limiting local transformations to functions with 16 or less inputs, so that fast truth table manipulations can be used. The proposed Boolean methods lessen the structural bias of algebraic methods, while still allowing for high speed and multiple iterations. Experimental results on area reduction of K-LUT networks, compared to heavily optimized versions of the same networks, show an average additional reduction of 5.4% in LUT count while preserving delay.

4.1 Introduction

The traditional way of decomposing and factoring logic networks uses algebraic methods. These represent the logic of each node as a sum of products (SOP) and apply algebraic methods to find factors or divisors. Kerneling or two-cube division is used to derive candidate divisors. These methods can be extremely fast if implemented properly, but they are biased because they rely on an SOP representation of the logic functions, from which only algebraic divisors are extracted. A long-time goal has been to develop similarly fast methods for finding and using good Boolean divisors, independent of any SOP form.

We present a new type of Boolean method, which uses as its underlying computation, a fast method for disjoint-support decomposition (DSD). This approach was influenced by the efficient BDD-based computation of complete maximum DSDs proposed in [6], but it has been made faster by using truth tables and sacrificing

---

R. Brayton (✉)
Department of EECS, University of California, Berkeley, CA, USA
e-mail: brayton@eecs.berkeley.edu

This work is based on "Boolean factoring and decomposition of logic networks", Robert Brayton, Alan Mishchenko, and Satrajit Chatterjee, in Proceedings of the 2008 IEEE/ACM international Conference on Computer-Aided Design, (2008) ACM, 2008.

S.P. Khatri, K. Gulati (eds.), *Advanced Techniques in Logic Synthesis, Optimizations and Applications*, DOI 10.1007/978-1-4419-7518-8_4,

completeness for speed. However, this heuristic, in practice, tends to find the maximum DSDs. This fast computation is extended for finding Boolean divisors for the case of non-disjoint-support decomposition.

Methods based on these ideas can be seen as a type of Boolean rewriting of logic networks, analogous to rewriting of AIG networks [27]. AIG rewriting has been successful, partly because it can be applied many times due to its extreme speed. Because of this, many iterations can be used, spreading the area of change and compensating for the locality of AIG-based transforms. Similar effects can be observed with the new methods.

This chapter is organized as follows. Section 4.2 provides the necessary background on DSD as well as a cut-based view of logic networks. Section 4.3 shows new results on extending DSD methods to non-disjoint decompositions. Section 4.4 looks at reducing the number of LUTs on top of a high-quality LUT mapping and high-effort resynthesis. Section 4.5 compares the proposed decomposition to Boolean matching. Section 4.6 presents experimental results. Section 4.7 concludes the chapter and reviews future work.

4.2 Background

A *Boolean network* is a directed acyclic graph (DAG) with nodes corresponding to logic gates and directed edges corresponding to wires connecting the gates. We use the terms *Boolean networks*, *logic networks*, and *circuits* interchangeably. We use the term *K-LUT network* to refer to Boolean networks whose nodes are K-input lookup tables (K-LUTs).

A node n has zero or more *fanins*, i.e., nodes that are driving n, and zero or more *fanouts*, i.e., nodes driven by n. The *primary inputs* (PIs) are nodes of the network without fanins. The *primary outputs* (POs) are a specified subset of nodes of the network.

An *And-Inverter Graph* (AIG) is a Boolean network whose nodes are 2-input ANDs. Inverters are indicated by a special attribute on the edges of the network.

A cut of node n is a set of nodes, called leaves, such that

1. each path from any PI to n passes through at least one cut node and
2. for each cut node, there is at least one path from a PI to n passing through the cut node and not passing through any other cut nodes.

Node n is called the root of C. A *trivial cut* of node n is the cut $\{n\}$ composed of the node itself. A non-trivial cut is said to *cover* all the nodes found on the paths from the leaves to the root, including the root but excluding the leaves. A trivial cut does not cover any nodes. A cut is *K-feasible* if the number of its leaves does not exceed K. A cut C_1 is said to be *dominated* if there is another cut C_2 of the same node such that $C_2 \subset C_1$.

A *cover* of an AIG is a subset R of its nodes such that for every $n \in R$, there exists exactly one non-trivial cut $C(n)$ associated with it such that

1. if n is a PO, then $n \in R$,
2. if $n \in R$, then for all $p \in C(n)$ either $p \in R$ or p is a PI, and
3. if n is not a PO, then $n \in R$ implies there exists $p \in R$ such that $n \in C(p)$.

The last requirement ensures that all nodes in R are "used."

In this chapter, we sometimes use an AIG accompanied with a cover to represent a logic network. This is motivated by our previous work on AIG rewriting and technology mapping. The advantage of viewing a logic network as a cover of an AIG is that different covers of the AIG (and thus different network structures) can be easily enumerated using fast cut enumeration.

The *logic function* of each node $n \in R$ of a cover is simply the Boolean function of n computed in terms of $C(n)$, the cut leaves. This can be extracted easily as a truth table using the underlying AIG between the node and its cut. The truth table computation can be performed efficiently as part of the cut computation. For practical reasons, the cuts in this chapter are limited to at most 16 inputs.[1]

A completely specified Boolean function F *essentially depends* on a variable if there exists an input combination such that the value of the function changes when the variable is toggled. The *support* of F is the set of all variables on which function F essentially depends. The supports of two functions are *disjoint* if they do not contain common variables. A set of functions is *disjoint* if their supports are pair-wise disjoint.

A *decomposition* of a completely specified Boolean function is a Boolean network with one PO that is functionally equivalent to the function. A *disjoint-support decomposition* (DSD – also called simple disjunctive decomposition) is a decomposition in which the set of nodes of the resulting network are disjoint. Because of the disjoint supports of the nodes, the DSD is always a tree (each node has one fanout). The set of leaf variables of any sub-tree of the DSD is called a *bound* set, the remaining variables a *free* set. A *single* disjoint decomposition of a function consists of one block with a bound set as inputs and a single output feeding into another block with the remaining (free) variables as additional inputs. A *maximal DSD* is one in which each node cannot be further decomposed by DSD.

It is known that internal nodes of a maximal DSD network can be of three types: AND, XOR, and PRIME. The AND and XOR nodes may have any number of inputs, while PRIME nodes have support three or more and only a trivial DSD. For example, a 2:1 MUX is a prime node. A DSD is called *simple* if it does not contain prime nodes.

Theorem 4.1 *[4] For a completely specified Boolean function, there is a unique maximal DSD (up to the complementation of inputs and outputs of the nodes).*

There are several algorithms for computing the maximal DSD [6, 23, 38]. *Our implementation follows* [6] *but uses truth tables instead of BDDs for functional manipulation.*

[1] The fast methods of this chapter are based on bit-level truth table manipulations and 16 is a reasonable limit for achieving speed for this.

Definition 4.1 Given a maximal DSD, the set of all leaf variables of any sub-tree of the DSD are said to be in a separate block.

Example 4.1 $f = xy + z$ has x, y and x, y, z in separate blocks but not x, z.

4.3 General Non-disjoint Decompositions

A general decomposition has the form

$$F(x) = \hat{H}(g_1(a,b), \cdots, g_k(a,b), b, c)$$

If $|b|$=0, the decomposition is called *disjoint* or *disjunctive* and if $k = 1$, it is called *simple*.

Definition 4.2 A function F has an (a, b)-decomposition if it can be written as $F(x) = H(D(a,b), b, c)$ where (a, b, c) is a partition of the variables x, $|a| > 1$, and D is a single-output function.

An (a, b)-decomposition is simple but, in general, non-disjoint. It can also be written as

$$F(x) = \hat{H}(g_1(a,b), \ldots, g_{|b|+1}(a,b), c)$$

where $g_1(a,b) = D, g_2 = b_1, g_{|b|+1} = b_{|b|}$. In this case, it is non-simple but disjoint. For general decompositions, there is an elegant theory [35] on their existence. Kravets and Sakallah [19] applied this to constructive decomposition using support-reducing decompositions along with a pre-computed library of gates.

In our case, since $|a| > 1$, an (a, b)-decomposition is support reducing. Although less general, the advantage of (a, b)-decompositions is that their existence can be tested much more efficiently (as far as we know) by using cofactoring and the fast DSD algorithms of this chapter. Recent works use ROBDDs to test the existence of decompositions, with the variables (a, b) ordered at the top of the BDD (e.g., see [37] for an easy-to-read description).

The variables a are said to be in a separate block and form the *bound* set, the variables c are the *free* set, and the variables b are the *shared* variables. If $|b| = 0$, the decomposition is a DSD. $D(a, b)$ is called a *divisor* of F.

A particular cofactor of F with respect to b may be independent of the variables a; however, we still consider that a is (trivially) in a separate block in this cofactor. We call such cofactors, bound set *independent* cofactors, or *bsi*-cofactors; otherwise *bsd*-cofactors.

Example 4.2 If $F = deb + \bar{b}c$, then $F_{\bar{b}} = c$ is independent of the bound variable set $a = (d, e)$ i.e., it is a *bsi*-cofactor.

A version of the following theorem can be found in [38] as Proposition 2 where $t = 1$.

Theorem 4.2 *A function $F(a,b,c)$ has an (a,b)-decomposition if and only if each of the $2^{|b|}$ cofactors of F with respect to b has a DSD structure in which the variables a are in a separate block.*

Proof **If**: Suppose F has an (a,b)-decomposition. Then $F(x) = H(D(a,b),b,c)$. Consider a cofactor $F_{b^j}(x)$ with respect to a minterm of b^j, say $b^j = b_1\bar{b}_2\bar{b}_3b_4$, for $k=4$. This sets $b = 1,0,0,1$, giving rise to the function,

$$F_{b^j}(a,c) = H(D(a,1,0,0,1),1,0,0,1,c) \equiv H_{b^j}(D_{b^j}(a),c).$$

Thus this cofactor has a DSD with a separated. Note that if b^j is bsi, then H_{b^j} may not depend on D_{b^j}.

Only if: Suppose all the b cofactors have DSDs with variables a in separate blocks. Thus $F_{b^j}(a,c) = H_j(D_j(a),c)$ for some functions H_j and D_j. We can take $D_j(a) = 0$ if b^j is bsi. The Shannon expansion gives $F(a,b,c) = \sum_{j=0}^{2^{|b|}-1} b^j H_j(D_j(a),c)$. Define $D(a,b) = \sum_{j=0}^{2^{|b|}-1} b^j D_j(a)$ and note that $b^j H_j(D_j(a),c) = b^j H_j\left(\sum_{m=0}^{2^{|b|}-1} b^m D_m(a),c\right)$. Thus, $F(a,b,c) = \sum_{j=0}^{2^{|b|}-1} b^j H_j\left(\sum_{m=0}^{2^{|b|}-1} b^m D_m(a),c\right) = H(D(a,b),b,c)$. **QED**

In many applications, such as discussed in Section 4.4, the shared variables b are selected first, and the bound set variables a are found next using Theorem 4.2 to search for a largest set a that can be used.

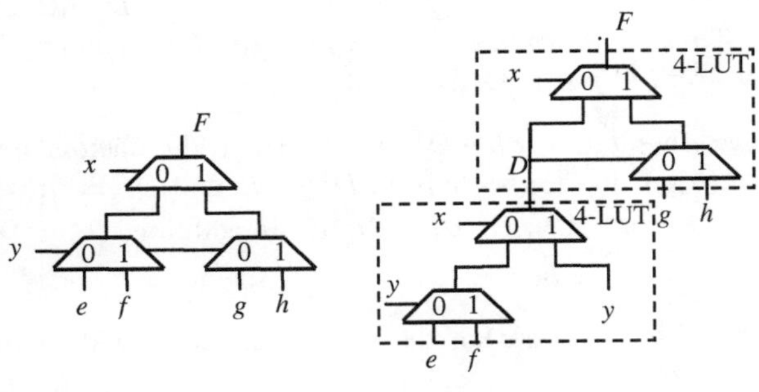

Fig. 4.1 Mapping 4:1 MUX into two 4-LUTs

Example 4.3 Consider the problem of decomposing a 4:1 MUX into two 4-LUTs. A structural mapper starting from the logic structure shown in Fig. 4.1 on the left would require three 4-LUTs, each containing a 2:1 MUX. To achieve a more compact mapping, we find an (a,b)-decomposition where $a = (e,f,y)$ and $b = x$. The free variables are $c = (g,h)$. This leads to cofactors $F_{\bar{x}} = \bar{y}e + yf$ and

$F_x = \bar{y}g + yh$. Both $F_{\bar{x}}$ and F_x have $a = (e, f, y)$ in a separate block.[2] Thus, $D_0 = \bar{y}e + yf$ and $D_1 = y$, while $H_0 = D_0$ and $H_1 = \bar{D}_1 g + D_1 h$. Thus we can write $F = \bar{x}H_0 + xH_1 = \bar{x}(D_0) + x(\bar{D}_1 g + D_1 h)$. Replacing D_0 and D_1 with $D = \bar{x}(\bar{y}e + yf) + x(y)$, we have $F = \bar{x}D + x(\bar{D}g + Dh)$. This leads to the decomposition shown on the right of Fig. 4.1. As a result, a 4:1 MUX is realized by two 4-LUTs.

We will use the notation $f \cong g$ to denote that f equals g up to complementation. The following theorem characterizes the set of all (a, b)-divisors of a function F.

Theorem 4.3 *Let F have an (a, b)-decomposition with an associated divisor $D(a, b) = \sum_{j=0}^{2^{|b|}-1} D_j(a)$. Then $\widehat{D}(a, b) = \sum_{j=0}^{2^{|b|}-1} b^j \widehat{D}_j(a)$ is also an (a, b)-divisor if and only if $\widehat{D}_j(a) \cong D_j(a), \forall j \in J$, where J is the set of indices of the bsd-cofactors of F.*

Proof **If:** Define $\widehat{D}_j(a) \cong D_j(a), j \in J$ and consider a function $\widehat{D}(a, b) = \sum_{j=0}^{2^{|b|}-1} b^j \widehat{D}_j(a)$ where $\widehat{D}_j(a), j \notin J$ is an arbitrary function. Since $F = H(D(a, b), b, c) = \sum_{j=0}^{2^{|b|}-1} b^j H_{bj}(D_j(a), c)$, we can define

$$\widehat{H}(\widehat{D}(a, b), b, c) = \sum_{j \in J_1} b^j H_{bj}(\widehat{D}_j(a), c) + \sum_{j \in J_2} b^j H_{bj}(\overline{\widehat{D}_j}(a), c)$$

where J_1 is the set of indices where $D_j = \widehat{D}_j$ and J_2 is the set of indices where $D_j = \overline{\widehat{D}_j}$. Clearly, $F(a, b, c) = \widehat{H}(\widehat{D}(a, b), b, c)$ and therefore $\widehat{D}(a, b)$ is an (a, b)-divisor of F.

Only if: Assume that $F = H(D(a, b), b, c)$ and $F = \widehat{H}(\widehat{D}(a, b), b, c)$. Cofactoring each with respect to b^j, $j \in J$, yields $F_{bj} = H_{bj}(D_{bj}(a), c)$ and $F_{bj} = \widehat{H}_{bj}(\widehat{D}_{bj}(a), c)$. Thus $F_{bj}(a, c)$ has a DSD with respect to a, and by Theorem 4.1, $D_{bj}(a) \cong \widehat{D}_{bj}(a)$ for b^j, $j \in J$. **QED**

Example 4.4 $F = ab + \bar{b}c = (ab + \bar{a}\bar{b})b + \bar{b}c$ The bsd-cofactors are $\{b\}$ and the bsi-cofactors are $(\bar{b}\}$. F has (a, b)-divisors $D^1 = ab$ and $D^2 = (ab + \bar{a}\bar{b})$, which agree in the bsd-cofactor b, i.e., $D_b^1(a) = D_b^2(a)$. In addition, $D^3 = \overline{D^1} = \bar{a} + \bar{b}$ is a divisor because $F = \overline{D^3} + \bar{b}c$.

In contrast to the discussion so far, the next result deals with finding common Boolean divisors among a *set* of functions.

Definition 4.3 A set of functions, $\{F_1, \cdots, F_n\}$ is said to be *(a, b)-compatible* if each has an (a, b)-divisor, and $\forall j \in J_1 \cap \cdots \cap J_n$, $D_{bj}^1(a) \cong D_{bj}^2(a)$, where J_i is the set of *bsd* b-cofactors of F_i.

---

[2] In $F_x = \bar{y}g + yh$, the DSD is a trivial one in which each input (leaf variable) is a separate block. Since the variables (e, f) do not appear in F_x, they can be considered as part of the separate block containing y. Thus $a = (e, f, y)$ appears in separate blocks of both $F_{\bar{x}}$ and F_x.

Note that compatibility is not transitive, but if $\{F_1, F_2, F_3\}$ is pair-wise (a, b)-compatible, then the set $\{F_1, F_2, F_3\}$ is (a, b)-compatible and by the next theorem, they all share a common (a, b)-divisor.

Theorem 4.4 [3] *There exists a common (a, b)-divisor of $\{F_1, \cdots, F_n\}$ if and only if the set $\{F_1, \cdots, F_n\}$ is pair-wise (a, b)-compatible.*

Proof For simplicity, we show the proof for $n = 2$.

If: Suppose F_1 and F_2 are (a, b)-compatible. Then $F_1(a, b, c) = H_1(D^1(a, b), b, c)$ and $F_2(a, b, c) = H_2(D^2(a, b), b, c)$ and $D^1_{b^j}(a) \cong D^2_{b^j}(a)$ for all bsd $\{b^j\}$ for both F_1 and F_2. Define $\tilde{D}_{b^j}(a) = D^1_{b^j}(a)$ for such b^j. If b^j is bsd for F_1 and bsi for F_2, let $\tilde{D}_{b^j}(a) = D^1_{b^j}(a)$. If b^j is bsd for F_2 and bsi for F_1 let $\tilde{D}_{b^j}(a) = D^2_{b^j}(a)$. Otherwise, let $\tilde{D}_{b^j}(a) = 0$. Clearly, by Theorem 4.3, $\tilde{D}(a, b) = \sum_{j=0}^{2^{|b|}-1} b^j \tilde{D}_{b^j}(a)$ is an (a, b)-divisor of both F_1 and F_1.

Only if: Suppose a common (a, b)-divisor exists, i.e., $F_1(a, b, c) = H_1(\tilde{D}(a, b), b, c)$ and $F_2(a, b, c) = H_2(\tilde{D}(a, b), b, c)$. Then both F_1 and F_1 have (a, b)-divisors such that $D^1_{b^j}(a) \cong D^2_{b^j}(a)$ for $j \in J_1 \cap J_2$, namely, $D^1 = D^2 = \tilde{D}$. **QED**

Thus a common divisor of two functions with shared variable b can be found by cofactoring with respect to b, computing the maximum DSDs of the cofactors, and looking for variables a for which the associated cofactors are compatible.

4.4 Rewriting K-LUT networks

This section presents a new method for rewriting K-LUT networks based on the ideas of Section 4.3, followed by a discussion of some implementation details and experimental results.

4.4.1 Global View

The objective is to rewrite a local window of a K-LUT mapped network. The window consists of a root node, n, and a certain number of transitive fanin (TFI) LUTs. The TFI LUTs are associated with a cut C. The local network to be rewritten consists of the LUT for n plus all LUTs between C and n. Our objective is to decompose the associated function of n, $f_n(C)$, expressed using the cut variables, into a smaller number of LUTs. For convenience, we denote this local network N_n.

An important concept is the *maximum fanout free cone* (MFFC) of N_n. This is defined as the set of LUTs in N_n, which are only used in computing $f_n(C)$. If node

[3] As far as we know, there is no equivalent theorem in the literature.

n were removed, then all of the nodes in MFFC(n) could be removed also. We want to re-decompose N_n into fewer K-LUTs taking into account that LUTs not in MFFC(n) must remain since they are shared with other parts of the network. Since it is unlikely that an improvement will be found when a cut has a small MFFC(n), we only consider cuts with no more than S shared LUTs. In our implementation S is a user-controlled parameter that is set to 3 by default.

Given n and a cut C for n, the problem is to find a decomposition of $f_n(C)$ composed of the minimum number N of K (or less) input blocks. For those N_n where there is a gain (taking into account the duplication of the LUTs not in the MFFC), we replace N_n with its new decomposition.

The efficacy of this method depends on the following:

- The order in which the nodes n are visited.
- The cut selected for rewriting the function at n.
- Not using decompositions that worsen delay.
- Creating a more balanced decomposition.
- Pre-processing to detect easy decompositions[4].

4.4.2 Cut Computation

The cut computation for resynthesis differs from traditional cut enumeration for technology mapping in several ways:

- Mapping is applied to an AIG, while resynthesis is applied to a mapped network.
- Mapping considers nodes in a topological order, while resynthesis may be applied to selected nodes, for example, nodes on a critical path, or nodes whose neighbors have changed in the last pass of resynthesis.
- During mapping, the size of a cut's MFFC is not important (except during some types of area recovery), while in resynthesis the size of a cut's MFFC is the main criterion to judge how many LUTs may be saved after resynthesis.
- Mapping for K-LUTs requires cuts of size K whereas for an efficient resynthesis of a K-LUT network, larger cuts need to be computed (typically up to 16 inputs).

Given these observations, complete or partial cut enumeration used extensively in technology mapping is not well-suited for resynthesis. Resynthesis requires a different cut computation strategy that is top-down: cuts are computed separately for each node starting with the trivial cut of the node and new cuts are obtained by expanding existing cuts towards primary inputs.

The pseudo-code of the cut computation is given in Fig. 4.2. The procedure takes the root node (*root*), for which the cuts are being computed, the limit on the cut size (N), the limit on the number of cuts (M), and the limit on the number of nodes duplicated when the computed cut is used for re-synthesized (S). The last

[4] For example, it can be a MUX decomposition of a function with at most $2K - 1$ inputs with cofactors of input size $K - 2$ and K.

```
cutset cutComputationByExpansion ( root, N, M, S )
{
        markMffcNode( root );
        cutset = { {node} };
        for each cut in cutset {
                for each leaf of cut
                        if ( !nodeIsPrimaryInput( leaf ) )
                                expandCutUsingLeaf( cut, leaf, N, S, cutset );
        }
        filterCutsWithDSDstructure( cutset );
        if (|cutset| > M ) {
                sortCutsByWeight( cutset );
                cutset = selectCutsWithLargerWeight( cutset, M );
        }
        return cutset;
}
expandCutUsingLeaf ( cut, leaf, N, S, cutset )
{
        // check if the cut is duplication-feasible
        if ( !nodeBelongsToMffc(leaf) && numDups(cut) == S )
                return;
        // derive the new cut
        cut_new = (cut \ leaf ) ⋃ nodeFanins( leaf );
        // check if the cut is N-feasible
        if (numLeaves(cut_new)> N )
                return;
        // check if cut_new is equal to or dominated by any cut in cutset
        // (also, remove the cuts in cutset that are contained in cut_new)
        if ( cutsetFilter( cutset, cut_new ) )
                return;
        // add cut_new to cutset to be expanded later
        cutset = cutset ⋃ cut_new;
}
```

Fig. 4.2 Cut computation for resynthesis

number is the maximum number of nodes that can be covered by a cut, which are not in the MFFC of the given node. Obviously, these nodes will be duplicated when the cut's function is expressed in terms of the cut leaves, using a set of new K-feasible nodes after decomposition.

For each cut computed, two sets of nodes are stored: the set of leaves and the set of nodes covered by the cut (these are the nodes found on the paths between the leaf nodes and the root, excluding the leaves and including the root). Additionally, each cut has a count of covered nodes that are not in the MFFC of the root. The value of this is returned by procedure **numDups**.

Procedure **expandCutUsingLeaf** tries to expand the cut by moving a leaf to the set of covered nodes and adding the leaf's fanins to the set of leaves. If the leaf belongs to the MFFC of the root and the number of duplicated nodes of the cut has already saturated, the new cut is not constructed. If the resulting cut has more leaves than is allowed by the limit, the new cut is not used. If the cut is equal or dominated by any of the previously computed cuts, it is also skipped. Finally, if none of the above conditions holds, the cut is appended to the resulting cutset. Later in the cut computation, this cut, too, is used to derive other cuts.

The above procedure converges because of the limits on the cut size (N) and the number of duplicated nodes (S), resulting in a set of cuts that are potentially useful for resynthesis. A cut is actually used for resynthesis if both of the following conditions are met: (a) the cut does not have a DSD-structure and (b) the cut has a high enough weight. These conditions are described below.

4.4.3 Cuts with a DSD Structure

Some cuts are not useful for resynthesis because the network structure covered by these cuts cannot be compacted. These are the cuts covering disjoint-support-decomposable parts of the structure. For example, Fig. 4.3 shows a cut $\{a, b, c, d\}$ of node n. This cut covers node m whose fanins, c and d, are disjoint-support with respect to the rest of the covered network structure which depends on a and b.

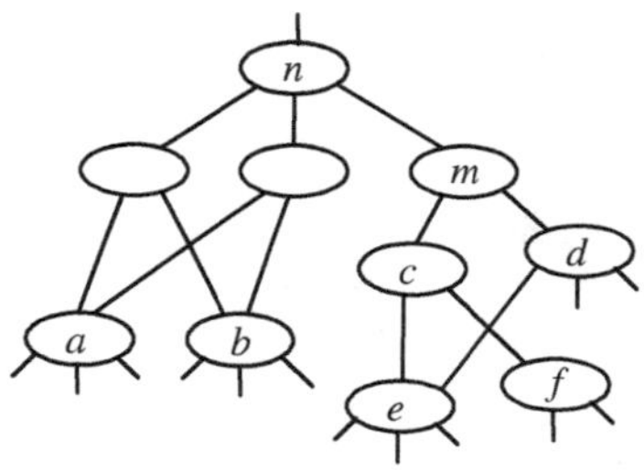

Fig. 4.3 Illustration of a cut with DSD-structure

Cuts with a DSD-structure can be efficiently identified by a dedicated procedure. These cuts can be skipped when attempting resynthesis, but they cannot be left out during the cut computation because expanding them can lead to other useful cuts. For example, expanding cut $\{a, b, c, d\}$ shown in Fig. 4.3 with respect to leaf c leads to a cut $\{a, b, d, e, f\}$, which does not have a DSD structure because node d depends on node e.

4.4.4 Cut Weight

Some cuts should be skipped during resynthesis because they have little or no potential for improvement and should not be tried, given a tight runtime budget. To filter out useless cuts and prioritize other cuts, the notion of *cut weight* is introduced. The weight is computed for all cuts and only a given number (M) of cuts with high enough weight is selected for resynthesis.

The weight of a cut is defined as follows:

$$\textbf{cutWeight}(c) = [\textbf{numCovered}(c) - \textbf{numDups}(c)]/\textbf{numLuts}(c)$$

where **numLuts**(c) = ceiling[(**numLeaves**(c)-1)/(K–1)].

The procedures **numCovered** and **numDups** return the total number of covered nodes and the number of covered nodes not belonging to the MFFC of the root, respectively. Procedure **numLeaves** return the number of cut leaves. Finally, **numLuts** computes the minimum number of K-LUTs needed to implement a cut of size *c*, provided that the cut's function has a favorable decomposition.

Intuitively, the weight of a cut shows how likely the cut will be useful in resynthesis: the more nodes it saves and the less LUTs possibly needed to realize it, the better the cut. The cuts whose weights are less than or equal to 1.0 can be skipped because they give no improvement. If resynthesis with zero-cost replacements is allowed, only the cuts with weight less than 1.0 are skipped.

4.4.5 Decomposition and Network Update

Figure 4.4 gives an overall pseudo-code of the proposed approach to resynthesis:

```
resynthesisByLutPacking( network, parameters )
{
    // apply resynthesis to each node
    for each node of network in topological order {
        cutset = cutComputationByExpansion( node, parameters );
        for each cut in cutset in decreasing order of cut weight {
            // derive function of the cut as a truth table
            F = cutComputeFunction( cut );
            // iteratively decompose the function
            while ( suppSize(F) > K ) {
                boundset = findSupportReducingBoundSet( F, K );
                if ( boundset == NONE )
                    break;
                F = functionDecompose( F, boundset );
            }
            // if F is decomposed, update and move to the next node
            if ( boundset != NONE ) {
                networkUpdate( network );
                break;
            }
        }
    }
}
```

Fig. 4.4 Overall pseudo-code of the proposed resynthesis

4.4.6 Finding the Maximum Support-Reducing Decomposition

The proposed algorithm works by cofactoring the non-decomposable blocks of the DSD of F and using Theorem 4.2 to find a Boolean divisor and bound variables a. The approach is heuristic and may not find a decomposition with the minimum number (N) of K-input blocks. The heuristic tries to extract a maximum support-reducing block at each step, based on the idea that support reduction leads to a good decomposition. In fact, any optimum implementation of a network into K-LUTs must be support reducing if any fanin of a block is support reducing for that block. However, in general, it may not be the most support-reducing.

The approach is to search for common bound sets of the cofactor DSDs where cofactoring is tried with respect to subsets of variables in the support of F. If all subsets are attempted, the proposed greedy approach reduces the associated block to a minimum number of inputs. However, in our implementation, we heuristically trade-off the quality of the decomposition found versus the runtime spent in exploring cofactoring sets. A limit is imposed on (a) the number of cofactoring variables, and (b) the number of different variable combinations tried. Our experiments show that the proposed heuristic approach usually finds a solution with a minimum number of blocks.

The pseudo-code in Fig. 4.5 shows how the DSD structures of the cofactors can be exploited to compute a bound set leading to the maximum support reduction.

The procedure **findSupportReducingBoundSet** takes a completely-specified function F and the limit K on the support size of the decomposed block. It returns a good bound set, that is, a bound set leading to the decomposition with a maximal support-reduction. If a support-reducing decomposition does not exist, the procedure returns NONE.

First, the procedure derives the DSD tree of the function itself. The tree is used to compute the set of all feasible bound sets whose sizes do not exceed K. Bound sets of larger size are not interesting because they cannot be implemented using K-LUTs. For each of the bound sets found, decomposition with a single output and no shared variables is possible. If a bound set of size K exists, it is returned. If such a bound set does not exist (for example, when the function has no DSD), the second best would be to have a bound set of size $K-1$. Thus, the computation enters a loop, in which cofactoring of the function with respect to several variables is tried, and common support-reducing bound sets of the cofactors are explored.

When the loop is entered, cofactoring with respect to one variable is tried first. If the two cofactors of the function have DSDs with a common bound set of size $K-1$, it is returned. In this case, although the decomposed block has K variables, the support of F is only reduced by $K-2$ because the cofactoring variable is shared and the output of the block is a new input. If there is no common bound set of size $K-1$, the next best outcome is one of the following:

1. There is a bound set of size $K-2$ of the original function.
2. There is a common bound set of size $K-2$ of the two cofactors with respect to the cofactoring variable.

```
varset findSupportReducingBoundSet( function F, int K )
{
    // derive DSD for the function
    DSDtree Tree = performDSD( F );
    // find K-feasible bound-sets of the tree
    varset BSets[0] = findKFeasibleBoundSets( F, Tree, K );
    // check if a good bound-set is already found
    if ( BSets[0] contains bound-set B of size K )
        return B;
    if ( BSets[0] contains bound-set B of size K -1 )
        return B;
    // cofactor F with respect to sets of variables and look for the largest
    // support-reducing bound-set shared by all the cofactors
    for ( int V = 1; V ≤ K − 2; V++ ) {
        // find the set including V cofactoring variables
        varset cofvars = findCofactoringVarsForDSD( F, V );
        // derive DSD trees for the cofactors and compute
        // common K-feasible bound-sets for all the trees
        set of varsets BSets[V] = {∅};
        for each cofactor Cof of function F with respect to cofvars {
          DSDtree Tree = performDSD( Cof );
          set of varsets BSetsC =
                      computeBoundSets(Cof, Tree, K-V);
          BSets[V] = mergeSets( BSets[V], BSetsC, K-V );
        }
        // check if at least one good bound-set is already found
        if ( BSets[V] contains bound-set B of size K-V )
          return B;
        // before trying to use more shared variables, try to find
        // bound-sets of the same size with fewer shared variable
        for ( int M = 0; M ≤ V; M++ )
          if ( BSets[M] contains bound-set B of size K-V-1 )
              return B;
    }
    return NONE;
}
```

Fig. 4.5 Computing a good support-reducing bound set

3. There is a common bound set of size $K-2$ of the four cofactors with respect to two variables.

The loop over M at the bottom of Fig. 4.2 tests for outcomes (1) and (2). If these are impossible, V is incremented and the next iteration of the loop is performed, which is the test for the outcome (3).

In the next iteration over V, cofactoring with respect to two variables is attempted and the four resulting cofactors are searched for common bound sets. The process is continued until a bound set is found, or the cofactoring with respect to K–2 variables is tried without success. When V exceeds K–2 (say, V is K–1), the decomposition is not support-reducing, because the composition function depends on shared K–1 variables plus the output of the decomposed block. In other words, the decomposition takes away K variables from the composition function and returns K variables. In this case, NONE is returned, indicating that there is no support-reducing decomposition.

Example 4.5 Consider the decomposition of function F of the 4:1 MUX shown in Fig. 4.1 (left). Assume $K = 4$. This function does not have a non-trivial DSD; its DSD is composed of one prime block. The set of K-feasible bound sets is trivial in this case: $\{ \{\oslash\}, \{a\}, \{b\}, \{c\}, \{d\}, \{x\}, \{y\} \}$. Clearly, none of these bound sets has size K or K–1. The above procedure enters the loop with $V = 1$. Suppose x is chosen as the cofactoring variable. The cofactors are $F_{\bar{x}} = \bar{y}a + yb$ and $F_x = \bar{y}c + yd$. The K–1-feasible bound sets are $\{\{\oslash\}, \{a\}, \{b\}, \{y\}, \{a, b, y\}\}$, and $\{\{\oslash\}, \{c\}, \{d\}, \{y\}, \{c, d, y\}\}$. A common bound set $\{a, b, y\}$ of size K-1 exists. The loop terminates and this bound set is returned, resulting in the decomposition in Fig. 4.1 (right).

4.4.7 Additional Details

In this section, we briefly present several aspects of the proposed algorithm that were not described above to keep that discussion simple.

4.4.7.1 Using Timing Information to Filter Candidate Bound Sets

Timing information, which can be taken into account by the algorithm, includes the arrival times of the leaves of the cut and the required time of the root. During decomposition, a lower bound on the root's required time is computed and further decomposition is not performed if this bound exceeds the required time. In this case, the resynthesis algorithm moves on to the next cut of the root node. If the node has no more cuts, the algorithm moves to the next node in a topological order.

The use of the timing information prunes the search space and therefore leads to faster resynthesis. However, the gain in area may be limited due to skipping some of the resynthesis opportunities as not feasible under the timing constraints.

4.4.7.2 Restricting Bound Sets for Balanced Decompositions

Experiments have shown that allowing all variables in the bound sets may lead to unbalanced decompositions. Therefore, the variables that are allowed to be in the bound sets are restricted to those that are in the transitive fanins cones of the prime blocks. If the DSD trees of the cofactors do not have prime blocks, all variables are allowed in the bound sets.

4.4.7.3 Opportunistic MUX-Decomposition

To further improve quality and runtime, a different type of decomposition is considered along with the DSD-based decomposition presented above. This decomposition attempts to cofactor the function with respect to every variable and checks the sizes of cofactors. If the smaller of the two cofactors has support size less than K–2 and the larger one has size less than K, the function can be implemented using two K-LUTs, one of which subsumes the smaller block together with the MUX, while another is used as the larger block. This type of decomposition can often be found faster than a comparable DSD.

4.5 Comparison with Boolean Matching

The proposed Boolean decomposition applied to rewriting of LUT networks is similar to Boolean matching, which checks if a function can be implemented using a LUT structure.

Boolean matchers can be compared using the following criteria:

- An average *runtime* needed to match a given K-variable function with a given N-variable LUT structure ($K \leq N$).
- An average *success rate*. This is the percentage of functions successfully matched, out of the total number of functions, for which a match exists.

The following approaches to Boolean matching are known:

(1) *Structural approach.* [11, 29] Structural LUT-based technology mappers fall into this category. This approach is incomplete because it requires the subject graph of a Boolean function to match the LUT structure. It is fast but has a low success rate. The actual rate may depend strongly on the function and the structure used. It our experiments with 16-input functions matched into structures composed of 6-LUTs, the success rate was about 30%.
(2) *SAT-based approach.* [17, 21, 24, 36] This complete method is guaranteed to find a match if it exists. It is also flexible because it can be customized easily to different LUT structures. However, it is time-consuming and takes seconds or minutes to solve 10-variable instances. When the problem is hard and a runtime limit is used, this method becomes incomplete with the success rate depending on the problem type. For example, [24] reports a success rate of 50% for 10-variable functions with a 60 s timeout.
(3) *NPN-equivalence-based approach.* [1, 8, 9, 14]. This potentially complete method pre-computes all NPN-classes of functions that can be implemented using the given architecture. Although this method is relatively fast, it does not scale well for functions above 9–12 inputs because of the large number of NPN classes. It is restricted to a particular LUT structure and may not be feasible for large/complex architectures with too many NPN-classes.
(4) *Functional approach.* [18, 25, 26, 39] This approach attempts to apply a Boolean decomposition algorithm to break the function down into K-input blocks. The runtime strongly depends on the number of inputs of the function, the type of decomposition used, and how the decomposability check is implemented. The completeness is typically compromised by the heuristic nature of

bound set selection [18] or fixing the variable order in the BDD [39] when looking for feasible decompositions. Exhaustively exploring all bound sets, as proposed in [26], can only be done for relatively small functions (up to 12 inputs).

The algorithm proposed in this chapter belongs to the last category. It uses properties of Boolean functions, such as DSD, to guide what K-input blocks to extract. The completeness of this method is guaranteed by Theorem 4.1. In practice the completeness requires performing DSD for a large number of cofactors of the function. Therefore, our current implementation limits the number of cofactors considered, making the method fast but incomplete.

4.6 Experimental Results

The proposed algorithm is implemented in ABC [5] and is available as command *lutpack*. Experiments targeting 6-input LUTs were run on an Intel Xeon 2-CPU 4-core computer with 8 GB of RAM. The resulting networks were verified using the combinational equivalence checker in ABC (command *cec*) [28].

The following ABC commands are included in the scripts which were used to collect experimental results, which targeted area minimization while preserving delay:

- *resyn* is a logic synthesis script that performs five iterations of AIG rewriting [27] trying to improve area without increasing depth.
- *resyn2* is a script that performs 10 iterations of a more diverse set of AIG rewritings than those of *resyn*.
- *choice* is a script that allows for accumulation of structural choices; *choice* runs *resyn* followed by *resyn2* and collects three snapshots of the network during this process: the original, the final, and the one after *resyn*, resulting in a circuit with structural choices.
- *if* is an efficient FPGA mapper using priority cuts [31], fine-tuned for area recovery (after a minimum delay mapping), and using subject graphs with structural choices[5].
- *imfs* is an area-oriented resynthesis engine for FPGAs [30] based on changing a logic function at a node by extracting don't cares from a surrounding window and using Boolean resubstitution to rewrite the node function using possibly new inputs.
- *lutpack* is the new resynthesis described in this section.

[5] The mapper was run with the following settings: at most 12 6-input priority cuts are stored at each node; five iterations of area recovery included three with area flow and two with exact local area.

The benchmarks used in this experiment are 20 large public benchmarks from the MCNC and ISCAS'89 suites used in previous work on FPGA mapping [11, 22, 29][6].

The following four experiments were performed:

- "Baseline" = (*resyn; resyn2; if*). This is a typical run of technology-independent synthesis followed by the default FPGA mapping.
- "Choices" = *resyn; resyn2; if;* $(choice;\ if)^4$.
- "imfs" = *resyn; resyn2; if;* $(choice;\ if;\ imfs)^4$.
- "imfs+lutpack" = *resyn; resyn2; if;* $(choice;\ if;\ imfs)^4$; $(lutpack)^2$.

We use exponentiation to denote iteration, e.g., $(com1;\ com2)^3$ means iterate (*com1; com2*) three times.

Table 4.1 lists the number of primary inputs ("PIs"), primary outputs ("POs"), registers ("Reg"), area calculated as the number of 6-LUTs ("LUT") and delay calculated as the depth of the 6-LUT network ("Level"). The ratios in the tables are the ratios of geometric averages of values reported in the corresponding columns.

Table 4.1 Evaluation of resynthesis after technology mapping for FPGAs ($K = 6$)

Designs	PI	PO	Reg	Baseline		Choices		imfs		imfs+lutpack	
				LUT	Level	LUT	Level	LUT	Level	LUT	Level
alu4	14	8	0	821	6	785	5	558	5	453	5
apex2	39	3	0	992	6	866	6	806	6	787	6
apex4	9	19	0	838	5	853	5	800	5	732	5
bigkey	263	197	224	575	3	575	3	575	3	575	3
clma	383	82	33	3323	10	2715	9	1277	8	1222	8
des	256	245	0	794	5	512	5	483	4	480	4
diffeq	64	39	377	659	7	632	7	636	7	634	7
dsip	229	197	224	687	3	685	2	685	2	685	2
ex1010	10	10	0	2847	6	2967	6	1282	5	1059	5
ex5p	8	63	0	599	5	669	4	118	3	108	3
elliptic	131	114	1122	1773	10	1824	9	1820	9	1819	9
frisc	20	116	886	1748	13	1671	12	1692	12	1683	12
i10	257	224	0	589	9	560	8	548	7	547	7
pdc	16	40	0	2327	7	2500	6	194	5	171	5
misex3	14	14	0	785	5	664	5	517	5	446	5
s38417	28	106	1636	2684	6	2674	6	2621	6	2592	6
s38584	12	278	1452	2697	7	2647	6	2620	6	2601	6
seq	41	35	0	931	5	756	5	682	5	645	5
spla	16	46	0	1913	6	1828	6	289	4	263	4
tseng	52	122	385	647	7	649	6	645	6	645	6
Geomean				1168	6.16	1103	5.66	716	5.24	677	5.24
Ratio1				1.000	1.000	0.945	0.919	0.613	0.852	0.580	0.852
Ratio2						1.000	1.000	0.649	0.926	0.614	0.926
Ratio3								1.000	1.000	0.946	1.000

The columns *Baseline* and *Choices* have been included to show that repeated re-mapping has a dramatic impact over conventional mapping (*Baseline*). However, the main purpose of the experiments is to demonstrate the additional impact of

[6] In the above set, circuit s298 was replaced by i10 because the former contains only 24 6-LUTs.

command *lutpack* after a very strong synthesis flow. Thus we only focus on the last line of the table, which compares *lutpack* against the strongest result obtained using other methods (*imfs*). Given the power of *imfs*, it is somewhat unexpected that *lutpack* can achieve an *additional* 5.4% reduction in area[7].

This additional area reduction speaks for the orthogonal nature of *lutpack* over *imfs*. While *imfs* tries to reduce area by analyzing alternative resubstitutions at each node, it cannot efficiently compact large fanout-free cones that may be present in the mapping. The latter is done by *lutpack*, which iteratively collapses fanout-free cones with up to 16 inputs and re-derives new implementations using the minimum number of LUTs.

The runtime of one run of *lutpack* was less than 20 s for any of the benchmarks reported in Table 4.1. The total runtime of the experiments was dominated by *imfs*. Table 4.1 illustrates only two passes of *lutpack* used for final processing; several iterations of *lutpack* in the inner loop, e.g., $(choice;\ if;\ imfs;\ lutpack)^4$, often show additional gains.

4.7 Conclusions and Future Work

This chapter presented an algorithm for Boolean decomposition of logic functions targeting area-oriented resynthesis of K-LUT structures. The new algorithm, *lutpack*, is based on cofactoring and disjoint-support decomposition and is much faster than previous solutions relying on BDDs and Boolean satisfiability. The algorithm achieved an additional 5.4% reduction in area, when applied to a network obtained by iterating high-quality technology mapping and another type of powerful resynthesis based on don't cares and windowing.

Future work might include the following:

- Extracting a common Boolean divisor from a pair of functions (using Theorem 4.4).
- Computing all decompositions of a function and using them to find common Boolean divisors among all functions of a logic network.
- Improving the DSD-based analysis, which occasionally fails to find a feasible match and is the most time-consuming part.
- Exploring other data structures for cofactoring and DSD decomposition for functions with more than 16 inputs. This would improve the quality of resynthesis.

Acknowledgments This work was supported in part by SRC contracts 1361.001 and 1444.001, NSF contract CCF-0702668, and the California MICRO Program with industrial sponsors Actel, Altera, Calypto, IBM, Intel, Intrinsity, Magma, Mentor Graphics, Synopsys (Synplicity), Tabula, and Xilinx. The authors thank Stephen Jang, for his masterful experimental evaluation of the

---

[7] Some readers may suspect that the result after *imfs* can be improved on easily. We invite them to take the benchmarks and come up with a better result than that recorded in the *imfs+lutpack* column.

proposed algorithm, Slawomir Pilarski and Victor Kravets, for their careful reading and useful comments, and to an anonymous reviewer for pointing out critical errors in the original manuscript. This work was performed while the third author was at Berkeley.

References

1. Abdollahi, A., Pedram, M.: A new canonical form for fast Boolean matching in logic synthesis and verification. In: Proceedings of DAC '05, pp. 379–384.
2. Actel Corporation: ProASIC3 flash family FPGAs datasheet.
3. Altera Corporation: Stratix II device family data sheet.
4. Ashenhurst, R.L.: The decomposition of switching functions. Proceedings of International Symposium on the Theory of Switching, Part I (Annals of the Computation Laboratory of Harvard University, Vol. XXIX), Harvard University Press, Cambridge, 1959, pp. 75–116.
5. Berkeley Logic Synthesis and Verification Group, ABC: A System for Sequential Synthesis and Verification, Release 70911. http://www.eecs.berkeley.edu/~ alanmi/abc/ (2010)
6. Bertacco, V., Damiani, M.: The disjunctive decomposition of logic functions. In: Proceedings ICCAD '97, pp. 78–82.
7. Brayton, R., McMullen, C.: The decomposition and factorization of Boolean expressions. In: Proceedings ISCAS '82, pp. 29–54.
8. Chai, D., Kuehlmann, A.: Building a better Boolean matcher and symmetry detector. In: Proceedings DATE '06, pp. 1079–1084.
9. Chatterjee, S., Mishchenko, A., Brayton, R.: Factor cuts. In: Proceedings ICCAD '06, pp. 143–150.
10. Chatterjee, S., Mishchenko, A., Brayton, R., Wang, X., Kam, T.: Reducing structural bias in technology mapping. In: Proceedings ICCAD '05, pp. 519–526.
11. Chen, D., Cong, J.: DAOmap: A depth-optimal area optimization mapping algorithm for FPGA designs. In: Proceedings ICCAD '04, pp. 752–757.
12. Cong, J., Wu, C., Ding, Y.: Cut ranking and pruning: Enabling a general and efficient FPGA mapping solution. In: Proceedings FPGA'99, pp. 29–36.
13. Curtis, A.: New approach to the design of switching circuits. Van Nostrand, Princeton, NJ (1962)
14. Debnath, D., Sasao, T.: Efficient computation of canonical form for Boolean matching in large libraries. In: Proceedings ASP-DAC '04, pp. 591–596.
15. Farrahi, A., Sarrafzadeh, M.: Complexity of lookup-table minimization problem for FPGA technology mapping. IEEE TCAD, **13**(11), 319–1332 (Nov. 1994)
16. Files, C., Perkowski, M.: New multi-valued functional decomposition algorithms based on MDDs. IEEE TCAD, **19**(9), 1081–1086 (Sept. 2000)
17. Hu, Y., Shih, V., Majumdar, R., He, L.: Exploiting symmetry in SAT-based Boolean matching for heterogeneous FPGA technology mapping. In: Proceedings ICCAD '07.
18. Kravets, V.N.: Constructive multi-level synthesis by way of functional properties. PhD thesis. University of Michigan, 2001.
19. Kravets, V.N., Sakallah, K.A.: Constructive library-aware synthesis using symmetries. In: Proceedings of DATE, pp. 208–213 (March 2000)
20. Lehman, E., Watanabe, Y., Grodstein, J., Harkness, H.: Logic decomposition during technology mapping. IEEE TCAD, **16**(8), 813–833 (1997)
21. Ling, A., Singh, D., Brown, S.: FPGA technology mapping: A study of optimality. In: Proceedings of DAC '05, pp. 427–432.
22. Manohara-rajah, V., Brown, S.D., Vranesic, Z.G.: Heuristics for area minimization in LUT-based FPGA technology mapping. In: Proceedings of IWLS '04, pp. 14–21
23. Matsunaga, Y.: An exact and efficient algorithm for disjunctive decomposition. In: Proceedings of SASIMI '98, pp. 44–50

24. Minkovich, K., Cong, J. An improved SAT-based Boolean matching using implicants for LUT-based FPGAs. In: Proceedings of FPGA'07.
25. Mishchenko, A., Sasao, T.: Encoding of Boolean functions and its application to LUT cascade synthesis. In: Proceedings of IWLS '02, pp. 115–120.
26. Mishchenko, A., Wang, X., Kam, T.: A new enhanced constructive decomposition and mapping algorithm. In: Proceedings of DAC '03, pp. 143–148.
27. Mishchenko, A., Chatterjee, S., Brayton, R.: "DAG-aware AIG rewriting: A fresh look at combinational logic synthesis. In: Proceedings of DAC'06, pp. 532–536.
28. Mishchenko, A., Chatterjee, S., Brayton, R., Een, N.: "Improvements to combinational equivalence checking," In: Proceedings of ICCAD '06, pp. 836–843.
29. Mishchenko, A., Chatterjee, S., Brayton, R.: Improvements to technology mapping for LUT-based FPGAs. IEEE TCAD, **26**(2), 240–253 (Feb 2007)
30. Mishchenko, A., Brayton, R., Jiang, J.-H.R., Jang, S.: Scalable don't care based logic optimization and resynthesis. In: Proceedings of FPGA'09, pp. 151–160.
31. Mishchenko, A., Cho, S., Chatterjee, S., Brayton, R.: Combinational and sequential mapping with priority cuts. In: Proceedings of ICCAD '07.
32. Mishchenko, A., Chatterjee, S., Brayton, R.: Fast Boolean matching for LUT structures. ERL Technical Report, EECS Deptartment, UC Berkeley.
33. Pan, P., Lin, C.-C.: A new retiming-based technology mapping algorithm for LUT-based FPGAs. In: Proceedings of FPGA '98, pp. 35–42
34. Perkowski, M., Marek-Sadowska, M., Jozwiak, L., Luba, T., Grygiel, S., Nowicka, M., Malvi, R., Wang, Z., Zhang, J.S.: Decomposition of multiple-valued relations. In: Proceedings of ISMVL'97, pp. 13–18
35. Roth, J.P., Karp, R.: Minimization over Boolean graphs. IBM Journal of Research and Development **6**(2), 227–238 (1962)
36. Safarpour, S., Veneris, A., Baeckler, G., Yuan, R.: Efficient SAT-based Boolean matching for FPGA technology mapping.' In: Proceedings of DAC '06
37. Sawada, H., Suyama, T., Nagoya, A.: Logic synthesis for lookup tables based FPGAs using functional decomposition and support minimization. In: Proceedings of ICCAD, pp. 353–358, (1995)
38. Sasao, T., Matsuura, M.: DECOMPOS: An integrated system for functional decomposition. In: Proceedings of IWLS'98, pp. 471–477
39. Vemuri, N., Kalla, P., Tessier, R.: BDD-based logic synthesis for LUT-based FPGAs. ACM TODAES **7**, 501–525 (2002)
40. Wurth, B., Schlichtmann, U., Eckl, K., Antreich, K.: Functional multiple-output decomposition with application to technology mapping for lookup table-based FPGAs. ACM Transactions on Design Automation of Electronic Systems **4**(3), 313–350 (1999)

Chapter 5
Ashenhurst Decomposition Using SAT and Interpolation

Hsuan-Po Lin, Jie-Hong Roland Jiang, and Ruei-Rung Lee

Abstract Functional decomposition is a fundamental operation in logic synthesis to facilitate circuit transformation. Since the first formulation by Ashenhurst in 1959, functional decomposition has received much attention and has been generalized and studied to some extent. Recent practical approaches to functional decomposition relied on the well-studied data structure binary decision diagram (BDD), which, however, is known to suffer from the memory explosion problem and thus not scalable to decompose large Boolean functions. In these BDD-based approaches, variable partitioning, a crucial step in functional decomposition, has to be specified a priori and often restricted to few bound set variables. Moreover, non-disjoint decomposition requires substantial sophistication in formulation. This report shows that, when Ashenhurst decomposition (the simplest and preferable functional decomposition) is considered, both single- and multiple-output decomposition can be computed with satisfiability solving, Craig interpolation, and functional dependency. Variable partitioning can be automated and integrated into the decomposition process without the bound set size restriction. The computation naturally extends to non-disjoint decomposition. Experimental results show that the proposed method can effectively decompose functions with up to 300 input variables.

5.1 Introduction

Functional decomposition [1, 6, 11] aims at decomposing a Boolean function into a network of smaller sub-functions. It is a fundamental operation in logic synthesis and has various applications to FPGA synthesis, minimization of circuit communication complexity, circuit restructuring, and other contexts. The most widely applied area is perhaps FPGA synthesis, especially for the look-up table (LUT)-based FPGA

J.-H.R. Jiang (✉)
National Taiwan University, Taipei Taiwan
e-mail: jhjiang@cc.ee.ntu.edu.tw

This work is based on an earlier work: To SAT or not to SAT: Ashenhurst decomposition in a large scale, in Proceedings of the 2008 IEEE/ACM International Conference on Computer-Aided Design, ISBN ISSN:1092-3152, 978-1-4244-2820-5 (2008) © ACM, 2008.

S.P. Khatri, K. Gulati (eds.), *Advanced Techniques in Logic Synthesis, Optimizations and Applications*, DOI 10.1007/978-1-4419-7518-8_5,

architecture, where each LUT can implement an arbitrary logic function with up to five or six inputs. Because of the input-size limitation of each LUT, a Boolean function to be realized using LUTs has to be decomposed into a network of sub-functions each conforming to the input-size requirement. Since FPGAs became a viable design style and highly optimized BDD packages were available, BDD-based functional decomposition [3, 13] has been intensively studied over the previous two decades. A comprehensive introduction to this subject is available in [19].

Most prior work on functional decomposition used BDD as the underlying data structure. By ordering variables in some particular way, BDD can be exploited for the computation of functional decomposition. Despite having been a powerful tool, BDD poses several limitations: First, BDDs are very sensitive to variable ordering and suffer from the notorious memory explosion problem. In representing a Boolean function, a BDD can be of large size (in the worst case, exponential in the number of variables). It is even more so when special variable ordering rules need to be imposed on BDDs for functional decomposition. Therefore it is typical that a function under decomposition can have just a few variables. Second, variable partitioning needs to be specified a priori and cannot be automated as an integrated part of the decomposition process. In order to effectively enumerate different variable partitions and keep BDD sizes reasonably small, the set of bound set variables cannot be large. Third, for BDD-based approaches, non-disjoint decomposition cannot be handled easily. In essence, decomposability needs to be analyzed by cases exponential in the number of joint (or common) variables. Finally, even though multiple-output decomposition [22] can be converted to single-output decomposition [9], BDD sizes may grow largely in this conversion.

The above limitations motivate the need for new data structures and computation methods for functional decomposition. We show that, when Ashenhurst decomposition [1] is considered, these limitations can be overcome through satisfiability (SAT)-based formulation. Ashenhurst decomposition is a special case of functional decomposition, where, as illustrated in Fig. 5.1, a function $f(X)$ is decomposed into two sub-functions $h(X_H, X_C, x_g)$ and $g(X_G, X_C)$ with $f(X) = h(X_H, X_C, g(X_G, X_C))$. For general functional decomposition, the function g can be a functional vector $(g_1, \ldots, g_k)$ instead. It is this simplicity that makes Ashenhurst decomposition particularly attractive in practical applications.

The enabling techniques of our method, in addition to SAT solving, include Craig interpolation [5] and functional dependency [10]. Specifically, the decomposability of function f is formulated as SAT solving, the derivation of function g is by Craig interpolation, and the derivation of function h is by functional dependency.

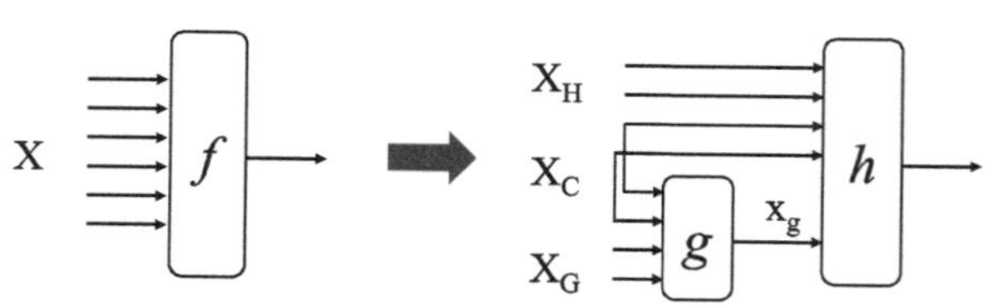

Fig. 5.1 Ashenhurst decomposition

Compared with BDD-based methods, the proposed algorithm is advantageous in the following aspects. First, it does not suffer from the memory explosion problem and is scalable to large functions. Experimental results show that Boolean functions with more than 300 input variables can be decomposed effectively. Second, variable partitioning need not be specified a priori and can be automated and derived on the fly during decomposition. Hence the size of the bound set variables X_G need not be small. Third, it works for non-disjoint decomposition naturally. Finally, it is easily extendable to multiple-output decomposition. Nonetheless, a limitation of the method is its expensive generalization to functional decomposition beyond Ashenhurst's special case.

A scalable decomposition method may be beneficial to modern VLSI design. For example, the dominating interconnect delays in nanometer IC design may be reduced by proper decomposition at the functional level; complex system realization using FPGAs or 3D ICs may require a design being decomposed at the chip level. On the other hand, the scalability of the proposed method may provide a global view on how a large function can be decomposed. Accordingly, hierarchical and chip-level logic decomposition might be made feasible in practice. In addition, our results may possibly shed light on scalable Boolean matching for heterogeneous FPGAs as well as topologically constrained logic synthesis [20].

5.2 Previous Work

Aside from BDD-based functional decomposition [19], we compare some related work using SAT. In bi-decomposition [14], a function f is written as $f(X) = h(g_1(X_A, X_C), g_2(X_B, X_C))$ under variable partition $X = \{X_A|X_B|X_C\}$, where function h is known a priori and is of special function types (namely, two-input OR, AND, and XOR gates) while functions g_1 and g_2 are the unknown to be computed. In contrast, the complication of Ashenhurst decomposition $f(X) = h(X_H, X_C, g(X_G, X_C))$ comes from the fact that both functions h and g are unknown. The problem needs to be formulated and solved differently while the basic technique used is similar to that in [14].

FPGA Boolean matching, see, e.g., [4], is a subject closely related to functional decomposition. In [15], Boolean matching was achieved with SAT solving, where quantified Boolean formulas were converted into CNF formulas. The intrinsic exponential explosion in formula sizes limits the scalability of the approach. Our method may provide a partial solution to this problem, at least for some special PLB configurations.

5.3 Preliminaries

As conventional notation, sets are denoted by upper-case letters, e.g., S; set elements are in lower-case letters, e.g., $e \in S$. The cardinality of S is denoted by $|S|$. A partition of a set S into $S_i \subseteq S$ for $i = 1, \ldots, k$ (with $S_i \cap S_j = \emptyset, i \neq j$, and $\bigcup_i S_i = S$) is denoted by $\{S_1|S_2|\ldots|S_k\}$. For a set X of Boolean variables, its set of valuations (or truth assignments) is denoted by $[\![X]\!]$, e.g., $[\![X]\!] = \{(0,0), (0,1), (1,0), (1,1)\}$ for $X = \{x_1, x_2\}$.

5.3.1 Functional Decomposition

Definition 5.1 Given a completely specified Boolean function f, variable x is a *support variable* of f if $f_x \neq f_{\neg x}$, where f_x and $f_{\neg x}$ are the positive and negative cofactors of f on x, respectively.

Definition 5.2 A set $\{f_1(X), \ldots, f_m(X)\}$ of completely specified Boolean functions is *(jointly) decomposable* with respect to some variable partition $X = \{X_H|X_G|X_C\}$ if every function f_i, $i = 1, \ldots, m$, can be written as

$$f_i(X) = h_i(X_H, X_C, g_1(X_G, X_C), \ldots, g_k(X_G, X_C))$$

for some functions h_i, g_1, ..., g_k with $k < |X_G|$. The decomposition is called *disjoint* if $X_C = \emptyset$ and *non-disjoint* otherwise.

It is known as *single-output decomposition* for $m = 1$ and *multiple-output decomposition* for $m > 1$. Note that, in multiple-output decomposition, functions $h_1, \ldots, h_m$ share the same functions $g_1, \ldots, g_k$. For $k = 1$, the decomposition is known as the so-called *Ashenhurst decomposition* [1].

Note that, for $|X_G| = 1$, there is no successful decomposition because of the violation of the criterion $k < |X_G|$. On the other hand, the decomposition trivially holds if $X_C \cup X_G$ or $X_C \cup X_H$ equals X. The corresponding variable partition is called *trivial*. We are concerned about decomposition under non-trivial variable partition and furthermore focus on Ashenhurst decomposition.

The decomposability of a set $\{f_1, \ldots, f_m\}$ of functions under the variable partition $X = \{X_H|X_G|X_C\}$ can be analyzed through the so-called *decomposition chart*, consisting of a set of matrices, one for each member of $[\![X_C]\!]$. The rows and columns of a matrix are indexed by $\{1, \ldots, m\} \times [\![X_H]\!]$ and $[\![X_G]\!]$, respectively. For $i \in \{1, \ldots, m\}$, $a \in [\![X_H]\!]$, $b \in [\![X_G]\!]$, and $c \in [\![X_C]\!]$, the entry with row index (i, a) and column index b of the matrix of c is of value $f_i(X_H = a, X_G = b, X_C = c)$.

Proposition 5.1 (Ashenhurst [1], Curtis [6], and Karp [11]) *A set $\{f_1, \ldots, f_m\}$ of Boolean functions is decomposable as*

$$f_i(X) = h_i(X_H, X_C, g_1(X_G, X_C), \ldots, g_k(X_G, X_C))$$

for $i = 1, \ldots, m$ under variable partition $X = \{X_H|X_G|X_C\}$ if and only if, for every $c \in [\![X_C]\!]$, the corresponding matrix of c has at most 2^k column patterns (i.e., at most 2^k different kinds of column vectors).

5.3.2 Functional Dependency

Definition 5.3 Given a Boolean function $f : \mathbb{B}^m \to \mathbb{B}$ and a vector of Boolean functions $G = (g_1(X), \ldots, g_n(X))$ with $g_i : \mathbb{B}^m \to \mathbb{B}$ for $i = 1, \ldots, n$, over the same set of variable vector $X = (x_1, \ldots, x_m)$, we say that f *functionally depends* on G if there exists a Boolean function $h : \mathbb{B}^n \to \mathbb{B}$, called the *dependency function*, such that $f(X) = h(g_1(X), \ldots, g_n(X))$. We call functions f, G, and h the *target function*, *base functions*, and *dependency function*, respectively.

Note that functions f and G are over the same domain in the definition; h need not depend on all of the functions in G.

The necessary and sufficient condition of the existence of the dependency function h was given in [8]. Moreover a SAT-based computation of functional dependency was presented in [10]. It forms an important ingredient in part of our formulation.

5.3.3 Propositional Satisfiability and Interpolation

Let $V = \{v_1, \ldots, v_k\}$ be a finite set of Boolean variables. A *literal* l is either a Boolean variable v_i or its negated form $\neg v_i$. A *clause* c is a disjunction of literals. Without loss of generality, we shall assume that there are no repeated or complementary literals in the same clause. A *SAT instance* is a conjunction of clauses, i.e., in the so-called *conjunctive normal form* (CNF). An *assignment* over V gives every variable v_i a Boolean value either true or false. A SAT instance is *satisfiable* if there exists a satisfying assignment such that the CNF formula evaluates to true. Otherwise it is *unsatisfiable*. Given a SAT instance, the *satisfiability (SAT) problem* asks whether it is satisfiable or not. A SAT solver is a designated program to solve the SAT problem.

5.3.3.1 Refutation Proof and Craig Interpolation

Definition 5.4 Assume literal v is in clause c_1 and $\neg v$ in c_2. A *resolution* of clauses c_1 and c_2 on variable v yields a new clause c containing all literals in c_1 and c_2 except for v and $\neg v$. The clause c is called the *resolvent* of c_1 and c_2 and variable v the *pivot variable*.

Proposition 5.2 *A resolvent c of c_1 and c_2 is a logical consequence of $c_1 \wedge c_2$, that is, $c_1 \wedge c_2$ implies c.*

Theorem 5.1 (Robinson [18]) *For an unsatisfiable SAT instance, there exists a sequence of resolution steps leading to an empty clause.*

Theorem 5.1 can be easily proved by Proposition 5.2 since an unsatisfiable SAT instance must imply a contradiction. Often only a subset of the clauses, called an *unsatisfiable core*, of the SAT instance participate in the resolution steps leading to an empty clause.

Definition 5.5 A refutation proof Π of an unsatisfiable SAT instance S is a directed acyclic graph (DAG) $\Gamma = (N, A)$, where every node in N represents a clause which is either a root clause in S or a resolvent clause having exactly two predecessor nodes and every arc in A connects a node to its ancestor node. The unique leaf of Π corresponds to the empty clause.

Theorem 5.2 (Craig Interpolation Theorem [5]) *Given two Boolean formulas φ_A and φ_B, with $\varphi_A \wedge \varphi_B$ unsatisfiable, then there exists a Boolean formula ψ_A referring only to the common variables of φ_A and φ_B such that $\varphi_A \Rightarrow \psi_A$ and $\psi_A \wedge \varphi_B$ is unsatisfiable.*

The Boolean formula ψ_A is referred to as the *interpolant* of φ_A with respect to φ_B. Some modern SAT solvers, e.g., MiniSat [7], are capable of constructing an interpolant from an unsatisfiable SAT instance [16]. Detailed exposition on how to construct an interpolant from a refutation proof in linear time can be found in [12, 16, 17]. Note that the so-derived interpolant is in a circuit structure, which can then be converted into the CNF as discussed below.

5.3.3.2 Circuit-to-CNF Conversion

Given a circuit netlist, it can be converted to a CNF formula in such a way that the satisfiability is preserved. The conversion is achievable in linear time by introducing some intermediate variables [21].

5.4 Main Algorithms

We show that Ashenhurst decomposition of a set of Boolean functions $\{f_1, \ldots, f_m\}$ can be achieved by SAT solving, Craig interpolation, and functional dependency. Whenever a non-trivial decomposition exists, we derive functions h_i and g automatically for $f_i(X) = h_i(X_H, X_C, g(X_G, X_C))$ along with the corresponding variable partition $X = \{X_H | X_G | X_C\}$.

5.4.1 Single-Output Ashenhurst Decomposition

We first consider Ashenhurst decomposition for a single function $f(X) = h(X_H, X_C, g(X_G, X_C))$.

5.4.1.1 Decomposition with Known Variable Partition

Proposition 5.1 in the context of Ashenhurst decomposition of a single function can be formulated as satisfiability solving as follows.

Proposition 5.3 *A completely specified Boolean function $f(X)$ can be expressed as $h(X_H, X_C, g(X_G, X_C))$ for some functions g and h if and only if the Boolean* *formula*

$$(f(X_H^1, X_G^1, X_C) \not\equiv f(X_H^1, X_G^2, X_C)) \wedge$$
$$(f(X_H^2, X_G^2, X_C) \not\equiv f(X_H^2, X_G^3, X_C)) \wedge$$
$$(f(X_H^3, X_G^3, X_C) \not\equiv f(X_H^3, X_G^1, X_C)) \tag{5.1}$$

is unsatisfiable, where a superscript i in Y^i *denotes the ith copy of the instantiation of variables* Y.

Observe that formula (5.1) is satisfiable if and only if there exists more than two distinct column patterns in some matrix of the decomposition chart. Hence its unsatisfiability is exactly the condition of Ashenhurst decomposition.

Note that, unlike BDD-based counterparts, the above SAT-based formulation of Ashenhurst decomposition naturally extends to non-disjoint decomposition. It is because the unsatisfiability checking of formula (5.1) essentially tries to assert that under every valuation of variables X_C the corresponding matrix of the decomposition chart has at most two column patterns. In contrast, BDD-based methods have to check the decomposability under every valuation of X_C separately.

Whereas the decomposability of function f can be checked through SAT solving of formula (5.1), the derivations of functions g and h can be realized through Craig interpolation and functional dependency, respectively, as shown below.

To derive function g, we partition formula (5.1) into two sub-formulas

$$\varphi_A = f(X_H^1, X_G^1, X_C) \not\equiv f(X_H^1, X_G^2, X_C) \text{ and} \tag{5.2}$$

$$\varphi_B = (f(X_H^2, X_G^2, X_C) \not\equiv f(X_H^2, X_G^3, X_C)) \wedge$$
$$(f(X_H^3, X_G^3, X_C) \not\equiv f(X_H^3, X_G^1, X_C)) \tag{5.3}$$

Figure 5.2 shows the corresponding circuit representation of formulas (5.2) and (5.3). The circuit representation can be converted into a CNF formula in linear time [21] and thus can be checked for satisfiability.

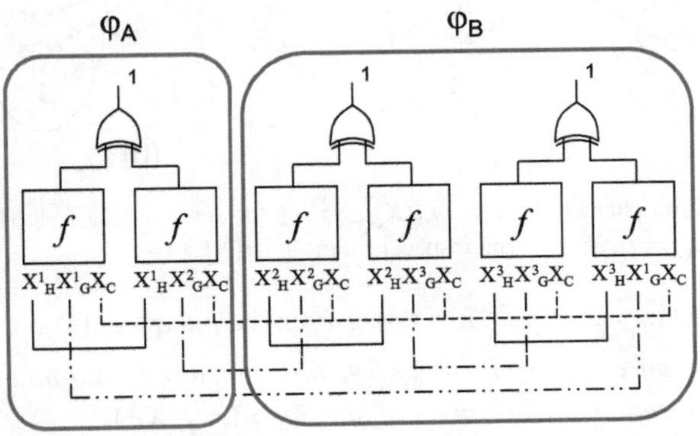

Fig. 5.2 Circuit representing the conjunction condition of formulas (5.2) and (5.3)

Lemma 5.1 *For function $f(X)$ decomposable under Ashenhurst decomposition with variable partition $X = \{X_H|X_G|X_C\}$, the interpolant ψ_A with respect to φ_A of formula (5.2) and φ_B of formula (5.3) corresponds to a characteristic function such that,*

(i) for φ_A satisfiable under some $c \in [\![X_C]\!]$, $\psi_A(b_1, b_2, c) = 1$ with $b_1 \in [\![X_G^1]\!]$ and $b_2 \in [\![X_G^2]\!]$ if and only if the column vectors indexed by b_1 and b_2 in the matrix of c of the decomposition chart of f are different;
(ii) for φ_A unsatisfiable under some $c \in [\![X_C]\!]$, there is only one column pattern in the matrix of c of the decomposition chart of f; and
(iii) for unsatisfiable φ_A, variables X_G are not the support variables of f and thus $\{X_H|X_G|X_C\}$ is a trivial variable partition for f.

Figure 5.3a illustrates the relation characterized by interpolant $\psi_A(X_G^1, X_G^2, c)$ for some $c \in [\![X_C]\!]$. The left and right sets of gray dots denote the elements of $[\![X_G^1]\!]$ and $[\![X_G^2]\!]$, respectively. For function f to be decomposable, there are at most two equivalence classes for the elements of $[\![X_G^i]\!]$ for $i = 1, 2$. In the figure, the two clusters of elements in $[\![X_G^i]\!]$ signify two equivalence classes of column patterns indexed by $[\![X_G^i]\!]$. An edge (b_1, b_2) between $b_1 \in [\![X_G^1]\!]$ and $b_2 \in [\![X_G^2]\!]$ denotes that b_1 is not in the same equivalence class as b_2, i.e., $\psi_A(b_1, b_2, c) = 1$. For example, p and r in the figure are in different equivalence classes and $\psi_A(p, r, c) = 1$, whereas p and q are in the same equivalence class and $\psi_A(p, q, c) = 0$. Essentially the set of such edges is characterized by the equivalence relation $\psi_A(X_G^1, X_G^2, c)$. So every element in one equivalence class of $[\![X_G^1]\!]$ is connected to every element in the other equivalence class of $[\![X_G^2]\!]$, and vice versa, in Fig. 5.3a.

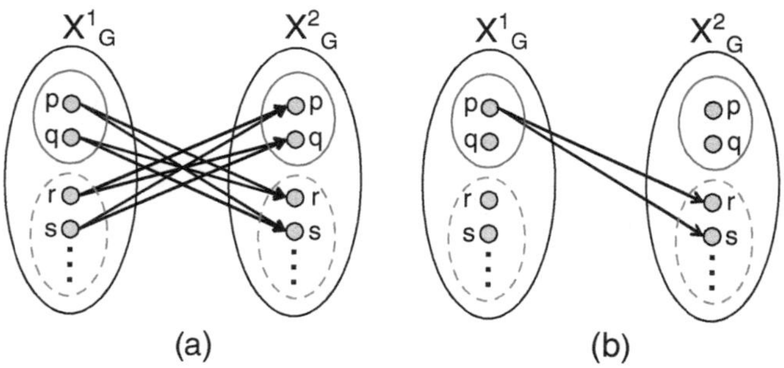

Fig. 5.3 (**a**) Relation characterized by $\psi_A(X_G^1, X_G^2, c)$ for some $c \in [\![X_C]\!]$; (**b**) relation after cofactoring $\psi_A(X_G^1 = p, X_G^2, c)$ with respect to some $p \in [\![X_G^1]\!]$

We next show how to extract function g from the interpolant ψ_A.

Lemma 5.2 *For an arbitrary $a \in [\![X_G^1]\!]$, the cofactored interpolant $\psi_A(X_G^1 = a, X_G^2, X_C)$ is a legal implementation of function $g(X_G^2, X_C)$.*

After renaming X_G^2 to X_G, we get the desired $g(X_G, X_C)$.

Consider Fig. 5.3. After cofactoring $\psi_A(X_G^1, X_G^2, c)$ with respect to $p \in [\![X_G^1]\!]$, all the edges in Fig. 5.3a will disappear except for the ones connecting p with the elements in the other equivalence class of $[\![X_G^2]\!]$ as shown in Fig. 5.3b. Hence $\psi_A(p, X_G^2, c)$ can be used as an implementation of g function.

So far we have successfully obtained function g by interpolation. Next we need to compute function h. The problem can be formulated as computing functional dependency as follows. Let $f(X)$ be our target function; let function $g(X_G, X_C)$ and identity functions $\iota_x(x) = x$, one for every variable $x \in X_H \cup X_C$, be our base functions. So the computed dependency function corresponds to our desired h. Since functional dependency can be formulated using SAT solving and interpolation [10], it well fits in our computation framework.

Remark 5.1 For disjoint decomposition, i.e., $X_C = \emptyset$, we can simplify the derivation of function h, without using functional dependency.

Given two functions $f(X)$ and $g(X_G)$ with variable partition $X = \{X_H|X_G\}$, we aim to find a function $h(X_H, x_g)$ such that $f(X) = h(X_H, g(X_G))$, where x_g is the output variable of function $g(X_G)$. Let $a, b \in [\![X_G]\!]$ with $g(a) = 0$ and $g(b) = 1$. Then by Shannon expansion

$$h(X_H, x_g) = (\neg x_g \wedge h_{\neg x_g}(X_H)) \vee (x_g \wedge h_{x_g}(X_H))$$

where $h_{\neg x_g}(X_H) = f(X_H, X_G = a)$ and $h_{x_g}(X_H) = f(X_H, X_G = b)$. The derivation of the offset and onset minterms is easy because we can pick an arbitrary minterm c in $[\![X_G]\!]$ and see if $g(c)$ equals 0 or 1. We then perform SAT solving on either $g(X_G)$ or $\neg g(X_G)$ depending on the value $g(c)$ to derive another necessary minterm.

The above derivation of function h, however, does not scale well for decomposition with large $|X_C|$ because we may need to compute $h(X_H, X_C = c, x_g)$, one for every valuation $c \in [\![X_C]\!]$. There are $2^{|X_C|}$ cases to analyze. Consequently when common variables exist, functional dependency may be a better approach to computing h.

The correctness of the so-derived Ashenhurst decomposition follows from Lemma 5.2 and Proposition 5.1, as the following theorem states.

Theorem 5.3 *Given a function f decomposable under Ashenhurst decomposition with variable partition $X = \{X_H|X_G|X_C\}$, then $f(X) = h(X_H, X_C, g(X_G, X_C))$ for functions g and h obtained by the above derivation.*

5.4.1.2 Decomposition with Unknown Variable Partition

The previous construction assumes that a variable partition $X = \{X_H|X_G|X_C\}$ is given. We show how to automate the variable partition within the decomposition process of function f. A similar approach was used in [14] for bi-decomposition of Boolean functions.

For each variable $x_i \in X$ we introduce two control variables α_{x_i} and β_{x_i}. In addition we instantiate variable X into six copies X^1, X^2, X^3, X^4, X^5, and X^6. Let

$$\varphi_A = (f(X^1) \not\equiv f(X^2)) \wedge \bigwedge_i ((x_i^1 \equiv x_i^2) \vee \beta_{x_i}) \tag{5.4}$$

and

$$\begin{aligned}\varphi_B = {} & (f(X^3) \not\equiv f(X^4)) \wedge (f(X^5) \not\equiv f(X^6)) \wedge \\ & \bigwedge_i (((x_i^2 \equiv x_i^3) \wedge (x_i^4 \equiv x_i^5) \wedge (x_i^6 \equiv x_i^1)) \vee \alpha_{x_i}) \wedge \\ & \bigwedge_i (((x_i^3 \equiv x_i^4) \wedge (x_i^5 \equiv x_i^6)) \vee \beta_{x_i})\end{aligned} \tag{5.5}$$

where $x_i^j \in X^j$ for $j = 1, \ldots, 6$ are the instantiated versions of $x_i \in X$. Observe that $(\alpha_{x_i}, \beta_{x_i}) = (0,0)$, $(0,1)$, $(1,0)$, and $(1,1)$ indicate that $x_i \in X_C$, $x_i \in X_G$, $x_i \in X_H$, and x_i can be in either of X_G and X_H, respectively.

In SAT solving the conjunction of formulas (5.4) and (5.5), we make *unit assumptions* [7] on the control variables. Similar to [14] but with a subtle difference, we introduce the following *seed variable partition* to avoid trivial variable partition and to avoid $|X_G| = 1$. For the unit assumption, initially we specify three distinct variables with one, say, x_j, in X_H and two, say, x_k, x_l, in X_G and specify all other variables in X_C. That is, we have $(\alpha_{x_j}, \beta_{x_j}) = (1,0)$, $(\alpha_{x_k}, \beta_{x_k}) = (0,1)$, $(\alpha_{x_l}, \beta_{x_l}) = (0,1)$, and $(\alpha_{x_i}, \beta_{x_i}) = (0,0)$ for $i \neq j, k, l$.

Lemma 5.3 *For an unsatisfiable conjunction of formulas (5.4) and (5.5) under a seed variable partition, the final conflict clause consists of only the control variables, which indicates a valid non-trivial variable partition.*

If the conjunction of formulas (5.4) and (5.5) is unsatisfiable under a seed variable partition, then the corresponding decomposition (indicated by the final conflict clause) is successful. Otherwise, we should try another seed variable partition. For a given function $f(X)$ with $|X| = n$, the existence of non-trivial Ashenhurst decomposition can be checked with at most $3 \cdot C_3^n$ different seed partitions.

Rather than just looking for a valid variable partition, we may further target one that is more balanced (i.e., $|X_H|$ and $|X_G|$ are of similar sizes) and closer to disjoint (i.e., $|X_C|$ is small) by enumerating different seed variable partitions. As SAT solvers usually refer to a small unsatisfiable core, the returned variable partition is desirable because $|X_C|$ tends to be small. Even if a returned unsatisfiable core is unnecessarily large, the corresponding variable partition can be further refined by modifying the unit assumption to reduce the unsatisfiable core and reduce $|X_C|$ as well. The process can be iterated until the unsatisfiable core is minimal.

After automatic variable partition, functions g and h can be derived through a construction similar to the foregoing one. The correctness of the overall construction can be asserted.

Theorem 5.4 *For a function f decomposable under Ashenhurst decomposition, we have $f(X) = h(X_H, X_C, g(X_G, X_C))$ for functions g and h and a non-trivial variable partition $X = \{X_H | X_G | X_C\}$ derived from the above construction.*

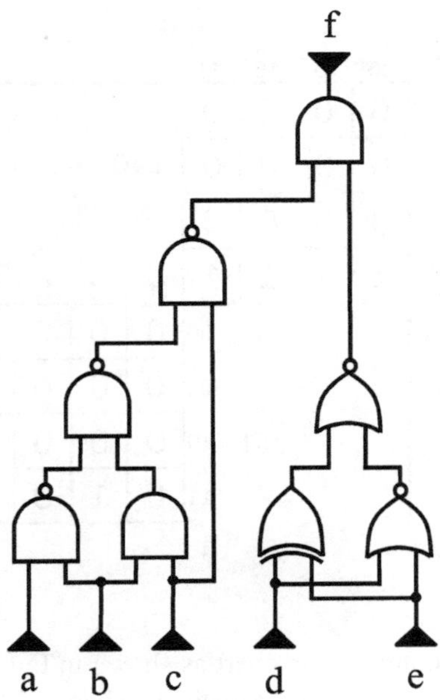

Fig. 5.4 Circuit to be decomposed

Example 5.1 To illustrate the computation, consider the circuit of Fig. 5.4. The first step is to derive a valid variable partition. To exclude trivial partition, suppose we force variables a and b in X_G and d in X_H. Then the assignments along with the assignments of the other variables, i.e., c and e, in X_C form a seed variable partition. These conditions can be specified by unit assumption setting control variables $(\alpha_a, \beta_a) = (\alpha_b, \beta_b) = (0, 1)$, $(\alpha_d, \beta_d) = (1, 0)$, and $(\alpha_c, \beta_c) = (\alpha_e, \beta_e) = (0, 0)$. Solving the conjunction of formulas (5.4) and (5.5) under the unit assumption results in an unsatisfiable result. It indicates that the seed partition is valid. Furthermore suppose the returned conflict clause is $(\alpha_a \vee \alpha_b \vee \alpha_c \vee \beta_c \vee \beta_d \vee \beta_e)$. It corresponds to a valid partition suggesting that $c \in X_C$, $a, b \in X_G$, and $d, e \in X_H$. For illustration convenience, the decomposition chart of the circuit under this variable partition is given in Fig. 5.5.

Given a valid variable partition, the second step is to derive the corresponding g function. In turn, an interpolant can be derived from the unsatisfiability proof of the conjunction of formulas (5.2) and (5.3). Suppose the derived interpolant is

$$\psi_A = \neg a^1 b^1 \neg c^1 a^2 \neg b^2 \neg c^2 \vee a^1 \neg b^1 \neg c^1 \neg a^2 \neg c^2 \vee a^1 \neg c^1 \neg a^2 b^2 \neg c^2 \vee \\ \neg b^1 c^1 \neg a^2 b^2 c^2 \vee a^1 c^1 \neg a^2 b^2 c^2 \vee \neg a^1 b^1 c^1 \neg b^2 c^2 \vee \neg a^1 b^1 c^1 a^2 c^2$$

Then the Boolean relation characterized by the interpolant can be depicted in Fig. 5.6a, where the solid and dashed circles indicate different column patterns in the decomposition chart of Fig. 5.5. Note that, when $c = 0$, there is only one

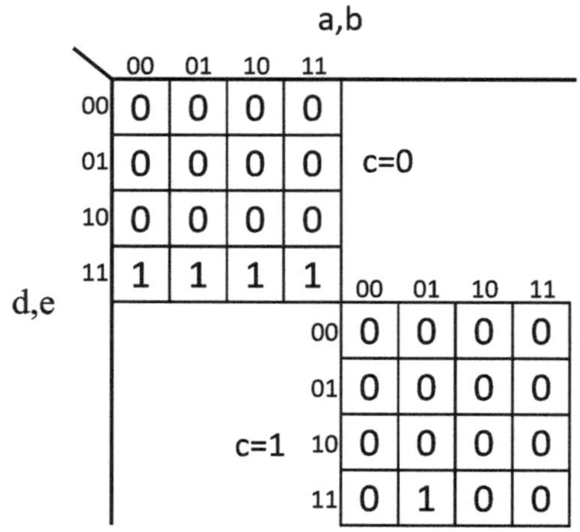

Fig. 5.5 Decomposition chart

column pattern in the decomposition chart as shown in Fig. 5.5. In effect both formulas (5.2) and (5.3) are themselves unsatisfiable when $c = 0$. Hence the interpolant under $c = 0$ is unconstrained and can be arbitrary. On the other hand, when $c = 1$, the interpolant corresponds to the Boolean relation characterizing different column patterns of the decomposition chart as indicated in Fig. 5.6. By cofactoring the interpolant with $(a^1 = 0, b^1 = 0)$, we obtain a legal implementation of function $g(a, b, c)$ after renaming variables a^2 to a, b^2 to b, and c^1 and c^2 to c. Note that the

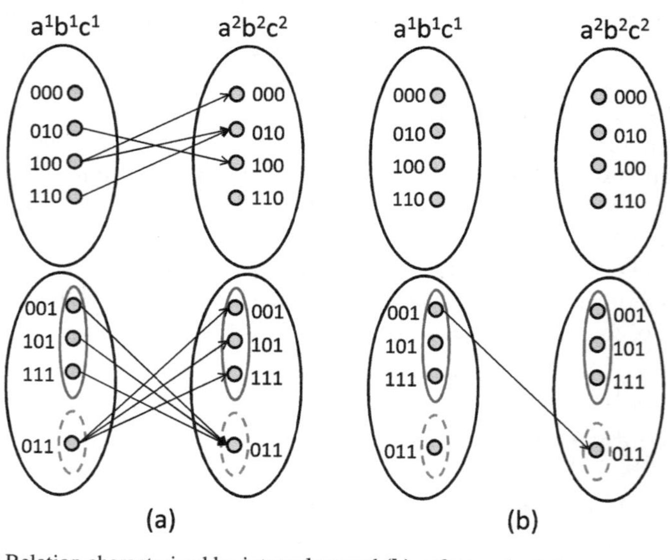

Fig. 5.6 **(a)** Relation characterized by interpolant and **(b)** cofactored relation

derivation of the g function is not unique, which depends on the cofactoring values of (a^1, b^1).

Finally the third step is to derive the h function using functional dependency computation. In the computation, as shown in Fig. 5.7a the base functions include the obtained g function and identity functions each representing a variable in $X_H \cup X_C$. Furthermore the original f function in Fig. 5.4 is considered as the target function. Under such arrangement, the computed dependency function is what we desire for the h function. In this example the derived h function is shown in Fig. 5.7b. Therefore after Ashenhurst decomposition, $f(a, b, c, d, e)$ can be re-expressed by $h(d, e, c, g(a, b, c))$ with g and h functions derived above.

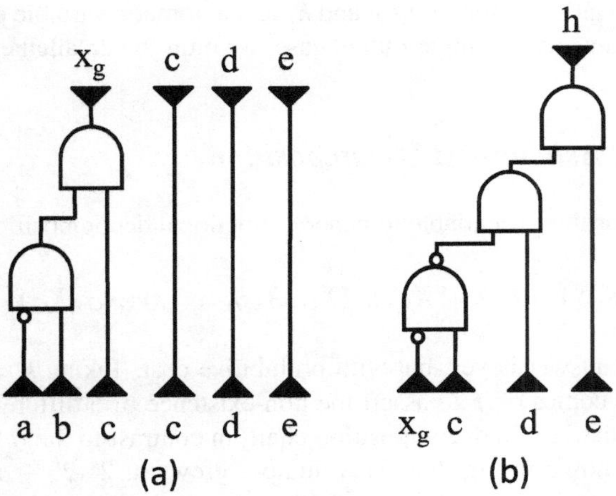

Fig. 5.7 **(a)** Base functions for functional dependency computation and **(b)** computed dependency function

5.4.2 Multiple-Output Ashenhurst Decomposition

So far we considered single-output Ashenhurst decomposition for a single function f. We show that the algorithm is extendable to multiple-output Ashenhurst decomposition for a set $\{f_1, \ldots, f_m\}$ of functions.

Proposition 5.1 in the context of Ashenhurst decomposition of a set of functions can be formulated as satisfiability solving as follows.

Proposition 5.4 *A set* $\{f_1(X), \ldots, f_m(X)\}$ *of completely specified Boolean functions can be expressed as*

$$f_i(X) = h_i(X_H, X_C, g(X_G, X_C))$$

for some functions h_i *and* g *with* $i = 1, \ldots, m$ *if and only if the Boolean formula*

$$\left(\bigvee_i f_i(X_H^1, X_G^1, X_C) \not\equiv f_i(X_H^1, X_G^2, X_C)\right) \wedge$$
$$\left(\bigvee_i f_i(X_H^2, X_G^2, X_C) \not\equiv f_i(X_H^2, X_G^3, X_C)\right) \wedge$$
$$\left(\bigvee_i f_i(X_H^3, X_G^3, X_C) \not\equiv f_i(X_H^3, X_G^1, H_C)\right) \quad (5.6)$$

is unsatisfiable.

Since the derivation of functions g and h_i and automatic variable partitioning are essentially the same as the single-output case, we omit the detailed exposition.

5.4.3 Beyond Ashenhurst Decomposition

Is the above algorithm extendable to general functional decomposition, namely,

$$f(X) = h(X_H, X_C, g_1(X_G, X_C), \ldots, g_k(X_G, X_C))$$

for $k > 1$? The answer is yes, but with prohibitive cost. Taking $k = 2$, for example, we need 20 copies of f to assert the non-existence of 5 different column patterns for every matrix of a decomposition chart, in contrast to the 6 for Ashenhurst decomposition shown in Fig. 5.2. This number grows in $2^k(2^k + 1)$. Aside from this duplication issue, the derivation of functions $g_1, \ldots, g_k$ and h may involve several iterations of finding satisfying assignments and performing cofactoring. The number of iterations varies depending on how the interpolation is computed and can be exponential in k. Therefore we focus mostly on Ashenhurst decomposition.

5.5 Experimental Results

The proposed approach to Ashenhurst decomposition was implemented in C++ within the ABC package [2] and used MiniSAT [7] as the underlying solver. All the experiments were conducted on a Linux machine with Xeon 3.4 GHz CPU and 6 GB RAM.

Large ISCAS, MCNC, and ITC benchmark circuits were chosen to evaluate the proposed method. Only large transition and output functions (with no less than 50 inputs in the transitive fanin cone) were considered. We evaluated both single-output and two-output Ashenhurst decompositions. For the latter, we decomposed simultaneously a pair of functions with similar input variables. For a circuit, we heuristically performed pairwise matching among its transition and output functions for decomposition. Only function pairs with joint input variables no less than 50 were decomposed. Note that the experiments target the study of scalability, rather than comprehensiveness as a synthesis methodology.

Tables 5.1 and 5.2 show the decomposition statistics of single-output and two-output decompositions, respectively. In these tables, circuits to be decomposed are listed in column 1. Columns 2 and 3 list the numbers of instances (i.e., functions for single-output decomposition and function pairs for two-output decomposition) with no less than 50 inputs and the ranges of the input sizes of these instances, respectively. Column 4 lists the numbers of instances that we cannot find any successful variable partition within 60 s or within 1500 seed variable partitions. Column 5 lists the numbers of instances that are decomposable but spending over 30 s in SAT solving for the derivation of function g or h. Columns 6 and 7 list the numbers of successfully decomposed instances and the ranges of the input sizes of these instances, respectively. Columns 8 and 9 list the average numbers of tried seed partitions in 60 s and the average rates hitting valid seed partitions. Column 10 shows the average CPU times spending on decomposing an instance. Finally, Column 11 shows the memory consumption. As can be seen, our method can effectively decompose functions or function pairs with up to 300 input variables.

Table 5.1 Single-output Ashenhurst decomposition

Circuit	#func	#var	#fail	#SAT_TO	#succ	#var_succ	#VP_avg	rate_valid-VP	time_avg (s)	mem (Mb)
b14	153	50–218	0	108	45	50–101	1701	0.615	144.22	90.01
b15	370	143–306	0	51	319	143–306	1519	0.917	96.62	107.20
b17	1009	76–308	0	148	861	76–308	1645	0.904	87.12	125.84
C2670	6	78–122	0	1	5	78–122	1066	0.835	83.80	58.91
C5315	20	54–67	0	4	16	54–67	3041	0.914	50.90	51.34
C7552	36	50–194	0	2	34	50–194	1350	0.455	64.38	36.65
s938	1	66–66	0	0	1	66–66	3051	0.726	19.03	24.90
s1423	17	51–59	0	0	17	51–59	3092	0.723	13.66	25.34
s3330	1	87–87	0	0	1	87–87	3336	0.599	58.30	27.75
s9234	13	54–83	0	0	13	54–83	3482	0.857	37.86	35.33
s13207	3	212–212	0	0	3	212–212	569	0.908	70.26	50.62
s38417	256	53–99	6	72	178	53–99	1090	0.523	103.33	136.04
s38584	7	50–147	0	0	7	50–147	1120	0.924	47.13	51.56

Table 5.2 Two-output Ashenhurst decomposition

Circuit	#pair	#var	#fail	#SAT_TO	#succ	#var_succ	#VP_avg	rate_valid-VP	time_avg (s)	mem (Mb)
b14	123	50–223	18	65	40	50–125	1832	0.568	96.86	226.70
b15	201	145–306	0	31	170	145–269	1176	0.845	113.86	224.07
b17	583	79–310	0	88	495	79–308	676	0.824	103.12	419.35
C2670	5	78–123	0	1	4	78–123	254	0.724	66.95	55.71
C5315	11	56–69	0	2	9	56–69	370	0.594	59.20	60.05
C7552	21	56–195	0	2	19	56–141	188	0.465	89.57	78.67
s938	1	66–66	0	0	1	66–66	3345	0.720	61.24	34.77
s1423	14	50–67	0	0	14	50–67	3539	0.591	55.34	45.66
s3330	1	87–87	0	0	1	87–87	1278	0.423	66.83	47.43
s9234	12	54–83	0	0	12	54–83	2193	0.708	48.11	55.15
s13207	3	212–228	0	0	3	212–228	585	0.700	93.36	118.03
s38417	218	53–116	13	30	175	53–116	689	0.498	109.06	319.48
s38584	9	50–151	0	0	9	50–151	1656	0.713	46.17	207.78

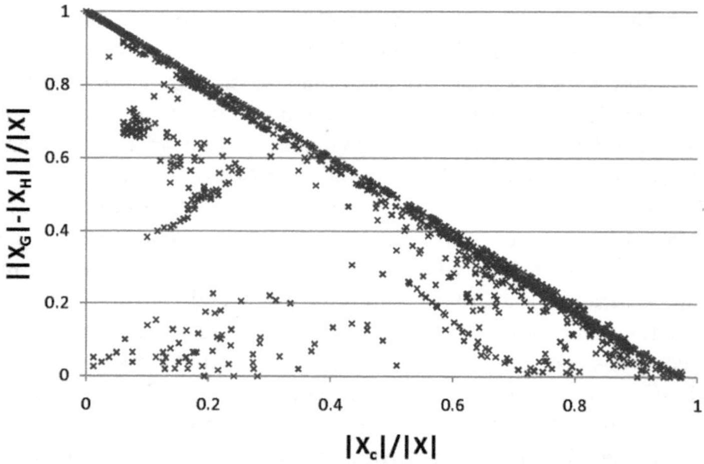

Fig. 5.8 Best variable partition found in 60 s – without minimal UNSAT core refinement

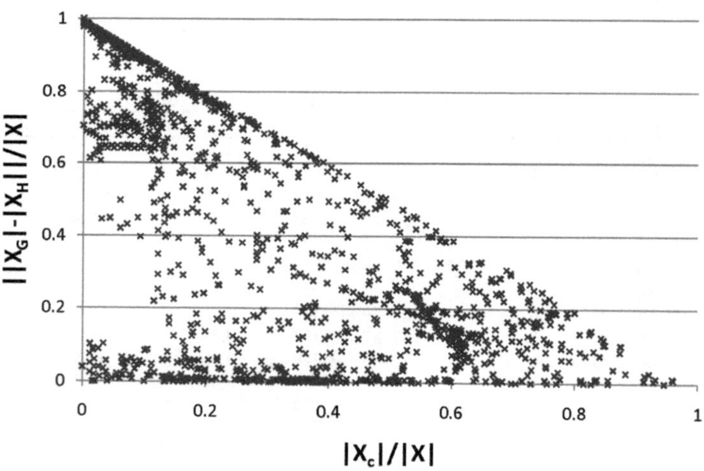

Fig. 5.9 Best variable partition found in 60 s – with minimal UNSAT core refinement

We measure the quality of a variable partition in terms of disjointness, indicated by $|X_C|/|X|$, and balancedness, indicated by $||X_G| - |X_H||/|X|$. The smaller the values are, the better a variable partition is. Figures 5.8 and 5.9 depict, for each decomposition instance, the quality of best variable partition found within 60 s[1] in terms of the above two metrics, with emphasis on disjointness. A spot on these two figures corresponds to a variable partition for some decomposition instance. Figs. 5.8 and 5.9 show the variable partition data *without* and *with* further minimal

[1] The search for a best variable partition may quit before 60 s if both disjointness and balancedness cannot be improved in consecutive 1500 trials.

unsatisfiable (UNSAT) core refinement[2], respectively. Since a final conflict clause returned by a SAT solver may not reflect a minimal UNSAT core, very likely we can further refine the corresponding variable partition. Suppose the variable partition is $X = \{X_H|X_G|X_C\}$ before the refinement. We iteratively and greedily try to move a common variable of X_C into X_G or X_H, if available, making the new partition more balanced as well. The iteration continues until no such movement is possible. On the other hand, for a variable x with control variables $(\alpha_x, \beta_x) = (1, 1)$, indicating x can be placed in either of X_H and X_G, we put it in the one such that the final partition is more balanced. Comparing Figs. 5.8 and 5.9, we see that minimal UNSAT core refinement indeed can substantially improve the variable partition quality. Specifically, the improvement is 42.37% for disjointness and 5.74% for balancedness.

Figure 5.10 compares the qualities of variable partitioning under four different efforts. In the figure, "1st" denotes the first-found valid partition and "$t$sec" denotes the best found valid partition in t seconds. The averaged values of $|X_C|/|X|$ and $||X_G| - |X_H||/|X|$ with and without minimal UNSAT core refinement are plotted. In our experiments, improving disjointness is preferable to improving balancedness. These two objectives, as can be seen, are usually mutually exclusive. Disjointness can be improved at the expense of sacrificing balancedness and vice versa. The figure reveals as well the effectiveness of the minimal UNSAT core refinement in

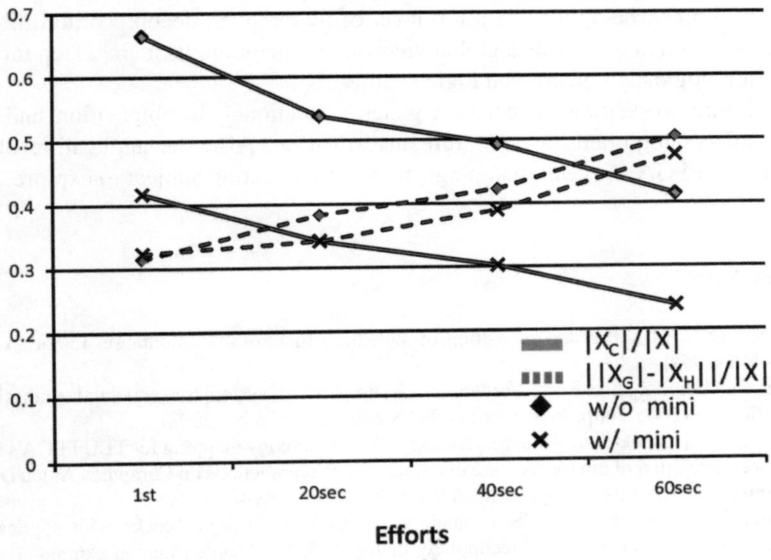

Fig. 5.10 Variable partition qualities under four different efforts

[2] For every decomposition instance, the UNSAT core refinement is applied only once to the best found variable partition. The CPU times listed in Tables 5.1 and 5.2 include those spent on such refinements.

improving disjointness. It is interesting to note that, on average, 1337 seed partitions are tried in 60 s, in contrast to 3 seed partitions tried to identify the first valid one.

Practical experience suggests that the AIG sizes and levels of the composition functions g and h are typically much larger than those of the original function f by an order of magnitude, despite the reduction of support variables. How to minimize interpolants effectively becomes an important subject for our method to directly benefit logic synthesis.

5.6 Chapter Summary

A new formulation of Ashenhurst decomposition was proposed based on SAT solving, Craig interpolation, and functional dependency. Traditionally difficult non-disjoint and multiple-output decompositions can be handled naturally. Moreover variable partition need not be specified a priori and can be embedded in the decomposition process. It allows effective enumeration over a wide range of partition choices, which is not possible before. Although Ashenhurst decomposition is a special case of functional decomposition, its simplicity is particularly attractive and preferable.

Because of its scalability to large designs as justified by experimental results, our approach can be applied at a top level of hierarchical decomposition in logic synthesis, which may provide a global view on optimization. It can be a step forward toward topologically constrained logic synthesis.

For future work, how to perform general functional decomposition and how to minimize interpolants await future investigation. Also the application of our approach to FPGA Boolean matching can be an interesting subject to explore.

References

1. Ashenhurst, R.L.: The decomposition of switching functions. Computation Laboratory **29**, 74–116 (1959)
2. Berkeley Logic Synthesis and Verification Group: ABC: A system for sequential synthesis and verification (2005). http://www.eecs.berkeley.edu/~alanmi/abc/(2008)
3. Chang, S.C., Marek-Sadowska, M., Hwang, T.T.: Technology mapping for TLU FPGA's based on decomposition of binary decision diagrams. IEEE Transactions on Computer-Aided Design of Integrated Circuits and Systems **15**(10), 1226–1236 (1996)
4. Cong, J., Hwang, Y.Y.: Boolean matching for LUT-based logic blocks with applications to architecture evaluation and technology mapping. IEEE Transactions on Computer-Aided Design of Integrated Circuits and Systems **20**(9), 1077–1090 (2001)
5. Craig, W.: Linear reasoning: A new form of the Herbrand-Gentzen theorem. Journal of Symbolic Logic **22**(3), 250–268 (1957)
6. Curtis, A.: A New Approach to the Design of Switching Circuits. Van Nostrand, Princeton, NJ (1962)
7. Eén, N., Söensson, N.: An extensible SAT-solver. In: Proceedings of International Conference on Theory and Applications of Satisfiability Testing pp. 502–518. Santa Margherita Ligure, Italy (2003)

8. Jiang, J.H.R., Brayton, R.K.: Functional dependency for verification reduction. In: Proceedings of the International Conference on Computer Aided Verification, pp. 268–280. Boston, MA, USA (2004)
9. Jiang, J.H.R., Jou, J.Y., Huang, J.D.: Compatible class encoding in hyper-function decomposition for FPGA synthesis. In: Proceedings of the Design Automation Conference, pp. 712–717. San Francisco, CA, USA (1998)
10. Jiang, J.H.R., Lee, C.C., Mishchenko, A., Huang, C.Y.: To SAT or Not to SAT: Scalable exploration of functional dependency. IEEE Transactions on Computers **59**(4), 457–467 (2010)
11. Karp, R.M.: Functional decomposition and switching circuit design. Journal of the Society for Industrial and Applied Mathematics **11**(2), 291–335 (1963)
12. Krajicek, J.: Interpolation theorems, lower bounds for proof systems, and independence results for bounded arithmetic. Journal of Symbolic Logic **62**(2), 457–486 (1997)
13. Lai, Y.T., Pan, K.R., Pedram, M.: OBDD-based function decomposition: Algorithms and implementation. IEEE Transactions on Computer Aided Design of Integrated Circuits and Systems **15**(8), 977–990 (1996)
14. Lee, R.R., Jiang, J.H.R., Hung, W.L.: Bi-decomposing large Boolean functions via interpolation and satisfiability solving. In: Proceedings of the Design Automation Conference, pp. 636–641. Anaheim, CA, USA (2008)
15. Ling, A., Singh, D., Brown, S.: FPGA technology mapping: A study of optimality. In: Proceedings of the Design Automation Conference, pp. 427–432. San Diego, CA, USA (2005)
16. McMillan, K.L.: Interpolation and SAT-based model checking. In: Proceedings of the International Conference on Computer Aided Verification, pp. 1–13. Boulder, CO, USA (2003)
17. Pudlak, P.: Lower bounds for resolution and cutting plane proofs and monotone computations. Journal of Symbolic Logic **62**(3), 981–998 (1997)
18. Robinson, J.A.: A machine-oriented logic based on the resolution principle. Journal of the ACM **12**(1), 23–41 (1965)
19. Scholl, C.: Functional Decomposition with Applications to FPGA Synthesis. Dordrecht, The Netherlands (2001)
20. Sinha, S., Mishchenko, A., Brayton, R.K.: Topologically constrained logic synthesis. In: Proceedings of the International Conference on Computer Aided Design, pp. 679–686. San Jose, CA, USA (2002)
21. Tseitin, G.: On the complexity of derivation in propositional calculus. In: A.O. Slisenko (ed.) Studies in Constructive Mathematics and Mathematical Logic, Part II, vol. 8, p. 280. Serial Zap. Nauchn. Sem. LOMI. Nauka, Leningrad (1968)
22. Wurth, B., Schlichtmann, U., Eckl, K., Antreich, K.: Functional multiple-output decomposition with application to technology mapping for lookup table-based FPGAs. ACM Transactions on Design Automation of Electronic Systems **4**(3), 313–350 (1999)

Chapter 6
Bi-decomposition Using SAT and Interpolation

Ruei-Rung Lee, Jie-Hong Roland Jiang, and Wei-Lun Hung

Abstract Boolean function bi-decomposition is a fundamental operation in logic synthesis. A function $f(X)$ is bi-decomposable under a variable partition X_A, X_B, X_C on X if it can be written as $h(f_A(X_A, X_C), f_B(X_B, X_C))$ for some functions h, f_A, and f_B. The quality of a bi-decomposition is mainly determined by its variable partition. A preferred decomposition is disjoint, i.e., $X_C = \emptyset$, and balanced, i.e., $|X_A| \approx |X_B|$. Finding such a good decomposition reduces communication and circuit complexity and yields simple physical design solutions. Prior BDD-based methods may not be scalable to decompose large functions due to the memory explosion problem. Also as decomposability is checked under a fixed variable partition, searching a good or feasible partition may run through costly enumeration that requires separate and independent decomposability checkings. This chapter proposes a solution to these difficulties using interpolation and incremental SAT solving. Preliminary experimental results show that the capacity of bi-decomposition can be scaled up substantially to handle large designs.

6.1 Introduction

Functional decomposition [1, 7] is a fundamental operation on Boolean functions that decomposes a large function f on variables X into a set of small subfunctions $h, g_1, \ldots, g_m$ with $f(X) = h(g_1(X), \ldots, g_m(X))$, often $m < |X|$. It plays a pivotal role in the study of circuit and communication complexity and has important applications on multilevel and FPGA logic synthesis. Extensive research has been published on this subject, see, e.g., [15] for an introduction.

When $m = 2$, the decomposition is known as *bi-decomposition* [4, 5, 11, 14, 16], the simplest non-trivial case, yet the most widely applied since a logic netlist is often

---

J.-H.R. Jiang (✉)
National Taiwan University, Taipei, Taiwan
e-mail: jhjiang@cc.ee.ntu.edu.tw

This work is based on an earlier work: Bi-decomposing large Boolean functions via interpolation and satisfiability solving, in Proceedings of the 45th Annual Design Automation Conference, ISBN ISSN:0738-100X , 978-1-60558-115-6 (2008) © ACM, 2008. DOI= http://doi.acm.org/10.1145/1391469.1391634

S.P. Khatri, K. Gulati (eds.), *Advanced Techniques in Logic Synthesis, Optimizations and Applications*, DOI 10.1007/978-1-4419-7518-8_6,

expressed as a network of two-fanin gates. A primary issue of bi-decomposition is *variable partition*. For $f(X) = h(f_A(X_A, X_C), f_B(X_B, X_C))$, the variable partition $\{X_A, X_B, X_C\}$ on X (i.e., X_A, X_B, X_C are pairwise disjoint and $X_A \cup X_B \cup X_C = X$) mainly determines the decomposition quality. A particularly desirable bi-decomposition is *disjoint*, i.e., $|X_C| = 0$, and *balanced*, i.e., $|X_A| \approx |X_B|$. Figure 6.1 shows the general circuit topology of bi-decomposition. An ideal bi-decomposition reduces circuit and communication complexity and in turn simplifies physical design. Effective approaches to bi-decomposition can be important not only in large-scale circuit minimization but also in early design closure if combined well with physical design partitioning.

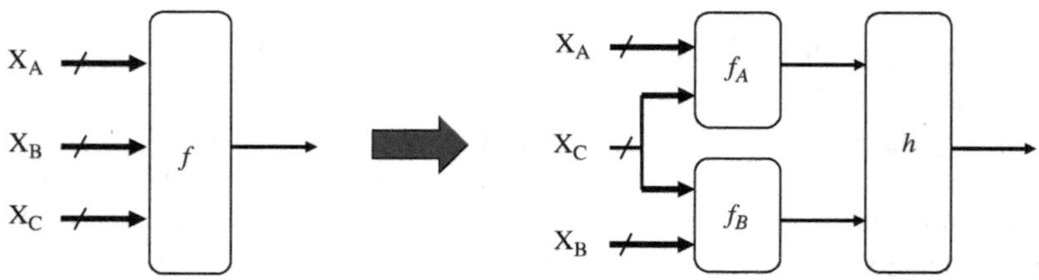

Fig. 6.1 Bi-decomposition

Modern approaches to bi-decomposition, such as [11], were based on BDD data structure for its powerful capability supporting various Boolean manipulations. They are, however, not scalable to handle large Boolean functions due to the common memory explosion problem. Furthermore, the variable partition problem cannot be solved effectively. Because decomposability is checked under a fixed variable partition, searching a good or even feasible partition may run through costly enumeration that requires separate and independent decomposability checkings.

To overcome these limitations, this chapter proposes a solution based on satisfiability (SAT) solving. The formulation is motivated by the recent work [9], where a SAT-based formulation made the computation of functional dependency scalable to large designs. Our main results include (1) a pure SAT-based solution to bi-decomposition, (2) subfunction derivation by interpolation and cofactoring, and (3) automatic variable partitioning by incremental solving under unit assumptions. Thereby the scalability of bi-decomposition and the optimality of variable partition can be substantially improved. Experiments show promising results on the scalability of bi-decomposition and the optimality of variable partitioning.

6.2 Previous Work

There has been several bi-decomposition methods developed over the past two decades. Among them, the efforts [4, 16] laid the early formulation of (AND, OR, and XOR) bi-decomposition. Bi-decomposition through irredundant sum-of-products expression and positive polarity Reed–Muller expression was studied in

[14]. Its capacity is limited due to the expensiveness of representing functions in these expressions. Later BDD-based improvements were actively pursued. For example, a BDD-based approach to bi-decomposing multiple-output incompletely specified functions was proposed in [11]. On the other hand, logic bi-decomposition may have different optimization objectives, e.g., support minimization [18], timing minimization [5]. An effective BDD-based disjoint-support decomposition was proposed in [3]. The computation complexity is polynomial in the BDD size, and the decomposition is not restricted to bi-decomposition. However, the decomposition was done with respect to the BDD structure for a fixed variable ordering. As BDD-based methods suffer from the memory explosion problem, they are not scalable to decompose large Boolean functions.

In comparison with the closest work [11], in addition to the scalability and variable partitioning issues, we focus on *strong decomposition* (namely, X_A and X_B cannot be empty), whereas [11] gave a more general approach allowing *weak decomposition* (namely, X_A or X_B can be empty). Moreover, as don't cares are better handled in BDD than in SAT, they were exploited in [11] for logic sharing.

6.3 Preliminaries

As conventional notation, sets are denoted in upper-case letters, e.g., S; set elements are in lower-case letters, e.g., $e \in S$. The cardinality of S is denoted as $|S|$. A partition of a set S into $S_i \subseteq S$ for $i = 1, \ldots, k$ (with $S_i \cap S_j = \emptyset, i \neq j$, and $\bigcup_i S_i = S$) is denoted as $\{S_1|S_2|\ldots|S_k\}$. For a set X of Boolean variables, its set of valuations (or truth assignments) is denoted as $[\![X]\!]$, e.g., $[\![X]\!] = \{(0,0), (0,1), (1,0), (1,1)\}$ for $X = \{x_1, x_2\}$.

6.3.1 Bi-Decomposition

Definition 6.1 Given a completely specified Boolean function f, variable x is a *support variable* of f if $f_x \neq f_{\neg x}$, where f_x and $f_{\neg x}$ are the positive and negative cofactors of f on x, respectively.

Definition 6.2 An *incompletely specified function* F is a 3-tuple (q, d, r), where the completely specified functions q, d, and r represent the onset, don't care set, and offset functions, respectively.

Definition 6.3 A completely specified function $f(X)$ is $\langle \text{OP} \rangle$ *bi-decomposable*, or simply $\langle \text{OP} \rangle$*-decomposable*, under variable partition $X = \{X_A|X_B|X_C\}$ if it can be written as $f(X) = f_A(X_A, X_C) \ \langle \text{OP} \rangle \ f_B(X_B, X_C)$, where $\langle \text{OP} \rangle$ is some binary operator. The decomposition is called *disjoint* if $X_C = \emptyset$ and *non-disjoint* otherwise.

Note that bi-decomposition trivially holds if $X_A \cup X_C$ or $X_B \cup X_C$ equals X. The corresponding variable partition is called *trivial*. We are concerned about non-trivial bi-decomposition. In the sequel, a binary operator ⟨OP⟩ can be OR2, AND2, and XOR. Essentially OR2-, AND2-, and XOR-decompositions form the basis of all types of bi-decompositions because any bi-decomposition is simply one of the three cases with some proper complementation on f, f_A, and/or f_B.

6.3.2 Propositional Satisfiability

Let $V = \{v_1, \ldots, v_k\}$ be a finite set of Boolean variables. A *literal* l is either a Boolean variable v_i or its negation $\neg v_i$. A *clause* c is a disjunction of literals. Without loss of generality, we shall assume that there are no repeated or complementary literals appearing in the same clause. A *SAT instance* is a conjunction of clauses, i.e., in the so-called *conjunctive normal form* (CNF). In the sequel, a clause set $C = \{c_1, \ldots, c_k\}$ shall mean to be the CNF formula $c_1 \wedge \cdots \wedge c_k$. An *assignment* over V gives every variable v_i a Boolean value either 0 or 1. A SAT instance is *satisfiable* if there exists a satisfying assignment such that the CNF formula evaluates to 1. Otherwise it is *unsatisfiable*.

6.3.2.1 Refutation Proof and Craig Interpolation

Definition 6.4 Assume literal v is in clause c_1 and $\neg v$ in c_2. A *resolution* of clauses c_1 and c_2 on variable v yields a new clause c containing all literals in c_1 and c_2 except for v and $\neg v$. The clause c is called the *resolvent* of c_1 and c_2 and variable v the *pivot variable*.

Theorem 6.1 (Robinson [13]) *For an unsatisfiable SAT instance, there exists a sequence of resolution steps leading to an empty clause.*

Often only a subset of the clauses of a SAT instance participates in the resolution steps leading to an empty clause.

Definition 6.5 A *refutation proof* Π of an unsatisfiable SAT instance C is a directed acyclic graph (DAG) $\Gamma = (N, A)$, where every node in N represents a clause which is either a root clause in C or a resolvent clause having exactly two predecessor nodes, and every arc in A connects a node to its ancestor node. The unique leaf of Π corresponds to the empty clause.

Modern SAT solvers (e.g., MiniSat [8]) are capable of producing a refutation proof from an unsatisfiable SAT instance.

Theorem 6.2 (Craig Interpolation Theorem [6]) *For any two Boolean formulas ϕ_A and ϕ_B with $\phi_A \wedge \phi_B$ unsatisfiable, then there exists a Boolean formula $\phi_{A'}$ referring only to the common input variables of ϕ_A and ϕ_B such that $\phi_A \Rightarrow \phi_{A'}$ and $\phi_{A'} \wedge \phi_B$ is unsatisfiable.*

The Boolean formula $\phi_{A'}$ is referred to as the *interpolant* of ϕ_A and ϕ_B. We shall assume that ϕ_A and ϕ_B are in CNF. So a refutation proof of $\phi_A \wedge \phi_B$ is available from a SAT solver. How to construct an interpolant circuit from a refutation proof in linear time can be found in, e.g., [10].

6.3.3 Circuit to CNF Conversion

Given a circuit netlist, it can be converted to a CNF formula in such a way that the satisfiability is preserved [17]. The conversion is achievable in linear time by introducing extra intermediate variables. In the sequel, we shall assume that the clause set of a Boolean formula ϕ (similarly $\neg\phi$) is available from such conversion.

6.4 Our Approach

6.4.1 OR Bi-decomposition

We show that OR2-decomposition can be achieved using SAT solving. Whenever a non-trivial OR2-decomposition exists, we obtain a feasible variable partition and the corresponding subfunctions f_A and f_B.

6.4.1.1 Decomposition of Completely Specified Functions

Decomposition with Known Variable Partition

Given a function $f(X)$ and a non-trivial variable partition $X = \{X_A|X_B|X_C\}$, we study if f can be expressed as $f_A(X_A, X_C) \vee f_B(X_B, X_C)$ for some functions f_A and f_B.

The following proposition lays the foundation of OR2-decomposition.

Proposition 6.1 (Mischenko et al. [11]) *A completely specified function $f(X)$ can be written as $f_A(X_A, X_C) \vee f_B(X_B, X_C)$ for some functions f_A and f_B if and only if the quantified Boolean formula*

$$f(X) \wedge \exists X_A.\neg f(X) \wedge \exists X_B.\neg f(X) \tag{6.1}$$

is unsatisfiable.

It can be restated as follows.

Proposition 6.2 *A completely specified function $f(X)$ can be written as $f_A(X_A, X_C) \vee f_B(X_B, X_C)$ for some functions f_A and f_B if and only if the Boolean formula*

$$f(X_A, X_B, X_C) \wedge \neg f(X'_A, X_B, X_C) \wedge \neg f(X_A, X'_B, X_C) \tag{6.2}$$

is unsatisfiable, where variable set Y' is an instantiated version of variable set Y.

By renaming quantified variables, the quantifiers of Formula (6.1) can be removed. That is, Formula (6.1) can be rewritten as the quantifier-free formula of (6.2) because existential quantification is implicit in satisfiability checking. Note that the complementations in Formulas (6.1) and (6.2) need not be computed. Rather, the complementations can be achieved by adding inverters in the corresponding circuit before circuit-to-CNF conversion or alternatively by asserting the corresponding variables to be false in SAT solving.

The decomposability of a function under OR2-decomposition can be explained through the visualization of a decomposition chart, which consists of $2^{|X_C|}$ two-dimensional tables corresponding to the valuations of X_C, whereas each table has its columns and rows indexed by the valuations of X_A and X_B, respectively. An entry in the decomposition chart corresponds to the function value under the corresponding valuations of X_A, X_B, and X_C. A function is decomposable if and only if, for every 1-entry in a table of the decomposition chart, its situated column and row cannot contain 0-entries. (It is because this 1-entry after bi-decomposition must result from $f_A = 1$ or $f_B = 1$ under the corresponding valuation of X. Moreover $f_A = 1$ and $f_B = 1$ in turn make the entire column and row, respectively, valuated to 1.)

To illustrate, Fig. 6.2 shows that function $f(a, b, c, d) = \neg ab \vee c\neg d$ under variable partition $X_A = \{a, b\}$, $X_B = \{c, d\}$, and $X_C = \emptyset$ is OR2-decomposable with $f_A(X_A) = \neg ab$ and $f_B(X_B) = c\neg d$. In contrast, Fig. 6.3 shows that function $f(a, b, c, d) = \neg ab \vee c\neg d \vee a\neg bc$ under the same variable partition is not OR2-decomposable because of the problematic 1-entry of $(a, b, c, d) = (1, 0, 1, 1)$. In this case, Formula (6.2) is satisfiable. Note that, in Formula (6.2), $f(X_A, X_B, X_C)$

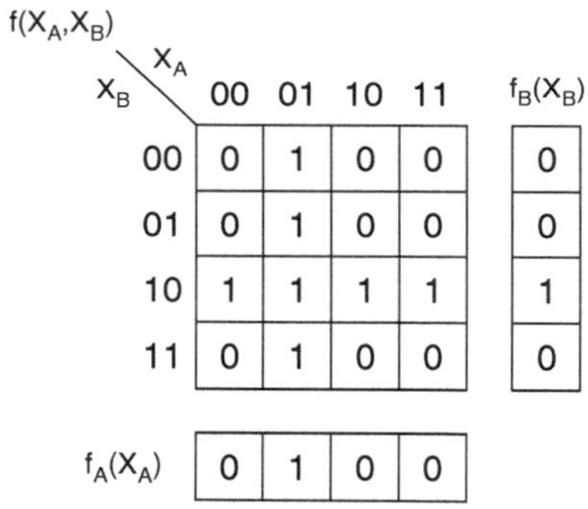

Fig. 6.2 A function decomposable subject to OR2-decomposition

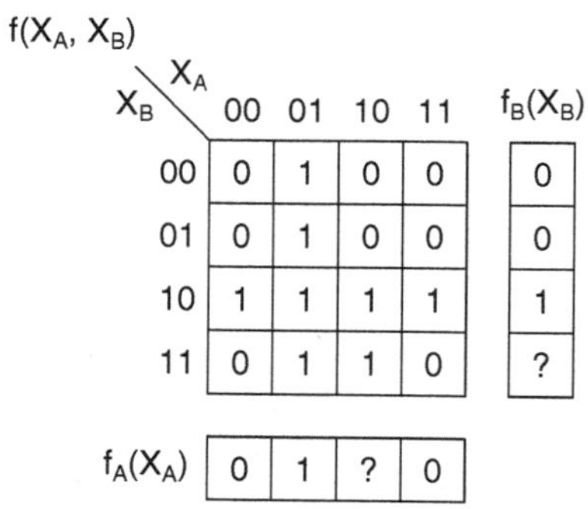

Fig. 6.3 A function not decomposable subject to OR2-decomposition

asserts that there exists a 1-entry in the decomposition chart; $\neg f(X'_A, X_B, X_C)$ and $\neg f(X_A, X'_B, X_C)$ assert that there exist 0-entries in the row and column where the 1-entry situates, respectively. The satisfiability indicates that the function $f(a,b,c,d) = \neg ab \vee c\neg d \vee a\neg bc$ is not decomposable under the specified variable partition.

A remaining problem to be resolved is how to derive f_A and f_B. We show that they can be obtained through interpolation from a refutation proof of Formula (6.2). Consider partitioning the clause set of Formula (6.2) into two subsets C_A and C_B with C_A the clause set of

$$f(X_A, X_B, X_C) \wedge \neg f(X'_A, X_B, X_C) \tag{6.3}$$

and C_B the clause set of

$$\neg f(X_A, X'_B, X_C) \tag{6.4}$$

Then the corresponding interpolant corresponds to an implementation of f_A. Figure 6.4 depicts the construction to derive f_A according to the above formulas.

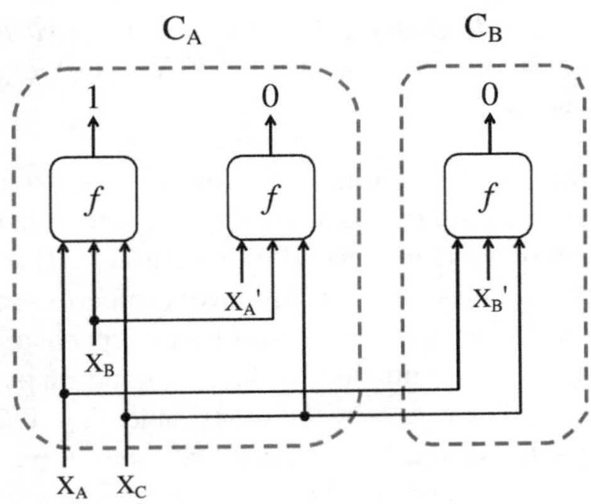

Fig. 6.4 The construction of f_A

On the other hand, to derive f_B we perform a similar computation, but now with C_A the clause set of

$$f(X_A, X_B, X_C) \wedge \neg f_A(X_A, X_C) \tag{6.5}$$

and C_B the clause set of

$$\neg f(X'_A, X_B, X_C) \tag{6.6}$$

Then the corresponding interpolant corresponds to an implementation of f_B. The construction to derive f_B is shown in Fig. 6.5.

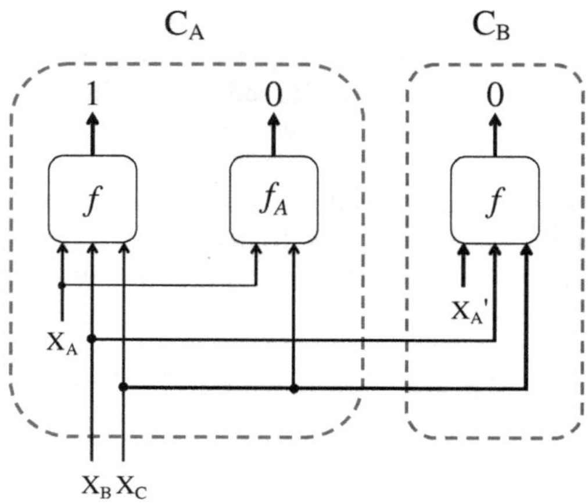

Fig. 6.5 The construction of f_B

We show the correctness of the above construction.

Theorem 6.3 *For any* OR2*-decomposable function f under variable partition $X = \{X_A|X_B|X_C\}$, we have $f(X) = f_A(X_A, X_C) \vee f_B(X_B, X_C)$ for f_A and f_B derived from the above construction.*

Proof To show that the interpolant obtained from the unsatisfiable conjunction of Formulas (6.3) and (6.4) indeed corresponds to f_A, observe that X_A and X_C variables are the only common variables shared by Formulas (6.3) and (6.4). Moreover, Formula (6.3) (respectively, Formula (6.4)), for every valuation on X_C, characterizes the set of valuations on X_A that must be in the onset (respectively, offset) of f_A. As the interpolant is an over-approximation of the onset and disjoint from the offset, it is a valid implementation of f_A. On the other hand, X_B and X_C variables are the only common variables shared by Formulas (6.5) and (6.6). They, respectively, corresponds to the care onset and care offset of f_B, where f_A sets the don't care condition of f_B. Therefore, the interpolant obtained from the unsatisfiable conjunction of Formulas (6.5) and (6.6) is a valid implementation of f_B.

Remark 6.1 An interpolant itself is in fact a netlist composed of OR2 and AND2 gates [10]. The "bi-decomposed" netlist, however, may contain some amount of redundancy; moreover, variable partitioning is not used in its derivation.

Decomposition with Unknown Variable Partition

The previous construction assumes that a variable partition $X = \{X_A|X_B|X_C\}$ is given. We further automate variable partition in the derivation of f_A and f_B as follows. For each variable $x_i \in X$, we introduce two control variables α_{x_i} and β_{x_i}. In addition we instantiate variables X into X' and X''. Let C_A be the clause set of

$$f(X) \wedge \neg f(X') \wedge \bigwedge_i ((x_i \equiv x'_i) \vee \alpha_{x_i}) \tag{6.7}$$

and C_B be the clause set of

$$\neg f(X'') \wedge \bigwedge_i ((x_i \equiv x''_i) \vee \beta_{x_i}) \tag{6.8}$$

where $x' \in X'$ and $x'' \in X''$ are the instantiated versions of $x \in X$. Observe that $(\alpha_{x_i}, \beta_{x_i}) = (0,0), (0,1), (1,0)$, and $(1,1)$ indicate $x_i \in X_C, x_i \in X_B, x_i \in X_A$, and x_i can be in either of X_A and X_B, respectively.

In SAT solving the conjunction of Formulas (6.7) and (6.8), we make *unit assumptions* [8] on the control variables. Under an unsatisfiable unit assumption, the SAT solver will return a final conflict clause consisting of only the control variables. Notice that every literal in the conflict clause is of positive phase because the conflict arises from a subset of the control variables set to 0. It reveals that setting to 0 the control variables present in the conflict clause is sufficient, making the whole formula unsatisfiable. Hence setting to 1 the control variables absent from the conflict clause cannot affect the unsatisfiability. The more the control variables can be set to 1, the better the bi-decomposition is because $|X_C|$ is smaller. In essence, this final conflict clause indicates a variable partition X_A, X_B, X_C on X. For example, the conflict clause $(\alpha_{x_1} + \beta_{x_1} + \alpha_{x_2} + \beta_{x_3})$ indicates that the unit assumption $\alpha_{x_1} = 0$, $\beta_{x_1} = 0$, $\alpha_{x_2} = 0$, and $\beta_{x_3} = 0$ results in unsatisfiability. It in turn suggests that $x_1 \in X_C$, $x_2 \in X_B$, and $x_3 \in X_A$.

To see how the new construction works, imagine setting all the control variables to 0. As SAT solvers tend to refer to a small subset of the clauses relevant to a refutation proof, it may return a conflict clause with just a few literals. It in effect conducts a desirable variable partition. This perception, unfortunately, is flawed in that SAT solvers are very likely to return a conflict clause that consists of all the control variables reflecting the trivial variable partition $X_C = X$. In order to avoid trivial variable partitions, we initially specify two distinct variables x_a and x_b to be in X_A and X_B, respectively, and all other variables in X_C, that is, having $(\alpha_{x_a}, \beta_{x_a}) = (1,0)$, $(\alpha_{x_b}, \beta_{x_b}) = (0,1)$, and $(\alpha_{x_i}, \beta_{x_i}) = (0,0)$ for $i \neq a, b$ in the unit assumption. We call such an initial variable partition as a *seed variable partition*. If the conjunction of Formulas (6.7) and (6.8) is unsatisfiable under a seed partition, then the corresponding bi-decomposition is successful. As SAT solvers often refer to a small unsatisfiable core, the returned variable partition is desirable because $|X_C|$ tends to be small. Otherwise, if the seed partition fails, we should try another one. For a given function $f(X)$ with $|X| = n$, the existence of non-trivial OR2-decomposition can be checked with at most $(n-1) + \cdots + 1 = n(n-1)/2$ different seed partitions. On the other hand, we may enumerate different variable partitions using different seed partitions to find one that is more balanced and closer to disjoint. Even from a successful seed partition, we may further refine the returned variable partition by reducing the corresponding unsatisfiable core. The process can be iterated until the unsatisfiable core is minimal.

Lemma 6.1 *For an unsatisfiable conjunction of Formulas (6.7) and (6.8) under a seed variable partition, the final conflict clause contains only the control variables, which indicates a valid non-trivial variable partition.*

Proof The values of control variables are specified in the unit assumption as if they are in the first decision level. In solving an unsatisfiable instance, both 0 and 1 valuations of any other variable must have been tried and failed, and only the control variables are not valuated in both cases. Because unit assumption causes unsatisfiability, the final learned clause indicates the conflict decisions made in the first decision level. On the other hand, as discussed earlier this clause corresponds to a valid variable partition, which is non-trivial since $|X_A|, |X_B| \geq 1$ due to the seed variable partition.

Theorem 6.4 asserts the correctness of the construction.

Theorem 6.4 *For any* OR2*-decomposable function* f*, we have* $f(X) = f_A(X_A, X_C) \vee f_B(X_B, X_C)$ *for* f_A*,* f_B *and a non-trivial variable partition* $X = \{X_A|X_B|X_C\}$ *derived from the above construction.*

Proof Given an unsatisfiable conjunction of Formulas (6.7) and (6.8) under a seed variable partition, by Lemma 6.1 the final learned clause indicates which variables assigned in X_C are unnecessary and can be placed in X_A or X_B instead. The resultant partition is indeed a non-trivial variable partition. Under the obtained variable partition, Theorem 6.3 is applicable, and functions f_A and f_B can be obtained.

One might speculate about whether $(\alpha_x, \beta_x) = (1, 1)$ is possible as it tends to suggest that x can be in either of X_A and X_B. To answer this question, we study the condition that x_i can be in either of X_A and X_B.

Lemma 6.2 (Saso and Butler [14]) *If* f *is bi-decomposable under some variable partition, then the cofactors* f_x *and* $f_{\neg x}$ *for any variable* x *are both bi-decomposable under the same variable partition.*

Proof The lemma follows from the fact that cofactor and Boolean operations, such as $\wedge, \vee, \neg$, commute.

The converse, however, is not true. The following theorem gives the condition that x can be in either of X_A and X_B.

Theorem 6.5 *Let* $X = \{X_a|X_b|X_C|\{x\}\}$ *for some* $x \in X$*. A function* $f = f_A(X_A, X_C) \vee f_B(X_B, X_C)$ *can be bi-decomposed under variable partition* $\{X_a \cup \{x\}|X_b|X_C\}$ *as well as under variable partition* $\{X_a|X_b \cup \{x\}|X_C\}$ *if and only if both* f_x *and* $f_{\neg x}$ *are themselves* OR2*-decomposable under variable partition* $\{X_a|X_b|X_C\}$ *and also* $(f_x \not\equiv 1 \wedge f_{\neg x} \not\equiv 1) \Rightarrow (f_x \equiv f_{\neg x})$ *under every* $c \in [\![X_C]\!]$.

Proof ($\Longrightarrow$) For bi-decomposable f, the bi-decomposability of f_x and $f_{\neg x}$ follows from Lemma 6.2. On the other hand, assume by contradiction that $(f_x \not\equiv 1 \wedge f_{\neg x} \not\equiv 1) \Rightarrow (f_x \equiv f_{\neg x})$ does not hold under every $c \in [\![X_C]\!]$, i.e., $(f_x \not\equiv 1 \wedge f_{\neg x} \not\equiv 1) \wedge (f_x \not\equiv f_{\neg x})$ under some $c \in [\![X_C]\!]$. Since $f_x \not\equiv f_{\neg x}$, assume without loss of generality that there exists some $m_a \in [\![X_a]\!]$ and $m_b \in [\![X_b]\!]$ such that $f_x(m_a, m_b, c) = 0$ and $f_{\neg x}(m_a, m_b, c) = 1$. Then, $f_{\neg x}(m_a, X_b, c) \equiv 1$ since x can be in X_a and f is OR2-decomposable. Also, $f_{\neg x}(m_a, X_b, c) \equiv 1$ implies $f_{\neg x}(X_a, X_b, c) \equiv 1$ since x can be in X_b and f is OR2-decomposable. It contradicts with the assumption that $f_{\neg x}(X_a, X_b, c) \not\equiv 1$.

($\Longleftarrow$) Consider under every $c \in [\![X_C]\!]$. For $f_x \equiv f_{\neg x}$, x is not a support variable of f and thus can be redundantly placed in X_A or X_B. On the other hand, for $f_x \not\equiv f_{\neg x}$, at least one of f_x and $f_{\neg x}$ is a tautology. If $f_x \equiv 1$, f is OR2-decomposable no matter $x \in X_A$ or $x \in X_B$ since $f_{\neg x}$ is OR2-decomposable. Similarly, if $f_{\neg x} \equiv 1$, f is OR2-decomposable. Hence x can be in either of X_A and X_B in the OR2-decomposition of f.

That is, under every $c \in [\![X_C]\!]$ either x is not a support variable of f, or f_x or $f_{\neg x}$ equals constant 1. It is interesting to note that only the former can make $(\alpha_x, \beta_x) = (1, 1)$. Whenever the latter happens, (α_x, β_x) equals $(0, 1)$, $(1, 0)$, or $(0, 0)$ if the solver is unable to identify a *minimal* unsatisfiable core. To see it, consider $f_x \equiv 1$ (similar for $f_{\neg x} \equiv 1$) and $f_{\neg x} \not\equiv f_x$ under some $c \in [\![X_C]\!]$. If $(\alpha_x, \beta_x) = (1, 1)$, Formula (6.2) reduces to

$$
\begin{aligned}
&(\exists x.f(X_a, X_b, c, x)) \wedge \neg(\forall x.f(X'_a, X_b, c, x)) \wedge \\
&\neg(\forall x.f(X_a, X'_b, c, x)) \\
&= 1 \wedge \neg f_{\neg x}(X'_a, X_b, c) \wedge \neg f_{\neg x}(X_a, X'_b, c)
\end{aligned}
$$

which is satisfiable because $f_{\neg x} \not\equiv 1$ under c. Hence $(\alpha_x, \beta_x) = (1, 1)$ only if x is not a support variable of f.

6.4.1.2 Decomposition of Incompletely Specified Functions

Proposition 6.2 can be generalized for incompletely specified functions as follows.

Proposition 6.3 *Given an incompletely specified function $F = (q, d, r)$, there exists a completely specified function f with $f(X) = f_A(X_A, X_C) \vee f_B(X_B, X_C)$, $q(X) \Rightarrow f(X)$, and $f(X) \Rightarrow \neg r(X)$ if and only if the Boolean formula*

$$
q(X_A, X_B, X_C) \wedge r(X'_A, X_B, X_C) \wedge r(X_A, X'_B, X_C) \tag{6.9}
$$

is unsatisfiable.

The derivations of f_A and f_B can be computed in a way similar to the aforementioned construction. We omit the detailed exposition to save space.

6.4.2 AND Bi-decomposition

Proposition 6.4 (Saso and Butler [14]) *A function f is* AND2*-decomposable if and only if $\neg f$ is* OR2*-decomposable.*

By decomposing $\neg f$ as $f_A \vee f_B$, we obtain $f = \neg f_A \wedge \neg f_B$. Hence our results on OR2-decomposition are convertible to AND2-decomposition.

6.4.3 XOR Bi-decomposition

6.4.3.1 Decomposition of Completely Specified Functions

Decomposition with Known Variable Partition

We formulate the XOR-decomposability in the following proposition, which differs from prior work [16] and is more suitable for SAT solving.

Proposition 6.5 *A function f can be written as $f(X) = f_A(X_A, X_C) \oplus f_B(X_B, X_C)$ for some functions f_A and f_B under variable partition $X = \{X_A|X_B|X_C\}$ if and only if*

$$(f(X_A, X_B, X_C) \equiv f(X_A, X'_B, X_C)) \wedge (f(X'_A, X_B, X_C) \not\equiv f(X'_A, X'_B, X_C)) \tag{6.10}$$

is unsatisfiable. Furthermore, $f_A = f(X_A, \mathbf{0}, X_C)$ and $f_B = f(\mathbf{0}, X_B, X_C) \oplus f(\mathbf{0}, \mathbf{0}, X_C)$ or alternatively $f_A = \neg f(X_A, \mathbf{0}, X_C)$ and $f_B = f(\mathbf{0}, X_B, X_C) \oplus \neg f(\mathbf{0}, \mathbf{0}, X_C)$.

Proof We show that the proposition is true for every valuation $c \in [\![X_C]\!]$.

($\Longrightarrow$) Observe that, for f is XOR-decomposable under variable partition $X = \{X_A|X_B|X_C\}$, then either $f(a_1, X_B, c) \equiv f(a_2, X_B, c)$ or $f(a_1, X_B, c) \equiv \neg f(a_2, X_B, c)$ for any $a_1, a_2 \in [\![X_A]\!]$ and also either $f(X_A, b_1, c) \equiv f(X_A, b_2, c)$ or $f(X_A, b_1, c) \equiv \neg f(X_A, b_2, c)$ for any $b_1, b_2 \in [\![X_B]\!]$. If Formula (6.10) is satisfiable, the property is violated.

($\Longleftarrow$) If Formula (6.10) is unsatisfiable, then for every $c \in [\![X_C]\!]$ either $f(a_1, X_B, c) \equiv f(a_2, X_B, c)$ or $f(a_1, X_B, c) \equiv \neg f(a_2, X_B, c)$ for any $a_1, a_2 \in [\![X_A]\!]$ and also either $f(X_A, b_1, c) \equiv f(X_A, b_2, c)$ or $f(X_A, b_1, c) \equiv \neg f(X_A, b_2, c)$ for any $b_1, b_2 \in [\![X_B]\!]$. We can simply choose $f(X_A, \mathbf{0}, c)$ as $f_A(X_A, c)$. However, in choosing function $f(\mathbf{0}, X_B, c)$ as $f_B(X_B, c)$, we need to make $f_A(\mathbf{0}, c) \oplus f_B(\mathbf{0}, c)$ consistent with $f(\mathbf{0}, \mathbf{0}, c)$. Hence we have $f_B(X_B, c) = f(\mathbf{0}, X_B, c) \oplus f_A(\mathbf{0}, c)$. On the other hand, for any $f = g \oplus h$, $f = \neg g \oplus \neg h$. We can alternatively let $f_A(X_A, c) = \neg f(X_A, \mathbf{0}, c)$ and $f_B(X_B, c) = f(\mathbf{0}, X_B, c) \oplus \neg f_A(\mathbf{0}, c)$.

Since the arguments are universally true for every $c \in [\![X_C]\!]$, the theorem follows. Accordingly interpolation is not needed in computing f_A and f_B in XOR-decomposition.

The decomposability of a function under XOR-decomposition can be analyzed using a decomposition chart, similar to that under OR2-decomposition. A function is XOR-decomposable if and only if any two columns (rows) in each table of its decomposition chart are either identical or complementary to each other. It is due to

$f(X_A, X_B)$

$X_B \backslash X_A$	00	01	10	11	$f_B(X_B)$
00	1	0	1	1	0
01	1	0	1	1	0
10	0	1	0	0	1
11	1	0	1	1	0
$f_A(X_A)$	1	0	1	1	

Fig. 6.6 A function decomposable subject to XOR-decomposition

the fact that, for every variable valuation, $f_A \oplus f_B = \neg f_A$ if $f_B = 1$ and $f_A \oplus f_B = f_A$ if $f_B = 0$ (likewise $f_A \oplus f_B = \neg f_B$ if $f_A = 1$ and $f_A \oplus f_B = f_B$ if $f_A = 0$).

To illustrate, Fig. 6.6 shows that function $f(a, b, c, d) = a\neg c \vee ad \vee \neg b\neg c \vee \neg bd \vee \neg abc\neg d$ under variable partition $X_A = \{a, b\}$, $X_B = \{c, d\}$, and $X_C = \emptyset$ is XOR-decomposable with $f_A(X_A) = a \vee \neg b$ and $f_B(X_B) = c\neg d$. Observe that, in this figure, any two columns (respectively, rows) in decomposition chart are either identical or complementary to each other, that is, either $f(a_1, X_B) \equiv f(a_2, X_B)$ or $f(a_1, X_B) \equiv \neg f(a_2, X_B)$ for any $a_1, a_2 \in [\![X_A]\!]$ (respectively, $f(X_A, b_1) \equiv f(X_A, b_2)$ or $f(X_A, b_1) \equiv \neg f(X_A, b_2)$ for any $b_1, b_2 \in [\![X_B]\!]$).

Decomposition with Unknown Variable Partition

The XOR-decomposition of Proposition 6.5 assumes that a variable partition is given. We further automate variable partition as follows. For each variable $x_i \in X$, we introduce two control variables α_{x_i} and β_{x_i}. In addition we instantiate variables X into X', X'', and X'''. We modify Formula (6.10) as

$$
\begin{aligned}
&(f(X) \equiv f(X')) \wedge (f(X'') \not\equiv f(X''')) \wedge \\
&\bigwedge_i (((x_i \equiv x_i'') \wedge (x_i' \equiv x_i''')) \vee \alpha_{x_i}) \wedge \\
&\bigwedge_i (((x_i \equiv x_i') \wedge (x_i'' \equiv x_i''')) \vee \beta_{x_i})
\end{aligned} \tag{6.11}
$$

By Formula (6.11), an automatic variable partition can be obtained from a seed partition, similar to what we have in OR2-decomposition.

The correctness of the construction is asserted as follows.

Theorem 6.6 *For any XOR-decomposable function f, we have $f(X) = f_A(X_A, X_C) \oplus f_B(X_B, X_C)$ for f_A, f_B, along with a non-trivial variable partition $X = \{X_A|X_B|X_C\}$ derived from the above construction.*

Proof The proof relies on Proposition 6.5 and is similar to that of Theorem 6.4. We omit the detailed exposition.

To see whether $(\alpha_x, \beta_x) = (1, 1)$ is possible or not for some variable x, we study the condition that x can be in either of X_A and X_B.

Theorem 6.7 *Let $X = \{X_a|X_b|X_C|\{x\}\}$ for some $x \in X$. A function $f = f_A(X_A, X_C) \oplus f_B(X_B, X_C)$ can be bi-decomposed under variable partition $\{X_a \cup \{x\}|X_b|X_C\}$ as well as under variable partition $\{X_a|X_b \cup \{x\}|X_C\}$ if and only if both f_x and $f_{\neg x}$ are themselves* XOR*-decomposable under variable partition $\{X_a|X_b|X_C\}$ and also $(f_x \equiv f_{\neg x}) \vee (f_x \equiv \neg f_{\neg x})$ under every $c \in [\![X_C]\!]$.*

Proof ($\Longrightarrow$) For f bi-decomposable, f_x and $f_{\neg x}$ are bi-decomposable as well by Lemma 6.2. On the other hand, assume by contradiction that $(f_x \equiv f_{\neg x}) \vee (f_x \equiv \neg f_{\neg x})$ does not hold under every $c \in [\![X_C]\!]$, i.e., $(f_x \not\equiv f_{\neg x}) \wedge (f_x \not\equiv \neg f_{\neg x})$ under some $c \in [\![X_C]\!]$. Hence assume without loss of generality that $f_x(m_{a_1}, m_{b_1}, c) = 0$, $f_x(m_{a_2}, m_{b_2}, c) = 0$, and $f_{\neg x}(m_{a_1}, m_{b_1}, c) = 0$, $f_{\neg x}(m_{a_2}, m_{b_2}, c) = 1$ for $m_{a_1}, m_{a_2} \in [\![X_a]\!]$, $m_{b_1}, m_{b_2} \in [\![X_b]\!]$. Two cases $f_x(m_{a_1}, m_{b_2}, c) = 0$ and $f_x(m_{a_1}, m_{b_2}, c) = 1$ need to be analyzed. If $f_x(m_{a_1}, m_{b_2}, c) = 0$, then $f_x(m_{a_1}, m_{b_2}, c) = f_x(m_{a_2}, m_{b_2}, c)$ infers $f_x(m_{a_2}, m_{b_1}, c) = 0$, $f_{\neg x}(m_{a_1}, m_{b_2}, c) = 0$, and $f_{\neg x}(m_{a_2}, m_{b_1}, c) = 1$ for $x \in X_A$ due to the XOR-decomposability of f. It contradicts with the XOR-decomposability of f under $x \in X_B$. Likewise $f_x(m_{a_1}, m_{b_2}, c) = 1$ leads to a contradiction. Thus $(f_x \equiv f_{\neg x}) \vee (f_x \equiv \neg f_{\neg x})$ under every $c \in [\![X_C]\!]$.

($\Longleftarrow$) Consider under every $c \in [\![X_C]\!]$. For f_x and $f_{\neg x}$ XOR-decomposable with $f_x \equiv f_{\neg x}$ or $f_x \equiv \neg f_{\neg x}$, Formula (6.10) is unsatisfiable for either case of $x \in X_A$ and $x \in X_B$. That is, f is XOR-decomposable no matter $x \in X_A$ or $x \in X_B$.

Under the flexible partition for variable x, Formula (6.10) reduces to

$$\begin{aligned}(\exists x.f(X_a, X_b, X_C, x) \equiv \exists x.f(X_a, X_b', X_C, x)) \wedge \\ (\exists x.f(X_a', X_b, X_C, x) \not\equiv \exists x.f(X_a', X_b', X_C, x))\end{aligned} \tag{6.12}$$

If $f_x \equiv f_{\neg x}$, the unsatisfiability of Formula (6.10) implies the unsatisfiability of Formula (6.12). On the other hand, if $f_x \equiv \neg f_{\neg x}$, $\exists x.f$ is a constant-1 function and thus Formula (6.12) is unsatisfiable. Hence $(\alpha_x, \beta_x) = (1, 1)$ is possible even if x is a support variable of f. In this case, we can first decompose f as $f = x \oplus f_{\neg x}$ or equivalently $\neg x \oplus f_x$. Moreover, it can be generalized as follows.

Corollary 6.1 *For an* XOR*-decomposable function f, suppose x_i, for $i = 1, \ldots, k$, are the support variables of f with $(\alpha_{x_i}, \beta_{x_i}) = (1, 1)$ after variable partition. Then f can be decomposed as*

$$f = x_1 \oplus \cdots \oplus x_k \oplus f_{\neg x_1 \cdots \neg x_k} \tag{6.13}$$

Further, for $X_a = \{x \mid (\alpha_x, \beta_x) = (1, 0)\}$, $X_b = \{x \mid (\alpha_x, \beta_x) = (0, 1)\}$, and $X_C = \{x \mid (\alpha_x, \beta_x) = (0, 0)\}$, then

$$f = x_1 \oplus \cdots \oplus x_k \oplus f_A(X_a, X_C) \oplus f_B(X_b, X_C) \tag{6.14}$$

with f_A and f_B derived from the previous construction.

Proof Since $f_x \equiv \neg f_{\neg x}$ implies $(f_l)_x \equiv \neg(f_l)_{\neg x}$ for some literal $l \not\equiv x, \neg x$, we can successively decompose f as Formula (6.13). Moreover, for $(\alpha_{x_i}, \beta_{x_i}) = (1, 1), i = 1, \ldots, k$, the unsatisfiability of Formula (6.11) is independent of variables x_i. Thus, $f_{\neg x_1 \cdots \neg x_k} = f_A(X_a, X_C) \oplus f_B(X_b, X_C)$.

6.4.4 Implementation Issues

When disjoint variable partitioning is concerned, it corresponds to computing a *minimum* unsatisfiable core. Incremental SAT solving is useful in finding a good minimal unsatisfiable core, see, e.g., [12]. In our implementation, a variable of X_C is greedily moved to either of X_A and X_B favoring the small one. The process iterates until no more reduction can be made on X_C.

When balanced variable partitioning is concerned, SAT solvers usually tend to make decisions in a descending priority order based on variable IDs. From empirical experience, this bias makes variable partition unbalanced. To overcome, we interleave the variable IDs of X' and those of X'' of Formulas (6.7) and (6.8) for OR2- and AND2-decomposition and interleave those of Formula (6.11) for XOR-decomposition. For example, assume that variables x_i', x_i'', x_{i+1}', and x_{i+1}'' are originally of IDs 100, 200, 101, and 201, respectively. We rename them to 100, 200, 201, and 101, respectively. This shuffling makes variable partitioning more balanced.

6.5 Experimental Results

The algorithms were implemented in C++ in ABC [2] with MiniSAT [8] as the underlying solver. All experiments were conducted on a Linux machine with Xeon 3.4 GHz CPU and 6 Gb RAM.

Two sets of experiments were designed to demonstrate the scalability of bi-decomposition and the optimality of variable partitioning. Only circuits containing output functions with large support sizes (≥30) were chosen from the ISCAS, ITC, and LGSYNTH benchmark suites.[1] To show the efficiency of decomposing large functions, Table 6.1 shows the results of OR2- and XOR-decompositions on the output functions of the listed circuits. As can be seen, functions with up to 200 inputs, such as `i2`, can be decomposed effectively. It may not be the case using BDD-based methods.

To measure the quality of a variable partition, we use two metrics: $|X_C|/|X|$ for disjointness and $||X_A| - |X_B||/|X|$ for balancedness. The smaller they are, the better a partition is. In particular, we prefer disjointness to balancedness since the former yields better variable reduction. Experience suggests that $|X_C|/|X|$ very often can be maximally reduced within the first few enumerations while keeping $||X_A| - |X_B||/|X|$ as low as possible. Figure 6.7 shows how these two values

---

[1] Sequential circuits are converted to combinational ones by replacing register inputs and outputs with primary outputs and inputs, respectively.

Table 6.1 Bi-decomposition of PO functions

				OR2-decomposition				XOR-decomposition			
Circuit	#In	#Max	#Out	#Dec	#Slv	Time (s)	Mem (Mb)	#Dec	# Slv	Time (s)	Mem (Mb)
b04c	76	38	74	49	3878	12.26	19.35	49	2714	28.82	20.02
b07c	49	42	57	14	12985	27.59	22.3	39	601	5.43	18.72
b12c	125	37	127	80	12526	25.14	23.32	84	4862	19.22	26.93
C1355	41	41	32	0	26240	354	20.32	–	–	TO	–
C432	36	36	7	7	102	13.15	18.54	0	3654	197.81	17.46
C880	60	45	26	16	222	8.36	20.72	11	4192	83.08	18.72
comp	32	32	3	0	1488	2.61	15.86	1	1014	13.69	16.9
dalu	75	75	16	1	26848	352.87	24.14	16	210	26.59	19.68
e64	65	65	65	0	45760	17.98	22.91	0	45760	388.18	24.37
i2	201	201	1	1	1	1.07	18.6	1	34	2.16	18.59
i3	132	32	6	6	82	0.96	16.32	0	1986	9.28	16.36
i4	192	47	6	4	6	0.58	16.08	0	4326	60.04	16.54
k2	45	45	45	33	1071	17.51	22.33	33	612	5.29	20.71
my_adder	33	33	17	0	3656	2.61	18.05	16	577	4.92	17.32
o64	130	130	1	1	1	0.36	16.17	0	8385	623.43	16.12
pair	173	53	137	119	4429	20.63	21.56	101	5676	41.81	21.61
rot	135	63	107	49	19927	65.97	23.21	46	4975	59.23	21.96
s1423c	91	59	79	26	42744	121.49	27.17	68	7281	161.98	20.25
s3330c	172	87	205	60	2941	9.42	23.09	71	3135	16.45	21.87
s3384c	226	48	209	76	12685	28.16	30.21	147	2467	24.95	21.33
s6669c	322	49	294	101	24423	198.14	29.13	176	3120	279.03	22.87
s938c	66	66	33	1	5985	2.81	19.86	33	426	4.49	16.28
too_large	38	36	3	3	22	9.89	19.87	2	629	33.38	18.4

#In: number of PIs; #Max: maximum number of support vars in POs; #Out: number of POs; #Dec: number of decomposable POs; #Slv: number of SAT solving runs; TO: time out at 1500 s

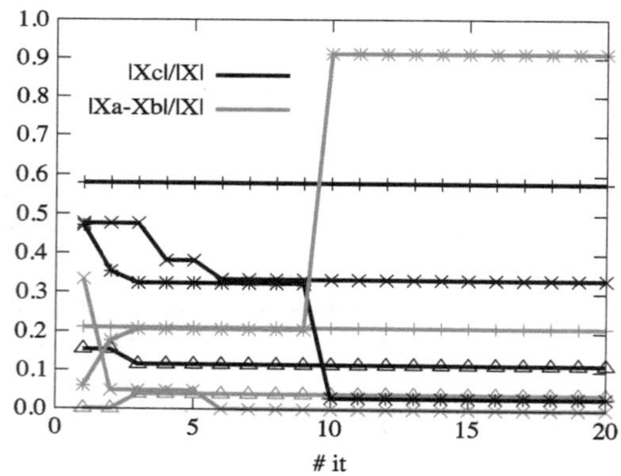

Fig. 6.7 $|X_C|/|X|$ and $||X_A| - |X_B||/|X|$ in the enumeration of variable partitions in OR2-decomposition

change in enumerating different variable partitions under OR2-decomposition on some sample circuits, where every variable partition corresponds to two markers in the same symbol, one in black and the other in gray.

It is interesting to note that OR2- and XOR-decompositions exhibit very different characteristics in variable partitioning. Figures 6.8 and 6.9 show the difference. In these two plots, a marker corresponds to a first found valid variable partition in decomposing some function. As can be seen, the decomposition quality is generally good in OR2-decomposition, but not in XOR-decomposition. This phenomenon is because XOR-decomposable circuits, e.g., arithmetic circuits, possess some regular structures in their functionality. This regularity makes disjointness and balancedness mutually exclusive in variable partitioning.

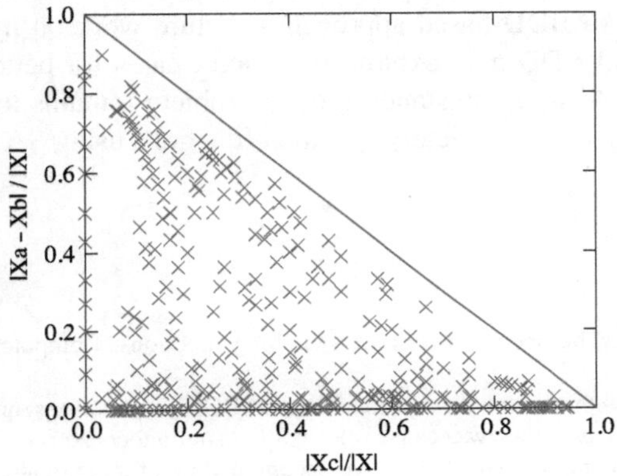

Fig. 6.8 Variable partition in OR2-decomposition

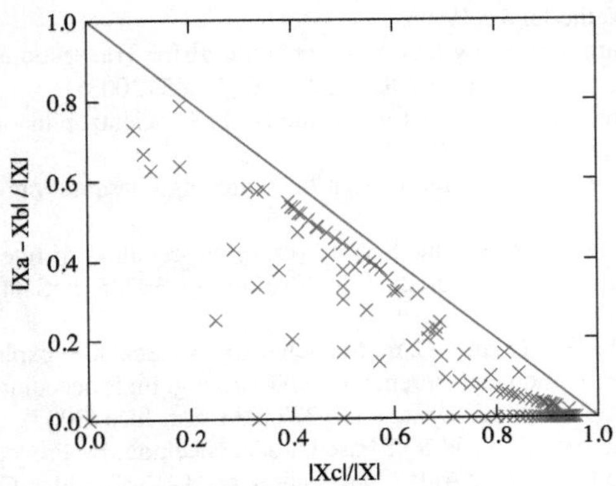

Fig. 6.9 Variable partition in XOR-decomposition

6.6 Summary

We showed that the bi-decomposition of a Boolean function can be achieved through SAT solving. Interpolation (respectively, cofactoring) turned out playing an essential role in the computation of OR2- and AND2-decompositions (respectively,

XOR-decomposition). The formulation much extends the capacity of bi-decomposition for large functions. In addition, we automated the search of disjoint (respectively, balanced) variable partition by finding a minimal unsatisfiable core with incremental SAT solving under unit assumptions (respectively, by shuffling variable IDs for an unbiased decision order). Experiments show promising results on the scalability of bi-decomposition and the optimality of variable partitioning.

Although the SAT-based method has its strengths in dealing with large functions and in automating variable partitioning, it is weak in handling don't cares when compared with BDD-based approaches. Future work on hybrid approaches combining SAT and BDD may exploit more don't cares for better decomposition of large functions. Also, an outstanding open problem remains to be solved is the XOR-decomposition of incompletely specified functions using SAT solving.

References

1. Ashenhurst, R.L.: The decomposition of switching functions. Computation Laboratory **29**, 74–116 (1959)
2. Berkeley Logic Synthesis and Verification Group: ABC: A system for sequential synthesis and verification (2005). http://www.eecs.berkeley.edu/~alanmi/abc/ (2007)
3. Bertacco, V., Damiani, M.: The disjunctive decomposition of logic functions. In: Proceedings of the International Conference on Computer Aided Design, pp. 78–82. San Jose (1997)
4. Bochmann, D., Dresig, F., Steinbach, B.: A new decomposition method for multilevel circuit design. In: Proceedings of the European Design Automation Conference, pp. 374–377. Amsterdam, The Netherlands (1991)
5. Cortadella, J.: Timing-driven logic bi-decomposition. IEEE Transactions on Computer Aided Design of Integrated Circuits and Systems **22**(6), 675–685 (2003)
6. Craig, W.: Linear reasoning: A new form of the Herbrand-Gentzen theorem. Journal of Symbolic Logic **22**(3), 250–268 (1957)
7. Curtis, A.: A New Approach to the Design of Switching Circuits. Van Nostrand, Princeton, NJ (1962)
8. Eén, N., Söensson, N.: An extensible SAT-solver. In: Proceedings of International Conference on Theory and Applications of Satisfiability Testing, pp. 502–518. Santa Margherita Ligure, Italy (2003)
9. Lee, C.C., Jiang, J.H.R., Huang, C.Y., Mishchenko, A.: Scalable exploration of functional dependency by interpolation and incremental SAT solving. In: Proceedings of the International Conference on Computer Aided Design, pp. 227–233. San Jose (2007)
10. McMillan, K.L.: Interpolation and SAT-based model checking. In: Proceedings of the International Conference on Computer Aided Verification, pp. 1–13. Boulder, CO, USA (2003)
11. Mishchenko, A., Steinbach, B., Perkowski, M.A.: An algorithm for bi-decomposition of logic functions. In: Proceedings of the Design Automation Conference, pp. 103–108. Las Vegas, Nevada, USA (2001)
12. Oh, Y., Mneimneh, M., Andraus, Z., Sakallah, K., Markov, I.: Amuse: A minimally unsatisfiable subformula extractor. In: Proceedings of the Design Automation Conference, pp. 518–523. San Diego, CA, USA (2004)
13. Robinson, J.A.: A machine-oriented logic based on the resolution principle. Journal of the ACM **12**(1), 23–41 (1965)
14. Sasao, T., Butler, J.: On bi-decomposition of logic functions. In: Proceedings of the International Workshop on Logic Synthesis Tahoe City, CA, USA (1997)

15. Scholl, C.: Functional Decomposition with Applications to FPGA Synthesis. Kluwer Dordrecht, The Netherlands (2001)
16. Steinbach, B., Wereszczynski, A.: Synthesis of multi-level circuits using EXOR-gates. In: Proceedings of IFIP Workshop on Applications of the Reed-Muller Expansion in Circuit Design, pp. 161–168. Makuhari, Japan (1995)
17. Tseitin, G.: On the complexity of derivation in propositional calculus. In: A.O. Slisenko (ed.) Studies in Constructive Mathematics and Mathematical Logic, Part II, vol. 8, p. 280. Serial Zap. Nauchn. Sem. LOMI. Nauka, Leningrad (1968)
18. Yamashita, S., Sawada, H., Nagoya, A.: New methods to find optimal nondisjoint bidecompositions. In: Proceedings of the Design Automation Conference, pp. 59–68. San Francisco, CA, USA (1998)

Part II
Boolean Satisfiability

In the area of Boolean satisfiability, four research works are presented. The first chapter in this category studies the use equivalence checking formulas to compare unsatisfiability proofs built by a conflict-driven SAT-solver. The details of a SAT-sweeping approach that exploits local observability don't-cares (ODCs) to increase the number of vertices merged for logic optimization are presented in the second chapter. The next chapter proposes an approximation of MIN ONE SAT problem (which is the satisfying assignment with minimum number of ones). The last chapter presents a solution to MaxSAT (the problem of statisfying the maximum number of clauses in the original formula) using unsatisfiable cores.

Chapter 7
Boundary Points and Resolution

Eugene Goldberg and Panagiotis Manolios

Abstract We use the notion of boundary points to study resolution proofs. Given a CNF formula F, an $l(x)$-boundary point is a complete assignment falsifying only clauses of F having the same literal $l(x)$ of variable x. An $l(x)$-boundary point p mandates a resolution on variable x. Adding the resolvent of this resolution to F eliminates p as an $l(x)$-boundary point. Any resolution proof has to eventually eliminate all boundary points of F. Hence one can study resolution proofs from the viewpoint of boundary point elimination. We use equivalence checking formulas to compare proofs of their unsatisfiability built by a conflict-driven SAT-solver and very short proofs tailored to these formulas. We show experimentally that in contrast to proofs generated by this SAT-solver, almost every resolution of a specialized proof eliminates a boundary point. This implies that one may use the share of resolutions eliminating boundary points as a metric of proof quality. We argue that obtaining proofs with a high value of this metric requires taking into account the formula structure. We show that for any unsatisfiable CNF formula there always exists a proof consisting only of resolutions eliminating cut boundary points (which are a relaxation of the notion of boundary points). This result enables building resolution SAT-solvers that are driven by elimination of cut boundary points.

This chapter is an extended version of the conference paper [9].

7.1 Introduction

Resolution-based SAT-solvers [3, 6, 10, 12, 13, 15, 16] have achieved great success in numerous applications. However, the reason for this success and, more generally, the semantics of resolution are not well understood yet. This obviously impedes

E. Goldberg (✉)
Northeastern University, Baston, MA, USA
e-mail: eigold@ccs.neu.edu

This work is based on "Boundary Points and Resolution", Eugene Goldberg, in Proceedings of the 2009 the 12th International Conference on Theory and Applications of Satisfiability Testing ACM, 2009.

S.P. Khatri, K. Gulati (eds.), *Advanced Techniques in Logic Synthesis, Optimizations and Applications*, DOI 10.1007/978-1-4419-7518-8_7,

progress in SAT-solving. In this chapter, we study the relation between the resolution proof system [2] and boundary points [11]. The most important property of boundary points is that they mandate particular resolutions of a proof. So by studying the relation between resolution and boundary points one gets a deeper understanding of resolutions proofs, which should lead to building better SAT-solvers.

Given a CNF formula F, a non-satisfying complete assignment p is called an $l(x)$-boundary point, if it falsifies only the clauses of F that have the same literal $l(x)$ of variable x. The name is due to the fact that for satisfiable formulas the set of such points contains the boundary between satisfying and unsatisfying assignments. If F is unsatisfiable, for every $l(x)$-boundary point p there is a resolvent of two clauses of F on variable x that eliminates p. (That is, after adding such a resolvent to F, p is not an $l(x)$-boundary point anymore.) On the contrary, for a non-empty satisfiable formula F, there is always a boundary point that cannot be eliminated by adding a clause implied by F.

To prove that a CNF formula F is unsatisfiable it is sufficient to eliminate all its boundary points. In the resolution proof system, one reaches this goal by adding to F resolvents. If formula F has an $l(x)$-boundary point, a resolution proof has to have a resolution operation on variable x. The resolvents of a resolution proof eventually eliminate all boundary points. We will call a resolution mandatory if it eliminates a boundary point of the initial formula F that has not been eliminated by adding the previous resolvents. (In [9] such a resolution was called boundary.)

Intuitively, one can use the Share of Mandatory Resolutions (SMR) of a proof as a metric of proof quality. The reason is that finding mandatory resolutions is not an easy task. (Identification of a boundary point is computationally hard, which implies that finding a mandatory resolution eliminating this point is not easy either.) However, finding mandatory resolutions becomes much simpler if one knows subsets of clauses of F such that resolving clauses of these subsets eliminate boundary points. (An alternative is to try to guess these subsets heuristically.) Intuitively, such subsets have a lot to do with the structure of the formula. So the value of SMR may be used to gauge how well the resolution proof built by a SAT-solver follows the structure of the formula.

We substantiate the intuition above experimentally by comparing two kinds of proofs for equivalence checking formulas. (These formulas describe equivalence checking of two copies of a combinational circuit.) Namely, we consider short proofs of linear size particularly tailored for equivalence checking formulas and much longer proofs generated by a SAT-solver with conflict-driven learning. We show experimentally that the share of boundary resolution operations in high-quality specialized proofs is much greater than in proofs generated by the SAT-solver.

Generally speaking, it is not clear yet if for any irredundant unsatisfiable formula there is a proof consisting only of mandatory resolutions. However, as we show in this chapter, for any unsatisfiable formula F there always exists a proof where each resolution eliminates a *cut boundary point*. The latter is computed with respect to the CNF formula F_T consisting of clauses of F and their resolvents that specify a cut T of a resolution graph describing a proof. (Formula F_T is unsatisfiable for any cut T.) The notion of a cut boundary point is a relaxation of that of a boundary point computed with respect to the initial formula F. Point p that is boundary for F_T may not be boundary for F. Proving that resolution is complete with respect to elimination of cut boundary points enables building SAT-solvers that are driven by cut boundary point elimination. Another important observation is that the metric that computes the share of resolutions eliminating cut boundary points is more robust than SMR metric above. Namely, it can be also applied to the formulas that do not have proofs with a 100% value of the SMR metric.

The contributions of this chapter are as follows. First, we show that one can view resolution as elimination of boundary points. Second, we introduce the SMR metric that can be potentially used as a measure of proof quality. Third, we give some experimental results about the relation between SMR metric and proof quality. Fourth, we show that resolution remains complete even when it is restricted to resolutions eliminating cut boundary points.

This chapter is structured as follows. Section 7.2 introduces main definitions. Some properties of boundary points are given in Section 7.3. Section 7.4 views a resolution proof as a process of boundary point elimination. A class of equivalence checking formulas and their short resolution proofs are described in Section 7.5. Experimental results are given in Section 7.6. Some relevant background is recalled in Section 7.7. In Section 7.8 we show that for any unsatisfiable formula there is proof consisting only of resolutions eliminating cut boundary points. Conclusions and directions for future research are listed in Section 7.9.

7.2 Basic Definitions

Definition 7.1 *A literal* of a Boolean variable x (denoted as $l(x)$) is a Boolean function of x. The identity and negation functions (denoted as x and $\overline{x}$, respectively) are called the *positive literal* and *negative literal* of x, respectively. We will denote $l(x)$ as just l if the identity of variable x is not important.

Definition 7.2 *A clause* is the disjunction of literals where no two (or more) literals of the same variable can appear. *A CNF formula* is the conjunction of clauses. We will also view a CNF formula *as a set of clauses*. Denote by ***Vars(**F**)*** (respectively, ***Vars(**C**)***) the set of variables of CNF formula F (respectively, clause C).

Definition 7.3 Given a CNF formula $F(x_1, \ldots, x_n)$, *a complete assignment* p (also called *a point*) is a mapping $\{x_1, \ldots, x_n\} \rightarrow \{0, 1\}$. Given a complete assignment p and a clause C, denote by $C(p)$ the value of C when its variables are assigned by p. A clause C is *satisfied* (respectively, *falsified*) by p if $C(p) = 1$ (respectively, $C(p) = 0$).

Definition 7.4 Given a CNF formula F, *a satisfying assignment* p is a complete assignment satisfying every clause of F. *The satisfiability problem (SAT)* is to find a satisfying assignment for F or to prove that such an assignment does not exist.

Definition 7.5 Let F be a CNF formula and p be a complete assignment. Denote by ***Unsat(**p, F**)*** the set of all clauses of F falsified by p.

Definition 7.6 Given a CNF formula F, a complete assignment p is called an $l(x_i)$-*boundary point*, if *Unsat*$(p,F) \neq \emptyset$ and every clause of *Unsat*(p,F) contains literal $l(x_i)$.

Example 7.1 Let F consist of five clauses: $C_1 = x_2$, $C_2 = \overline{x}_2 \vee x_3$, $C_3 = \overline{x}_1 \vee \overline{x}_3$, $C_4 = x_1 \vee \overline{x}_3$, $C_5 = \overline{x}_2 \vee \overline{x}_3$. Complete assignment $p_1 = (x_1 = 0, x_2 = 0, x_3 = 1)$ falsifies only clauses C_1, C_4. So *Unsat*$(p_1, F) = \{C_1, C_4\}$. There is no literal shared by all clauses of *Unsat*(p_1,F). Hence p_1 is not a boundary point. On the other hand, $p_2 = (x_1 = 0, x_2 = 1, x_3 = 1)$ falsifies only clauses C_4, C_5 that share literal $\overline{x}_3$. So p_2 is a $\overline{x}_3$-boundary point.

7.3 Properties

In this section, we give some properties of boundary points.

7.3.1 Basic Propositions

In this section, we prove the following propositions. The set of boundary points contains the boundary between satisfying and unsatisfying assignments (Proposition 7.1). A CNF formula without boundary points is unsatisfiable (Proposition 7.2). Boundary points come in pairs (Proposition 7.3).

Definition 7.7 Denote by *Bnd_pnts(F)* the set of all boundary points of a CNF formula F. We assume that an $l(x_i)$-boundary point p is specified in *Bnd_pnts*(F) as the pair $(l(x_i),p)$. So the same point p may be present in *Bnd_pnts*(F) more than once (e.g., if p is an $l(x_i)$-boundary point and an $l(x_j)$-boundary point at the same time).

Proposition 7.1 *Let F be a satisfiable formula whose set of clauses is not empty. Let p_1 and p_2 be two complete assignments such that (a) $F(p_1) = 0$, $F(p_2) = 1$; (b) p_1 and p_2 are different only in the value of variable x_i. Then p_1 is an $l(x_i)$-boundary point.*

Proof Assume the contrary, i.e., *Unsat*(p_1,F) contains a clause C of F that does not have variable x_i. Then p_2 falsifies C too and so p_2 cannot be a satisfying assignment. A contradiction.

Proposition 7.1 means that *Bnd_pnts*(F) contains the boundary between satisfying and unsatisfying assignments of a satisfiable CNF formula F.

Proposition 7.2 *Let F be a CNF formula that has at least one clause.*
If Bnd_pnts$(F) = \emptyset$, then F is unsatisfiable.

Proof Assume the contrary, i.e., *Bnd_pnts*$(F) = \emptyset$ and F is satisfiable. Since F is not empty, one can always find two points p_1 and p_2 such that $F(p_1) = 0$ and $F(p_2) = 1$ and that are different only in the value of one variable x_i of F. Then according to Proposition 7.1, p_1 is an $l(x_i)$-boundary point. A contradiction.

Proposition 7.3 *Let p_1 be an $l(x_i)$-boundary point for a CNF formula F. Let p_2 be the point obtained from p_1 by changing the value of x_i. Then p_2 is either a satisfying assignment or a $\overline{l(x_i)}$-boundary point.*

Proof Reformulating the proposition, one needs to show that $Unsat(p_2,F)$ either is empty or contains only clauses with literal $\overline{l(x_i)}$. Assume that contrary, i.e., $Unsat(p_2,F)$ contains a clause C with no literal of x_i. (All clauses with $l(x_i)$ are satisfied by p_2.) Then C is falsified by p_1 too and so p_1 is not an $l(x_i)$-boundary point. A contradiction.

Definition 7.8 Proposition 7.3 means that for unsatisfiable formulas every x_i-boundary point has the corresponding $\overline{x}_i$-boundary point (and vice versa). We will call such a pair of points *twin boundary points in variable* $\boldsymbol{x_i}$.

Example 7.2 The point $p_2 = (x_1 = 0, x_2 = 1, x_3 = 1)$ of Example 7.1 is a $\overline{x}_3$-boundary point. The point $p_3 = (x_1 = 0, x_2 = 1, x_3 = 0)$ obtained from p_2 by flipping the value of x_3 falsifies only clause $C_2 = \overline{x}_2 \vee x_3$. So p_3 is an x_3-boundary point.

7.3.2 Elimination of Boundary Points by Adding Resolvents

In this section, we prove the following propositions. Clauses of a CNF formula F falsified by twin boundary points can be resolved (Proposition 7.4). Adding such a resolvent to F eliminates these boundary points (Proposition 7.5). Adding the resolvents of a resolution proof eventually eliminates all boundary points (Proposition 7.6). An $l(x_i)$-boundary point can be eliminated only by a resolution on variable x_i (Proposition 7.7). If formula F has an $l(x_i)$-boundary point, any resolution proof that F is unsatisfiable has a resolution on variable x_i (Proposition 7.8).

Definition 7.9 Let C_1 and C_2 be two clauses that have opposite literals of variable x_i (and no opposite literals of any other variable). The *resolvent* C of C_1 and C_2 is the clause consisting of all the literals of C_1 and C_2 but the literals of x_i. The clause C is said to be obtained by *a resolution operation* on variable x_i. C_1 and C_2 are called *the parent clauses* of C.

Proposition 7.4 *Let p_1 and p_2 be twin boundary points of a CNF formula F in variable x_i. Let C_1 and C_2 be two arbitrary clauses falsified by p_1 and p_2, respectively. Then (a) C_1, C_2 can be resolved on variable x_i; (b) $C(p_1) = 0$, $C(p_2) = 0$ where C is the resolvent of C_1 and C_2.*

Proof Since $C_1(p_1) = 0$, $C_2(p_2) = 0$, and p_1 and p_2 are twin boundary points in x_i, C_1 and C_2 have opposite literals of variable x_i. Since p_1 and p_2 are different only in the value of x_i, clauses C_1 and C_2 cannot contain opposite literals of a variable other than x_i. (Otherwise, p_1 and p_2 had to be different in values of at least two variables.) Since p_1 and p_2 are different only in the value of x_i, they both set to 0 all the literals of C_1 and C_2 but literals of x_i. So the resolvent C of C_1 and C_2 is falsified by p_1 and p_2.

Example 7.3 Points $p_2 = (x_1 = 0, x_2 = 1, x_3 = 1)$ and $p_3 = (x_1 = 0, x_2 = 1, x_3 = 0)$ from Examples 7.1 and 7.2 are twin boundary points in variable x_3. $Unsat(p_2, F) = \{C_4, C_5\}$ and $Unsat(p_3, F) = \{C_2\}$. For example, $C_4 = x_1 \vee \overline{x}_3$ can be resolved with $C_2 = \overline{x}_2 \vee x_3$ on variable x_3. Their resolvent $C = x_1 \vee \overline{x}_2$ is falsified by both p_2 and p_3.

Proposition 7.5 *Let p_1 and p_2 be twin boundary points in variable x_i and C_1 and C_2 be clauses falsified by p_1 and p_2, respectively. Then adding the resolvent C of C_1 and C_2 to F eliminates the boundary points p_1 and p_2. That is, pairs (x_i, p_1) and $(\overline{x}_i, p_2)$ are not in the set $Bnd\_pnts(F \wedge C)$ (here we assume that p_1 is an x_i-boundary point and p_2 is a $\overline{x}_i$-boundary point of F).*

Proof According to Proposition 7.4, any clauses C_1 and C_2 falsified by p_1 and p_2, respectively, can be resolved on x_i and p_1 and p_2 falsify the resolvent C of C_1 and C_2. Since clause C does not have a literal of x_i, p_1 is not an x_i-boundary point and p_2 is not a $\overline{x}_i$-boundary point of $F \wedge C$.

Proposition 7.6 *If a CNF formula F contains an empty clause, then $Bnd\_pnts(F) = \emptyset$.*

Proof For any complete assignment p, the set $Unsat(p, F)$ contains the empty clause of F. So p cannot be an l-boundary point.

Proposition 7.6 works only in one direction, i.e., if $Bnd\_pnts(F) = \emptyset$, it does not mean that F contains an empty clause. Proposition 7.6 only implies that, given an unsatisfiable formula F for which $Bnd\_pnts(F)$ is not empty, the resolvents of any resolution proof of unsatisfiability of F eventually eliminate all the boundary points.

Proposition 7.7 *Let F be a CNF formula and p be an $l(x_i)$-boundary point of F. Let C be the resolvent of clauses C_1 and C_2 of F that eliminates p (i.e., $(l(x_i), p)$ is not in $Bnd\_pnts(F \wedge C)$). Then C is obtained by resolution on variable x_i. In other words, an $l(x_i)$-boundary point can be eliminated only by adding to F a resolvent on variable x_i.*

Proof Assume the contrary, i.e., adding C to F eliminates p and C is obtained by resolving C_1 and C_2 on variable x_j, $j \neq i$. Since C eliminates p as an $l(x_i)$-boundary point, it is falsified by p and does not contain $l(x_i)$. This means that neither C_1 nor C_2 contains variable x_i. Since C is falsified by p, one of the parent clauses, say clause C_1, is falsified by p too. Since C_1 does not contain literal $l(x_i)$, p is not an $l(x_i)$- boundary point of F. A contradiction.

Proposition 7.8 *Let p be an $l(x_i)$-boundary point of a CNF formula F. Then any resolution derivation of an empty clause from F has to contain a resolution operation on variable x_i.*

Proof According to Proposition 7.6, every boundary point of F is eventually eliminated in a resolution proof. According to Proposition 7.7, an $l(x_i)$-boundary point can be eliminated only by adding to F a clause produced by resolution on variable x_i.

7.3.3 Boundary Points and Redundant Formulas

In this section, we prove the following propositions. A clause C of a CNF formula F that has a literal l and is not falsified by an l-boundary point of F is redundant in F (Proposition 7.9). If F does not have any $l(x_i)$-boundary points, all clauses depending on variable x_i can be removed from F (Proposition 7.10). If p is an $l(x_i)$-boundary point of F, it is also an $l(x_i)$-boundary point of every unsatisfiable subset of clauses of F.

Definition 7.10 A clause C of a CNF formula F is called *redundant* if $F \setminus \{C\} \to C$.

Proposition 7.9 *Let C be a clause of a CNF formula F. Let l be a literal of C. If no l-boundary point of F falsifies C, then C is redundant.*

Proof Assume the contrary, i.e., C is not redundant. Then there is an assignment p such that C is falsified and all the other clauses of F are satisfied. Then p is an l-boundary point. A contradiction.

Importantly, Proposition 7.9 works only in one direction. That is, the fact that a clause C is redundant in F does not mean that no boundary point of F falsifies C. Let CNF formula $F(x_1, x_2)$ consist of four clauses: $\overline{x}_1$, x_1, $x_1 \vee \overline{x}_2$, $x_1 \vee x_2$. Although the clause x_1 is redundant in F, $p = (x_1 = 0,\ x_2 = 0)$ is an x_1-boundary point falsifying x_1 (and $x_1 \vee x_2$). The resolvent of clauses x_1 and $\overline{x}_1$ eliminates p as a boundary point.

Proposition 7.10 *Let a CNF formula F have no $l(x_i)$-boundary points. Then removing the clauses containing x_i or $\overline{x}_i$ from F does not change F (functionally).*

Proof Let C be a clause of F with a literal $l(x_i)$. Then according to Proposition 7.9 C is redundant in F and so its removal does not change the Boolean function specified by F. Removing C from F cannot produce an $l(x_i)$-boundary point in $F \setminus \{C\}$. So Proposition 7.9 can be applied again to any of the remaining clauses with x_i or $\overline{x}_i$ (and so on).

Proposition 7.11 *Let F be an unsatisfiable formula. Let p be an $l(x_i)$-boundary point of F and F' be an unsatisfiable subset of clauses of F. Then p is an $l(x_i)$-boundary point of F'.*

Proof Since $F' \subseteq F$ then $Unsat(p,F') \subseteq Unsat(p,F)$. Since F' is unsatisfiable, $Unsat(p,F') \neq \emptyset$.

7.4 Resolution Proofs and Boundary Points

In this section, we view construction of a resolution proof as a process of boundary point elimination and give a metric for measuring proof quality.

7.4.1 Resolution Proof as Boundary Point Elimination

First, we define the notion of a resolution proof [2] and a boundary resolution.

Definition 7.11 Let F be an unsatisfiable formula. Let $R_1, \ldots, R_k$ be a set of clauses such that (a) each clause R_i is obtained by resolution operation where a parent clause is either a clause of F or the resolvent of a previous resolution operation; (b) clauses R_i are numbered in the order they are derived; (c) R_k is an empty clause. Then the set of resolutions that produced the resolvents $R_1, \ldots, R_k$ is called *a resolution proof*. We assume that this proof is *irredundant*, i.e., removal of any non-empty subset of these k resolvents breaks condition (a).

Definition 7.12 Let $\{R_1, \ldots, R_k\}$ be the set of resolvents forming a resolution proof that a CNF formula F is unsatisfiable. Denote by F_i the CNF formula that is equal to F for $i = 1$ and to $F \cup \{R_1, \ldots, R_{i-1}\}$ for $i = 2, \ldots, k$. We will say that the ith resolution (i.e., one that produces resolvent R_i) is *non-mandatory* if $Bnd\_pnts(F_i)$ = $Bnd\_pnts(F_{i+1})$. Otherwise (i.e., if $Bnd\_pnts(F_i) \subset Bnd\_pnts(F_{i+1})$, because adding a clause cannot create a boundary point), ith resolution is called *mandatory*. So a resolution operation is mandatory if adding R_i to F_i eliminates a boundary point.

In Section 7.3, we showed that eventually all the boundary points of a CNF formula F are removed by resolvents. Importantly, an $l(x_i)$-boundary point mandates a resolution on x_i. Besides, as we showed in Section 7.3.3, even redundant clauses can be used to produce new resolvents eliminating boundary points. It is important because all clauses derived by resolution (e.g., conflict clauses generated by modern SAT-solvers) are redundant. So the derived clauses are as good as the original ones for boundary point elimination.

A natural question arises about the role of non-mandatory resolutions. When answering this question it makes sense to separate redundant and irredundant formulas. (A CNF formula F is said to be *irredundant* if no clause of F is redundant, see Definition 7.10.) For a redundant formula, one may have to use non-mandatory resolutions. (In particular, a heavily redundant formula may not have boundary points at all. Then every resolution operation is non-mandatory.) For irredundant formulas the situation is different.

Proposition 7.12 *Let F be an irredundant formula of m clauses. Then F has at least d boundary points where d is the number of literals in F.*

Proof Let C be a clause of F. Then there is a complete assignment p falsifying C and satisfying the clauses of $F \setminus \{C\}$. This assignment is an l-boundary point where l is a literal of C.

7.4.2 SMR Metric and Proof Quality

Intuitively, to efficiently build a short proof for an unsatisfiable CNF formula F, a resolution-based SAT-solver has to find mandatory resolutions as soon as possible. Otherwise, a lot of non-mandatory resolutions may be generated that would not have been necessary had mandatory resolutions been derived early. (In particular, as we show in experiments, an entire proof may consist only of mandatory resolutions.)

This implies that the Share of Mandatory Resolutions (SMR) of a proof can be used as a proof quality metric. Generation of proofs with a high value of SMR most likely requires a good knowledge of the formula structure. The reason is as follows. Finding a mandatory resolution suggests identification of at least one boundary point this resolution eliminates. But detection of boundary points is hard. (Finding an $l(x_i)$-boundary point of formula F reduces to checking the satisfiability of the set of clauses $F \setminus \{$ the clauses of F with $l(x_i)\}$, see Section 7.6.) Identification of mandatory resolutions without looking for boundary points of F requires the knowledge of "special" subsets of clauses of the current formula F. (These special subsets should contain clauses whose resolutions produce resolvents eliminating boundary points.) Intuitively, such subsets can be identified if the formula structure is known. This intuition is substantiated experimentally in Section 7.6.

The simplest example of information about the formula structure is to identify a small unsatisfiable core. Even if the initial unsatisfiable CNF formula F to be solved is irredundant, an unsatisfiable subformula of F inevitably appears due to the addition of new clauses. (In particular, one can view an empty clause as the smallest unsatisfiable subformula of the final CNF formula.) Let F_1 be an unsatisfiable subformula of F. Then no $l(x_i)$-boundary point exists if x_i is in $Vars(F) \setminus Vars(F_1)$. (The set of clauses falsified by any point p contains at least one clause C of F_1 and $x_i \notin Vars(C)$.) So any resolution on a variable of $Vars(F) \setminus Vars(F_1)$ is non-mandatory.

The appearance of unsatisfiable subformulas may lead to increasing the share of non-mandatory resolutions in the final proof. For example, instead of deriving an empty clause from F_1, the SAT-solver may first derive some clauses having variables of $Vars(F_1)$ from clauses of $F \setminus F_1$. It is possible since clauses of $F \setminus F_1$ may contain variables of $Vars(F_1)$. When deriving such clauses the SAT-solver may use (non-mandatory) resolutions on variables of $Vars(F) \setminus Vars(F_1)$, which leads to redundancy of the final proof.

Unfortunately, we do not know yet if, given an unsatisfiable irredundant CNF formula, there is always a proof consisting only of mandatory resolutions (and so having a 100% value of SMR metric). Hence a low value of the SMR metric for a CNF formula F may mean that the latter does not have a "natural" proof in the resolution proof system (see Section 7.8). However, as we show in Section 7.8, for any unsatisfiable formula there always exists a proof where each resolution eliminates a *cut boundary point*. So, for measuring proof quality one can also use the share of cut mandatory resolutions (i.e., resolutions eliminating cut boundary points). This metric should be more robust than SMR in the sense that it should work even for formulas that do not have proofs with a 100% value of SMR metric.

7.5 Equivalence Checking Formulas

In this section, we introduce the formulas we use in the experimental part of this chapter. These are the formulas that describe equivalence checking of two copies of a combinational circuit. In Section 7.5.1 we show how such formulas are constructed. In Section 7.5.2, we build short proofs of unsatisfiability particularly tailored for equivalence checking formulas.

7.5.1 Building Equivalence Checking Formulas

Let N and N^* be two single-output combinational circuits. To check their functional equivalence one constructs a circuit called *a miter* (we denote it as $Miter(N, N^*)$). It is a circuit that is satisfiable (i.e., its output can be set to 1) if and only if N and N^* are not functionally equivalent. (N and N^* are not functionally equivalent if there is an input assignment for which N and N^* produce different output values.) Then a CNF formula F_{Miter} is generated that is satisfiable if and only if $Miter(N, N^*)$ is satisfiable. In our experiments, we use a miter of two identical copies of the same circuit. Then $Miter(N, N^*)$ is always unsatisfiable and so is CNF formula F_{Miter}.

Example 7.4 Figure 7.1 shows the miter of copies N and N^* of the same circuit. Here g_1, g_1^* are OR gates, g_2, g_2^* are AND gates, and h is an XOR gate (implementing modulo-2 sum). Note that N and N^* have the same set of input variables but different intermediate and output variables. Since $g_2 \oplus g_2^*$ evaluates to 1 if and only if $g_2 \neq g_2^*$, and N and N^* are functionally equivalent, the circuit $Miter(N, N^*)$ evaluates only to 0.

A CNF formula F_{Miter} whose satisfiability is equivalent to that of $Miter(N, N^*)$ is formed as $F_N \wedge F_N^* \wedge F_{\text{xor}} \wedge h$. Here F_N and F_N^* are formulas specifying the functionality of N and N^*, respectively. The formula F_{xor} specifies the functionality of the XOR gate h, and the unit clause h forces the output of $Miter(N, N^*)$ to be set to 1. Since, in our case, the miter evaluates only to 0, the formula F_{Miter} is unsatisfiable.

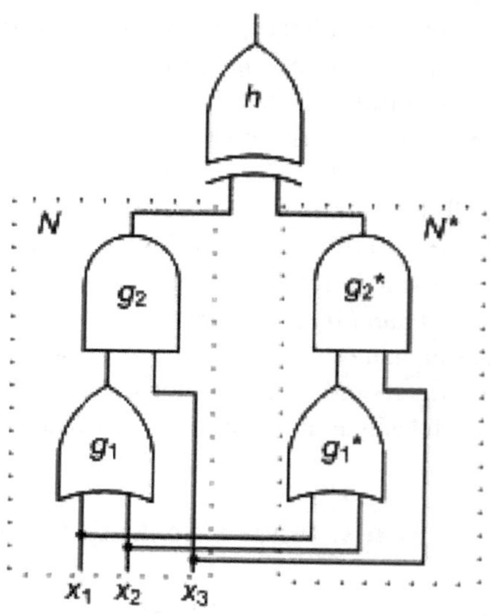

Fig. 7.1 Circuit $Miter(N, N^*)$

Formulas F_N and F_N^* are formed as the conjunction of subformulas describing the gates of N and N^*. For instance, $F_N = F_{g_1} \wedge F_{g_2}$ where, for example, $F_{g_1} = (x_1 \vee x_2 \vee \overline{g}_1) \wedge (\overline{x}_1 \vee g_1) \wedge (\overline{x}_2 \vee g_1)$ specifies the functionality of an OR gate. Each clause of F_{g_1} rules out some inconsistent assignments to the variables of gate g_1. For example, the clause $(x_1 \vee x_2 \vee \overline{g}_1)$ rules out the assignment $x_1 = 0,\ x_2 = 0,\ g_1 = 1$.

7.5.2 Short Proofs for Equivalence Checking Formulas

For a CNF formula F_{Miter} describing equivalence checking of two copies N, N^* of the same circuit, there is a short resolution proof that F_{Miter} is unsatisfiable. This proof is linear in the number of gates in N and N^*. The idea of this proof is as follows. For every pair g_i, g_i^* of the corresponding gates of N and N^*, the clauses of CNF formula $Eq(g_i, g_i^*)$ specifying the equivalence of variables g_i and g_i^* are derived where $Eq(g_i, g_i^*) = (\overline{g}_i \vee g_i^*) \wedge (g_i \vee \overline{g}_i^*)$. These clauses are derived according to topological levels of gates g_i, g_i^* in *Miter*(N, N^*). (The topological level of a gate g_i is the length of the longest path from an input to gate g_i measured in the number of gates on this path.) First, clauses of $Eq(g_i, g_i^*)$ are derived for all pairs of gates g_i, g_i^* of topological level 1. Then using previously derived $Eq(g_i, g_i^*)$, same clauses are derived for the pairs of gates g_j, g_j^* of topological level 2 and so on. Eventually, the clauses of $Eq(g_s, g_s^*)$ relating the output variables g_s, g_s^* of N and N^* are derived. Resolving the clauses of $Eq(g_s, g_s^*)$ and the clauses describing the XOR gate, the clause $\overline{h}$ is derived. Resolution of $\overline{h}$ and the unit clause h of F_{Miter} produces an empty clause.

Example 7.5 Let us explain the construction of the proof using the CNF F_{Miter} from Example 7.4. Gates g_1, g_1^* have topological level 1 in *Miter*(N, N^*). So first, the clauses of $Eq(g_1, g_1^*)$ are obtained. They are derived from the CNF formulas F_{g_1} and $F_{g_1^*}$ describing gates g_1 and g_1^*. That the clauses of $Eq(g_1, g_1^*)$ can be derived from $F_{g_1} \wedge F_{g_1^*}$ just follows from the completeness of resolution and the fact that $Eq(g_1, g_1^*)$ is implied by the CNF formula $F_{g_1} \wedge F_{g_1^*}$. (This implication is due to the fact that F_{g_1} and $F_{g_1^*}$ describe two functionally equivalent gates with the same set of input variables.) More specifically, the clause $\overline{g}_1 \vee g_1^*$ is obtained by resolving the clause $x_1 \vee x_2 \vee \overline{g}_1$ of F_{g_1} with the clause $\overline{x}_1 \vee g_1^*$ of $F_{g_1^*}$ and then resolving the resolvent with the clause $\overline{x}_2 \vee g_1^*$ of $F_{g_1^*}$. In a similar manner, the clause $g_1 \vee \overline{g}_1^*$ is derived by resolving the clause $x_1 \vee x_2 \vee \overline{g}_1^*$ of $F_{g_1^*}$ with clauses $\overline{x}_1 \vee g_1$ and $\overline{x}_2 \vee g_1$ of F_{g_1}.

Then the clauses of $Eq(g_2, g_2^*)$ are derived (gates g_2, g_2^* have topological level 2). $Eq(g_2, g_2^*)$ is implied by $F_{g_2} \wedge F_{g_2^*} \wedge Eq(g_1, g_1^*)$. Indeed, g_2 and g_2^* are functionally equivalent gates that have the same input variable x_3. The other input variables g_1 and g_1^* are identical too due to the presence of $Eq(g_1, g_1^*)$. So the clauses of $Eq(g_2, g_2^*)$ can be derived from clauses of $F_{g_2} \wedge F_{g_2^*} \wedge Eq(g_1, g_1^*)$ by resolution. Then the clause $\overline{h}$ is derived as implied by $F_{\text{xor}} \wedge Eq(g_2, g_2^*)$ (an XOR gate produces output 0 when its input variables have equal values). Resolution of h and $\overline{h}$ produces an empty clause.

7.6 Experimental Results

The goal of experiments was to compare the values of SMR metric (see Section 7.4.2) for two kinds of proofs of different quality. In the experiments we used formulas describing the equivalence checking of two copies of combinational circuits. The reason for using such formulas is that one can easily generate high-quality specialized proofs of their unsatisfiability (see Section 7.5). In the experiments we compared these short proofs with the ones generated by a well-known SAT-solver Picosat, version 913 [3].

We performed the experiments on a Linux machine with Intel Core 2 Duo CPU with 3.16 GHz clock frequency. The time limit in all experiments was set to 1 h. The formulas and specialized proofs we used in the experiments can be downloaded from [17].

Given a resolution proof R of k resolutions that a CNF formula F is unsatisfiable, computing the value of SMR metric of R reduces to k SAT-checks. In our experiments, these SAT-checks were performed by a version of DMRP-SAT [8]. Let F_i be the CNF formula $F \cup \{R_1, \ldots, R_{i-1}\}$ where $\{R_1, \ldots, R_{i-1}\}$ are the resolvents generated in the first $i - 1$ resolutions. Let C_1 and C_2 be the clauses of F_i that are the parent clauses of the resolvent R_i. Let C_1 and C_2 be resolved on variable x_j. Assume that C_1 contains the positive literal of x_j. Checking if ith resolution eliminates an x_j-boundary point can be performed as follows. First, all the clauses with a literal of x_j are removed from F_i. Then one adds to F_i the unit clauses that force the assignments setting all the literals of C_1 and all the literals of C_2 but the literal $\overline{x}_j$ to 0. Denote the resulting CNF formula by G_i.

If G_i is satisfiable then there is a complete assignment p that is falsified by C_1 and maybe by some other clauses with literal x_j. So p is an x_j-boundary point of F_i. Since p falsifies all the literals of C_2 but $\overline{x}_j$, it is falsified by the resolvent of C_1 and C_2. So the satisfiability of G_i means that ith resolution eliminates p and so this resolution is mandatory. If G_i is unsatisfiable, then no x_j-boundary point is eliminated by ith resolution. All boundary points come in pairs (see Proposition 7.3). So no $\overline{x}_j$-boundary point is eliminated by ith resolution either. Hence the unsatisfiability of G_i means that the ith resolution is non-mandatory.

Table 7.1 shows the value of SMR metric for the short specialized proofs. The first column gives the name of the benchmark circuit whose self-equivalence is

Table 7.1 Computing value of SMR metric for short specialized proofs

Name	#Vars	#Clauses	#Resolutions	SMR (%)	Time (s)
c432	480	1,333	1,569	95	1.4
9symml	480	1,413	1,436	100	0.6
mlp7	745	2,216	2,713	100	1.3
c880	807	2,264	2,469	100	3.8
alu4	2,369	7,066	8,229	96	43
c3540	2,625	7,746	9,241	97	137
x1	4,381	12,991	12,885	97	351
dalu	4,714	13,916	15,593	84	286

described by the corresponding CNF formula. The size of this CNF formula is given in the second and third columns. The fourth column of Table 7.1 gives the size of the proof (in the number of resolutions). The fifth column shows the value of SMR metric, and the last column of Table 1 gives the run time of computing this value. These run times can be significantly improved if one uses a faster SAT-solver and tunes it to the problem of computing the SMR metric. (For example, one can try to share conflict clauses learned in different SAT-checks.)

Looking at Table 7.1 one can conclude that the specialized proofs have a very high value of SMR metric (almost every resolution operation eliminates a boundary point). The only exception is the dalu formula (84%). The fact that the value of SMR metric for dalu and some other formulas is different from 100% is probably due to the fact that the corresponding circuits have some redundancies. Such redundancies would lead to redundancy of CNF formulas specifying the corresponding miters, which would lower the value of SMR metric.

The values of SMR metric for the proofs generated by Picosat are given in Table 7.2. The second column gives the size of resolution proofs generated by Picosat. When computing the size of these proofs we removed the obvious redundancies. Namely, the derivation of the conflict clauses that did not contribute to the derivation of an empty clause was ignored. The third column shows the value of SMR metric, and the last column gives the run time of computing this value. In the case the computation did not finish within the time limit, the number in parentheses shows the percent of the resolution operations processed before the computation terminated.

Table 7.2 Computing value of SMR metric for proofs generated by Picosat

Name	#Resolutions	SMR (%)	Run time (s) (% of proof finished)
c432	19, 274	41	75
9symml	12, 198	47	28
mlp7	7, 253, 842	60	>1 h (1.2)
c880	163, 655	17	>1 h (72)
alu4	672, 293	41	>1 h (12)
c3540	3, 283, 170	29	>1 h (2.7)
x1	92, 486	45	>1 h (84)
dalu	641, 714	33	>1 h (6.9)

Table 7.2 shows that the size of the proofs generated by Picosat is much larger than that of specialized proofs (Table 7.1, fourth column). Importantly, the value of SMR metric we give for the formulas for which computation was terminated due to exceeding the time limit is higher than it should be. Typically, the later a resolution occurs in a resolution proof, the more likely it is that this resolution is non-mandatory. So the early termination of SMR metric computation ignored resolutions with the highest chances to be non-mandatory.

The intuition above is confirmed by the results of Table 7.3. To reach later resolutions we sampled four largest resolution proofs; that is, we checked only every kth resolution whether it was mandatory. The value of k is shown in the second

Table 7.3 Using sampling to compute SMR metric for Picosat proofs

Name	Sampling rate	SMR (%)	Proof processed (%)
mlp7	100	18	11
alu4	10	29	36
c3540	100	11	26
dalu	10	37	26

column. The next column gives the value of SMR metric computed over the set of sampled resolutions. The part of the proof covered by sampling within the 1-h time limit is shown in the last column. It is not hard to see that taking into account later resolutions significantly reduced the value of SMR metric for three proofs out of four.

Summing up, one can conclude that for the formulas we considered in experiments, the proofs of poorer quality (generated by Picosat) have lower values of SMR metric. This substantiates the reasoning of Section 7.4 that not taking into account the formula structure leads to generation of proofs with a low value of SMR metric. On the contrary, picking "right" subsets of clauses to be resolved with each other, i.e., closely following the formula structure, leads to generation of proofs with a high value of SMR metric.

7.7 Some Background

The notion of boundary points was introduced in [11] where they were called essential points. (We decided to switch to the term "boundary point" as more precise.) Boundary points were used in [11] to help a SAT-solver prune the search space. If the subspace x_i=0 does not contain a satisfying assignment or an x_i-boundary point, one can claim that the symmetric subspace x_i=1 cannot contain a satisfying assignment either (due to Proposition 7.3). The same idea of search pruning was independently described in [14] and implemented in the SAT-solver Jerusat. The ideas of search pruning introduced in [11] were further developed in [5].

In [7], we formulate two proof systems meant for exploring the 1-neighborhood of clauses of the formula to be solved. The union of the 1-neighborhoods of these clauses is essentially a superset approximation of the set of boundary points. To prove that a formula is unsatisfiable it is sufficient to eliminate all boundary points (Proposition 7.2). The proof systems of [7] show that one can eliminate all boundary points without generation of an empty clause. So resolution can be viewed as a special case of boundary point elimination.

The results of this chapter can also be considered as an approach to improving automatizability of resolution [4]. General resolution is most likely non-automatizable [1]. This means that finding short proofs cannot be done efficiently in general resolution. A natural way to mitigate this problem is to look for restricted versions of general resolution that are "more automatizable," i.e., that facilitate finding good proofs. Intuitively, mandatory resolutions is a tiny part of the set of all possible resolutions. So the restriction of resolutions to mandatory ones (or cut mandatory ones, see Section 7.8) can be viewed as a way to make it easier to find good proofs.

7.8 Completeness of Resolution Restricted to Boundary Point Elimination

In this section, we show that, given a CNF formula (irredundant or not), there always exists a proof consisting only of resolutions that eliminate so-called cut boundary points. (This result was not published in [9].) The importance of this result is that it enables SAT-solvers to use (cut) boundary point elimination for building resolution proofs.

7.8.1 Cut Boundary Points

Let F be a CNF formula and resolvents $R_1, \ldots, R_k$ form a proof R that F is unsatisfiable. So far we considered elimination of boundary points of the original CNF formula F (by adding resolvents R_i). Now we introduce the notion of a cut boundary point.

Denote by G_R a DAG (called *a resolution graph*) specified by proof R. The nodes of G_R correspond to the clauses of F (the sources of G_R) and resolvents $R_1, \ldots, R_k$ (the empty clause R_k being the sink of G_R). Graph G_R has an edge directed from n_1 to n_2 if and only if n_2 corresponds to a resolvent and n_1 corresponds to a parent clause of this resolvent.

Denote by T *a cut* of G_R, i.e., a set of nodes such that every path from a source to the sink of G_R has to go through a node of T. Denote by F_T the CNF formula that consists of the clauses corresponding to the nodes of cut T. (If cut T consists of the sources of G_R, then F_T is the initial formula F.) Formula F_T is unsatisfiable for any cut T. Indeed, the resolutions corresponding to the nodes located between the nodes of T and this sink of G_R form a proof that F_T is unsatisfiable. In terms of Definition 7.12, F_T is a subset of F_i for some $i \leq k$ where $F_i = F \cup R_1 \cup \ldots \cup R_i$. (Since F_i is redundant, one can remove some clauses of F_i without breaking its unsatisfiability.)

Definition 7.13 Let T be a cut of a proof G_R that a CNF formula F is unsatisfiable. Let p be an l-boundary point for F_T (i.e., $p \in Bnd\_pnts(F_T)$). We will call p an l-boundary point with respect to cut T or just *cut boundary point*.

Note that $Vars(F_i) = Vars(F)$ and $Vars(F_T) \subseteq Vars(F)$. For the sake of simplicity, we will assume that if p is a boundary point for F_T, the variables of $Vars(F) \setminus Vars(F_T)$ are assigned in p (even though these assignments cannot affect satisfying or falsifying a clause of F_T). Importantly, since F_T is only a subset of F_i, a point p that is l-boundary for F_T may not be such for F_i.

Note that if F is an irredundant CNF formula, all resolution graphs G_R have the same cut T consisting only of the nodes that are sources of G_R. (For such a cut, $F_T = F$.)

7.8.2 The Completeness Result

In this section, we describe the procedure *gen_proof* that, given an unsatisfiable CNF formula F, builds a resolution proof where every resolution eliminates a cut boundary point. This means that the resolution proof system remains complete even if it is restricted only to the resolution operations that eliminate cut boundary points.

The pseudocode of *gen_proof* is shown in Fig. 7.2. In the outer *while* loop, *gen_proof* processes variables of F one by one. First, a variable x_i to be processed is chosen by the function *pick_variable*. Then all $l(x_i)$-boundary points are eliminated by *gen_proof* in the inner *while* loop. (Finding and elimination of $l(x_i)$-boundary points are performed by procedures *find_bp* and *elim_bp*, respectively.)

```
gen_proof(F)
{/* F_{x_i} - the clauses of F with x_i
  F_{x̄_i} - the clauses of F with x̄_i */
while (true)
  {x_i = pick_variable(F);
  while (true)
   {(ans,p) ← find_bp(F,x_i);
   if (ans == failure)
     {F = F \ (F_{x_i} ∪ F_{x̄_i});
     break;} /* leave the inner loop */
   (C,C_{x_i},C_{x̄_i}) ← elim_bp(F_{x_i},F_{x̄_i},p);
   update_proof(G_R,C,C_{x_i},C_{x̄_i});
   if (empty_clause(C))
     return(rem_redundant(G_R));
   F = F ∪ C; }}}
```

Fig. 7.2 *gen_proof* generates a proof where every resolution eliminates a cut boundary point

To eliminate an $l(x_i)$-boundary point, the resolvent C of parent clauses C_{x_i} and $C_{\overline{x}_i}$ is generated. Here $C_{x_i} \in F_{x_i}$ and $C_{\overline{x}_i} \in F_{\overline{x}_i}$ where F_{x_i} and $F_{\overline{x}_i}$ are the clauses of F having literals x_i and $x_{\overline{x}_i}$, respectively. If C is an empty clause, *gen_proof* stops and returns the resolution graph G_R specifying a proof that F is unsatisfiable. (G_R may contain nodes corresponding to resolvents that are not on a path leading from a source to the empty clause. So, the procedure *rem_redundant* is called to remove the parts of G_R corresponding to redundant resolution operations.) Otherwise, a new $l(x_i)$-boundary point is looked for. When all $l(x_i)$-boundary points are eliminated, the clauses F_{x_i} and $F_{\overline{x}_i}$ are removed from F and *proof_gen* leaves the inner loop.

That the *proof_gen* procedure is sound follows from the soundness of the resolution proof system. The *proof_gen* procedure is also complete. The number of $l(x_i)$-boundary points is finite and monotonically decreases in the process of adding resolvents C to F. So after a finite number of steps, all $l(x_i)$-boundary points are eliminated from the current formula F. At this point, the clauses of F containing

x_i or $\overline{x}_i$ are removed from F, which does affect its unsatisfiability (see Proposition 7.10). After processing all the variables, an empty clause is inevitably derived (because the resulting formula has no variables and has to be functionally equivalent to the original formula F that is unsatisfiable).

Let us show that every resolution operation of the proof G_R produced by *gen_proof* eliminates a cut boundary point. Let C be a resolvent on variable x_i produced from parent clauses C_{x_i} and $C_{\overline{x}_i}$ in the inner loop of the *gen_proof* procedure. By construction, C eliminates an $l(x_i)$-boundary point p of the current formula F. The nodes corresponding to F form a cut of G_R (because every resolvent produced in the future is a descendent of clauses of F). The elimination of redundant nodes of G_R by the procedure *rem_redundant* of *gen_proof* does not change much. If the resolvent C turns out to be redundant, then whether or not adding C to F eliminates a cut boundary point is irrelevant. If the resolvent C stays in the proof, it still eliminates the same $l(x_i)$ boundary p (according to Proposition 7.11).

Summarizing, each resolution of the proof specified by the graph G_R built by *gen_proof* eliminates a cut boundary point. (The cut corresponding to a resolution is specified by the formula F at the time this resolution was performed minus some redundant clauses identified by *rem_redundant*.)

7.8.3 Boundary Points as Complexity Measure

The existence of an $l(x_i)$-boundary point of a CNF formula F implies dependency of F on variable x_i. In this subsection, we argue that in the context of resolution proofs, $Bnd\_pnts(F)$ is a more precise complexity measure than $Vars(F)$. Besides, we introduce the notion of natural resolution proofs.

Let resolution graph G_R specify a proof that a CNF formula F is unsatisfiable. Such a proof can be represented as a sequence of unsatisfiable formulas $F_{T_1}, \ldots, F_{T_m}$ corresponding to cuts $T_1, \ldots, T_m$ of G_R. We assume that $F_{T_1} = F$ and F_{T_m} consists only of an empty clause and that $T_i \leq T_j$ if $i \leq j$. (The partial order $T_i \leq T_j$ holds iff there is no path from a source to the sink of G_R such that a node of T_j appears on this path before a node of T_i.)

It is natural to expect that formulas F_{T_i} are getting easier to solve as the value of i grows and eventually the trivial formula F_m is derived. Note that in terms of the number of variables this is true because $T_i \leq T_j \rightarrow Vars(T_i) \subseteq Vars(T_j)$. However, in terms of boundary points this is, in general, not true. For example, it is possible that $Bnd\_pnts(F_{T_i})$ is not a subset of $Bnd\_pnts(F_{T_j})$ even though $T_i \leq T_j$. The reason is that F_{T_j} is obtained from F_{T_i} by adding and *removing* some clauses. The removal of a clause C from F_{T_i} may lead to appearance of a new $l(x_s)$-boundary point p where $x_s \notin Vars(C)$. This means that F_{T_j} in some aspect depends on x_s stronger than F_{T_i} (because a resolution on variable x_s is mandated by p).

The observation above implies that boundary points provide a more precise way to measure formula complexity than just computing the set of variables on which the formula depends. It would be interesting to find classes of formulas for which there exist resolution proofs where the complexity of formulas F_{T_i} monotonically reduces in terms of $Bnd\_pnts(F_{T_i})$. Intuitively, this is possible if the proof structure follows the natural structure of the formula. So such proofs can be called natural. On the other hand, there may be a class of formulas for which natural resolution proofs do not exist (due to the fact that the structure of any resolution proof does not agree with that of a formula from this class).

7.9 Conclusions and Directions for Future Research

We show that a resolution proof can be viewed as the process of boundary point elimination. We introduce the SMR metric that is the percent of resolutions of the proof that eliminate boundary points (mandatory resolutions). This metric can be used for gauging proof quality. We experimentally show that short specialized proofs for equivalence checking formulas have high values of SMR metric. On the other hand, values of this metric for proofs generated by a SAT-solver with conflict-driven learning are low. As we argue in Section 7.4, this may be attributed to not taking into account the formula structure 1.

The idea of treating resolution as boundary proof elimination has many interesting directions for research. Here are a few of them.

1. Studying the quality of proofs consisting only of resolutions eliminating cut boundary points. (We showed the existence of such proofs for every CNF formula but have not made any claims about their quality.)
2. Studying further the relation between the value of SMR metric for resolution proofs obtained by SAT-solvers and their ability to take into account the formula structure.
3. Building SAT-solvers based on the idea of (cut) boundary point elimination.
4. Finding the classes of formulas for which "natural" proofs exist, i.e., proofs for which the complexity of cut CNF formulas monotonically decreases (in terms of cut boundary points).

References

1. Alekhnovich, M., Razborov, A.: Resolution is not automatizable unless w[p] is tractable. SIAM Journal on Computing **38**(4), 1347–1363 (2008)
2. Bachmair, L., Ganzinger, H.: Resolution theorem proving. In: A. Robinson, A. Voronkov (eds.) Handbook of Automated Reasoning, vol. I, chap. 2, pp. 19–99. North-Holland Elsevier Science (2001)
3. Biere, A.: Picosat essentials. Journal of Substance Abuse Treatment **4**(2–4), 75–97 (2008)
4. Bonet, L., Pitassi, T., Raz, R.: On interpolation and automatization for Frege systems. SIAM Journal of Computing **29**(6), 1939–1967 (2000)
5. Babic, D., Bingham, J., Hu, A.: Efficient sat solving: Beyond supercubes. In: Design Automation Conference, pp. 744–749. Anaheim, California, USA (2005)
6. Eén, N., Sörensson, N.: An extensible sat-solver. In: Proceedings of SAT, pp. 502–518. Santa Margherita Ligure, Italy (2003)

7. Goldberg, E.: Proving unsatisfiability of CNFs locally. Journal of Automated Reasoning **28**(4), 417–434 (2002)
8. Goldberg, E.: A decision-making procedure for resolution-based sat-solvers. In: Proceedings of SAT-08, pp. 119–132. Guangzhou, China (2008)
9. Goldberg, E.: Boundary points and resolution. In: Proceedings of SAT-09, pp. 147–160. Swansea, Wales, United Kingdom (2009)
10. Goldberg, E., Novikov, Y.: Berkmin: A fast and robust sat-solver. Discrete Applied Mathematics **155**(12), 1549–1561 (2007). Doi: http://dx.doi.org/10.1016/j.dam.2006.10.007
11. Goldberg, E., Prasad, M., Brayton, R.: Using problem symmetry in search based satisfiability algorithms. In: Proceedings of DATE '02, pp. 134–141. Paris, France (2002)
12. Marques-Silva, J., Sakallah, K.: Grasp—A new search algorithm for satisfiability. In: Proceedings of ICCAD-96, pp. 220–227. Washington, DC (1996)
13. Moskewicz, M., Madigan, C., Zhao, Y., Zhang, L., Malik, S.: Chaff: Engineering an efficient sat solver. In: Design Automation Conference-01, pp. 530–535. New York, NY (2001). Doi: http://doi.acm.org/10.1145/378239.379017
14. Nadel, A.: Backtrack search algorithms for propositional logic satisfiability: Review and innovations. Master's thesis, The Hebrew University (2002)
15. Zhang, H.: Sato: An efficient propositional prover. In: Proceedings of CADE-97, pp. 272–275. Springer, London (1997)
16. Ryan, L.: The siege sat-solver. http://www.cs.sfu.ca/~cl/software/siege (2010)
17. Goldberg, E.: Benchmarks. http://eigold.tripod.com/benchmarks/book2010.tar.gz (2010)

Chapter 8
SAT Sweeping with Local Observability Don't-Cares

Qi Zhu, Nathan B. Kitchen, Andreas Kuehlmann, and Alberto Sangiovanni-Vincentelli

Abstract Boolean reasoning is an essential ingredient of electronic design automation. AND-INVERTER graphs (AIGs) are often used to represent Boolean functions but have a high degree of structural redundancy. SAT sweeping is a method for simplifying an AIG by systematically merging graph vertices from the inputs toward the outputs using a combination of structural hashing, simulation, and SAT queries. Due to its robustness and efficiency, SAT sweeping provides a solid algorithm for Boolean reasoning in functional verification and logic synthesis. In previous work, SAT sweeping merges two vertices only if they are functionally equivalent. In this chapter we present a significant extension of the SAT-sweeping algorithm that exploits local observability don't-cares (ODCs) to increase the number of vertices merged. We use a novel technique to bound the use of ODCs and thus the computational effort to find them, while still finding a large fraction of them. Our reported results based on a set of industrial benchmark circuits demonstrate that the use of ODCs in SAT sweeping results in significantly more graph simplification with great benefit for Boolean reasoning with a moderate increase in computational effort.

8.1 Introduction

Boolean reasoning is a key part of many tasks in computer-aided circuit design, including logic synthesis, equivalence checking, and property checking. Circuit graphs such as AND-INVERTER graphs (AIGs) [8] are often used to represent

Q. Zhu (✉)
Intel Corporation, Hillsboro, OR, USA
e-mail: qi.dts.zhu@intel.com

This work is based on an earlier work: SAT sweeping with local observability don't-cares, in Proceedings of the 43rd Annual Design Automation Conference, ISBN:1-59593-381-6 (2006) © ACM, 2006. DOI= http://doi.acm.org/10.1145/1146909.1146970.
This chapter is an extended version of [20]. The added content includes examples, a proof, details of the algorithm flow and implementation, discussion of applications, and explanation of the experimental results.

S.P. Khatri, K. Gulati (eds.), *Advanced Techniques in Logic Synthesis, Optimizations and Applications*, DOI 10.1007/978-1-4419-7518-8_8,

Boolean functions because their memory complexity compares favorably with other representations such as binary decision diagrams. In many Boolean reasoning problems, circuit graphs have a high degree of structural redundancy [9]. The redundancy can be reduced by the application of *SAT sweeping* [7]. SAT sweeping is a method for simplifying an AND-INVERTER graph by systematically merging graph vertices from the inputs toward the outputs using a combination of structural hashing, simulation, and SAT queries. Due to its robustness and efficiency, SAT sweeping provides a solid algorithm for Boolean reasoning in functional verification and logic synthesis.

In previous work, SAT sweeping merges two vertices only if they are functionally equivalent. However, functional differences between vertices are not always observable at the outputs of a circuit. This fact can be exploited to increase the number of vertices merged. In this chapter we present a significant extension of the SAT-sweeping algorithm that uses observability don't-cares (ODCs) for greater graph simplification. Taking observability into account not only increases the effectiveness of SAT sweeping but also increases its computational expense. In order to find a balance between effectiveness and efficiency, we introduce the notion of *local observability*, in which only paths of bounded length are considered.

When observability is taken into account, the equivalence-class refinement approach of the original SAT-sweeping algorithm cannot be used. We introduce a new method of comparing simulation vectors to identify merging candidates. Although the number of comparisons is theoretically quadratic, on average near-linear complexity is observed in our results.

SAT sweeping interleaved with SAT queries to check a property can be utilized as an efficient engine for equivalence and bounded property checking [7, 9] (see also Section 8.4.5). Similarly, SAT sweeping can be applied to perform the equivalence-class refinement in van Eijk's algorithm for sequential equivalence checking [4]. In each iteration it can prove or disprove the functional equivalences of the candidates. In all these applications, the use of ODCs for SAT sweeping described in this chapter can increase the number of vertices merged and thus improve their overall reasoning power.

This chapter is structured as follows: Previous work is discussed in Section 8.2. Section 8.3 introduces some preliminary concepts, including AIGs and SAT sweeping. Section 8.4 explains the details of SAT sweeping with local ODCs. Section 8.5 presents the experimental results, and Section 8.6 contains conclusions.

8.2 Previous Work

The combined application of random simulation and satisfiability queries for finding functionally equivalent circuit components has been proposed in multiple publications [2, 6, 10–12]. A particular implementation on AIGs that combines these methods with structural hashing and circuit rewriting was described in [7] for simplifying the transition relation for bounded model checking (BMC).

In [16] the use of observability don't-cares was proposed to simplify the functionality of internal circuit nodes by exploiting non-observability of their input assignments [16]. Using BDDs for their representation, the don't-cares are systematically computed from the outputs toward the inputs and then applied for node resynthesis. Recently, SAT-based methods have been suggested for the same application [13, 14]. To reduce the huge computational cost of computing ODCs, existing methods use compatible ODCs [3, 18] to avoid frequent recomputations or windowing [17] to limit the problem size. In contrast, the ODC-based SAT sweeping method presented in this chapter encodes observability conditions directly in the check for mergeability of two vertices. Random simulation is utilized to effectively reduce the number of merge candidates. Furthermore, our approach is made robust by limiting the path length considered for observability.

Other previous work utilizes ODCs for CNF-based SAT checking. In [15] additional clauses which encode ODC conditions are added to the CNF formula of the circuit with the goal to improve the solver performance. In [5], a similar approach is presented that also introduces additional literals. The aim of both these approaches is to speed up the SAT search, whereas our method is targeted to simplify the circuit representation itself. This simplification can be used in multiple application domains, including equivalence checking [9], property verification [7], or logic synthesis.

8.3 Preliminaries

For completeness, this section briefly outlines the concept of AND-INVERTER graphs and the SAT-sweeping algorithm.

8.3.1 AND-INVERTER *Graphs*

Let $C = (V, E)$ denote an AIG with the set of vertices $V = \{v_0\} \cup X \cup G$ and set of directed edges $E \subseteq V \times V$. Vertex v_0 represents the logical constant 0, X is the set of inputs, G is the set of AND gates, and $O \subseteq G$ is the set of outputs. v_0 and all inputs $v \in X$ have no predecessors, whereas the gates $g \in G$ have exactly two incoming edges denoted by $right(g)$ and $left(g)$. Let $ref(e)$ denote the source vertex of edge e and $FO(v)$ be the set of successor vertices of v, i.e., $v' \in FO(v) \Leftrightarrow (v, v') \in E$. $TFO(v)$ refers to the set of vertices in the transitive fanout of v. Moreover, let $other(v', v)$ denote the incoming edge of vertex $v' \in FO(v)$ which does not come from v, i.e.,

$$other(v', v) = \begin{cases} right(v') \text{ if } v = ref(left(v')) \\ left(v') \text{ if } v = ref(right(v')) \end{cases} \tag{8.1}$$

For the definition of the AIG semantics, we assume that each input $v \in X$ is assigned a Boolean variable x_v and each gate $v \in G$ computes the conjunction of the incoming edge functions. For each edge $e \in E$ an attribute $inv(e)$ is used to indicate whether the function of the source vertex is complemented. More formally, the functions of a vertex $v \in V$ and an edge $e \in E$ of an AIG are defined as follows:

$$f(v) = \begin{cases} 0 & \text{if } v = v_0 \\ x_v & \text{if } v \in X \\ f(\mathit{left}(v)) \wedge f(\mathit{right}(v)) & \text{otherwise} \end{cases} \tag{8.2}$$
$$f(e) = f(\mathit{ref}(e)) \oplus \mathit{inv}(e)$$

Figure 8.1 illustrates part of an AIG. Inverted edges are shown with dots. v_1, v_2, v_3, and v_4 are inputs; their functions are x_1, x_2, x_3, and x_4, respectively. The function of v_9 is $(\neg x_1 \wedge x_2) \wedge x_3 = \neg x_1 \wedge x_2 \wedge x_3$. The function of the inverted edge (v_{10}, v_{13}) is $\neg((\neg x_1 \wedge x_2) \wedge (x_2 \wedge x_3)) = \neg(\neg x_1 \wedge x_2 \wedge x_3) = x_1 \vee \neg x_2 \vee \neg x_3$. Note that $f((v_{10}, v_{13}))$ is also the complement of v_9's function, so this AIG is redundant.

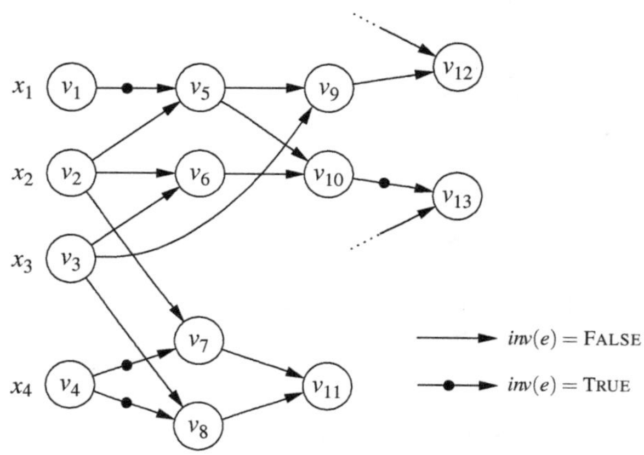

Fig. 8.1 Example of AND-INVERTER graph: $f(v_1) = x_1$, $f(v_2) = x_2$, $f(v_3) = x_3$, $f(v_4) = x_4$, $f(v_9) = (\neg x_1 \wedge x_2) \wedge x_3$; $f(v_{11}) = (x_2 \wedge \neg x_4) \wedge (x_3 \wedge \neg x_4)$; $f((v_{10}, v_{13})) = \neg(\neg x_1 \wedge x_2) \wedge (x_2 \wedge x_3)$

8.3.2 SAT Sweeping

Algorithm 2 outlines the basic flow of the SAT-sweeping algorithm as presented in [7]. The goal of the algorithm is to systematically identify all pairs of functionally equivalent vertices and merge them in the graph. For this the algorithm uses a classical partition refinement approach using a combination of functional simulation, SAT queries, and structural hashing. For simplicity, Algorithm 2 does not outline the handling of equivalence modulo complementation, i.e., the assignment of two vertices to the same equivalence class if their simulation vectors are identical or complements.

In Algorithm 2 we denote by $F(v)$ the simulation vector of vertex v. The value of $F(v)$ is computed by bitwise application of the semantics given in (8.2).

During the first iteration of the outer loop the randomly initialized input vectors are simulated to generate a first partitioning of the graph vertices into equivalence classes $\{V_1, V_2, \ldots, V_c\}$. To refine the initial partitioning, a SAT solver is applied to check for functional equivalence of the shallowest vertices in all shallow classes containing more than one vertex. If they are indeed functionally equivalent, they are merged. Otherwise, the simulation vectors are updated with the counterexample given by the satisfying input values. This ensures that this class is broken apart during the next iteration. The reader is referred to [7] for more implementation details on the SAT-sweeping algorithm.

Algorithm 2 Basic SAT Sweeping

```
{Given: AIG C = (V, E) with inputs X}
randomly initialize simulation vectors F(x) for all x ∈ X
Classes := {V}   {Initially all vertices in single class}
loop
   simulate C to update all simulation vectors F(v)
   refine all classes V_i s.t. ∀u, v ∈ V_i : F(u) = F(v)
   if ∀i : |V_i| = 1 ∨ (UsedResource > ResourceLimit) then
      return
   else
      for all shallow classes i with |V_i| > 1 do
         v := argmin_{v'∈V_i} LEVEL(v')
         u := argmin_{u'∈V_i\{v}} LEVEL(u')
         res := SAT-CHECK (f(u) ⊕ f(v))
         if res = SAT then
            extend simulation vectors F(x) by SAT counterexample
         else if res = UNSAT then {u and v equivalent}
            MERGE (u, v)
            remove u from V_i   {v is representative of u in V_i}
         end if
      end for
   end if
end loop
```

When SAT sweeping is applied to the AIG in Fig. 8.1, v_9 and v_{10} are identified as functionally equivalent and merged, resulting in the graph shown in Fig. 8.2. The

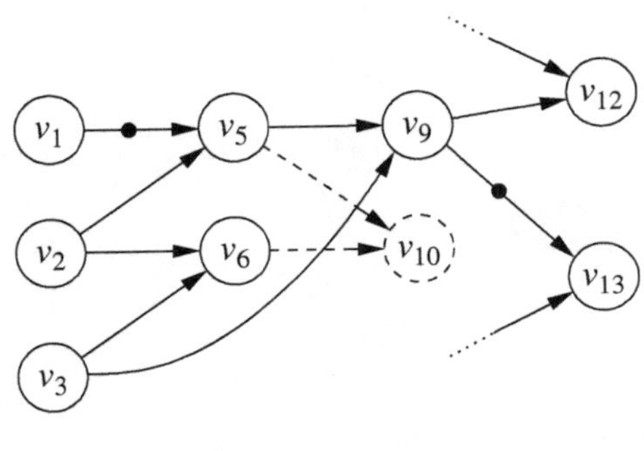

Fig. 8.2 Subgraph of AIG from Fig. 8.1 after SAT sweeping: v_{10} is merged onto v_9

outputs of v_{10} are moved to v_9. The rest of the AIG is unchanged; no other vertices can by merged by basic SAT sweeping.

8.4 SAT Sweeping with Observability Don't Cares

8.4.1 Motivating Example

Functional equivalence is a sufficient condition to prevent changes in overall circuit behavior when vertices are merged. However, it is not a necessary condition. Some functional differences between vertices are never observed at the outputs of the circuit. For example, in Fig. 8.1 v_6 and v_8 are not functionally equivalent, since $f(v_6) = x_2 \wedge x_3$ and $f(v_8) = x_3 \wedge \bar{x}_4$. However, their values only differ when $x_3 = 1$ and $x_2 = x_4$, which implies $f(v_7) = 0$. Since v_7 provides a controlling value to the input of v_{11}, this difference is not observable for v_8. Therefore, v_8 can safely be merged onto v_6.

8.4.2 Observability Don't Cares

For a given AIG, the *observability* $obs(v)$ of a vertex $v \in V$ that has no reconverging path in its transitive fanout, denoted by $TFO(v)$, is defined as follows:

$$obs(v) = \begin{cases} 1 & \text{if } v \in O \\ \bigvee_{v' \in FO(v)} \big(obs(v') \wedge f(other(v', v))\big) & \text{otherwise} \end{cases} \quad (8.3)$$

In other words, the value of vertex v is not observable iff for each of its fanouts either the value of the fanout vertex itself is not observable or its other input evaluates to 0, thus blocking the logical path.

In the presence of reconvergent paths beginning at v, the recursive computation of $obs(v)$ given in (8.3) may lead to incorrect results [18]. It may exclude input assignments for which a change of the value at v can propagate to an output through simultaneous switching of multiple paths. An example is shown in Fig. 8.3, where

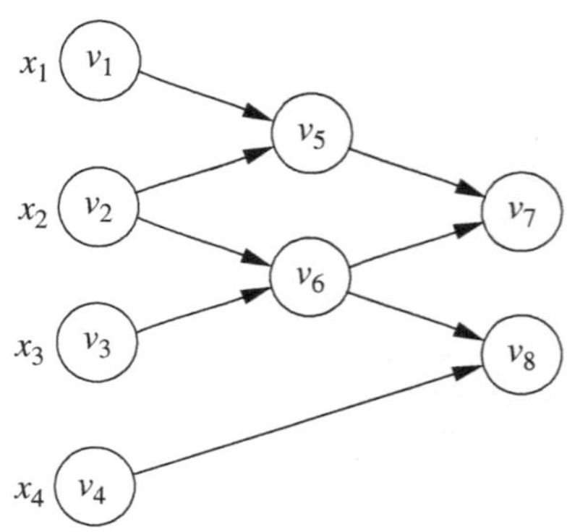

Fig. 8.3 Example of AIG for which (8.3) is incorrect

vertex v_2 has two paths to vertex v_7. Assuming $v_7, v_8 \in O$, the observability of v_2 according to (8.3) is computed as follows:

$$
\begin{aligned}
obs(v_2) &= (obs(v_5) \wedge x_1) \vee (obs(v_6) \wedge x_3) \\
&= (obs(v_7) \wedge f(v_6) \wedge x_1) \\
&\quad \vee \Big(\big((obs(v_7) \wedge f(v_5)) \vee (obs(v_8) \wedge f(v_4))\big) \wedge x_3\Big) \\
&= (1 \wedge x_2 \wedge x_3 \wedge x_1) \vee \Big(\big((1 \wedge x_1 \wedge x_2) \vee (1 \wedge x_4)\big) \wedge x_3\Big) \\
&= (1 \wedge x_2 \wedge x_3 \wedge x_1) \vee (1 \wedge x_1 \wedge x_2 \wedge x_3) \vee (1 \wedge x_4 \wedge x_3) \\
&= (x_1 \wedge x_2 \wedge x_3) \vee (x_3 \wedge x_4)
\end{aligned}
$$

The result implies that when $x_2 = x_4 = 0$, v_2 is not observable. This is incorrect; a change from $x_2 = 0$ to $x_2 = 1$ actually is observable when $x_1 = x_3 = 1$.

To avoid false ODCs of this kind, we use the following modified definition of $obs(v)$ which overapproximates the observability for reconverging structures:

$$
\begin{aligned}
obs(v) &= \overline{obs}(v, v) \\
\overline{obs}(v, u) &= \begin{cases} 1 & \text{if } v \in O \\ \bigvee_{v' \in FO(v)} \big(\overline{obs}(v', u) \wedge g(other(v', v), u)\big) & \text{otherwise} \end{cases} \qquad (8.4) \\
g(v, u) &= \begin{cases} 1 & \text{if } v = u \text{ or } v \in TFO(u) \\ f(v) & \text{otherwise} \end{cases}
\end{aligned}
$$

By using the auxiliary function $g(v, u)$, we sensitize paths from v conservatively: Effectively, we prevent any vertex in the transitive fanout of v from blocking the propagation of v's value. When we apply this definition to the example in Fig. 8.3, we get the following:

$$
\begin{aligned}
obs(v_2) &= \overline{obs}(v_2, v_2) \\
&= (\overline{obs}(v_5, v_2) \wedge x_1) \vee (\overline{obs}(v_6, v_2) \wedge x_3) \\
&= ((\overline{obs}(v_7, v_2) \wedge g(v_6, v_2)) \wedge x_1) \\
&\quad \vee \Big(\big((\overline{obs}(v_7, v_2) \wedge g(v_5, v_2)) \vee (\overline{obs}(v_8, v_2) \wedge x_4)\big) \wedge x_3\Big) \\
&= (1 \wedge 1 \wedge x_1) \vee \Big(\big((1 \wedge 1) \vee (1 \wedge x_4)\big) \wedge x_3\Big) \\
&= (1 \wedge 1 \wedge x_1) \vee (1 \wedge x_3) \vee (1 \wedge x_4 \wedge x_3) \\
&= x_1 \vee x_3 \vee x_4
\end{aligned}
$$

The result is an overapproximation to the exact observability of $(x_1 \wedge x_3) \vee (x_3 \wedge x_4)$.

The concept of *k-bounded observability* or *local observability* is based on limiting the length of the paths being considered for observability to reduce the effort for its computation. The approximate k-bounded observability, $obs(v, k)$, is defined as follows

$$obs(v,k) = \overline{obs}(v,v,k)$$

$$\overline{obs}(v,u,k) = \begin{cases} 1 & \text{if } v \in O \text{ or } k = 0 \\ \bigvee_{v' \in FO(v)} \big(\overline{obs}(v',u,k-1) \wedge g(other(v',v),u)\big) & \text{otherwise} \end{cases} \tag{8.5}$$

For example, for $k = 0$ vertex v is always observable, whereas $obs(v, 1)$ considers only the possible blockings at the immediate fanouts of v.

Clearly, every input assignment that results in $obs(v,k) = 0$ is an ODC for v. The idea of using local observability in SAT sweeping is to exploit the fact that v can be merged onto u if the functions of u and v are equal for all k-bounded observable input assignments, i.e., u and v have to be equal only if $obs(v,k) = 1$. More formally

Theorem 8.1 *For a given k, vertex v can be merged onto vertex u if, for every input assignment,*

$$f(u) = f(v) \vee obs(v,k) = 0.$$

Proof We begin by proving the unbounded version of the theorem using the definition of $obs(v)$ in (8.4).

Clearly, for input assignments with $f(u) = f(v)$, the values at the outputs do not change. For each remaining input assignment, suppose that $obs(v) = 0$. If we expand the recursion in (8.4) and distribute the conjunctions, we get a formula in disjunctive normal form with a disjunct for every path from v to an output. Formally, let $\mathcal{P}(v)$ be the set of paths from v to the vertices in O, where each path $p \in \mathcal{P}(v)$ is a sequence of edges. Then the expansion is

$$obs(v) = \bigvee_{p \in \mathcal{P}(v)} \bigwedge_{(y,z) \in p} g(other(z,y),v) \tag{8.6}$$

Since $obs(v) = 0$, the value of the disjunct for each path must be 0, so each path contains some edge (y,z) such that $g(other(z,y),v) = 0$. That is, the side input of z has value 0. Let w denote the side input; i.e., $w = other(z,y)$. By the definition of g, the vertex w cannot be in the transitive fanout of v or else $g(w,v)$ would be 1. Therefore, the value of w does not depend on v, and its value is not changed by merging v to u. Both before and after the merge, w blocks propagation along the path. Since every path from v to an output is blocked by some such w, the outputs must all have the same values after merging as before.

The argument using the bounded observability is the same. The only difference is that the expanded formula has only the first k terms from each conjunction in (8.6), so each path must be blocked within k levels from v. □

8.4.3 Algorithm

The criterion for merging two vertices exploiting observability don't-cares as given in Theorem 8.1 can be evaluated using a combination of random simulation and SAT queries, similar to the criterion of functional equivalence used in Algorithm 2. The challenge here is that the equivalence-class refinement method cannot be used, because the merging criterion of Theorem 8.1 is not an equivalence relation: It is neither symmetric nor transitive, so it is not possible to define equivalence classes of vertices such that each member of a class can be merged onto any other member. For example, in the graph in Fig. 8.1, v_8 can be merged onto v_6 as stated in Section 8.4.1, but v_6 cannot be merged onto v_8–There is an input assignment, (0,1,1,1), for which $f(v_6) = 1$ and $f(v_8) = 0$, while $f(v_5) = 1$, a non-controlling value. If v_6 were merged onto v_8, the value of $f(v_{10})$ would be changed for this assignment.

Instead of selecting candidate pairs for merging from equivalence classes, it is necessary to find independently for each vertex v the vertices onto which it may be merged. Possible candidates include vertices whose simulation vectors are identical to v's, but there may also be candidates whose simulation vectors differ from v's, as long as v is not observable for the input assignments resulting in the differing bits. In order to take observability into account when comparing simulation vectors, we compute *observability vectors* $OBS(v, k)$ from the simulation vectors $F(v)$ as follows:

$$OBS(v,k) = \begin{cases} [111\ldots11] & \text{if } v \in O \text{ or } k = 0 \\ \bigvee_{v' \in FO(v)} \big(OBS(v', k-1) \wedge F(other(v', v))\big) & \text{otherwise} \end{cases} \tag{8.7}$$

For example, in Fig. 8.1 if $F(v_2) = 110011$, $F(v_5) = 010010$, and $F(v_4) = 010101$, then the 1-bounded observability vector of v_3 is the following:

$$\begin{aligned} OBS(v_3, 1) &= (OBS(v_6, 0) \wedge F(v_2)) \\ &\vee (OBS(v_9, 0) \wedge F(v_5)) \\ &\vee (OBS(v_8, 0) \wedge \neg F(v_4)) \\ &= F(v_2) \vee F(v_5) \vee \neg F(v_4) \\ &= 111011 \end{aligned} \tag{8.8}$$

The bits of $OBS(v, k)$ represent the values of $obs(v, k)$ for the input assignments in the simulation vectors. When the bit at position i of $OBS(v, k)$ (denoted as $OBS(v, k)[i]$) is 1, the value of $F(v)[i]$ is observable, and it should be considered when comparing $F(u)$ to $F(v)$. When $OBS(v, k)[i] = 0$, we ignore any difference between $F(u)[i]$ and $F(v)[i]$. When $F(u)[i] = F(v)[i]$ or $OBS(v)[i] = 0$ for all i, we say that $F(u)$ is *compatible* with $F(v)$.

Note that for the bits of $OBS(v, k)$ we use an underapproximation of $obs(v, k)$ which may be incorrect for reconverging paths. However, as shown later in

Algorithm 3 SAT Sweeping with Local ODCs

```
{Given: AIG C = (V, E) with inputs X; bound k}
randomly initialize simulation vectors F(x) for all x ∈ X
simulate C to compute simulation vectors F(v) for all v ∈ V
compute observability vectors OBS(v, k) for all v ∈ V
for level L := 1 to L_max do
    build dictionary D for vertices at levels 1, ..., L
    for all v ∈ V such that LEVEL(v) = L do
        for all u ∈ SEARCH(D, F(v), OBS(v, k)) do
            if (F(u) ⇔ F(v)) ∨ ¬ OBS(v, k) = 1 {bitwise} then
                res := SAT-CHECK ((f(u) ⊕ f(v)) ∧ obs(v, k))
                if res = SAT then
                    add SAT counterexample to vectors F(x)
                    simulate C to update vectors F(v) and OBS(v, k)
                else if res = UNSAT then
                    MERGE (v, u)
                    simulate C to update vectors F(v) and OBS(v, k)
                    go to next v
                end if
            end if
        end for
    end for
end for
```

Algorithm 3, we use only the vectors as an initial filter for finding candidates for merging. Any false compatibilities are caught when we check the candidates against the criterion in Theorem 8.1.

In order to identify vertices with compatible simulation vectors, we build a dictionary of simulation vectors and utilize an efficient search routine that compares subsequences of the vectors using observability vectors as masks.

The overall flow of SAT sweeping with local ODCs is shown in Algorithm 3 and illustrated in Fig. 8.4. After initializing the simulation and observability vectors, we iterate through the vertices in order of their levels (i.e., their depths from the inputs). At each level, we rebuild the dictionary to contain the vertices at that level and below. For each vertex v, the SEARCH routine returns the vertices in the dictionary whose simulation vectors are compatible with $F(v)$. For each compatible candidate u, we check the criterion of Theorem 8.1 using a SAT query. If the SAT solver finds a violation of the criterion, the input assignment that distinguishes u and v is added to the simulation vectors. If the criterion holds, the vertices are merged. Because u and v are not functionally equivalent, the merge changes the functions of the vertices formerly in the transitive fanout of v. The functional differences are not themselves observable, but the change in graph structure can change the observability of other vertices. Therefore, we must update the simulation vectors.

The choice to iterate over the vertices in level order is motivated by a safety constraint. If a vertex v is merged onto a vertex u in its transitive fanout, a cycle will be created in the AIG. The original SAT-sweeping algorithm avoids

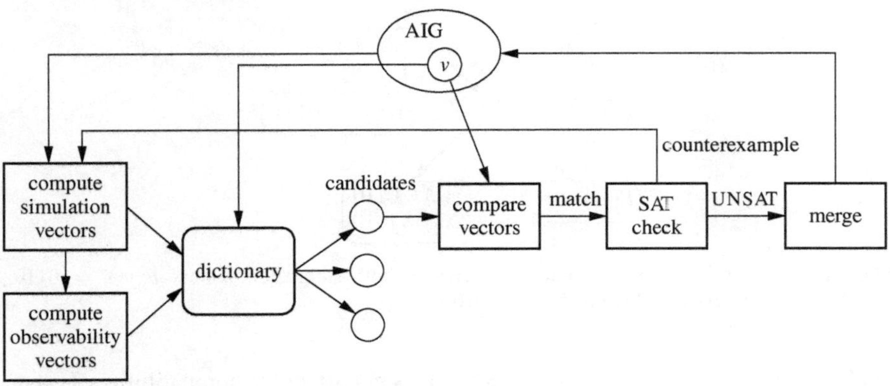

Fig. 8.4 Illustration of control and data flow for Algorithm 3

this case conservatively by comparing the levels of the vertices: v can be merged onto u only if $\text{LEVEL}(v) \geq \text{LEVEL}(u)$. We achieve the same effect by our order of iteration and by limiting the search for merging candidates to vertices at the current level and below. This choice also provides a benefit for efficiency: Because the search dictionary does not include all the vertices, the search cost is reduced.

The theoretical complexity of our algorithm is dominated by the SAT check, which can be exponential in the size of the circuit, regardless of the choice of observability bound k. In practice, a larger k usually leads to more merging candidates and larger SAT formulas due to more side inputs for the observability check. This results in significantly greater runtime, as shown in our experimental results in Section 8.5. Additionally, even if the SAT checks took linear time, our algorithm would have complexity cubic in the size of circuit, since for each vertex being swept the number of the merging candidates can potentially be of the order of the circuit size.

8.4.4 Implementation

To implement the search dictionary, we use a binary trie. For each level d of the trie, there is an associated bit index $i(d)$. Each internal node of the trie has two children. The two sub-tries rooted at the children of a node at level d contain AIG vertices whose simulation vectors have value 0 or 1 at bit position $i(d)$, respectively. Each leaf node corresponds to a bin of vertices. All the vertices in a bin have the same values for the sub-vectors indexed by the branching bit indices $i(1)$, $i(2)$, etc. Figure 8.5 illustrates the trie structure.

To search the trie for vertices with compatible simulation vectors, we use a recursive routine, shown in Algorithm 4. At level d, if $F(v)[i(d)]$ is observed, it is used

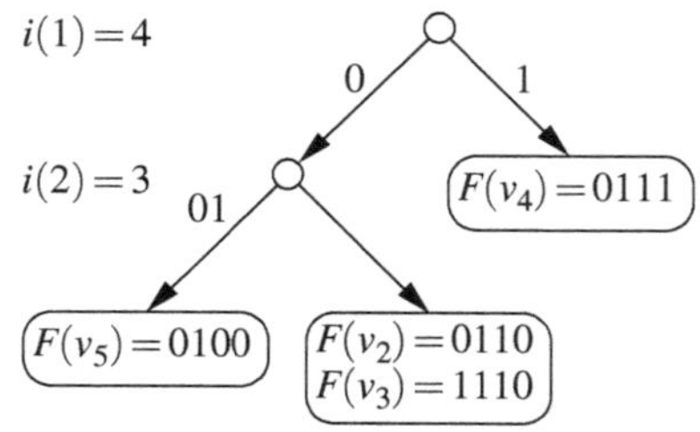

Fig. 8.5 Illustration of trie containing AIG vertices with simulation vectors $F(v_2) = 0110$, $F(v_3) = 1110$, $F(v_4) = 0111$, and $F(v_5) = 0100$

to choose a branch. If it is not observed, it does not affect the compatibility of other vectors with $F(v)$, so both branches are followed.

The efficiency of searching in the trie depends on the choice of branching bit indices. An uneven distribution of the vertices among the leaves will increase the complexity of the algorithm. In the extreme case, if all the vertices are in a single bin, then every pair of vertices is considered for merging, and the total number of SAT checks is quadratic in the graph size. To prevent this from happening, we set an upper limit on the size of the bins. Whenever a bin exceeds the size limit, we insert a new branch node. As a consequence, the number of branches along the path to the leaves increases. To keep the trie as shallow as possible for fast searches, the branching bits could be chosen so as to distribute the vertices evenly among the leaves.

Algorithm 4 Search(T, $F(v)$, $OBS(v,k)$, d)

```
{Given: (sub-)trie T; simulation vector F(v); observability vector OBS(v,k), depth d}
if T is a leaf then
    return V_T {vertices in bin of T}
else
    if OBS(v,k)[i(d)] = 1 then
        b := F(v)[i(d)]
        return SEARCH(T_b, F(v), OBS(v,k), d+1))
    else
        return SEARCH(T_0, F(v), OBS(v,k), d+1)) ∪ SEARCH(T_1, F(v), OBS(v,k), d+1))
    end if
end if
```

Alternatively, the branching bit indices may be chosen to minimize the number of don't care bits indexed during searches, since the number of leaves reached in a search is exponential in the number of unobservable bits indexed. In our implementation we chose this scheme, since we did not observe highly unbalanced distributions of vertices in our experiments. To minimize the number of unobservable bits in searches, we select the indices with a straightforward heuristic: For each index i, we compute the number of observed bits in the simulation vectors with index i, i.e., $\sum_v OBS(v,k)[i]$, and use the bits with the largest sums. Because we rebuild the trie for each level of the graph, the choice of branching bit indices can vary as the algorithm progresses.

The performance of SAT sweeping with local ODCs can be tuned by adjusting the maximum number of vertices stored in any leaf of the trie or several other parameters, including the initial number of simulation bits and the maximum number of backtracks per SAT query.

Updates to the simulation vectors are needed both when vertices are merged and when a SAT query proves that vertices are not mergeable. In the first case, the only vectors affected are those formerly in the transitive fanout of the merged vertex. (In fact, only k levels of transitive fanout are affected.) We take advantage of this fact by limiting vector propagation to the local fanout in order to minimize the cost of simulation vector updates.

We use another form of incremental simulation vector update when the SAT solver provides a new input assignment. In this case, the values propagated from previous input assignments are still valid, so there is no need to propagate them again. Instead, we update only the bits at the ends of the simulation vectors.

The efficiency of the algorithm is affected by the ability of the input assignments in the simulation vectors to sensitize paths through the graph. If the randomly selected inputs do not sensitize many paths, many of the simulation values will be unobservable. This increases the cost of search and the number of SAT checks needed. One strategy to avoid this problem is to populate the simulation vectors with input assignments that are known to sensitize paths in the graph. However, in our preliminary experiments using random long paths, this strategy did not significantly improve performance.

As presented in Section 17.5.1, our algorithm sweeps all vertices in the graph. In some applications, such as bounded model checking, different subsets of the vertices are swept at different times. In fact, additional vertices may be added to the graph after each sweep. To support this usage, our implementation allows a set of vertices to be explicitly specified and tries to merge only these vertices, not the entire graph.

8.4.5 Applications

This section describes how SAT sweeping with local ODCs interacts with other procedures in two application settings, bounded model checking and equivalence checking.

Figure 8.6 shows the application of SAT sweeping in bounded model checking for simplification of the unrolled frames. First, the design (typically given as a module in Verilog) is translated to an AND-INVERTER graph. Initially the graph contains only the first frame. It is simplified by SAT sweeping and the property of interest is checked by a SAT solver. Next, the second frame is unrolled from the simplified graph for the first. The SAT-sweeping engine tries to merge vertices in the second frame onto vertices from either of the first two frames, and the property is checked again. The third frame is unrolled in the same manner as the second and so on.

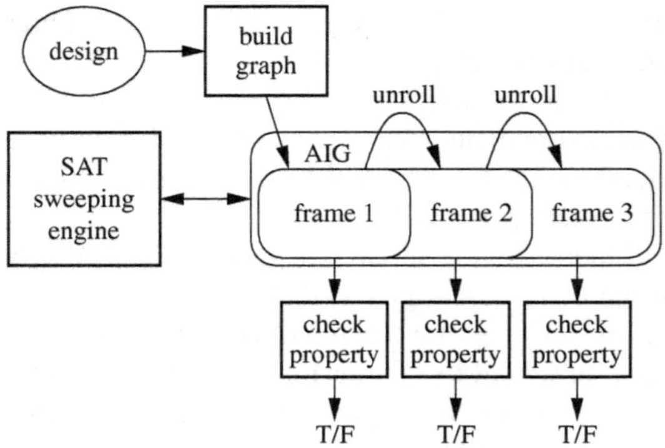

Fig. 8.6 Application of SAT sweeping to bounded model checking

The number of additional vertices needed for each frame decreases because of the simplifications from SAT sweeping, resulting in reduced effort for each property check.

SAT sweeping can also be applied to combinational equivalence checking, as shown in Fig. 8.7. The miter structure for the two designs is built as a single AIG, and the graph is simplified by SAT sweeping before pairs of outputs are checked for equivalence with a SAT solver. The best total runtime is often achieved by interleaving the SAT checks on the outputs with SAT sweeping, gradually increasing the resource limit on each procedure. Counterexamples from the equivalence checks can be used to augment the simulation vectors for SAT sweeping.

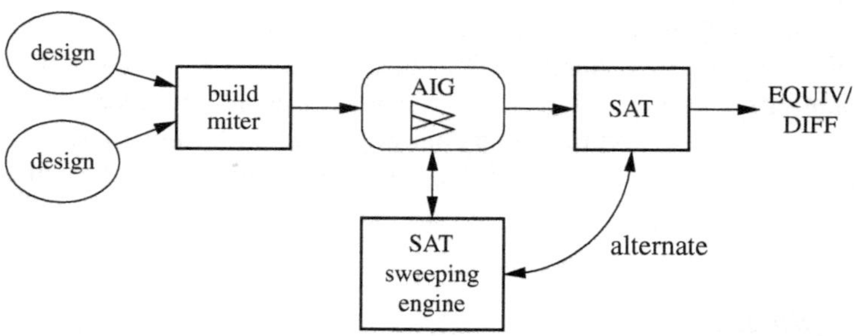

Fig. 8.7 Application of SAT sweeping to combinational equivalence checking

8.5 Results

We implemented SAT sweeping with local ODCs in OpenAccess [1] with the OpenAccess Gear toolkit [19]. For comparison, we also implemented basic SAT sweeping in the same framework. In this section we present the results of our experiments on input circuits from the IWLS 2005 benchmark set (http://iwls.org/iwls2005/benchmarks.html). In particular, we used IWLS benchmarks originating from the OpenCores repository, ranging in size from 200 to 48K AND vertices.

Our first set of experiments demonstrates the increased power of ODC-based SAT sweeping for graph simplification, the effect of varying levels of observability, and the scalability of the algorithm. In these experiments we applied ODC-based SAT sweeping to each design, varying the observability bound k from 1 to 5. As a preprocessing step, we applied basic SAT sweeping to remove functionally equivalent vertices, so that all reported merges are due to consideration of observability.

In our experiments, the initial length of the simulation vectors is set to 32 bits. The maximum bin size in the search dictionary is set to 16, i.e., at most 16 vertices are allowed in a single bin. We used MiniSat (http://www.minisat.se) for satisfiability checks.

Table 8.1 shows the number of merges for each benchmark and observability bound k. The first column gives the names of the designs. The second column lists the number of vertices in the original AIGs. The third column lists the number of vertices merged by basic SAT sweeping. The fourth to eighth columns report the number of merges for each value of k. The average percentage of vertices merged is also given. Table 8.2 reports the runtimes for these experiments. Figures 8.8 and 8.9 show plots of the results. In Fig. 8.8, the number of merges is shown as a percentage of the number of vertices in the original AIG. Figure 8.9 shows the runtimes relative to the runtime of basic SAT sweeping.

The results show that ODC-based SAT sweeping is able to merge many more vertices than basic SAT sweeping–On average, almost four times more when $k = 1$ and five times more when $k = 2$. Note that the greatest increases in the numbers of merges come with the first two levels of observability. For $k > 2$, the increases taper off quickly. The majority of the graph simplification is obtained at a moderate cost in runtime. In most cases, the runtime of ODC-based SAT sweeping with two

Table 8.1 Number of AIG vertices merged in OpenCores benchmarks (http://iwls.org/iwls2005/beachmarks.html) by basic SAT sweeping and SAT sweeping with observability to k levels

Design	AIG vertices	Number of merges for each k					
		Basic	1	2	3	4	5
steppermotordrive	215	8	20	24	27	30	31
ss_pcm	506	0	19	19	22	22	23
usb_phy	593	19	38	42	47	47	49
sasc	894	25	210	267	299	299	299
simple_spi	1183	38	266	348	383	387	388
i2c	1315	44	88	104	110	113	114
pci_spoci_ctrl	1467	125	275	352	420	439	456
systemcdes	3322	171	354	378	479	493	496
spi	4086	37	187	211	289	292	297
tv80	10026	360	945	1173	1378	1487	1609
systemcaes	13271	206	1609	2211	2464	2471	2483
ac97_ctrl	16553	49	1822	2096	2351	2403	2409
usb_funct	17755	407	1405	1680	1984	2040	2071
aes_core	22307	341	980	1032	1073	1102	1150
wb_conmax	49755	427	1538	3834	5304	5913	6189
Average % merged		2.73	10.3	12.8	14.8	15.3	15.6

Table 8.2 Runtime in seconds for basic SAT sweeping and SAT sweeping with observability to k levels on OpenCores benchmarks

	Runtime (s)					
Design	Basic	1	2	3	4	5
steppermotordrive	0.01	0.02	0.02	0.03	0.04	0.06
ss_pcm	0.01	0.02	0.04	0.04	0.05	0.06
usb_phy	0.02	0.05	0.07	0.09	0.09	0.10
sasc	0.05	0.08	0.11	0.13	0.99	1.00
simple_spi	0.09	0.17	0.24	0.27	1.35	1.46
i2c	0.07	0.29	0.39	0.41	0.88	1.04
pci_spoci_ctrl	0.32	0.61	0.76	0.93	1.05	1.42
systemcdes	0.51	1.12	1.48	1.72	2.16	2.24
spi	1.49	5.65	5.72	8.78	10.4	11.0
tv80	7.05	21.5	27.5	39.3	111	152
systemcaes	5.84	12.5	20.9	25.6	39.8	166
ac97_ctrl	5.74	11.3	18.1	22.3	32.9	35.1
usb_funct	14.1	29.8	53.2	71.8	99.6	149
aes_core	7.47	14.7	26.4	33.0	46.3	59.0
wb_conmax	21.9	58.8	115	169	196	237
Average relative time	1.00	2.40	3.48	4.41	8.39	11.3

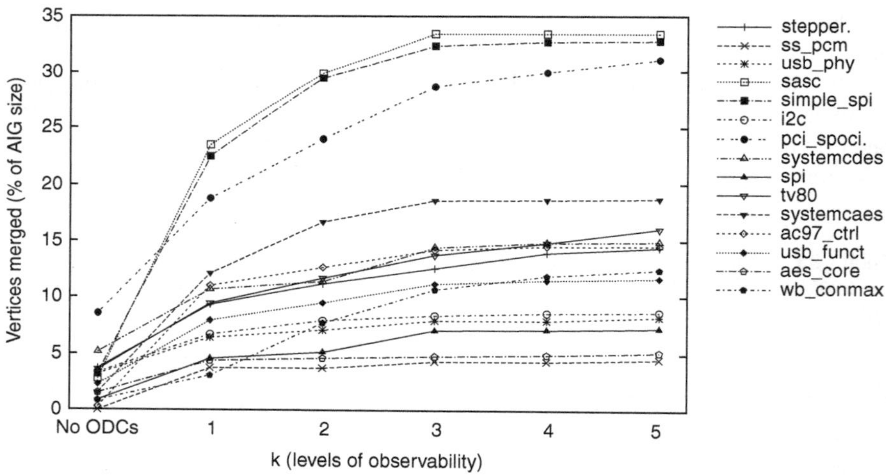

Fig. 8.8 Percentage of vertices merged in OpenCores benchmarks for each observability bound k

levels of observability is between two and four times that of basic SAT sweeping. Note also that the runtime increases linearly for low k where the greatest gains are obtained.

To determine how much optimization potential we sacrifice by our conservative approximation of observability given by (8.5), we compared our results with a SAT-sweeping approach that uses an exact computation of k-bounded observability based on Boolean difference. For $k = 2$, the number of vertices merged increased by less than 0.1% on average and at most 0.3%. For $k = 5$, the number of merges increased

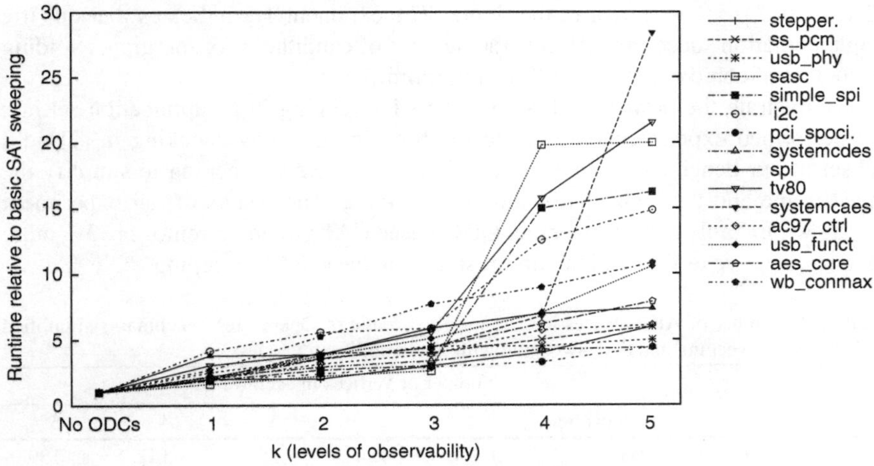

Fig. 8.9 Runtimes of SAT sweeping with ODCs relative to basic SAT sweeping for each observability bound k

by 0.5% on average and at most 3%. These results show that our approximation is very tight in practice.

Figure 8.10 shows how the runtime of ODC-based SAT sweeping increases with the size of the circuit graph for $k = 2$. To capture the trend, we applied nonlinear least squares fitting of the model ax^b to the data. The resulting curve,

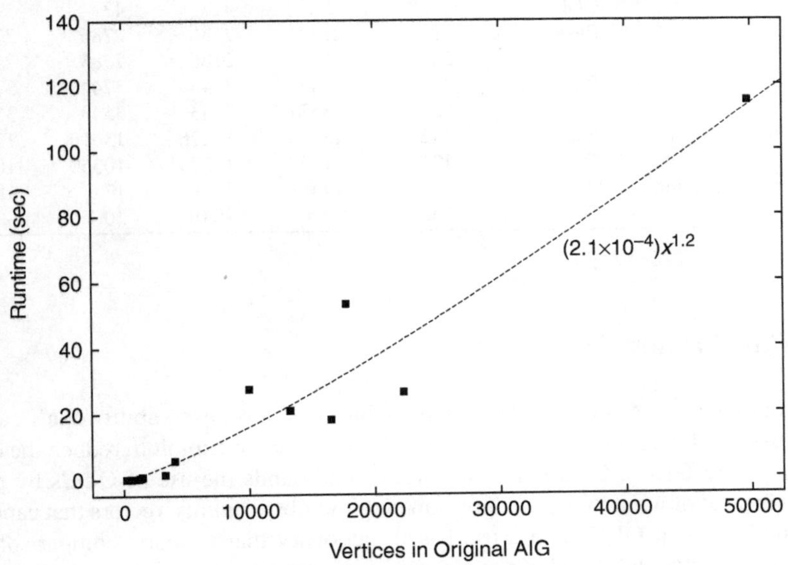

Fig. 8.10 Number of AIG vertices versus runtime of SAT sweeping with two levels of observability

$(2.1 \times 10^{-4})x^{1.2}$, is shown in the figure. The exponent 1.2 indicates that our trie implementation successfully limits the number of candidates for merging, avoiding the potential cubic complexity of the algorithm.

To illustrate the power of ODC-based SAT sweeping in an application setting, we performed experiments similar to the bounded property checking in [7] on a subset of our benchmarks. In each iteration, we ran SAT sweeping to simplify the latest frame and then unrolled it into a new frame. The results of our experiment are shown in Table 8.3. On average, ODC-based SAT sweeping removes 13% more vertices, relative to the original circuit size, than basic SAT sweeping.

Table 8.3 Number of AIG vertices in each unrolled frame of OpenCores benchmarks simplified by basic SAT sweeping and SAT sweeping with observability to two levels

		Number of vertices in each frame				
Design	Algorithm	1	2	3	4	5
steppermotordrive	Basic	177	156	152	142	139
	ODC	160	114	101	107	105
ss_pcm	Basic	398	383	363	361	361
	ODC	379	349	341	349	349
usb_phy	Basic	459	451	450	450	449
	ODC	436	429	429	429	426
sasc	Basic	734	491	487	476	468
	ODC	492	361	336	336	336
simple_spi	Basic	995	786	775	771	774
	ODC	685	552	541	530	527
i2c	Basic	1122	1084	1073	1076	1076
	ODC	1062	1005	1002	1004	1008
pci_spoci_ctrl	Basic	1255	886	704	689	687
	ODC	1028	559	417	422	397
systemcdes	Basic	2827	2787	2786	2786	2786
	ODC	2620	2592	2586	2583	2581
spi	Basic	3771	3748	3744	3744	3744
	ODC	3597	3557	3543	3543	3543
ac97_ctrl	Basic	14219	13730	13226	13010	12888
	ODC	12172	11369	10374	10360	10937
Average % reduction	Basic	3.5	14.0	16.6	17.6	18.0
	ODC	16.3	27.4	30.0	30.1	30.1

8.6 Conclusions

We proposed an extension of SAT sweeping that considers observability don't-cares. We introduced the notion of local observability, which we exploit to reduce the cost of computing ODCs. Ours is the first work that bounds the use of ODCs by path length instead of windows. Our algorithm employs observability vectors that capture information about ODCs and a trie-based dictionary that supports comparison of simulation vectors while taking don't-cares into account. Our experimental results show that SAT sweeping with two or three levels of observability finds several times more merges than basic SAT sweeping with a moderate increase in runtime and that it scales well with circuit size.

ODC-based SAT sweeping can be expected to have a great effect on various applications, such as equivalence and property checking. Miter structures in equivalence checking have many functionally equivalent points, which can be used as cutpoints for decomposition into subproblems in order to reduce the total complexity of the procedure. Our algorithm can provide more cutpoints while maintaining the locality of the subproblems because observability is tightly bounded. Similarly, for bounded model checking, ODC-based SAT sweeping can increase the number of merges during AIG compaction and thus increase its capacity and performance in comparison to non-ODC-based methods. These are just two examples of how ODC-based SAT sweeping is an effective method for increasing the power of Boolean reasoning.

References

1. Blanchard, T., Ferreri, R., Whitmore, J.: The OpenAccess coalition: The drive to an open industry standard information model, API, and reference implementation for IC design data. In: Proceedings of the International Symposium on Quality Electronic Design, pp. 69–74. San Jose, CA (2002)
2. Brand, D.: Verification of large synthesized designs. In: Proceedings of the IEEE/ACM International Conference on Computer Aided Design, pp. 534–537. Santa Clara, CA (1993)
3. Brayton, R.K.: Compatible observability don't cares revisited. In: Proceedings of the IEEE/ACM International Conference on Computer Aided Design, pp. 618–623. San Jose, CA (2001)
4. van Eijk, C.A.J.: Sequential equivalence checking without state space traversal. In: Proceedings of DATE, pp. 618–623. Paris, France (1998)
5. Fu, Z., Yu, Y., Malik, S.: Considering circuit observability don't cares in CNF satisfiability. In: Design, Automation and Test in Europe, pp. 1108–1113. Munich, Germany (2005)
6. Goldberg, E., Prasad, M.R., Brayton, R.K.: Using SAT in combinational equivalence checking. In: Proceedings of IEEE/ACM Design Automation and Test in Europe Conference and Exposition, pp. 114–121. Munich, Germany (2001)
7. Kuehlmann, A.: Dynamic transition relation simplification for bounded property checking. In: Digest of Technical Papers of the IEEE/ACM International Conference on Computer Aided Design, pp. 50–57. San Jose, CA (2004)
8. Kuehlmann, A., Krohm, F.: Equivalence checking using cuts and heaps. In: Proceedings of the 34th ACM/IEEE Design Automation Conference, pp. 263–268. Anaheim, CA (1997)
9. Kuehlmann, A., Paruthi, V., Krohm, F., Ganai, M.K.: Robust Boolean reasoning for equivalence checking and functional property verification. IEEE Transactions on Computer-Aided Design **21**(12), 1377–1394 (2002)
10. Kunz, W.: HANNIBAL: An efficient tool for logic verification based on recursive learning. In: Digest of Technical Papers of the IEEE/ACM International Conference on Computer Aided Design, pp. 538–543. Santa Clara, CA (1993)
11. Kunz, W., Stoffel, D., Menon, P.: Logic optimization and equivalence checking by implication analysis. IEEE Transactions on Computer-Aided Design of Integrated Circuits and Systems **16**(3), 266–281 (1997)
12. Lu, F., Wang, L.C., Cheng, K.T., Huang, R.C.Y.: A circuit SAT solver with signal correlation guided learning. In: Design Automation and Test in Europe, pp. 892–897. Munich, Germany (2003)

13. McMillan, K.L.: Don't-care computation using k-clause approximation. In: International Workshop on Logic Synthesis (IWLS'05). Lake Arrowhead, CA (2005)
14. Mishchenko, A., Brayton, R.K.: SAT-based complete don't-care computation for network optimization. In: Design, Automation and Test in Europe, pp. 412–417. Munich, Germany (2005)
15. Safarpour, S., Veneris, A., Drechsler, R., Lee, J.: Managing don't cares in Boolean satisfiability. In: Design, Automation and Test in Europe, p. 10260. Paris, France (2004)
16. Saldanha, A., Wang, A.R., Brayton, R.K., Sangiovanni-Vincentelli, A.L.: Multi-level logic simplification using don't cares and filters. In: Proceedings of the 26th ACM/IEEE Conference on Design Automation, pp. 277–282. Las Vegas, NV (1989)
17. Saluja, N., Khatri, S.P.: A robust algorithm for approximate compatible observability don't care (CODC) computation. In: Proceedings of the 41st Design Automation Conference, pp. 422–427. San Diego, CA (2004)
18. Savoj, H., Brayton, R.K.: The use of observability and external don't cares for the simplification of multi-level networks. In: Proceedings of the 27th ACM/IEEE Design Automation Conference, pp. 297–301. Orlando, FL (1990)
19. Xiu, Z., Papa, D.A., Chong, P., Albrecht, C., Kuehlmann, A., Rutenbar, R.A., Markov, I.L.: Early research experience with OpenAccess Gear: An open source development environment for physical design. In: Proceedings of the ACM International Symposium on Physical Design, pp. 94–100. San Francisco, CA (2005)
20. Zhu, Q., Kitchen, N., Kuehlmann, A., Sangiovanni-Vincentelli, A.: SAT sweeping with local observability don't-cares. In: Proceedings of the 43rd Design Automation Conference, pp. 229–234. San Francisco, CA (2006)

Chapter 9
A Fast Approximation Algorithm for MIN-ONE SAT and Its Application on MAX-SAT Solving

Lei Fang and Michael S. Hsiao

Abstract The current SAT solvers can provide an assignment for a satisfiable propositional formula. However, the capability for a SAT solver to return an "optimal" solution for a given objective function is severely lacking. MIN-ONE SAT is an optimization problem which requires the satisfying assignment with the minimal number of ONEs, and it can be easily extended to minimize an arbitrary linear objective function. While some research has been conducted on MIN-ONE SAT, the existing algorithms do not scale very well to large formulas. In this chapter, we propose a novel algorithm for MIN-ONE SAT that not only is efficient but also achieves a tight bound on the solution quality. Thus, it can handle very large and complex problem instances. The basic idea is to generate a set of constraints from the objective function to guide the search. The constraints are gradually relaxed to eliminate the conflicts with the original Boolean SAT formula until a solution is found. The experiments demonstrate that RelaxSAT is able to handle very large instances which cannot be solved by existing MIN-ONE algorithms; furthermore, RelaxSAT is able to obtain a very tight bound on the solution with one to two orders of magnitude speedup. We also demonstrated that RelaxSAT can be used to formulate and direct a new way to build a high-performance MAX-SAT solver.

9.1 Introduction

In the past decade, the formulation has been widely used to target a number of problems. With the help of modern-day SAT solvers, many hard EDA instances can now be handled. Although the recent successes of SAT have offered much promise, the current SAT solvers have their share of limitations. The heuristics embedded in the current SAT solvers generally steer the search to find the first solution as

---

L. Fang (✉)
Microsoft Corporation, Redmond, WA, USA
e-mail: lei.fang@microsoft.com

Based on Fang, L., Hsiao, M.S.; "A fast approximation algorithm for MIN-ONE SAT," Design, Automation and Test in Europe, 2008. DATE '08, pp.1087–1090, 10–14 March 2008. Doi: 10.1109/DATE.2008.4484921 © [2008] IEEE.

S.P. Khatri, K. Gulati (eds.), *Advanced Techniques in Logic Synthesis, Optimizations and Applications*, DOI 10.1007/978-1-4419-7518-8_9,

quickly as possible. Thus, for a satisfiable instance, one may argue that the solution returned by a SAT solver is an arbitrary one among the potentially huge number of satisfiable solutions. Furthermore, the "end-users" lack the ability to control the SAT solver to return a solution that they may favor. As a result, some applications are beyond the scope of the basic conventional SAT solver. For instance, consider the problem of generating an input sequence that can take a sequential circuit to satisfy some target such that the final state has the *least* Hamming distance from the initial state. When this sequence generation is modeled as a SAT problem on an unrolled instance of the circuit, the existing SAT solvers may be able to find a sequence that can satisfy the target objective, but they may have tremendous difficulty in obtaining the sequence that produces the final state with the least Hamming distance. Another related problem that requires an optimal solution is the problem of power-aware test generation, where the patterns generated should be generated with the minimum (or maximum) switching activities across two clock cycles [16]. In addition to these examples, other applications are emerging to demand a more intelligent SAT solver which can return an optimal (or close to optimal) solution among its solution set. **Objective functions** are commonly used to represent the optimality target. With appropriate simplification and normalization of the objective function, the objective function can often be represented as a linear function of the variable assignments. This puts the MIN-ONE SAT problem as an ideal candidate to address this class of problems.

MIN-ONE SAT can be viewed as a combination of the conventional SAT formulation and an optimization problem. It tries to find the best solution (minimal number of ONEs in the assignment) among all the possible solutions for a satisfiable Boolean formula. Note that the MIN-ONE SAT is not meaningful for unsatisfiable SAT formulas, since an unsatisfiable formula will also be MIN-ONE unsatisfiable. Formally, the MIN-ONE SAT problem is defined as follows:

> **MIN-ONE SAT Problem:** Given a formula, if it is satisfiable, find the variable assignment that contains the minimal number of ONEs.

Compared with the conventional SAT problem, MIN-ONE SAT demands more computation power because all the potential solutions need to be searched to identify the optimal one.

A naive approach to the MIN-ONE SAT problem is to compute all the possible solutions first and then identify the optimal one among all the solutions. This approach requires an all-solution SAT solver, so the overhead may render it to be infeasible for some instances. Another way is to treat the MIN-ONE SAT problem as a special case of problem [1, 18]. The basic idea and procedure behind the PBSAT approach will be explained here briefly: Using the PBSAT concept, the minimal-ONEs target can be viewed as searching the minimal sum of the assigned variable values. Suppose V_i is the value in the assignment of Boolean variable i, and suppose the total number of variables in the formula is N, then the to-be-minimized objective function for PBSAT is simply $minimize\ \sum_{i=1}^{N} V_i$. To find the minimal value, the PBSAT solver gradually tightens the minimal bound estimate until an optimum point is reached. In this approach it is easy to see that $\sum_{i=1}^{N} V_i$ can be replaced with any linear objective functions. More details about this technique will be covered in Section 14.3. In [15], a method called OPT-DLL is proposed to handle MIN-ONE SAT via a modified [11] search algorithm. Although the incorporation between DPLL algorithm and MIN-ONE SAT optimization performs very well on some benchmarks, it lacks the flexibility to process arbitrary linear objective functions. Meanwhile the performance of the DPLL algorithm may be degraded because of a forced decision order dictated by OPTSAT. The current MIN-ONE SAT algorithms are generally more suitable for small and medium-sized formulas. For large MIN-ONE SAT formulas, they often abort, unfortunately.

Due to the added complexity of the MIN-ONE SAT problem, performing a complete search to find the optimal solution is extremely hard. This observation motivated us to propose an efficient approximation algorithm for MIN-ONE SAT, which not only should be fast but also have low computational overhead and high-quality approximation of the solution. In our approach, a set of constraints from the objective function is automatically generated to guide the search. This set of constraints is gradually relaxed to eliminate any conflict(s) with the original Boolean SAT formula until a solution is found. To the best of our knowledge, this is the first approximation algorithm targeting specifically on MIN-ONE SAT to achieve a tight bound on the solution. The experimental results show that our approach can obtain a tight bound when compared with existing complete-search based methods, with one to two orders of magnitude speedup. Furthermore, we also demonstrate that the proposed RelaxSAT can be used to formulate and direct a new way to build a high-performance MAX-SAT solver.

The remainder of the chapter is organized as follows: First, previous algorithms will be discussed in Section 9.2. Then a novel approximation algorithm is proposed in Section 9.3. The experimental results are presented and explained in Section 9.4. Section 9.5 extends our algorithm to formulate a relaxation-based MAX-SAT solver. The conclusions and future works will be laid out in the last section.

9.2 Preliminaries

To help the readers better understand our proposed MIN-ONE SAT approximation algorithm, in this section we will first explain the terminology that will be used throughout this chapter, followed by the discussion of some of the existing MIN-ONE SAT algorithms.

The target formula is denoted as F_o. Suppose F_o has N variables, each of which is represented as A_i, $1 \leq i \leq N$. The value of A_i assigned by the MIN-ONE SAT solver is denoted as V_i. As mentioned before, the objective function of basic MIN-ONE SAT is

$$minimize \sum_{i=1}^{N} V_i$$

The above objective function can easily be extended to *minimize* $\sum_{i=1}^{N} c_i \times V_i$, where c_i denotes an integer coefficient corresponding to each variable value.

Listing 9.1 MIN-ONE SAT based on PBSAT

```
def MIN-ONE_SAT(F_o/*input formula*/)
begin
    i=0 /*iteration counter*/
    BE= InitBE();/*return the most pessimistic estimation*/
    while ()
        F_c=Bound_Transformation(BE);
        F_A=F_o · F_c
        Result=Call_SAT_Solver(F_A)
        if (Result == UNSATISFIABLE)
            return /*solution found*/
        else
            Update(BE); /*update based on current solution*/
end
```

We now describe a previous MIN-ONE SAT algorithm based on PBSAT. The complete algorithm is listed in Listing 9.1. To find the optimal solution, the PBSAT algorithm works in an iterative manner. In each iteration, a bound estimate BE is assigned to the objective function, ($\sum_{i=1}^{N} c_i \times V_i \leq BE$). Then, this inequality formula will be transformed into a Boolean formula F_c. There are various techniques to perform the transformation of the objective formula to a Boolean formula [13], based on the basic conversion of the inequality to a Boolean circuit. The transformation details will not be discussed here since it is out of the scope of this chapter. After obtaining F_c, a conventional SAT solver is invoked on the new formula F_A, where $F_A = F_o \cdot F_c$, whose satisfiability indicates whether the bound estimate can be achieved by the objective function. Initially, F_c is set to be a loose bound, usually set at a pessimistic estimate (e.g., set equal to N, the number of variables in the formula), and F_c is gradually tightened by decreasing the value of BE. If F_A is satisfiable, it indicates that a solution exists that satisfies both F_o and F_c, and the bound estimate, BE, formula F_c, and F_A will be updated accordingly. The algorithm stops when F_A becomes unsatisfiable. The unsatisfiability of F_A guarantees that no better solution exists and the previous bound estimate is the optimal solution.

A different MIN-ONE SAT algorithm has been proposed in [15] which is quite different from the PBSAT technique described above. In [15], the conventional DPLL algorithm is adapted to solve the SAT-related optimization problems, such as MAX-SAT [9], MIN-ONE, DISTANCE-SAT [5]. This algorithm is named as Optimization SAT (OPTSAT). Like the DPLL algorithm, OPTSAT performs a decision-based search through a recursive function. The only part that involves the optimization is in the decision heuristic. OPTSAT maintains a partial order P on all literals, and this order is used to make the next decision in the search. The order in P is defined in a way such that the objective function based on this order is monotonically increased or decreased. For example, given two variables $\{a, b\}$ there exists a P set as $\{\{\bar{a}, \bar{b}\}, \{\bar{a}, b\}, \{a, \bar{b}\}, \{a, b\}\}$. The order of this P set is ascending on the number of ONEs in the assignment. One can see that by forcing a partial order on the set of literals the solution space is searched in a specific sequence, where the potential solutions are ordered from the best to the worst. Thus, whenever a solution is obtained, the algorithm returns a solution that is guaranteed to be the optimal solution. OPTSAT achieves good performance in many benchmarks, but it also faces the problem that the forced decision order may degrade the solver performance in large formulas. It should also be noted that OPTSAT is not specifically for MIN-ONE SAT; it can solve other SAT optimization problems like MAX-SAT and DISTANCE-SAT.

Besides the above two MIN-ONE SAT algorithms, one may use an all-solution SAT solver as a base MIN-ONE solver as well. However, because the-state-of-art MIN-ONE SAT solvers are mainly built on the aforementioned algorithms, in this chapter we only compare our proposed algorithm with these two algorithms.

9.3 Our Approach

9.3.1 RelaxSAT

Our approach is fundamentally different from the existing OPTSAT and PBSAT algorithms. Instead, it is an approximation algorithm which means that the solution obtained may not be guaranteed to be optimal. It should be pointed out that our algorithm holds its value on fast solving because complete algorithms have to implicitly search all solutions to find the optimal one, which may become an excessive burden given the sheer size of solution set. Our approximation algorithm can potentially skip many search spaces, hence the computational complexity and resource usage can be significantly reduced. In addition, for many applications a low-cost near-optimal solution is more meaningful than the costly optimal solution, where our algorithm becomes a perfect fit. We call our approach as RelaxSAT because it uses a relaxation technique.

The basic idea of RelaxSAT is to generate a set of constraints from the objective function to guide the SAT solver. The detailed algorithm is shown in Listing 9.2. A set of constraints, S_c, is first generated in order to minimize (or maximize) the objection function. In other words, the constraint set defines a solution space within which the optimal value of the objective function is reached. The constraints are unit clauses that determine the corresponding variable assignments. For instance, in the basic MIN-ONE SAT, the initial constraint set, S_C, is the set of unit negative-literal clauses that forces each variable to be ZERO. Note that it is trivial to construct S_c when the objective function is a linear function of the variables. This constraint set can be represented as a conjunction of all such negative-literal clauses, resulting in a CNF constraint formula F_{sc}. It should be pointed out that F_{sc} contains at most N clauses when every variable is constrained. Then, $F_A = F_{sc} \cdot F_o$, and F_A is passed to a SAT solver to solve.

One may raise the concern that the constraint set, F_{sc}, may cause considerable conflict(s) with the original Boolean formula F_o. While this may be true, we want to take advantage of these conflicts to benefit our algorithm. Whenever one or more conflicts exist between F_{sc} and F_o, a relaxation procedure is invoked. This relaxation procedure analyzes the reason behind the conflict(s) through the trace information provided by the backbone SAT solver and removes the constraints that are most responsible to the conflicts. Generally speaking, the constraints that cause conflicts will be removed gradually, thereby *relaxing* the constraints along the way.

Listing 9.2 RelaxSAT algorithm

```
def RelaxSAT(F_o/*input formula*/)
begin
    i=0 /*iteration counter*/
    Sc= InitConstrantSet(Obj/*objective function*/);
    while ()
        FA=Fo · Fsc
        Result=Call_SAT_Solver(FA)
        if (Result == SATISFIABLE)
            return /*solution found*/
        else
            Sc=RelaxConstraints(Sc); /*Relaxation*/
end
```

Assume that F_o is satisfiable (otherwise, conventional SAT solvers are sufficient to determine its unsatisfiability), whenever F_A (i.e., $F_o \cdot F_{sc}$) is unsatisfiable, we can conclude that some clauses in F_{sc} are the reason that makes F_A unsatisfiable. In other words, whenever F_A is unsatisfiable, we can identify at least one unit clause in F_{sc} responsible for it. This allows us to formulate Theorem 9.1 as:

Theorem 9.1 *Let F_o be satisfiable. If $F_A = F_{sc} \cdot F_o$ is unsatisfiable, the UNSAT core [21] contains at least one clause from F_{sc}.*

Proof We prove this by contradiction. If the UNSAT core does not have any clause from F_{sc}, all the UNSAT core clauses will have to come from F_o. This leads to the conclusion that F_o is unsatisfiable, which conflicts with the assumption that F_o is satisfiable.

The relaxation procedure will be repeated until a solution is found. The number of iterations is bounded by the number of variables in F_o, as given in Theorem 9.2. Note that in both Theorems 9.1 and 9.2, the same symbol annotations as Listing 9.2 are used.

Theorem 9.2 *RelaxSAT will terminate within N iterations, where N is the number of variables in the formula.*

Proof From Theorem 9.1, in each iteration at least one constraint clause in F_{sc} will be removed. Since the maximal size of F_{sc} is N, so in the worst case, after N iterations F_{sc} will become empty, indicating that at this point $F_A = F_o$.

Theorem 9.2 gives an upper bound on the computational complexity of RelaxSAT. Although up to N SAT iterations may be needed, the search space for many of these iterations is generally very small, due to the constraints added. Details of the computational complexity will be furthered in Section 9.3.3.

9.3.2 Relaxation Heuristic

The basic principle of our relaxation heuristic is that the constraints that cause more conflicts should be relaxed first. The underlying assumption of this heuristic is that by removing those constraints that are more responsible for the conflict(s), we can reach a solution faster. Based on this principle, each constraint carries a *score* which represents how many conflicts it has involved. This conflict score is computed in each iteration through the trace file for the UNSAT core provided by the SAT solver[1] From these trace files, the implication tree that proves the formula unsatisfiable can be recovered [22]. Given the recovered implication tree, a backward search can be performed from the conflict to identify the conflict score of each constraint.

A simple example shown in Fig. 9.1 explains how the proposed heuristic works. In Fig. 9.1, two conflicts C1 and C2 rise due to the unit-clause constraints $(\bar{x})(\bar{y})(\bar{z})$. Based on a backward traversal, we identify that constraint $\bar{y}$ is involved in both conflicts, C1 and C2, while the other two only contribute to one conflict. Thus the conflict scores of $(\bar{x})(\bar{y})(\bar{z})$ are $\{1, 2, 1\}$. Using the obtained conflict scores, constraint $(\bar{y})$ will be chosen to be relaxed in this iteration.

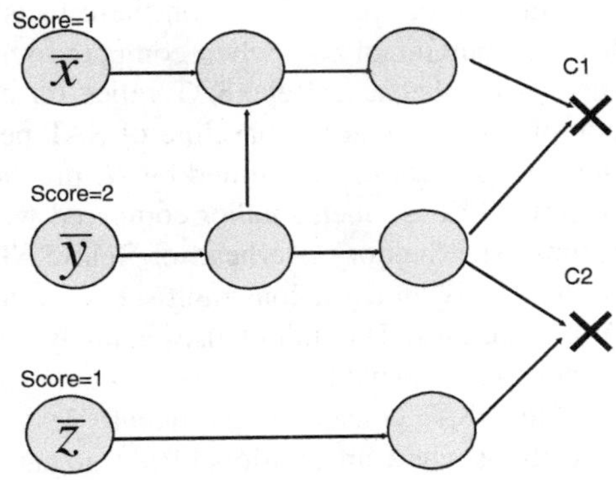

Fig. 9.1 Exploring space enlargement

It is not necessary to relax only one constraint at each iteration. To reduce the number of iterations, in our implementation we relax the top 10% of constraints on the conflict score list. For those variables with the same conflict scores, we break the tie by favoring on those variables that appear less frequently in the formula. It should be noted that if more constraints are relaxed in each iteration, the quality of the final solution may be degraded due to the decreased granularity in the search. For the MIN-ONE SAT extension, where each variable is associated with a coefficient, we pick the constraint that has the smallest coefficient and the highest conflict score in order to maintain minimal loss when trying to satisfy the objective function.

---

[1] Many modern SAT solvers are capable of providing the trace file for the unsatisfiable instances, such as ZChaff, MiniSAT [12].

A side benefit should be mentioned here. The returned solution from RelaxSAT on MIN-ONE SAT problems could have fewer ONEs than the constraint set specified in F_{sc}. For example, given a formula with 200 variables, if a solution is provided by RelaxSAT under a constraint set with 100 variables specified as ZEROS, the upper bound of ONEs would be $200 - 100 = 100$. However, during the search, the SAT solver may assign the (unconstrained) free variables to ZERO, leading to an even smaller number of ONEs.

The experiments in Section 9.4 demonstrate that our heuristic works very well to obtain a tight bound while keeping a low computational overhead.

9.3.3 Discussion on Computation Complexity

In this section, we will discuss briefly about the algorithm complexity. From Theorem 9.2, we know that in the worst case RelaxSAT may require N iterations. In each iteration the original Boolean formula F_o conjuncted with some unit-clause constraints F_{sc} will be solved. Although theoretically the SAT algorithm is NP complete, modern SAT solvers can solve many of the large instances without explosion in time or space. Based on these facts, conventional SAT solvers have much less computational cost when compared with existing MIN-ONE SAT solvers. In this regard, because RelaxSAT relies on conventional SAT, in each iteration its complexity stays at the baseline of SAT performance. Considering the maximum number of iterations is limited by N, practically the total computation complexity of RelaxSAT is much smaller compared with other existing MIN-ONE SAT algorithms. The memory overhead of RelaxSAT is also in line with conventional SAT because only an extra constraint set is introduced, which is simply a set of at most N unit clauses. The unit clauses actually help to reduce the search space of F_o by forcing some variables to a fixed value, especially in the earlier iterations. Generally RelaxSAT consumes significantly less resources compared with PBSAT-based algorithms where an additional Boolean network F_{sc} is needed, without any fixation of any variables.

9.4 Experimental Results

Our proposed RelaxSAT was implemented in C++ under 32-bit Linux, and all the experiments were conducted on a Intel Xeon 3.0G workstation with 2 G memory. ZChaff version 2004.11.15 simplified was chosen to serve as the backbone SAT solver in RelaxSAT due to the convenience on obtaining the trace information. There are three categories of benchmarks in our experiments. The first category is from the OPTSAT benchmarks, where OPTSAT outperformed some popular PBSAT-based solvers [15]. Since OPTSAT outperformed PBSAT for these benchmarks, in the first experiment we only compared RelaxSAT with OPTSAT. Note that the publicly distributed OPTSAT binary program can only solve the basic MIN-ONE SAT problem, so it is not compared in the second and third experiments, where the extended MIN-ONE SAT problem was targeted.

We report the results of the first experiment in Table 9.1. For each benchmark listed under the first column, the number of variables and clauses of each instance are listed in column 2. The third and fourth columns present the run time of OPTSAT and RelaxSAT, respectively, followed by a column of run time speedup. The speedup is defined as the ratio between the run time of OPTSAT and RelaxSAT. The objective function bound returned from OPTSAT and RelaxSAT is reported in the sixth and seventh columns, respectively. The last column shows the difference between the number of ONEs OPTSAT and RelaxSAT obtained. Note that whenever OPTSAT is able to solve the instance, it is guaranteed to be the optimal solution. In Table 9.1, OUT indicates OPTSAT times out after 3600 s and the corresponding run time speedup is recorded as INF. Whenever OPTSAT times out, the sixth and last columns are marked X.

Table 9.1 Experiment I: OPTSAT benchmarks

CNF	#Var/#Cls	OPTSAT time	RelaxSAT time	Speedup	OPTSAT min 1s	RelaxSAT min 1s	DIFF
c17.cnf	13/22	0.02	0.01	2.00	4	4	0
c880.cnf	469/1164	OUT	0.02	INF	X	198	X
2bitcomp_5	125/310	2.16	0.01	216.00	39	45	−6
2bitmax_6	252/766	333.88	0.11	3035.27	61	68	−7
3blocks	283/969	0.62	0.31	2.00	56	61	−5
4blocks	758/47820	OUT	6.34	INF	X	116	X
4blocksb	410/24758	1.06	0.68	1.56	66	66	0
qg1-08	512/148957	52.66	37.78	1.39	64	64	0
qg2-08	512/148957	31.4	16.64	1.89	64	64	0
qg3-08	512/10469	0.31	0.47	0.66	64	64	0
s27	72/141	0.02	0.01	2.00	24	24	0
s5378	12168/26527	OUT	97.89	INF	X	2796	X
sat_bw_large.a	459/4675	0.02	0.28	0.07	73	73	0
sat_bw_large.b	1087/13772	0.22	1.5	0.15	131	131	0
sat_logistics.a	728/5784	1.51	0.68	2.22	135	135	0
sat_logistics.b	757/6429	4.38	0.76	5.76	138	138	0
sat_rocket_ext.a	331/2246	1.4	0.16	8.75	65	65	0
sat_rocket_ext.b	351/2398	1.68	0.18	9.33	69	69	0
bmc-galileo-8	58074/294821	OUT	497.31	INF	X	12303	X
bmc-ibm-2	2810/11683	1013.13	10.64	95.22	940	1017	−77
bmc-ibm-3	14930/72106	22.37	92.83	0.24	6356	6362	−6

OUT=3600 s

In 12 of the 21 benchmarks, both OPTSAT and RelaxSAT found the optimal solutions. But RelaxSAT achieved a two to three times speedup. For example, in sat_rocket_ext.b, a minimum of 69 ONEs were obtained by both OPTSAT and RelaxSAT. However, RelaxSAT achieved a 9.33× speedup. For four of the instances, OPTSAT failed to finish in 3600 s while RelaxSAT can provide a solution in only a few seconds. While it is not possible to tell the quality of the result obtained by RelaxSAT for these four instances, the fact that a solution is obtained can be of great use in many applications. Note that the low computational cost also makes the approach attractive. RelaxSAT returned tight bounds on the remaining five benchmarks. For example, in 2bitmax_6, the solution achieved by RelaxSAT has seven more ONEs, but with more than 300× speedup. For very few simple instances, the run time is slightly increased due to the large number of iterations that RelaxSAT went through. These are the instances that both solvers can complete within a few seconds, but the iterative nature of RelaxSAT caused a higher cost for these simple instances.

The second experiment involves the benchmarks from the power-aware test pattern generation, where the goal is to generate a test pattern that can detect a target delay fault and simultaneously excite as much switching activities as possible to worsen the delay effect. As a brief background, increasing circuit activity pulls down the V_{DD} power source, thereby reduces the potential and increases delay. Since considering a complicated fault model is irrelevant with this chapter, in our experiments we only check the maximal switching activities of a sequential circuit in two consecutive clock cycles without any other constraints. In the experiment setup, we unroll each sequential circuit into a two-time-frame combinational circuit, signifying the circuit state across two clock cycles. The switching activities are monitored by a set of XNOR gates with the inputs from a pair of corresponding gates in two time frames. If the output of a monitor XNOR is ONE, it means that there is no switching activity on that gate across the two clock cycles. After transforming the two-time-frame combinational circuit and the monitor circuit into a Boolean formula, obviously the task of maximizing the switching activities now is equivalent to finding a MIN-ONE solution on the outputs of the XNOR gates. It should be emphasized that this problem is different from the basic MIN-ONE SAT because in basic MIN-ONE SAT *all* the variables have to be considered while in the switching activity maximization, only the outputs of the monitor circuits are considered. In other words, the objective function only involves a portion of the variables in the Boolean formula.

In this second experiment, RelaxSAT is compared with MiniSAT+ [13]. MiniSAT+ is a high-performance PBSAT solver which can be used to solve MIN-ONE SAT. The MiniSAT+ is chosen not only because it has been demonstrated to have high performance [19] but also due to the reason that its techniques favor those PBSAT problems where most of the constraints can be directly represented as single clauses. The usual MIN-ONE SAT instances fall into this category and can be benefited by using MiniSAT+, because all the constraints in MIN-ONE SAT are SAT clauses except the objective function.

The results of this second experiment are shown in Table 9.2. The eight columns of Table 9.2 are similar to Table 9.1, except that OPTSAT is replaced by MiniSAT+ due to the reason that OPTSAT is unable to handle the extended MIN-ONE SAT problems. Since MiniSAT+ works iteratively, whenever it cannot find the optimal solution within 3600 s, it reports the best solution obtained so far in the third column. A positive value in the last column indicates that RelaxSAT returned a *tighter* bound. Among the nine instances, RelaxSAT outperformed MiniSAT+ in five of them with better result and shorter run time. One can see that the run time of

Table 9.2 Experiment II: power-aware test generation benchmarks

CNF	#Var/#Cls	MiniSAT+ time	RelaxSAT time	Speedup	MiniSAT+ result	RelaxSAT result	DIFF
s510	972/2254	0.24	0.23	1.04	122	138	−16
s1494	2720/6812	9.25	2.15	4.30	342	421	−79
s5378	12168/26508	3600	20.85	172.66	1242	1075	167
s6669	13828/31152	3600	53.45	67.35	1635	1553	82
s15850	41880/89908	3600	402.51	8.94	6265	6259	6
s35932	72592/164716	3600	1131.21	3.18	8198	7456	742
b13	1448/3248	26.6	0.43	61.86	157	211	−54
b14	40392/98250	3600	458.99	7.84	5857	5785	72
b15	35688/87808	3600	281.97	12.77	6423	6624	−201

RelaxSAT is roughly 40× faster than that needed by MiniSAT+. For example, in s5378, RelaxSAT can return a solution containing 1075 non-switching gate in less than 21 s while MiniSAT+ generates 167 more non-switching gates even after 1 h. For some small instances, MiniSAT+ was able to find the global optimum, the solution obtained by RelaxSAT still presents a very tight margin.

In the third experiment, we evaluated RelaxSAT on a set of benchmarks whose objective functions have integer coefficients associated with each variable. Some of them were generated similarly as those in the second experiment. The only difference is that here the gate size is considered when computing the gate switching activities. Thus the objective functions will be slightly different, where the variables representing the outputs of the XNOR gates will have a positive coefficient corresponding to the gate size. Actually this is a more accurate model in the power consumption estimation, since the power consumed depends on the gate type and its load, which can be lumped to a gate size factor. The gate size information is obtained through the synthesized circuits. The rest of the benchmarks in this experiment are derived from the OPTSAT benchmarks with a randomly generated coefficient of each variable. So the objective function becomes *minimize* $\sum_{i=1}^{N} c_i \times V_i$. Note that c_i is a number between −20 and 20.

The layout of Table 9.3 is the same as Table 9.2. RelaxSAT outperformed MiniSAT+ on 6 of the total 12 benchmarks. For example, in s5378, RelaxSAT was able to achieve a much better solution (682 smaller in the objective function) at 190× less the cost. The remaining six instances are small ones except for s6669 and b15. For s6669 and b15, MiniSAT+ was able to find a better solution than RelaxSAT, although it could not produce the global optimum within 3600 s. The run times shown in the third column of these two circuits are the time needed to get a better solution than RelaxSAT. Here we take a closer look at b15 to discuss the strength and potential improvement of RelaxSAT. For b15, RelaxSAT was able to finish after 274 s with the result of 27748. Meanwhile MiniSAT+ discovered a better solution after more than 1800 s but could not step further until it finally timed out at 3600 s. One can see that although in this specific example RelaxSAT was not able to outperform MiniSAT+ in terms of the tightness of the bound, RelaxSAT holds the strength on a much smaller run time. In other words, MiniSAT+ needed

Table 9.3 Experiment III: objective functions with integer coefficients

CNF	#Var/#Cls	MiniSAT+ times	RelaxSAT time	Speedup Speedup	MiniSAT+ min 1s	RelaxSAT min 1s	DIFF
2bitcomp_5	125/310	0.94	0.01	94.00	−297	−203	−94
3blocks	283/969	3.39	0.34	9.97	−105	−81	−24
qg3-08	512/10469	1	0.39	2.56	−73	−29	−44
sat_bw_large.a	459/4675	0.34	0.2	1.70	−189	−189	0
sat_logistics.b	757/6429	3600	0.86	4186.05	−197	−269	72
bmc-ibm-2	2810/11683	3600	9.86	365.11	−616	−581	−35
s5378	12168/26508	3600	18.9	190.48	4570	3888	682
s6669	13828/31152	86.27*	180.65	0.48	6538	6732	−194
s13207	35088/74658	3600	243.07	14.81	14942	14248	694
s15850	41880/89908	3600	359.48	10.01	19394	19244	150
b14	40392/98250	3600	727.27	4.95	23174	21144	2030
b15	35688/87808	1821.3#	273.59	6.66	27392	27748	−356

* After 3600 s, the best result is 6386.
# After 3600 s, the best result is still 27392.

three times more execution time to outperform RelaxSAT by a small margin. This observation suggests that using RelaxSAT as a preprocessing step before MiniSAT+ may be beneficial, which could be a potential future extension of RelaxSAT.

The next experiment was set up to demonstrate the low memory overhead of RelaxSAT, compared with both OPTSAT and MiniSAT+. A total of ten large instances are selected to measure the peak memory consumption during the solving process. The results are presented in Figs. 9.2 and 9.3. In both figures the x-axis denotes the instance and y-axis lists the memory consumption in megabytes. In Fig. 9.2, five instances are selected from Experiment I and the memory consump-

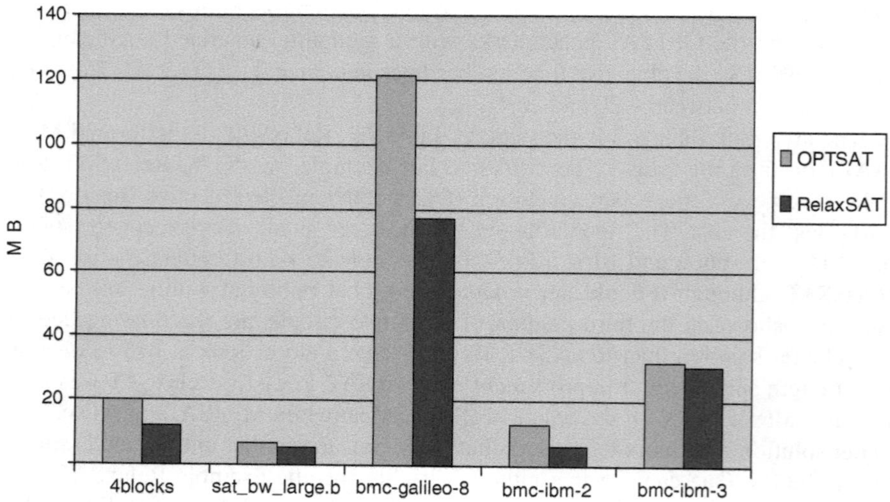

Fig. 9.2 Memory consumption: OPTSAT vs. RelaxSAT

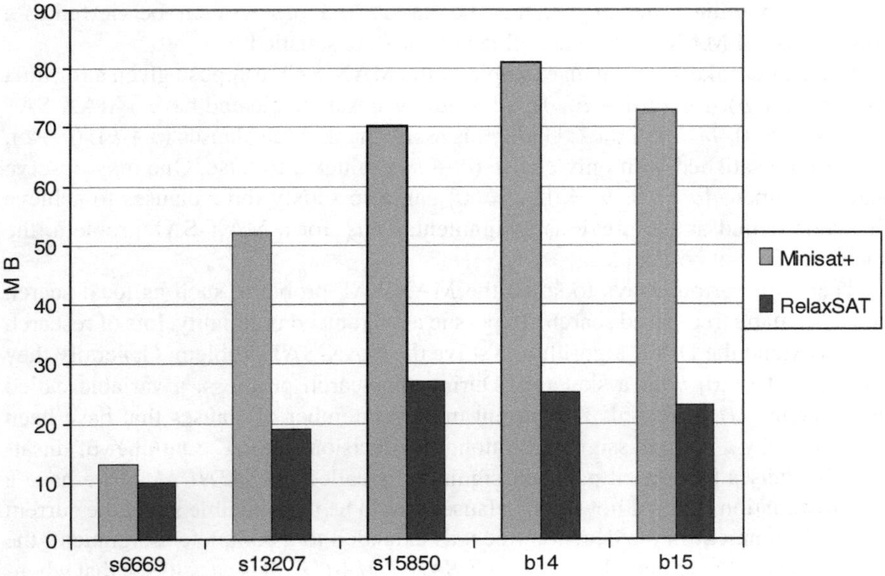

Fig. 9.3 Memory consumption: MiniSAT+ vs. RelaxSAT

tion is compared between OPTSAT and RelaxSAT. It is clear that RelaxSAT has a smaller memory cost especially on large instances like bmc-galileo-8. The memory cost of five instances from Experiment III is shown in Fig. 9.3. Similar to Fig. 9.2, RelaxSAT required a significantly smaller memory consumption compared with MiniSAT+. The reason is as we mentioned before RelaxSAT only has an additional constraint set while PBSAT-based algorithms need a large Boolean network of F_c.

9.5 Application Discussion: A RelaxSAT-Based MAX-SAT Solver

As is known, many difficult problems can be solved by MAX-SAT solver, such as Max-Cut, Max-Clique, and Binary Covering Problem [4, 7, 14, 17]. Due to its theoretical and practical importance, the MAX-SAT solver is receiving increased levels of attention. With our work on the MIN-ONE SAT solver, a new direction for MAX-SAT could be explored along a similar manner. In this section, we will describe how the proposed MIN-ONE SAT algorithm can benefit MAX-SAT solving. Before going further, MAX-SAT is formally defined as follows:

> **MAX-SAT Problem:** Given an unsatisfiable CNF formula, find a variable assignment that satisfies the maximum number of clauses.

Note that a clause is satisfied when at least one of its literals is evaluated to true. In other words, MAX-SAT optimizes the variable assignments to satisfy as many clauses as possible. One can see that the classic SAT problem can be viewed as a special case of MAX-SAT where all the clauses are satisfied.

First, let us take a look at an example of the MAX-SAT. Suppose given a formula $(a + b)(a + \bar{b})(\bar{a} + c)(\bar{a} + \bar{c})$. This formula is unsatisfiable and have a MAX-SAT solution $a = 0,\ b = 1,\ c = 1$. Under this assignment, three clauses $(a+b)$, $(\bar{a}+c)$, $(\bar{a} + \bar{c})$ are satisfied with only clause $(a + \bar{b})$ evaluated to false. One may observe that assignment $\{a = 0,\ b = 1,\ c = 0\}$ can also satisfy three clauses to achieve the same bound as the previous assignment. Thus, for a MAX-SAT problem, the solution may not be unique.

There are various ways to solve the MAX-SAT problem, such as local search and systematic tree-based search. To pursue a guaranteed optimality, lots of research turn to extend the DPLL algorithm to solve the MAX-SAT problem. Generally they are [2, 3, 10, 20] with a skeleton. During the search progress, a variable called upper bound (UB) records the current minimal number of clauses that have been unsatisfied by a partial assignment (along the decision tree). The number of unsatisfiable clauses by current partial assignment is called *UNSATNUM*. If we have a underestimation, UE, of how many clauses would be unsatisfiable when the current partial assignment in the decision tree was extended to a complete assignment, the best reachable lower bound would be $\mathit{UNSATNUM}+UE$. It is easy to see that whenever $UB <= \mathit{UNSATNUM} + UE$, there is no need to continue the search within the subspace under the current decision since no better solution would be found. Clearly this bound estimation helps to prune the search space effectively. On the other hand, when $UB > \mathit{UNSATNUM} + UE$ the branch-and-bound search strategy will continue under this subspace. The algorithm stops when the whole search space has been pruned and searched. There are other ways to solve MAX-SAT problems, like in [8] the author proposed an algorithm based on resolution.

Usually both DPLL-based MAX-SAT algorithm and MIN-ONE SAT-based MAX-SAT algorithm are complete. It is possible to modify them to search for sub-optimal solutions in order to save computational resources. Although the optimality of MAX-SAT is traded with the performance gain, the results achieved usually can justify the tradeoff.

As is known, MAX-SAT can be modeled and solved through MIN-ONE SAT, which is shown in Fig. 9.4. The framework can be partitioned into three steps. First, the target formula should be transformed into a new formula with auxiliary variables embedded into each clause. Then this new formula can be solved by a MIN-ONE SAT algorithm with the capability to find a solution on subset of variables. Finally, the MAX-SAT bound and the solution are derived from the MIN-ONE SAT solution of the transformed formula. Since the transformation and the final-step extraction are linear with the size of the clause, the computational complexity is determined by the MIN-ONE SAT solving in the second step in Fig. 9.4.

Here we present a simple example to see how this approach works. Given a formula $(x_1 + x_2)(x_1 + \bar{x_2})(\bar{x_1})$. This formula is first transformed to $(x_1 + x_2 + AX_1)(x_1 + \bar{x_2} + AX_2)(\bar{x_1} + AX_3)$, with the auxiliary variables as AX_1, AX_2, AX_3. We call the first formula as F_o and second formula as F_n. The MIN-ONE SAT solution of auxiliary variables in F_n is $\{AX_1 = 0,\ AX_2 = 0,\ AX_3 = 1\}$ with

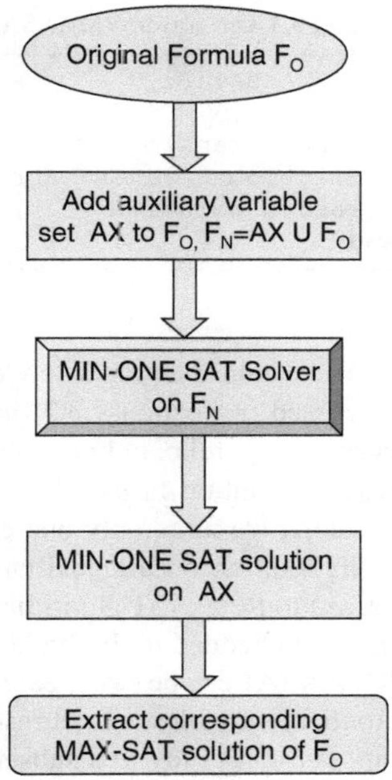

Fig. 9.4 MIN-ONE SAT-based MAX-SAT solving

associated assignment $\{x_1 = 1, x_2 = 0\}$. Note that the MIN-ONE solution is not unique in this example. It is easy to see the MAX-SAT bound equals two (maximal number of satisfied clauses are 2 with one clause left unsatisfied).

To explore the potential of our novel MIN-ONE SAT algorithm (RelaxSAT), we built a MAX-SAT solver on top of that. By taking advantage from the relaxation heuristic inside RelaxSAT, those clauses that potentially contribute most to the conflicts will be identified in order to ease the efforts to satisfy the rest of the formula. This local greedy scheme helps us to cover the possibly largest satisfiable part of the formula, which is exactly what the MAX-SAT solver attempts to do.

In the next section, we will introduce a new MAX-SAT solver () based on our previously proposed MIN-ONE SAT algorithm. From the preliminary experimental results our solver achieves a significant performance improvement over existing MAX-SAT solvers.

9.5.1 The New MAX-SAT Solver: RMAXSAT

Because our MAX-SAT solver is based on the RelaxSAT proposed in Section 9.5, we name it RMAXSAT. It consists of three components. The first component is formula transformation. A implementation is also embedded in RMAXSAT, and finally the MAX-SAT solution will be taken from the outputs of RelaxSAT and printed out. The pseudo code is presented in Listing 9.3. The program of RMAXSAT was implemented in C++ under Linux.

Listing 9.3 Our solver: RMAXSAT

```
def RMAXSAT(F)
begin
    F_n=transform(F)
  Solution=RelaxSAT(F_n)
  return solution //contains both assignment and bound
end
```

We would mention a few details about the implementation here. As previously discussed in Section 9.3.2, in the underlying MIN-ONE SAT algorithm, it is not necessary to relax only one constraint at a time, more constraints can be relaxed to reach a solution faster. However, in RMAXSAT, to achieve the bound as tight as possible, we relax only one constraint in each iteration. Because we want a MIN-ONE solution on the auxiliary variables only, the MIN-ONE SAT target function passed to RelaxSAT algorithm is $\sum_i^M AX_i$, where AX_i is the ith auxiliary variable that is embedded in the ith clause with a total of M clauses in the original formula. RMAXSAT can be easily extended to handle partial MAX-SAT problems, where the auxiliary variables are not inserted to every clause in the formula. By adding auxiliary variables only to a subset of clauses, those clauses without auxiliary variables will automatically be forced to be satisfied during the search for a solution.

Since the RMAXSAT inherits the search heuristics from RelaxSAT, it intends to leave those clauses that would most likely cause the conflicts alone (by setting their corresponding auxiliary variable to one) and satisfy the rest of the clauses in an iterative manner. Our experiments have shown that this conflict-guided relaxation heuristic works effectively in most cases.

First let us look at a case study of RMAXSAT. We have tested RMAXSAT on some equivalence checking problems to see if RMAXSAT can correctly identify the one clause that when removed can make the rest of the clauses satisfiable. Equivalence checking compares two different designs to check whether they are functionally equivalent or not. By adding the extra equivalence assertions, the satisfiability of the equivalence-checking instances determines the equivalence between the two designs. When the two original designs are indeed equivalent, the equivalence-checking instance should be unsatisfiable. In other words, an effective MAX-SAT solver should be able to identify the equivalence assertion clause, without which the rest of the clauses should be satisfiable.

Thus, in our preliminary setup, when an unsatisfiable instance derived from an equivalence-checking problem is fed to RMAXSAT, it is quite a surprise to us that RMAXSAT was able to quickly find a MAX-SAT solution that satisfies the clauses from the design without the clause from the equivalence assertion. This is extremely encouraging because logically the best way to make an unsatisfiable equivalence-checking instance to become satisfiable is by removing the equivalence assertion constraints. This observation gave us the hints that RMAXSAT might be a powerful tool to reveal some insightful reasoning encoded in the SAT instances. Next, we will discuss the performance of RMAXSAT for general MAX-SAT benchmarks.

9.5.2 Evaluation of MAX-SAT Solver

In this section some experiments were conducted to compare RMAXSAT with some existing MAX-SAT solvers. All the experiments were conducted on a PC workstation with Intel Xeon 3.2GHz CPU and 2GB memory.

First we cite the experimental results from [15], presented in Table 9.4. The reason we cite this table here is that we want to demonstrate that OPTSAT and MSAT+ are the state-of-the-art solvers which have a significant edge among the existing MAX-SAT solvers. Since the source code or executable files from most MAX-SAT solvers are not available to us, at current stage we first compare the performance between our RMAXSAT and OPTSAT, which is shown in Table 9.5. The comparison with MSAT+ is reported in Table 9.6. By evaluating on the same set of publicly available benchmarks, we can check to see if RMAXSAT is indeed a high-performance MAX-SAT solver.

Table 9.4 Comparison of existing MAX-SAT solvers

CNF	#C	BF	OPBDP	PBS4	MSAT+	OPTSAT
barrel3	941	0.23	2.04	0.88	0.12	0.9
barrel4	2034	0.65	47.59	11.67	0.34	21.19
barrel5	5382	21.42	MEM	MEM	24.01	177.11
barrel6	8930	213.6	MEM	–	95.56	896.45
barrel7	13764	SF	MEM	–	285.55	435.46
lmult0	1205	0.39	13.05	1.45	0.16	7.35
lmult2	3524	57.11	TIME	TIME	6.7	16.46
lmult4	6068	261.74	MEM	–	35.34	98.05
lmult6	8852	774.08	MEM	–	157.02	609.07
lmult8	11876	SF	MEM	–	297.32	704.08
qvar8	2272	0.62	MEM	17.67	2.95	36
qvar10	5621	2.21	MEM	234.97	55.54	156.44
qvar12	7334	6.2	MEM	–	36.8	74.49
qvar14	9312	SF	MEM	–	117.25	815.66
qvar16	6495	SF	MEM	–	51.33	117.31
c432	1114	131.06	TIME	7.22	0.24	7.6
c499	1869	TIME	TIME	100.41	0.8	4.59
c880	2589	TIME	TIME	320.96	5.54	38.91
c1355	3661	TIME	TIME	TIME	80.09	21.2
c1908	5095	TIME	MEM	TIME	58.01	165.99
c2670	6755	TIME	MEM	–	63.64	100.31
c3540	9325	TIME	MEM	–	242.02	786.33
u-bw.a	3290	7.81	TIME	249	209.03	178.18
u-bw.b	N/A	TIME	MEM	–	TIME	TIME
u-log.a	5783	TIME	MEM	TIME	59.65	179.3
u-log.b	6428	TIME	MEM	–	35.37	144.83
u-log.c	9506	TIME	MEM	–	383.65	731.87
u-rock.a	1691	13.29	TIME	41.29	206.56	6.26

TIME: Time out at 1800 s; MEM: Memory out at 1 GB; SF: Solver aborted unexpectedly; N/A: No solver could solve the instance; –: Solver returned incorrect result.

Table 9.5 RMAXSAT vs. OPTSAT

	OPTSAT		RMAXSAT			
CNF	#C	Run time	#C	Run time	Bound DIFF	Speedup
barrel3	941	0.335	941	0.02	0	16.75
barrel4	2034	2.26	2032	0.19	2	11.89
barrel5	5382	65.37	5380	2.11	2	30.98
barrel6	8930	257.48	8928	12.94	2	19.90
barrel7	13764	481.29	13763	17.63	1	27.30
lmult0	1205	2.36	1205	0.03	0	78.67
lmult2	3524	6.19	3523	0.1	1	61.90
lmult4	6068	49.74	6068	0.28	0	177.64
lmult6	8852	197.05	8852	7.2	0	27.37
lmult8	11876	787.3	11876	384.76	0	2.05
qvar8	2272	10.77	2272	0.18	0	59.83
qvar10	5621	49.18	5621	0.5	0	98.36
qvar12	7334	28.53	7334	0.95	0	30.03
qvar14	9312	188.09	9312	1.06	0	177.44
qvar16	6495	43.29	6495	1.2	0	36.08
c432	1114	2.41	1114	0.08	0	30.13
c499	1869	3.53	1869	0.3	0	11.77
c880	2589	14.43	2588	0.57	1	25.32
c1355	3661	10.88	3661	1.14	0	9.54
c1908	5095	61.04	5095	2.14	0	28.52
c2670	6755	43.14	6755	1.37	0	31.49
c3540	9325	353.27	9325	65.48	0	5.40
u-bw.a	3290	18.19	3285	0.27	5	67.37
u-bw.b	N/A	TIME	11480	1.14	N/A	N/A
u-log.a	5783	47.63	5782	0.22	1	216.50
u-log.b	6428	45.09	6428	0.23	0	196.04
u-log.c	9506	179.57	9506	0.39	0	460.44
u-rock.a	1691	1.59	1691	0.08	0	19.88
Average					0.56	72.54

In Table 9.4, the first column lists the names of the benchmark formula in the format of CNF. The maximal number of satisfied clauses is shown in the second column for each instance, which is the optimal bound if at least one of the deterministic MAX-SAT solver should return if it is able to finish. The running time of each MAX-SAT solver in the competition is presented in the following columns, in the order of BF [6], OPBDP [10], PBS4 [1], MSAT+ [13, 19]. Solver BF is a MAX-SAT solver, while OPBDP, PBS4, and MSAT+ are generic pseudo-Boolean solvers. MSAT+ is a pseudo-Boolean SAT solver based on MiniSat [12], and it is shown to be capable of solving large number of instances.

All of the deterministic MAX-SAT solvers can report the optimal bound if they could finish within the time and memory limits. The timeout limit is set at 1800 s while the memory consumption is limited within 1 GB. "TIME" in the table indicates that the corresponding solver times out. Similarly "MEM" indicates that memory is exhausted. Note that "SF" means the solver aborted unexpectedly. When no solver can solve the instance, the optimal bound in the second column is marked as

Table 9.6 RMAXSAT vs. MSAT+

	MSAT+		RMAXSAT			
CNF	#C	Run time	#C	Run time	Bound DIFF	Speedup
barrel3	941	0.04	941	0.02	0	2.23
barrel4	2034	0.04	2032	0.19	2	0.19
barrel5	5382	8.86	5380	2.11	2	4.20
barrel6	8930	27.45	8928	12.94	2	2.12
barrel7	13764	315.60	13763	17.63	1	17.90
lmult0	1205	0.05	1205	0.03	0	1.71
lmult2	3524	2.52	3523	0.1	1	25.20
lmult4	6068	17.93	6068	0.28	0	64.03
lmult6	8852	50.80	8852	7.2	0	7.06
lmult8	11876	332.46	11876	384.76	0	0.86
qvar8	2272	0.88	2272	0.18	0	4.90
qvar10	5621	17.46	5621	0.5	0	34.92
qvar12	7334	14.09	7334	0.95	0	14.84
qvar14	9312	27.04	9312	1.06	0	25.51
qvar16	6495	18.94	6495	1.2	0	15.78
c432	1114	0.08	1114	0.08	0	0.95
c499	1869	0.62	1869	0.3	0	2.05
c880	2589	2.05	2588	0.57	1	3.60
c1355	3661	41.10	3661	1.14	0	36.06
c1908	5095	21.33	5095	2.14	0	9.97
c2670	6755	27.37	6755	1.37	0	19.98
c3540	9325	108.73	9325	65.48	0	1.66
u-bw.a	3290	21.34	3285	0.27	5	79.03
u-bw.b	N/A	TIME	11480	1.14	N/A	N/A
u-log.a	5783	15.85	5782	0.22	1	72.03
u-log.b	6428	11.01	6428	0.23	0	47.88
u-log.c	9506	94.13	9506	0.39	0	241.36
u-rock.a	1691	52.46	1691	0.08	0	655.81
Average					0.56	51.55

"N/A." Finally "-" represents that the solver returns wrong results. It can be observed that solver BF, OPBDP, and PBS4 were not able to solve most of the instances, either due to the resource limit or due to the implementation bugs. On the other hand, MSAT+ and OPTSAT solved all the instances except for u-bw.b. For instance, in c3540, BF timed out, OPBDP ran out of memory, PBS4 returned wrong results, and both MSAT+ and OPTSAT finished in 242.02 and 786.33 s, respectively. The maximum number of clauses that could be satisfied is 9325. From this table we see that the two solvers, MSAT+ and OPTSAT, gave the best performance and capability edge over other three MAX-SAT solvers. Therefore, in the following experiments, we compare our proposed RMAXSAT with OPTSAT and MSAT+ separately.

Tables 9.5 and 9.6 share the same layout, where the first column exhibits the instance name, followed by the results from two solvers in the comparison. In Table 9.5, the second and the third columns report the bound and running time of OPTSAT, respectively, while the fourth and fifth columns show the bound and running time of RMAXSAT, respectively. The sixth column presents the bound difference between RMAXSAT and OPTSAT, computed as the bound obtained from RMAXSAT minus the bound obtained from OPTSAT. It should be noted that since RMAXSAT is based on the approximation algorithm of MIN-ONE SAT, the bound it returned may not be optimal. Finally, in the last column, the speedup is calculated as $\frac{\textit{Run time of RMAXSAT}}{\textit{Run time of OPTSAT}}$. We set the time-out limit the same as in Table 9.4, 1800 s.

One can see that the bounds returned by RMAXSAT are extremely tight. In all 27 instances, RMAXSAT reported optimal bounds on 18 of them. In the remaining nine instances, the difference with the optimal solution is at most 5; usually, the difference is only 1 or 2. Meanwhile the performance improvement is over 70× on average. For example, in qvar14, OPTSAT obtained the optimal MAX-SAT result of simultaneously satisfying 9312 clauses in 188.09 s, while RMAXSAT obtained the same MAX-SAT bound of 9312 in only 1.06 s. This is a speedup of 177.44×. In another example, RMAXSAT finished instance u-bw.b in just 1.14 s whereas all the other MAX-SAT solvers failed. With this result we know that in u-bw.b.cnf, at least 11480 clauses can be satisfied simultaneously, which means by removing only one of these clauses, the formula becomes satisfiable.

The results from Table 9.6 follow the same trend as in Table 9.5, and on average, over 50× performance improvement from RMAXSAT was achieved over MSAT+. For example, in the instance u-log.c, MSAT+ finished the instance in 94.13 s that simultaneously satisfied 9506 clauses. RMAXSAT, on the other hand, satisfied the same number of clauses in just 0.39 s. This is a 241× speedup. Based on the consistent results obtained in both experiments, it is safe to conclude that RMAXSAT is able to return tight bounds with over an order of magnitude reduction in the computational cost.

9.6 Conclusions and Future Works

In this chapter, a novel algorithm for MAX-SAT is proposed. A set of constraints automatically derived from the objective function is used to guide the search. Whenever the constraint-set causes conflicts with the original Boolean formula, those constraints most responsible to the conflicts will be identified and relaxed. The relaxation procedure continues until a solution is found. Our proposed algorithm has a low memory overhead and can provide a tight bound for the objective function. It is able to handle some large instances that the existing MIN-ONE SAT solvers failed. Meanwhile it can achieve one or two orders of magnitude run time reduction. Furthermore, a MAX-SAT solver built on top of RelaxSAT is presented to demonstrate the potential of our algorithm.

In the future, different relaxation heuristics can be explored, and a possible hybrid solution can also be investigated. Although RelaxSAT can provide a tight bound with a significant performance improvement, it may not guarantee a global optimum. To address this problem, we believe a hybrid solution of RelaxSAT and other complete MIN-ONE SAT algorithm can be very beneficial. One possible approach is to use the RelaxSAT as a preprocessing step in based algorithms where the bound estimation BE is passed from RelaxSAT. In this hybrid scheme, RelaxSAT would be able to provide a tight initial bound estimate quickly, allowing PBSAT-based algorithms to skip many unnecessary iterations.

References

1. Aloul, F., Ramani, A., Markov, I., Sakallah, K.: Pbs: A backtrack search pseudo-Boolean solver. In: Proceedings of the 5th International Conference on the Theory and Applications of Satisfiability Testing, pp. 46–353. (2002)
2. Alsinet, T., Manyà, F., Planes, J.: A max-sat solver with lazy data structures. In: Proceedings of Advances in Artificial Intelligence – IBERAMIA, pp. 334–342. (2004)
3. Alsinet, T., Manyk, F., Planes, J.: Improved branch and bound algorithms for max-sat. In: Proceedings of International Conference on the Theory and Applications of Satisfiability Testing (SAT), pp. 408–415. (2003)
4. Asano, T., Ono, T., Hirata, T.: Approximation algorithms for the maximum satisfiability problem. Nordic Journal of Computing **3**(4), 388–404 (1996)
5. Bailleux, O., Marquis, P.: Distance-sat: Complexity and algorithms. In: Proceedings of the 14th National Conference on Artificial Intelligence (AAAI-06) and the Collocated 11th Conference on Innovative Applications of Artificial Intelligence (IAAI-06), pp. 642–647. (1999)
6. Barth, P.: A Davis-Putnam enumeration algorithm for linear pseudo-Boolean optimization. Technical Report MPI-I-95–2-2003, Max Plank Institute for Computer Science (1995)
7. Bertoni, A., Campadelli, P., Carpentieri, M., Grossi, G.: A genetic model: Analysis and application to maxsat. Evolutionary Computation **8**(3), 291–309 (2000). Doi: http://dx.doi.org/10.1162/106365600750078790
8. Bonet, M.L., Levy, J., Manyà, F.: Resolution for max-sat. Artificial Intelligence **171**(8–9), 606–618 (2007). Doi: http://dx.doi.org/10.1016/j.artint.2007.03.001
9. Bonet, M.L., Levy, J.L., Manyà, F.: A complete calculus for max-sat. In: Proceedings of the 9th International Conference on the Theory and Applications of Satisfiability Testing, pp. 240–251. (2006)
10. Borchers, B., Furman, J.: A two-phase exact algorithm for MAX-SAT and weighted MAXSAT problems. Journal of Combinatorial Optimization **2**, 299–306 (1999)
11. Davis, M., Logemann, G., Loveland, D.W.: Machine program for theorem proving. Communications of the ACM **5**, 394–397 (1962)
12. Eén, N., Sörensson, N.: An extensible sat-solver. In: Proceedings of the 6th International Conference on the Theory and Applications of Satisfiability Testing, pp. 502–518. (2003)
13. Eén, N., Sörensson, N.: Translating pseudo-Boolean constraints into sat. Journal on Satisfiability, Boolean Modeling and Computation **2**, 1–26 (2006)
14. Fu, Z., Malik, S.: On solving the partial max-sat problem. Theory and Applications of Satisfiability Testing – SAT 2006 **4121**(4), 252–265 (2006)
15. Giunchiglia, E., Maratea, M.: Solving optimization problems with DLL. In: Proceedings of the 17th European Conference on Artificial Intelligence, pp. 377–381. (2006)
16. Kriplani, H., Najm, F., Hajj, I.: Worst case voltage drops in power and ground buses of CMOS VLSI circuits. PhD Thesis, University of Illinois at Urbana-Champaign (1994)
17. Menaï, M.E., Batouche, M.: An effective heuristic algorithm for the maximum satisfiability problem. Applied Intelligence **24**(3), 227–239 (2006). Doi: http://dx.doi.org/10.1007/s10489-006-8514-7
18. Sheini, H.M., Sakallah, K.A.: Pueblo: A modern pseudo-Boolean sat solver. In: Proceedings of Design, Automation and Test in Europe Conference, pp. 684–685. (2005)

19. Wedelin, D.: An algorithm for 0–1 programming with application to airline crew scheduling. In: Proceedings of the Second Annual European Symposium on Algorithms, pp. 319–330. (1994)
20. Xing, Z., Zhang, W.: Maxsolver: An efficient exact algorithm for (weighted) maximum satisfiability. Artificial Intelligence **164**(1–2), 47–80 (2005). Doi: http://dx.doi.org/10.1016/j.artint.2005.01.004
21. Zhang, L., Malik, S.: Extracting small unsatisfiable cores from unsatisfiable Boolean formulas. In: 6th International Conference on the Theory and Applications of Satisfiability Testing (2003)
22. Zhang, L., Malik, S.: Validating sat solvers using an independent resolution-based checker: Practical implementations and other applications. In: Proceedings of the 6th Design, Automation and Test in Europe Conference, pp. 10880–10885. (2003)

Chapter 10
Algorithms for Maximum Satisfiability Using Unsatisfiable Cores

Joao Marques-Sila and Jordi Planes

Abstract Many decision and optimization problems in electronic design automation (EDA) can be solved with Boolean satisfiability (SAT). These include binate covering problem (BCP), pseudo-Boolean optimization (PBO), quantified Boolean formulas (QBF), multi-valued SAT, and, more recently, maximum satisfiability (MaxSAT). The first generation of MaxSAT algorithms are known to be fairly inefficient in industrial settings, in part because the most effective SAT techniques cannot be easily extended to MaxSAT. This chapter proposes a novel algorithm for MaxSAT that improves existing state-of-the-art solvers by orders of magnitude on industrial benchmarks. The new algorithm exploits modern SAT solvers, being based on the identification of unsatisfiable subformulas. Moreover, the new algorithm provides additional insights between unsatisfiable subformulas and the maximum satisfiability problem.

10.1 Introduction

Boolean satisfiability (SAT) is used for solving an ever increasing number of decision and optimization problems in electronic design automation (EDA). These include model checking, equivalence checking, design debugging, logic synthesis, and technology mapping [8, 19, 34, 36]. Besides SAT, a number of well-known extensions of SAT also find application in EDA, including pseudo-Boolean optimization (PBO) (e.g., [27]), quantified Boolean formulas (QBF) (e.g., [13]), multi-valued SAT [26], and, more recently, maximum satisfiability (MaxSAT) [33].

MaxSAT is a well-known problem in computer science, consisting of finding the largest number of satisfied clauses in unsatisfiable instances of SAT. Algorithms for MaxSAT are in general not effective for large industrial problem instances, in part

J. Marques-Sila (✉)
University College Dublin, Dublin, Ireland
e-mail: jpms@ucd.ie

Based on Marques-Silva, J.; Planes, J.: "Algorithms for maximum satisfiability using unsatisfiable cores," Design, Automation and Test in Europe, 2008. DATE '08, pp. 408–413, 10–14 March 2008 Doi: 10.1109/DATE.2008.4484715 © [2008] IEEE.

S.P. Khatri, K. Gulati (eds.), *Advanced Techniques in Logic Synthesis, Optimizations and Applications*, DOI 10.1007/978-1-4419-7518-8_10,

because the most effective SAT techniques cannot be applied directly to MaxSAT [9] (e.g., unit propagation).

Motivated by the recent and promising application of MaxSAT in EDA (e.g., [33]) this chapter proposes a novel algorithm for MaxSAT, `msu4`, that performs particularly well for large industrial instances. Instead of the usual algorithms for MaxSAT, the proposed algorithm exploits existing SAT solver technology and the ability of SAT solvers for finding unsatisfiable subformulas. Despite building on the work of others, on the relationship between maximally satisfiable and minimally unsatisfiable subformulas [6, 16, 20, 21, 24], the approach outlined in this chapter is new, in that unsatisfiable subformulas are used for guiding the search for the solution to the MaxSAT problem. The `msu4` algorithm builds on recent algorithms for the identification of unsatisfiable subformulas, which find other significant applications in EDA [32, 37]. The `msu4` algorithm also builds on recent work on solving PBO with SAT [15], namely on techniques for encoding cardinality constraints as Boolean circuits obtained from BDDs. The `msu4` algorithm differs from the one in [16] in the way unsatisfiable subformulas are manipulated and in the overall organization of the algorithm.

Experimental results, obtained on representative EDA industrial instances, indicate that in most cases the new `msu4` algorithm is orders of magnitude more efficient than the best existing MaxSAT algorithms. The `msu4` also opens a new line of research that tightly integrates SAT, unsatisfiable subformulas, and MaxSAT.

The chapter is organized as follows. The next section provides a brief overview of MaxSAT and existing algorithms. Section 10.3 describes the `msu4` algorithm and proves the correctness of the proposed approach. Section 10.4 provides experimental results, comparing `msu4` with alternative MaxSAT algorithms. The chapter concludes in Section 10.6.

10.2 Background

This section provides definitions and background knowledge for the MaxSAT problem. Due to space constraints, familiarity with SAT and related topics is assumed and the reader is directed to the bibliography [10].

10.2.1 The MaxSAT Problem

The maximum satisfiability (MaxSAT) problem can be stated as follows. Given an instance of SAT represented in CNF, compute an assignment that maximizes the number of satisfied clauses. During the last decade there has been a growing interest on studying MaxSAT, motivated by an increasing number of practical applications, including scheduling, routing, bioinformatics, and EDA [33].

Despite the clear relationship with the SAT problem, most modern SAT techniques cannot be applied directly to the MaxSAT problem. As a result, most

MaxSAT algorithms are built on top of the standard DPLL [12] algorithm and so do not scale for industrial problem instances [16, 17, 22, 23].

The usual approach (most of the solvers in the MaxSAT competition [3, 4]) is based on a Branch and Bound algorithm, emphasizing the computation of lower bounds and the application of inference rules that simplify the instance [17, 22, 23]. Results from the MaxSAT competition [3] suggest that algorithms based on alternative approaches (e.g., by converting MaxSAT into SAT) do not perform well. As a result, the currently best performing MaxSAT solvers are based on branch and bound with additional inference rules.

More recently, an alternative, in general incomplete, approach to MaxSAT has been proposed [33]. The motivation for this alternative approach is the potential application of MaxSAT in design debugging and the fact that existing MaxSAT approaches do not scale for industrial problem instances.

10.2.2 Solving MaxSAT with PBO

One alternative approach for solving the MaxSAT problem is to use pseudo-Boolean optimization (PBO) (e.g., [24]). The PBO approach for MaxSAT consists of adding a new (*blocking*) variable to each clause. The blocking variable b_i for clause ω_i allows satisfying clause ω_i independently of other assignments to the problem variables. The resulting PBO formulation includes a cost function, aiming at minimizing the number of blocking variables assigned value 1. Clearly, the solution of the MaxSAT problem is obtained by subtracting from the number of clauses the solution of the PBO problem.

Example 10.1 Consider the CNF formula: $\varphi = (x_1)(x_2 + \bar{x}_1)(\bar{x}_2)$. The PBO MaxSAT formulation consists of adding a new blocking clause to each clause. The resulting instance of SAT becomes $\varphi_W = (x_1 + b_1)(x_2 + \bar{x}_1 + b_2)(\bar{x}_2 + b_3)$, where b_1, b_2, b_3 denote blocking variables, one for each clause. Finally, the cost function for the PBO instance is min $\sum_{i=1}^{3} b_i$.

Despite its simplicity, the PBO formulation does not scale for industrial problems, since the large number of clauses results in a large number of blocking variables, and corresponding larger search space. Observe that, for most instances, the number of clauses exceeds the number of variables. For the resulting PBO problem, the number of variables equals the sum of the number of variables and clauses in the original SAT problem. Hence, the modified instance of SAT has a much larger search space.

10.2.3 Relating MaxSAT with Unsatisfiable Cores

In recent years there has been work on relating minimum unsatisfiable and maximally satisfiable subformulas [16, 20, 21, 24]. problem.

This section summarizes properties on the relationship between unsatisfiable cores and MaxSAT, which are used in the next section for developing msu4. Let φ be an unsatisfiable formula with a number of unsatisfiable cores, which may or may not be disjoint. Note that two cores are disjoint if the cores have no identical clauses. Let $|\varphi|$ denote the number of clauses in φ.

Proposition 10.1 (MaxSAT upper bound) *Let φ contain K disjoint unsatisfiable cores. Then $|\varphi| - K$ denotes an upper bound on the solution of the MaxSAT problem.*

Furthermore, suppose blocking variables are added to clauses in φ such that the resulting formula φ_W becomes satisfiable.

Proposition 10.2 (MaxSAT lower bound) *Let φ_W be satisfiable, and let B denote the set of blocking variables assigned value 1. Then $|\varphi| - |B|$ denotes a lower bound on the solution of the MaxSAT problem.*

Clearly, the solution to the MaxSAT problem lies between any computed lower and upper bound.

Finally, it should be observed that the relationship of unsatisfiable cores and MaxSAT was also explored in [16] in the context of partial MaxSAT. This algorithm, msu1, removes one unsatisfiable core each time by adding a fresh set of blocking variables to the clauses in each unsatisfiable core. A possible drawback of the algorithm of [16] is that it can add multiple blocking variables to each clause, an upper bound being the number of clauses in the CNF formula [30]. In contrast, the msu4 algorithm adds at most one additional blocking variable to each clause. Moreover, a number of algorithmic improvements to the algorithm of [16] can be found in [30], i.e., msu2 and msu3. The proposed improvements include linear encoding of the cardinality constraints and an alternative approach to reduce the number of blocking variables used.

10.3 A New MaxSAT Algorithm

This section develops the msu4 algorithm by building on the results of Section 10.2.3. As shown earlier, the major drawback of using a PBO approach for the MaxSAT problem is the large number of blocking variables that have to be used (essentially one for each original clause). For most benchmarks, the blocking variables end up being significantly more than the original variables, which is reflected in the cost function and overall search space. The large number of blocking variables basically renders the PBO approach ineffective in practice.

The msu4 algorithm attempts to reduce as much as possible the number of necessary blocking variables, thus simplifying the optimization problem being solved. Moreover, msu4 avoids interacting with a PBO solver and instead is fully SAT based.

10.3.1 Overview

Following the results of Section 10.2.3, consider identifying disjoint unsatisfiable cores of φ. This can be done by iteratively computing unsatisfiable cores and adding blocking variables to the clauses in the unsatisfiable cores. The identification and blocking of unsatisfiable cores are done on a working formula φ_W. Eventually, a set of disjoint unsatisfiable cores is identified, and the blocking variables allow satisfying φ_W. From Proposition 10.2, this represents a lower bound on the solution of the MaxSAT problem. This lower bound can be refined by requiring fewer blocking variables to be assigned value 1. This last condition can be achieved by adding a cardinality constraint to φ[1].

The resulting formula can still be satisfiable, in which case a further refined cardinality constraint is added to φ_W. Alternatively, the formula is unsatisfiable. In this case, some clauses of φ without blocking variables may exist in the unsatisfiable core. If this is the case, each clause is augmented with a blocking variable, and a new cardinality constraint can be added to φ_W, which requires the number of blocking variables assigned value 1 to be less than the total number of new blocking clauses. Alternatively, the core contains no original clause without a blocking variable. If this is the case, then the highest computed lower bound is returned as the solution to the MaxSAT problem. The proof that this is indeed the case is given below.

In contrast with the algorithms in [16] and [30], the `msu4` algorithm is not exclusively based on enumerating unsatisfiable cores. The `msu4` algorithm also identifies satisfiable instances, which are then eliminated by adding additional cardinality constraints.

10.3.2 The Algorithm

Following the ideas of the previous section, the pseudocode for `msu4` is shown in Algorithm 5. The `msu4` algorithm works as follows. The main loop (lines 8–33) starts by identifying disjoint unsatisfiable cores. The clauses in each unsatisfiable core are modified so that any clause ω_i in the core can be satisfied by setting to 1 a new auxiliary variable b_i associated with ω_i. Consequently, a number of properties of the MaxSAT problem can be inferred. Let $|\varphi|$ denote the number of clauses, let ν_U represent the number of iterations of the main loop in which the SAT solver outcome is unsatisfiable, and let μ_{BV} denote the smallest of the number of blocking variables assigned value 1 each time φ_W becomes satisfiable. Then, an upper bound for the MaxSAT problem is $|\varphi| - \nu_U$, and a lower bound is $|\varphi| - \mu_{BV}$. Both the lower and the upper bounds provide approximations to the solution of the MaxSAT problem, and the difference between the two bounds provides an indication on the number of iterations. Clearly, the MaxSAT solution will require *at most* μ_{BV} blocking variables to be assigned value 1. Also, each time the SAT solver declares the CNF formula

[1] Encodings of cardinality constraints are studied, for example, in [15].

to be unsatisfiable, then the number of blocking variables that must be assigned value 1 can be increased by 1. Each time φ_W becomes satisfiable (line 25), a new cardinality constraint is generated (line 30), which requires the number of blocking variables assigned value 1 to be reduced given the current satisfying assignment (and so requires the lower bound to be increased, if possible). Alternatively, each time φ_W is unsatisfiable (line 12), the unsatisfiable core is analyzed. If there exist initial clauses in the unsatisfiable core, which do not have blocking variables, then additional blocking variables are added (line 17). Formula φ_W is updated accordingly by removing the original clauses and adding the modified clauses (line 18). A cardinality constraint is added to require at least one of the blocking clauses to be assigned value 1 (line 19). Observe that this cardinality constraint is in fact optional, but experiments suggest that it is most often useful. If φ_W is unsatisfiable, and no additional original clauses can be identified, then the solution to the MaxSAT problem has been identified (line 22). Also, if the lower bound and upper bound estimates become equal (line 32), then the solution to the MaxSAT problem has also been identified. Given the previous discussion, the following result is obtained.

Proposition 10.3 *Algorithm 5 gives the correct MaxSAT solution.*

Proof The algorithm iteratively identifies unsatisfiable cores and adds blocking variables to the clauses in each unsatisfiable core that do not yet have blocking variables (i.e., *initial clauses*), until the CNF formula becomes satisfiable. Each computed solution represents an upper bound on the number of blocking variables assigned value 1, and so it also represents a lower bound on the MaxSAT solution. For each computed solution, a new cardinality constraint is added to the formula (see line 30), requiring a smaller number of blocking variables to be assigned value 1. If the algorithm finds an unsatisfiable core containing no more initial clauses without blocking variables, then the algorithm can terminate and the last computed upper bound represents the MaxSAT solution. Observe that in this case the same unsatisfiable core C can be generated, even if blocking clauses are added to other original clauses without blocking clauses. As a result, the existing lower bound is the solution to the MaxSAT problem. Finally, note that the optional auxiliar constraint added in line 19 does not affect correctness, since it solely requires an existing unsatisfiable core not to be re-identified.

10.3.3 A Complete Example

This section illustrates the operation of the msu4 algorithm on a small example formula.

Example 10.2 Consider the following CNF formula:

$$\begin{aligned}\varphi = {} & \omega_1 \cdot \omega_2 \cdot \omega_3 \cdot \omega_4 \cdot \omega_5 \cdot \omega_6 \cdot \omega_7 \cdot \omega_8 \\ & (x_1)\,(\bar{x}_1 + \bar{x}_2)\,(x_2)\,(\bar{x}_1 + \bar{x}_3)\,(x_3)\,(\bar{x}_2 + \bar{x}_3) \\ & (x_1 + \bar{x}_4)\,(\bar{x}_1 + x_4)\end{aligned}$$

Algorithm 5 The msu4 algorithm

```
MSU4(φ)
▷ Clauses of CNF formula φ are the initial clauses
φ_W ← φ                                    ▷ Working formula, initially set to φ
μ_BV ← |φ|                                 ▷ Min blocking variables w/ value 1
ν_U ← 0                                    ▷ Iterations w/ unsat outcome
V_B ← ∅                                    ▷ IDs of blocking variables
UB ← |φ| + 1                               ▷ Upper bound estimate
LB ← 0                                     ▷ Lower bound estimate
while true
    do (st, φ_C) ← SAT(φ_W)
       ▷ φ_C is an unsat core if φ_W is unsat
       if st = UNSAT
         then
             φ_I = φ_C ∩ φ                 ▷ Initial clauses in core
             I ← {i | ω_i ∈ φ_I}
             V_B ← V_B ∪ I
             if |I| > 0
               then φ_N ← {ω_i ∪ {b_i} | ω_i ∈ φ_I}
                    φ_W ← (φ_W − φ_I) ∪ φ_N
                    φ_T ← CNF(Σ_{i∈I} b_i ≥ 1)
                    φ_W ← φ_W ∪ φ_T
               else ▷ Solution to MaxSAT problem
                    return LB
             ν_U ← ν_U + 1
             UB ← |φ| − ν_U                ▷ Refine UB
         else
             ν ← | blocking variables w/ value 1 |
             if μ_BV < ν
               then μ_BV ← ν
                    LB ← |φ| − μ_BV         ▷ Refine LB
             φ_T ← CNF(Σ_{i∈V_B} b_i ≤ μ_BV − 1)
             φ_W ← φ_W ∪ φ_T
       if LB = UB ▷ Solution to MaxSAT problem
         then return LB
```

Initially φ_W contains all the clauses in φ. In the first loop iteration, the core $\omega_1, \omega_2, \omega_3$ is identified. As a result, the new blocking variables b_1, b_2, and b_3 are added, respectively, to clauses ω_1, ω_2, and ω_3, and the CNF encoding of the cardinality constraint $b_1+b_2+b_3 \geq 1$ is also (optionally) added to φ_W. In the second iteration, φ_W is satisfiable, with $b_1 = b_3 = 1$. As a result, the CNF encoding of a new cardinality constraint, $b_1 + b_2 + b_3 \leq 1$, is added to φ_W. For the next iteration, φ_W is unsatisfiable and the clauses ω_4, ω_5, and ω_6 are listed in the unsatisfiable core. As a result, the new blocking variables b_4, b_5, and b_6 are added, respectively, to clauses ω_4, ω_5, and ω_6, and the CNF encoding of the cardinality constraint $b_4+b_5+b_6 \geq 1$ is also (optionally) added to φ_W. In this iteration, since the lower and the upper bounds become equal, then the algorithm terminates, indicating that two blocking variables need to be assigned value 1, and the MaxSAT solution is 6.

From the example, it is clear that the algorithm efficiency depends on the ability for finding unsatisfiable formulas effectively and for generating manageable cardinality constraints. In the implementation of msu4, the cardinality constraints were encoded either with BDDs or with sorting networks [15].

10.4 Experimental Results

The msu4 algorithm described in the previous section has been implemented on top of MiniSAT [14]. Version 1.14 of MiniSAT was used, for which an unsatisfiable core extractor was available. Two versions of msu4 are considered, one (v1) uses BDDs for representing the cardinality constraints and the other (v2) uses sorting networks [15].

All results shown below were obtained on a 3.0 GHz Intel Xeon 5160 with 4 GB of RAM running RedHat Linux. A time-out of 1000 s was used for all MaxSAT solvers considered. The memory limit was set to 2 GB. The MaxSAT solvers evaluated are the best performing solver in the MaxSAT evaluation [3], maxsatz [23], minisat+ [15] for the MaxSAT PBO formulation, and finally msu4. Observe that the algorithm in [16] targets partial MaxSAT, and so performs poorly for MaxSAT instances [3, 30].

In order to evaluate the new MaxSAT algorithm, a set of industrial problem instances was selected. These instances were obtained from existing unsatisfiable subsets of industrial benchmarks, obtained from the SAT competition archives and from SATLIB [7, 18]. The majority of instances considered was originally from EDA applications, including model checking, equivalence checking, and test-pattern generation. Moreover, MaxSAT instances from design debugging [33] were also evaluated. The total number of unsatisfiable instances considered was 691.

Table 10.1 shows the number of aborted instances for each algorithm. As can be concluded, for practical instances, existing MaxSAT solvers are ineffective. The use of the PBO model for MaxSAT performs better than maxsatz, but aborts more instances than either version of msu4. It should be noted that the PBO approach uses minisat+, which is based on a more recent version of MiniSAT than msu4.

Table 10.1 Number of aborted instances

Total	maxsatz	pbo	msu4 v1	msu4 v2
691	554	248	212	163

Figures 10.1, 10.2, and 10.3 show scatter plots comparing maxsatz, the PBO formulation, and msu4 v1 with msu4 v2. As can be observed, the two versions of msu4 are clearly more efficient than either maxsatz or minisat+ on the MaxSAT formulations. Despite the performance advantage of both versions of msu4, there are exceptions. With few outliers, maxsatz can only outperform msu4 v2 on instances where both algorithms take less than 0.1 s. In contrast, minisat+ can outperform msu4 v2 on a number of instances, in part because of the more recent version of MiniSAT used in minisat+.

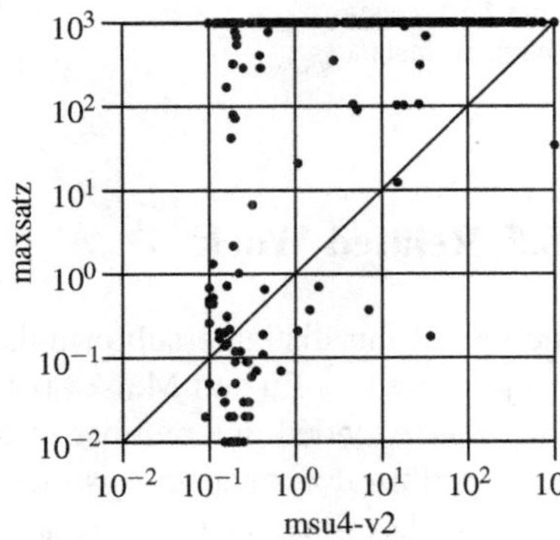

Fig. 10.1 Scatter plot: maxsatz vs. msu4-v2

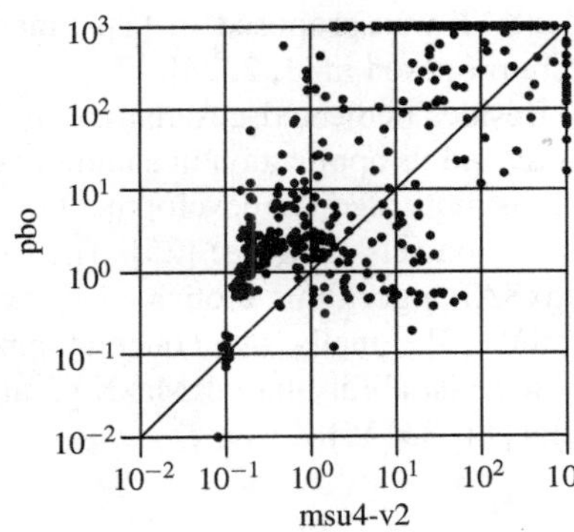

Fig. 10.2 Scatter plot: pbo vs. msu4-v2

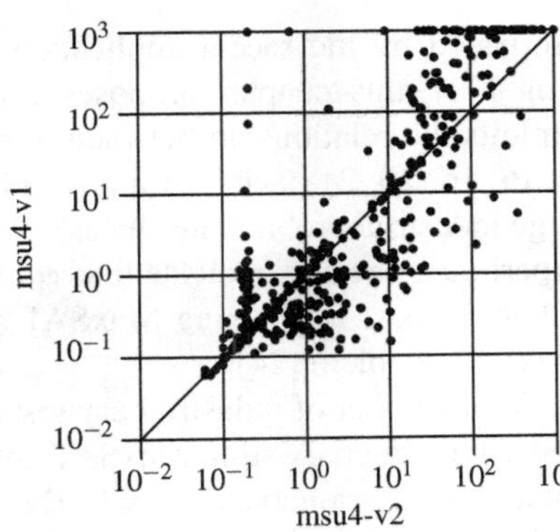

Fig. 10.3 Scatter plot: msu4-v1 vs. msu4-v2

Finally, Table 10.2 summarizes the results for design debugging instances [33]. As can be concluded, both versions of `msu4` are far more effective than either `maxsatz` or `minisat+` on the PBO model for MaxSAT.

Table 10.2 Design debugging instances

Total	maxsatz	pbo	msu4 v1	msu4 v2
29	26	21	3	3

10.5 Related Work

The use of unsatisfiable subformulas for solving (partial) MaxSAT problems was first proposed by Fu and Malik [16]. This algorithm is referred to as msu1.0. This work was extended in a number of different ways in our own work [29–31]. msu4, the algorithm described in this chapter, was first proposed in [31], whereas msu3 was first described in [30]. In addition, msu2 as well as different variations of msu1.0 (namely, msu1.1 and msu1.2) were proposed in [29]. There has been additional work on unsatisfiability-based MaxSAT [1, 28]. A new algorithm for partial MaxSAT was proposed in [1]. Finally, algorithms for weighted partial MaxSAT were proposed in [1, 2, 28].

Besides dedicated unsatisfiability-based algorithms for MaxSAT, this work has motivated its application in a number of areas. Unsatisfiability-based MaxSAT algorithms motivated the development of similar algorithms for computing the minimal correction sets (MCSes) [25]. The use of CNF encodings in unsatisfiability-based MaxSAT algorithms motivated work on improved encodings for cardinality constraints [5]. Finally, one concrete application area where the best solution is given by unsatisfiability-based MaxSAT algorithms is design debugging of digital circuits [11, 33, 35].

10.6 Conclusions

Motivated by the recent application of maximum satisfiability to design debugging [33], this chapter proposes a new MaxSAT algorithm, msu4, that further exploits the relationship between unsatisfiable formulas and maximum satisfiability [6, 16, 20, 21, 24]. The motivation for the new MaxSAT algorithm is to solve large industrial problem instances, including those from design debugging [33]. The experimental results indicate that msu4 performs in general significantly better than either the best performing MaxSAT algorithm [3] or the PBO formulation of the MaxSAT problem [24].

For a number of industrial classes of instances, which modern SAT solvers solve easily but which existing MaxSAT solvers are unable to solve, msu4 is able to find solutions in reasonable time. Clearly, msu4 is effective only for instances for which SAT solvers are effective at identifying small unsatisfiable cores and from which manageable cardinality constraints can be obtained.

Despite the promising results, additional improvements to msu4 are expected. One area for improvement is to exploit alternative SAT solver technology. msu4 is based on MiniSAT 1.14 (due to the core generation code), but more recent SAT solvers could be considered. Another area for improvement is considering alternative encodings of cardinality constraints, given the performance differences observed for the two encodings considered. Finally, the interplay between different algorithms based on unsatisfiable core identification (i.e., `msu1` [16] and `msu2` and `msu3` [30]) should be further developed.

References

1. Ansótegui, C., Bonet, M.L., Levy, J.: Solving (weighted) partial MaxSAT through satisfiability testing. In: International Conference on Theory and Applications of Satisfiability Testing, pp. 427–440. Swansea, UK (2009)
2. Ansótegui, C., Bonet, M.L., Levy, J.: A new algorithm for weighted partial MaxSAT. In: National Conference on Artificial Intelligence. Atlanta, USA (2010)
3. Argelich, J., Li, C.M., Manyá, F., Planes, J.: MaxSAT evaluation. http://www.maxsat07.udl.es/ (2008)
4. Argelich, J., Li, C.M., Manyá, F., Planes, J.: The first and second Max-SAT evaluations. Journal on Satisfiability, Boolean Modeling and Computation **4**, 251–278 (2008)
5. Asín, R., Nieuwenhuis, R., Oliveras, A., Rodríguez-Carbonell, E.: Cardinality networks and their applications. In: International Conference on Theory and Applications of Satisfiability Testing, pp. 167–180. Swansea, UK (2009)
6. de la Banda, M.G., Stuckey, P.J., Wazny, J.: Finding all minimal unsatisfiable sub-sets. In: International Conference on Principles and Practice of Declarative Programming, pp. 32–43. Uppsala, Sweden (2003)
7. Berre, D.L., Simon, L., Roussel, O.: SAT competition. http://www.satcompetition.org/ (2008)
8. Biere, A., Cimatti, A., Clarke, E., Zhu, Y.: Symbolic model checking without BDDs. In: Tools and Algorithms for the Construction and Analysis of Systems, pp. 193–207. (1999)
9. Bonet, M.L., Levy, J., Manyá, F.: Resolution for Max-SAT. Artificial Intelligence **171**(8–9), 606–618 (2007)
10. Bordeaux, L., Hamadi, Y., Zhang, L.: Propositional satisfiability and constraint programming: A comparative survey. ACM Computing Surveys **38**(4) (2006)
11. Chen, Y., Safarpour, S., Veneris, A.G., Marques-Silva, J.: Spatial and temporal design debug using partial MaxSAT. In: ACM Great Lakes Symposium on VLSI, pp. 345–350. Boston, USA (2009)
12. Davis, M., Logemann, G., Loveland, D.: A machine program for theorem-proving. Communications of the ACM **5**, 394–397 (1962)
13. Dershowitz, N., Hanna, Z., Katz, J.: Bounded model checking with QBF. In: International Conference on Theory and Applications of Satisfiability Testing, pp. 408–414. St. Andrews, UK (2005)
14. Een, N., Sörensson, N.: An extensible SAT solver. In: International Conference on Theory and Applications of Satisfiability Testing, pp. 502–518. Santa Margherita Ligure, Italy (2003)
15. Een, N., Sörensson, N.: Translating pseudo-Boolean constraints into SAT. Journal on Satisfiability, Boolean Modeling and Computation **2**, 1–26 (2006)
16. Fu, Z., Malik, S.: On solving the partial MAX-SAT problem. In: International Conference on Theory and Applications of Satisfiability Testing, pp. 252–265. Seattle, USA (2006)
17. Heras, F., Larrosa, J., Oliveras, A.: MiniMaxSat: a new weighted Max-SAT solver. In: International Conference on Theory and Applications of Satisfiability Testing, pp. 41–55. Lisbon, Portugal (2007)
18. Hoos, H., Stützle, T.: SAT lib. http://www.satlib.org/ (2008)
19. Kuehlmann, A., Paruthi, V., Krohm, F., Ganai, M.K.: Robust Boolean reasoning for equivalence checking and functional property verification. IEEE Transactions on CAD of Integrated Circuits and Systems **21**(12), 1377–1394 (2002)

20. Kullmann, O.: Investigations on autark assignments. Discrete Applied Mathematics **107**(1–3), 99–137 (2000)
21. Kullmann, O.: Lean clause-sets: generalizations of minimally unsatisfiable clause-sets. Discrete Applied Mathematics **130**(2), 209–249 (2003)
22. Li, C.M., Manyá, F., Planes, J.: Detecting disjoint inconsistent subformulas for computing lower bounds for Max-SAT. In: National Conference on Artificial Intelligence, pp. 86–91. Boston, USA (2006)
23. Li, C.M., Manyá, F., Planes, J.: New inference rules for Max-SAT. Journal of Artificial Intelligence Research **30**, 321–359 (2007)
24. Liffiton, M.H., Sakallah, K.A.: On finding all minimally unsatisfiable subformulas. In: International Conference on Theory and Applications of Satisfiability Testing, pp. 173–186. (2005)
25. Liffiton, M.H., Sakallah, K.A.: Generalizing core-guided max-sat. In: International Conference on Theory and Applications of Satisfiability Testing, pp. 481–494. Swansea, UK (2009)
26. Liu, C., Kuehlmann, A., Moskewicz, M.W.: CAMA: A multi-valued satisfiability solver. In: International Conference on Computer-Aided Design, pp. 326–333. San Jose, USA (2003)
27. Mangassarian, H., Veneris, A.G., Safarpour, S., Najm, F.N., Abadir, M.S.: Maximum circuit activity estimation using pseudo-Boolean satisfiability. In: Design, Automation and Testing in Europe Conference, pp. 1538–1543. Nice, France (2007)
28. Manquinho, V., Marques-Silva, J., Planes, J.: Algorithms for weighted Boolean optimization. In: International Conference on Theory and Applications of Satisfiability Testing, pp. 495–508. Swansea, UK (2009)
29. Marques-Silva, J., Manquinho, V.: Towards more effective unsatisfiability-based maximum satisfiability algorithms. In: International Conference on Theory and Applications of Satisfiability Testing, pp. 225–230. Guangzhou, China (2008)
30. Marques-Silva, J., Planes, J.: On using unsatisfiability for solving maximum satisfiability. http://arxiv.org/corr/ (2007)
31. Marques-Silva, J., Planes, J.: Algorithms for maximum satisfiability using unsatisfiable cores. In: Design, Automation and Testing in Europe Conference, pp. 408–413. Munich, Germany (2008)
32. McMillan, K.L.: Interpolation and SAT-based model checking. In: Computer-Aided Verification, pp. 1–13. Boulders, USA (2003)
33. Safarpour, S., Mangassarian, H., Veneris, A.G., Liffiton, M.H., Sakallah, K.A.: Improved design debugging using maximum satisfiability. In: Formal Methods in Computer-Aided Design, pp. 13–19. Austin, USA (2007)
34. Smith, A., Veneris, A.G., Ali, M.F., Viglas, A.: Fault diagnosis and logic debugging using Boolean satisfiability. IEEE Transactions on CAD of Integrated Circuits and Systems **24**(10), 1606–1621 (2005)
35. Sülflow, A., Fey, G., Bloem, R., Drechsler, R.: Using unsatisfiable cores to debug multiple design errors. In: ACM Great Lakes Symposium on VLSI, pp. 77–82. Orlando, USA (2008)
36. Wang, K.H., Chan, C.M.: Incremental learning approach and SAT model for Boolean matching with don't cares. In: International Conference on Computer-Aided Design, pp. 234–239. San Jose, USA (2007)
37. Zhang, L., Malik, S.: Validating SAT solvers using an independent resolution-based checker: Practical implementations and other applications. In: Design, Automation and Testing in Europe Conference. Munich, Germany (2003)

Part III
Boolean Matching

In Boolean matching, three research works are presented. By integrating graph-based, simulation-driven, and SAT-based techniques, the first chapter in this section makes Boolean matching feasible for large designs. The next chapter identifies equivalent signals between the original and the modified circuits by using fast functional and structural analysis techniques, and then uses a topologically guided dynamic matching algorithm to identify and hash reusable portions of logic. The last chapter in Boolean matching proposes an incremental learning-based algorithm and a SAT-based approach for Boolean matching.

Chapter 11
Simulation and SAT-Based Boolean Matching for Large Boolean Networks

Kuo-Hua Wang, Chung-Ming Chan, and Jung-Chang Liu

Abstract In this chapter, we explore the permutation-independent (P-equivalent) Boolean matching problem for large Boolean networks. Boolean matching is to check the equivalence of two target functions under input permutation and input/output phase assignment. We propose a matching algorithm seamlessly integrating Simulation and Boolean Satisfiability (S&S) techniques. Our algorithm first utilizes functional properties like unateness and symmetry to reduce the searching space. In the followed simulation phase, three types of input vector generation and checking are applied to match the inputs of two target functions. Moreover, for those inputs that cannot be distinguished in the simulation phase, we propose a recursive matching algorithm to find all feasible mapping solutions. Experimental results on large benchmarking circuits demonstrate that our matching algorithm is indeed very effective and efficient to solve Boolean matching for large Boolean networks.

11.1 Introduction

Logic simulation technique has been widely used in design verification and debugging over a long period of time. The major disadvantage of simulation comes from the fact that it is very time consuming and almost impossible to catch completely the functionality of a very large Boolean network. To solve this issue, Boolean satisfiability (SAT) technique was proposed and exploited in many industrial formal verification tools. In recent years, the technique of combining simulation and SAT (S&S) was popular and successfully applied in many verification and synthesis problems like equivalence checking [1] and logic minimization [2–5].

Boolean matching is to check whether two functions are equivalent or not under input permutation and input/output phase assignment (so-called *NPN-class*). The

K.-H. Wang (✉)
Fu Jen Catholic University, Taipei County, Taiwan
e-mail: khwang@csie.fju.edu.tw

This work is based on an earlier work: Simulation and SAT-based Boolean matching for large Boolean networks, in Proceedings of the 46th Annual Design Automation Conference, ISBN:978-1-60558-497-3 (2009) © ACM, 2009. DOI= http://doi.acm.org/10.1145/1629911.1630016

S.P. Khatri, K. Gulati (eds.), *Advanced Techniques in Logic Synthesis, Optimizations and Applications*, DOI 10.1007/978-1-4419-7518-8_11,

important applications of Boolean matching involve the verification of two circuits under unknown input correspondences, cell-library binding, and table look-up based FPGA's technology mapping. In the past decades, various Boolean matching techniques had been proposed and some of these approaches were discussed in the survey paper [6]. Among those previously proposed approaches, computing *signatures* [6, 7] and transforming into *canonical form* [8–10] of Boolean functions were the most successful techniques to solve Boolean matching. Recently, SAT technique was also applied for Boolean matching [11, 12]. Most of these techniques were proposed to handle completely specified functions, comparatively little research had focused on dealing with Boolean functions with don't cares [7, 12, 13].

The first and foremost issue for Boolean matching is the data structure for representing Boolean functions. As we know that many prior techniques used truth table, sum of products (SOPs), and Binary Decision Diagrams (BDD's) [14] to represent the target functions during the matching process, they suffered from the same memory explosion problem. It is clear that the required storage for truth table will grow exponentially as the number of input variables increases. In addition, many types of Boolean functions cannot be represented by SOPs for large input set and the memory space will explode while constructing their BDD's. Therefore, these Boolean matching techniques were constrained to apply for small to moderate Boolean networks (functions). To address the above issues, And-Inverter Graphs (AIGs) [15] had been utilized and successfully applied in verification and synthesis problems [4, 15].

In this chapter, we propose a permutation-independent (P-equivalent) Boolean matching algorithm for large Boolean functions represented by AIGs. The contributions of our matching algorithm involve three aspects:

- it is the first one seamlessly integrating simulation and SAT-based techniques to deal with Boolean matching;
- it applies functional properties such as unateness and symmetry to reduce the searching space quickly;
- it is complete by integrating a recursive matching algorithm which can find not only one feasible mapping solution but also all mapping solutions.

The remainder of this chapter is organized as follows. Section 11.2 gives a brief research background for our work. Section 11.3 shows our procedure of detecting functional properties. Some definitions and notations are given in Section 11.4. Section 11.5 presents our simulation strategy for distinguishing the input variables of Boolean functions. Sections 11.6 and 11.7 show our S&S-based matching algorithm with implementation issues and the experimental results, respectively. We summarize the chapter in Section 11.8.

11.2 Background

11.2.1 Boolean Matching

Consider two target Boolean functions $f(X)$ and $g(Y)$. Boolean matching is to check the equivalence of these two functions under input permutation and input/output phase assignment. To solve this problem, we have to search a *feasible mapping*

ψ such that $f(\psi(X)) = g(Y)$ (or $\bar{g}(Y)$). It is impractical to search all possible mappings because the time complexity is $O(2^{n+1} \cdot n!)$, where n is the number of input variables. Among those previously proposed techniques for Boolean matching, *signature* is one of the most effective approaches. Various signatures were defined to characterize input variables of Boolean functions. Since these signatures are invariant under the permutation or complementation of input variables, the input variables with different signatures can be distinguished to each other and many infeasible mappings can be pruned quickly. However, it had been proved that signatures have the inherent limitation to distinguish all input variables for those functions with *$\mathcal{G}$-symmetry* [16, 17].

11.2.2 Boolean Satisfiability

The Boolean Satisfiability (SAT) problem is to find a variable assignment to satisfy a given conjunctive normal form (CNF) or prove it is equal to the constant 0. Despite the fact that SAT problem is NP-complete, many advanced techniques like *non-chronological backtracking*, *conflict driven clause learning*, and *watch literals* have been proposed and implemented in state-of-the-art SAT solvers [18–20]. Therefore, SAT technique has been successfully applied to solve many EDA problems [21] over the past decade. Among these applications, combinational equivalence checking (CEC) is an important one of utilizing SAT solver to check the equivalence of two combinational circuits. The following briefly describes the concept of SAT-based equivalence checking. Consider two functions (circuits) f and g to be verified. A *miter* circuit with functionality $f \oplus g$ is constructed first and then transformed into a SAT instance (circuit CNF) by simple gate transformation rules. If this circuit CNF cannot be satisfied, then f and g are equivalent; otherwise, they are not equivalent.

11.2.3 And-Inverter Graph

And-Inverter Graph (AIG) is a directed acyclic graph which can be used as the structural representation of Boolean functions. It consists of three types of nodes: primary input, 2-input AND, and constant 0(1). The edges with INVERTER attribute denote the Boolean complementation. Due to the simple and regular structure of AIG, it is easy to transform a Boolean network (function) into AIG by simple gate transformation rules. Moreover, it is very fast to perform simulation on AIG with respect to (w.r.t.) a large set of input vectors at one time. However, it is unlike to the BDD which is a canonical form of Boolean functions for a specific input variable order. It may also have many functionally equivalent nodes in the graph. In the paper [13], SAT sweeping and structural hashing techniques were applied to reduce the graph size. More recently, Mishchenko et al. exploited SAT-based equivalence checking techniques to remove equivalent nodes while constructing an AIG, i.e., Functionally Reduced AIGs (FRAIGs) [22]. By our experimental observation, FRAIGs can represent many large Boolean functions which cannot be constructed as BDD's because of the memory explosion problem.

11.3 Detection of Functional Property Using S&S Approach

Consider a function $f(X)$ and an input $x_i \in X$. The *cofactor* of f w.r.t. x_i is $f_{x_i} = f(x_1, \ldots, x_i = 1, \ldots, x_n)$. The cofactor of f w.r.t. $\bar{x_i}$ is $f_{\bar{x_i}} = f(x_1, \ldots, x_i = 0, \ldots, x_n)$. A function f is *positive* (*negative*) *unate* in variable x_i if $f_{\bar{x_i}} \subseteq f_{x_i}$ ($f_{x_i} \subseteq f_{\bar{x_i}}$). Otherwise, it is *binate* in that variable. Given two inputs $x_i, x_j \in X$, the *non-equivalence symmetry* (*NE*), *equivalence symmetry* (*E*), and *single variable symmetry* (*SV*) of f w.r.t. x_i and x_j are defined and summarized in Table 11.1.

Table 11.1 Definition and S&S-based checking of functional properties

	Property		**S&S checking**		
Name	**Definition**	**Notation**	**Disj.**	**Removal_Condition**	**SAT_Check**
Positive unate	$f_{\bar{x_i}} \subseteq f_{x_i}$	$PU(x_i)$	x_i	$f(v_1) = 1, val(v_1, x_i) = 0$	$f_{\bar{x_i}} \cdot \bar{f}_{x_i} = 0$
Negative unate	$f_{x_i} \subseteq f_{\bar{x_i}}$	$NU(x_i)$	x_i	$f(v_1) = 1, val(v_1, x_i) = 1$	$\bar{f}_{\bar{x_i}} \cdot f_{x_i} = 0$
NE symmetry	$f_{\bar{x_i}x_j} = f_{x_i\bar{x_j}}$	$NE(x_i, x_j)$	x_i, x_j	$val(v_1, x_i) \neq val(v_1, x_j)$	$f_{\bar{x_i}x_j} \oplus f_{x_i\bar{x_j}} = 0$
E symmetry	$f_{\bar{x_i}\bar{x_j}} = f_{x_ix_j}$	$E(x_i, x_j)$	x_i, x_j	$val(v_1, x_i) = val(v_1, x_j)$	$f_{\bar{x_i}\bar{x_j}} \oplus f_{x_ix_j} = 0$
	$f_{\bar{x_i}\bar{x_j}} = f_{x_i\bar{x_j}}$	$SV(x_i, \bar{x_j})$	x_i	$val(v_1, x_j) = 0, \forall x_j \in X - \{x_i\}$	$f_{\bar{x_i}\bar{x_j}} \oplus f_{x_i\bar{x_j}} = 0$
Single variable	$f_{\bar{x_i}x_j} = f_{x_ix_j}$	$SV(x_i, x_j)$	x_i	$val(v_1, x_j) = 1, \forall x_j \in X - \{x_i\}$	$f_{\bar{x_i}x_j} \oplus f_{x_ix_j} = 0$
symmetry	$f_{\bar{x_i}\bar{x_j}} = f_{\bar{x_i}x_j}$	$SV(x_j, \bar{x_i})$	x_j	$val(v_1, x_i) = 0, \forall x_i \in X - \{x_j\}$	$f_{\bar{x_i}\bar{x_j}} \oplus f_{\bar{x_i}x_j} = 0$
	$f_{x_i\bar{x_j}} = f_{x_ix_j}$	$SV(x_j, x_i)$	x_j	$val(v_1, x_i) = 1, \forall x_i \in X - \{x_j\}$	$f_{x_i\bar{x_j}} \oplus f_{x_ix_j} = 0$

S&S approach is applied to check functional unateness and symmetry of target functions in our matching algorithm. Instead of enumerating all items (possible functional properties) and checking them directly, we exploit simulation to quickly remove impossible items. For those items that cannot be removed by simulation, SAT-based technique is applied to verify them. The generic functional property detection algorithm is shown in Fig. 11.1. A similar *NE-symmetry* detection pro-

Algorithm *S&S-Based-Functional-Property-Detection*(F, p)
Input: F is a Boolean network;
    p is the functional property to be checked;
Output: S is the set of items with property p;
Begin
  Initialize $S = \emptyset$ and add all possible items into T;
  Perform *Random Simulation* on F to remove impossible items in T;
  /* Refer to Table 11.1. */
  while ($T \neq \emptyset$) **do**
    repeat
      Take an item t_i away from T;
      $flag$ = *SAT-Based-Checking*(F, p, t_i);   /* Refer to Table. 11.1 */
      if ($flag$ is **FALSE**) $S = S \cup \{t_i\}$;   /* t_i is true property p. */
    until ($flag$ is **TRUE**);
    Perform *Guided Simulation* on F to remove impossible items in T
    based on counter examples by SAT solving;
  endwhile
  return S;
End

Fig. 11.1 The S&S-based functional property checking algorithm

cedure was proposed in the paper [5]. Consider a function f and some functional property p to be checked. Our detecting algorithm starts by using random simulation to remove impossible items as many as possible. If there still exist some unchecked items, it repeats taking an item and checking it with SAT-based technique until the taken item is a true functional property of f. Guided simulation will then be used to filter out the remaining impossible items. Rather than generating pure random vectors, guided simulation can generate simulation vectors based on counterexamples by SAT solving, i.e., solutions of the SAT instance.

In order to remove impossible items, we generate many pairs of random vectors (v_1, v_2)'s with Hamming distance 1 or 2 for simulation. The vector pairs with distance 1 are used to remove functional unateness and single variable symmetries, while the pairs with distance 2 are used to remove *NE* and *E* symmetries. Assume that v_1 and v_2 are disjoint on input x_i (and x_j) if their distance is 1 (2). Without loss of generality, let $f(v_1) \neq f(v_2)$ and $f(v_1) = 1$. The conditions for removing impossible functional properties and SAT-based equivalence checking are briefly summarized in Table 11.1, where the notation $val(v_1, x_i)$ denotes the value of x_i in the input vector v_1. The following example is given for illustration.

Example 11.1 Consider a function $f(x_1, x_2, x_3, x_4, x_5)$. Let two vectors v_1 = 00011 and v_2 = 10011 be disjoint on the first variable x_1. Assume that $f(v_1) = 1$ and $f(v_2) = 0$. It is obvious f is not positive unate in x_1 ($PU(x_1)$ can be removed) because of $val(v_1, x_1) = 0$. In addition, since $val(v_1, x_2) = val(v_1, x_3) = 0$ and $val(v_1, x_4) = val(v_1, x_5) = 1$, $SV(x_1, \bar{x_2})$, $SV(x_1, \bar{x_3})$, $SV(x_1, x_4)$, and $SV(x_1, x_5)$ can be removed accordingly. Consider the third input vector v_3 = 00101 with $f(v_3) = 0$. It is easy to see that v_1 and v_3 are disjoint on the variables x_3 and x_4. It fulfills the condition $val(v_1, x_3) \neq val(v_1, x_4)$. Consequently, $NE(x_3, x_4)$ can be removed. ■

11.4 Definitions and Notations

Let $P = \{X_1, X_2, \ldots, X_k\}$ be a *partition* of input set X, where $\bigcup_{i=1}^{k} X_i = X$ and $X_i \cap X_j = \emptyset$ for $i \neq j$. Each X_i is an *input group* w.r.t. P. The *partition size* of P is the number of subsets X_i's in P, denoted as $|P|$. The *group size* of X_i is the number of input variables in X_i, denoted as $|X_i|$.

Definition 11.1 Given two input sets X and Y with the same number of input variables, let $P_X = \{X_1, X_2, \ldots, X_k\}$ and $P_Y = \{Y_1, Y_2, \ldots, Y_k\}$ be two ordered input partitions of X and Y, respectively. A **mapping relation** $R = \{G_1, G_2, \ldots, G_k\}$ is a set of mappings between the input groups of P_X and P_Y, where $G_i = {X_i}^{Y_i}$ and $|X_i| = |Y_i|$. Each element $G_i \in R$ is a **mapping group** which maps X_i to Y_i. ■

Definition 11.2 Consider a mapping relation R and a mapping group $G_i = {X_i}^{Y_i} \in R$. The **mapping relation size** is the number of mapping groups in R, denoted as $|R|$. The **mapping group size** of G_i, denoted as $|G_i|$, is the group size of X_i (or Y_i), i.e., $|G_i| = |X_i| = |Y_i|$. ■

Definition 11.3 Consider two functions $f(X)$ and $g(Y)$. Let $G_i = X_i^{Y_i}$ be a mapping group in a mapping relation R. G_i is **unique** if and only if $|G_i| = 1$ or X_i (Y_i) is a *NE-symmetric* set of f (g). The mapping relation R is **unique** if and only if all mapping groups in R are unique. ■

Definition 11.4 Let v_i be an input vector w.r.t. the input set X. The **input weight** of v_i is the number of inputs with binary value 1. It is denoted as $\rho(v_i, X)$. ■

Definition 11.5 Consider a function $f(X)$ and a vector set V involving m distinct input vectors. The **output weight** of f w.r.t. V is the number of vectors v_i's in V such that $f(v_i) = 1$. It is denoted as $\sigma(f, V)$ and $0 \le \sigma(f, V) \le m$. ■

Example 11.2 Consider the Boolean function $f(X) = \bar{x_1}x_2 + x_2x_4 + x_1\bar{x_3}x_4$ and input vector set $V = \{v_1, v_2, v_3, v_4\}$. The Karnaugh map of f and set V are shown in Fig. 11.2. In this figure, $O = 1010$ is the output vector of f w.r.t. V, where each value in O indicates the output value of $f(v_i)$. We can see that the input weight of each $v_i \in V$ is $\rho(v_i, X) = 3$ and the output weight of f w.r.t. V is $\sigma(f, V) = 2$. ■

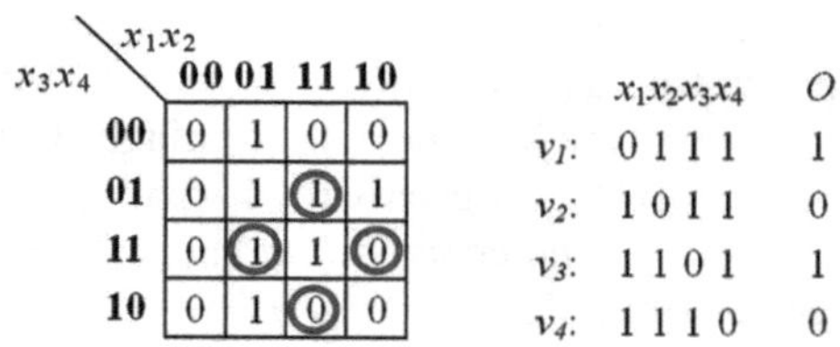

Fig. 11.2 The Karnaugh map of f and input/output vector sets V/O

11.5 Simulation Approach for Distinguishing Inputs

The idea behind our simulation approach is the same as the concept of exploiting signatures to quickly remove impossible input correspondences as many as possible. Consider two target functions $f(X)$ and $g(Y)$. Let $R = \{G_1, \ldots, G_i, \ldots, G_c\}$ be the current mapping relation. Without loss of generality, groups $G_1, \ldots, G_i$ are assumed non-unique while the remaining groups are unique. To partition a non-unique mapping group $G_i = X_i^{Y_i}$, for each input $x_j \in X_i$ $(y_j \in Y_i)$, we generate a vector v or a set of vectors V for simulation and use the simulation results as the signature of x_j (and y_j) w.r.t. f (and g). It can further partition G_i into two or more smaller mapping groups by means of these signatures. Suppose the mapping size of G_i is m, i.e., $|G_i| = |X_i| = |Y_i| = m$. In the following, we will propose three types of input vectors and show how to distinguish the input variables of X_i (and Y_i) in terms of the simulation results.

11.5.1 Type-1

We first generate c subvectors $v_1, \ldots, v_i, \ldots, v_c$, where v_i is a random vector with input weight 0 or m, i.e., $\rho(v_i, X_i) = \rho(v_i, Y_i) = 0$ or m. For each input variable $x_j \in X_i$ (and $y_j \in Y_i$), the subvector $\tilde{v}_i$ with input weight 1 or $m - 1$ can be obtained by complementing the value of x_j (and y_j) in v_i. The concatenated vector $v^j = v_1|\cdots|\tilde{v}_i|\cdots|v_m$ will then be used for simulating on f (and g) and its corresponding output value $f(v^j)$ $(g(v^j))$ can be viewed as the signature of x_j (y_j). Figure 11.3 demonstrates the vector set V_i used to partition G_i, where A_i is the set X_i or Y_i. Each vector (row) in V_i is dedicated to an input variable x_j (y_j) in X_i (Y_i). Using such a set V_i for simulation, in most cases, X_i (Y_i) can be partitioned into two subsets X_{i0} (Y_{i0}) and X_{i1} (Y_{i1}), where the signatures of input variables in these two sets are corresponding to the output value 0 and 1, respectively. Therefore, G_i can be divided into two mapping groups $G_{i0} = {X_{i0}}^{Y_{i0}}$ and $G_{i1} = {X_{i1}}^{Y_{i1}}$. Moreover, in our matching algorithm all non-unique mapping groups can be partitioned simultaneously. The following example is given for illustration.

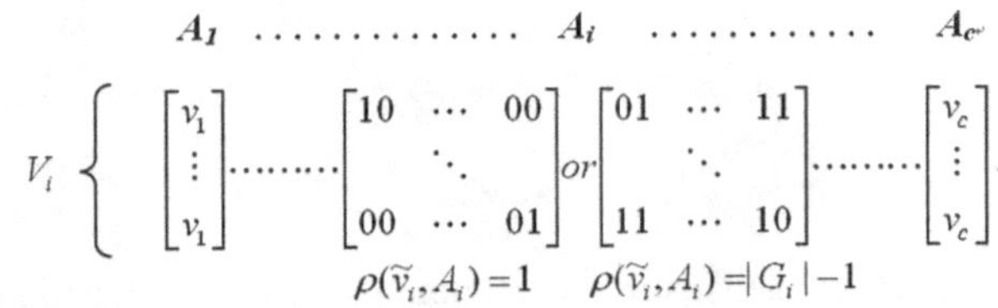

Fig. 11.3 Type-1 simulation vectors V_i for G_i

Example 11.3 Consider two functions $f(X)$, $g(Y)$, and the initial mapping relation $R = \{G_1, G_2\}$. Suppose that $G_1 = {X_1}^{Y_1}$ and $G_2 = {X_2}^{Y_2}$, where $X_1 = \{x_1, x_3\}$, $X_2 = \{x_2, x_4\}$, $Y_1 = \{y_2, y_3\}$, and $Y_2 = \{y_1, y_4\}$. To partition the mapping groups G_1 and G_2, we first generate two random vectors $v_1 = 00$ and $v_2 = 11$ for X_1 (Y_1) and X_2 (Y_2), respectively. Based on the method described as above, two input vector sets V_1 and V_2 are then generated for simulation. Figure 11.4 shows V_1, V_2, and the corresponding simulated output vectors $O_f = 1100$, $O_g = 1101$. By the output result of simulating V_1, we cannot partition G_1 since all its output values are equivalent to 1. To look at the output values simulated by V_2, we can further partition G_2 into two new mapping groups $\tilde{G}_2 = \{x_2\}^{\{y_4\}}$ and $\tilde{G}_3 = \{x_4\}^{\{y_1\}}$ corresponding to the output value 1 and 0, respectively. So we can get the new mapping relation $\tilde{R} = \{G_1, \tilde{G}_2, \tilde{G}_3\}$. ■

V	x_1x_3 / y_2y_3	x_2x_4 / y_1y_4	O_f	O_g
V_1	10	11	1	1
	01	11	1	1
V_2	00	01	1	0
	00	10	0	1

Fig. 11.4 Input/output vectors of $f(X)$ and $g(Y)$

11.5.2 Type-2

For each input $x_j \in X_i$ ($y_j \in Y_i$), a vector set V_j involving $|G_i| - 1$ vectors with input weight 2 (or $m-2$) will be generated. For simplicity, Fig. 11.5 only shows out the subvectors w.r.t. the input set X_i while the subvectors w.r.t. the remaining input sets can be generated like the initial subvectors v_i's of Type-1. Consider any vector v in V_j. We assign 1 (or 0) to the input variable x_j and one of the remaining inputs, while the other inputs are assigned 0 (or 1). After the simulation, the output weight $\sigma(f, V_j)$ ($\sigma(g, V_j)$) will be used as the signature of x_j (y_j) and input variables with different output weight can be distinguished to each other. Consequently, we can partition G_i into at most m groups because of $0 \leq \sigma(f, V_j) \leq m-1$. Besides, if some input $x_j \in X_i$ can uniquely map to an input $y_j \in Y_i$, we can further apply Type-1 checking to partition the set $X_i - \{x_j\}$ ($Y_i - \{y_j\}$) using the simulation results by the vector set V_j. We give an example for illustration.

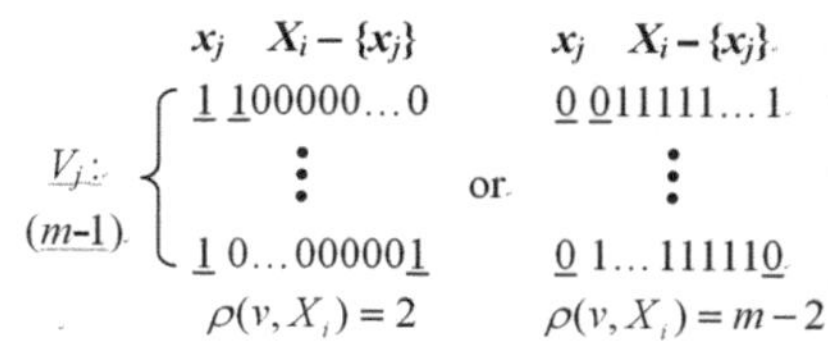

Fig. 11.5 Type-2 simulation vectors V_j

Example 11.4 Consider two functions $f(X)$, $g(Y)$, and initial mapping relation $R = \{G_1, G_2\}$, where $G_1 = {X_1}^{Y_1}$ and $G_2 = {X_2}^{Y_2}$. Suppose that $X_1 = \{x_1, x_2, x_3\}$, $Y_1 = \{y_2, y_3, y_4\}$, $X_2 = \{x_4\}$, and $Y_2 = \{y_1\}$. The input vector set $V = V_1 \cup V_2 \cup V_3$ is generated to partition G_1. These vector sets V_1, V_2, and V_3 are dedicated to x_1 (y_2), x_2 (y_3), and x_3 (y_4), respectively. Figure 11.6 shows the input vector sets V_i's, the output simulation vectors O_f, O_g, and their corresponding output weights. By the distribution of output weight, x_2 can uniquely map to y_2 because of $\sigma(f, V_2) = \sigma(g, V_1) = 1$. The input set $\{x_1, x_3\}$ can map to $\{y_3, y_4\}$ in the same way. Consequently, the original group G_1 can be partitioned into two groups $\tilde{G_1} = \{x_2\}^{\{y_2\}}$ and $\tilde{G_2} = \{x_1, x_3\}^{\{y_3, y_4\}}$. Moreover, since $\tilde{G_1}$ is unique, we can further partition $\tilde{G_2}$ using Type-1 checking method. By the simulation results of f on V_2 and of g on V_1, x_1 (x_3) can map to y_4 (y_3) because they have the same output value 0 (1). Finally, each input variable of f can uniquely map to one variable of g. ∎

11.5.3 Type-3

The third type of vectors can only be used for the mapping group $G_i = {X_i}^{Y_i}$, where X_i and Y_i involve several *NE-symmetric* sets. The idea is mainly based on functional symmetry that function f is invariant under the permutation of inputs in its *NE-symmetric* set. Suppose X_i consists of e symmetric set $S_1, S_2, \ldots, S_e$ each

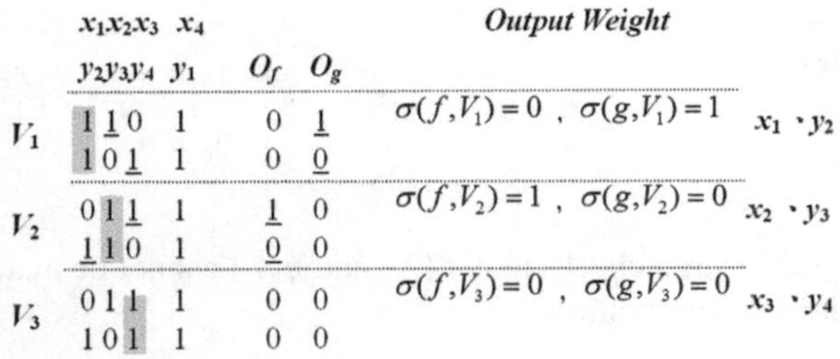

Fig. 11.6 Input/output vectors and output weight of Example 11.4

with k input variables. To partition the group G_i, two random vectors a_1 and a_2 with different input weight w_1 and w_2 will be generated, where $0 \leq w_1, w_2 \leq k$. For each symmetric set S_i, we then generate a vector v_i by assigning a_2 and a_1's to S_i and the remaining sets, respectively. Figure 11.7 shows only the weight distribution of vectors v_i's, where v_i is dedicated to S_i for simulation. As to the other mapping groups, the weights of their subvectors must be 0 or $|G_i|$ like we use in Type-1. By such a vector set for simulation, we can partition $S_1, S_2, \ldots, S_e$ into two groups of symmetric sets which have output simulation value 0 and 1, respectively. It is clear that at most $k \times (k + 1)$ combinations of w_1 and w_2 are required for this type of checking.

Fig. 11.7 Type-3 simulation vectors of G_i

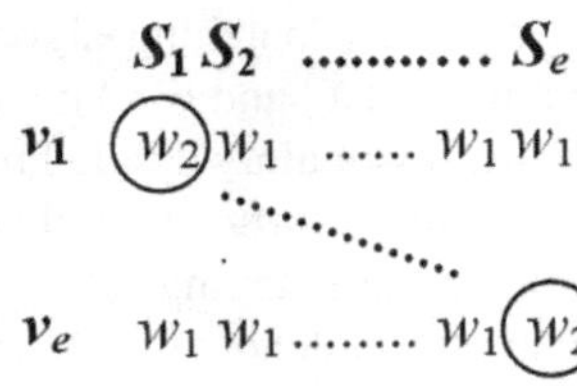

Example 11.5 Consider two functions $f(X)$, $g(Y)$, and initial mapping relation $R = \{G_1, G_2\}$, where $G_1 = X_1{}^{Y_1}$ and $G_2 = X_2{}^{Y_2}$. Suppose that $X_1 = \{x_1, x_2, x_3, x_5, x_6, x_7\}$, $Y_1 = \{y_2, y_3, y_4, y_5, y_6, y_7\}$, $X_2 = \{x_4, x_8\}$, and $Y_2 = \{y_1, y_8\}$. In the input set X_1, there are three *NE-symmetric* sets with size 2, i.e., $\{x_1, x_2\}$, $\{x_3, x_5\}$, and $\{x_6, x_7\}$. Y_1 also has three *NE-symmetric* sets $\{y_2, y_3\}$, $\{y_4, y_5\}$, and $\{y_6, y_7\}$. Figure 11.8 shows one set of input vectors for simulation and two output vectors of f and g. We first assign the input subvector of each symmetric set in X_1 (and Y_1) as 01, i.e., the input weight is 1. For each symmetric set to be checked, we change its input subvector to 11 with input weight 2. In this case, the input subvector of X_2 (and Y_2) is assigned as 11. According to the output vectors O_f and O_g, the set X_1 can be divided into two groups $\tilde{X_1} = \{x_1, x_2\}$ and $\tilde{X_2} = \{x_3, x_5, x_6, x_7\}$ which have output values 0 and 1, respectively. Similarly, Y_1 can be divided into two groups $\tilde{Y_1} = \{y_4, y_5\}$ and $\tilde{Y_2} = \{y_2, y_3, y_6, y_7\}$. Therefore, we can obtain a new mapping relation $\tilde{R} = \{\tilde{G_1}, \tilde{G_2}, G_2\}$, where $\tilde{G_1} = \tilde{X_1}^{\tilde{Y_1}}$ and $\tilde{G_2} = \tilde{X_2}^{\tilde{Y_2}}$. So $\tilde{G_1}$ becomes a unique mapping group by such a partitioning. ■

Fig. 11.8 Input/output vectors in Example 11.5

	x_1x_2 y_2y_3	x_3x_5 y_4y_5	x_6x_7 y_6y_7	x_4x_8 y_1y_8	Output O_f	O_g
v_1	11	01	01	11	0	1
v_2	01	11	01	11	1	0
v_3	01	01	11	11	1	1

Our matching algorithm checks that $f(X)$ and $g(Y)$ cannot be matched successfully by the following observation.

Observation 11.5.1 *Consider a non-unique mapping group $G_i = X_i^{Y_i}$. Through the simulation and checking steps as we described above, let P_{X_i} and P_{Y_i} be the resultant partitions w.r.t. to X_j and Y_j, respectively. For each group $A \in P_{X_i}$, there is a corresponding group $B \in P_{Y_i}$. The following shows two situations that $f(X)$ and $g(Y)$ cannot match each other:*

- $|P_{X_j}| \neq |P_{Y_j}|$, *i.e., their partition sizes are different.*
- $|A| \neq |B|$, *i.e., their set sizes are different.* ■

11.6 S&S-Based Boolean Matching Algorithm

11.6.1 Our Matching Algorithm

Our Boolean matching algorithm is shown in Fig. 11.9. It can match two target functions $f(X)$ and $g(Y)$ under a *threshold*, i.e., the maximum number of simulation rounds. It is mainly divided into three phases: *Initialization*, *Simulation*, and *Recursion*. In the *Initialization* phase, it exploits functional unateness and symmetries to initialize the mapping relation R and computes the maximum number of mapping groups in R, denoted as $MaxSize$. Since the input variables in the *NE-symmetric* set can be permuted without affecting the functionality, $MaxSize$ is equal to the number of *NE-symmetric* sets adding the number of non-symmetric inputs. It is the upper bound used to check the first terminating condition of the second phase. In the *Simulation* phase, it improves the mapping relation R by the *Simulate-and-Update* procedure implementing the simulation approach described in Section 11.5. This step is repeated until it can find a unique mapping relation R with $|R| = MaxSize$ or no improvement can be made to R under the *threshold* bound. If this phase is ended by the second terminating condition, it calls the *Recursive-Matching* procedure and enters into the *Recursion* phase to search the feasible mapping relation. Otherwise, the *SAT-Verify* procedure exploiting SAT-based technique is called to verify if two target functions are matched under the unique mapping relation R searched by the *Simulation* phase.

11.6.2 Recursive-Matching Algorithm

Figure 11.10 shows our recursive matching algorithm. Consider target functions f, g, a mapping relation R, and the size bound $MaxSize$ of R. It starts by checking if R is unique and so to verify R using SAT technique. For the case that R is not

```
Algorithm S&S-Based-Boolean-Matching(f(X), g(Y), threshold)
Input: f and g are target functions;
       threshold is the maximum number of simulation rounds;
Output: ∅ or ψ, i.e., the feasible mapping relation for f and g;
Begin
   ψ = ∅;   cnt = 0;
   R = Initial-Mapping(f,g);                          // Phase 1: Initialization
   MaxSize = #NE_Classes + #Non_Symm_Inputs;
   while (|R| < MaxSize and cnt < threshold) do      // Phase 2: Simulation
      NewR = Simulate-And-Update(f,g,R);
      if (NewR = ∅) return ∅;
      if (|R| == |NewR|) then
         cnt = cnt + 1;     // no improvement on R
      else
         R = NewR;  cnt = 0;
      endif
   endwhile
   if (|R| < MaxSize) then
      ψ = Recursive-Matching(f,g,R,MaxSize);         // Phase 3: Recursion
   else
      if (SAT-Verify(f,g,R) is TRUE) ψ = R;
   endif
   return ψ;
End
```

Fig. 11.9 Simulation and SAT-based Boolean matching algorithm

```
Algorithm Recursive-Matching(f(X),g(Y),R,MaxSize)
Input: f and g are target functions;
       R is the current mapping relation;
       MaxSize is the maximum mapping relation size for R;
Output: ∅ or ψ, i.e., the feasible mapping relation for f and g;
Begin
   if (|R| == MaxSize) then   // the terminating condition
      if (SAT-Verify(f,g,R) is TRUE) return ψ;
      return ∅;
   endif
   G_i = the smallest non-unique mapping group in R;
   ψ = ∅; Choose an input x_j ∈ X_i;
   for each possible mapping relation T w.r.t. G_i do
      tmpR = R ∪ T − {G_i};
      NewR = Simulate-And-Update(f,g,tmpR);
      if (NewR ≠ ∅)
         ψ = Recursive-Matching(f,g,NewR,MaxSize);
      // comment the next line if to find all mapping solutions;
      if (ψ ≠ ∅) return ψ;
   endfor
   return ψ;
End
```

Fig. 11.10 Recursive-matching algorithm

unique, the smallest non-unique mapping group $G_i = {X_i}^{Y_i}$ in R will be selected for partitioning. Consider the mapping of an input $x_j \in X_i$ to an input $y_k \in Y_i$. It will partition G_i into two mapping groups $A = \{x_j\}^{\{y_k\}}$ and $B = (X_i - \{x_j\})^{Y_i - \{y_k\}}$. With such a partitioning, we can derive a new mapping relation $tmpR = R \cup T - \{G_i\}$ with mapping relation size $|R| + 1$, where $T = \{A, B\}$ is derived from G_i.

Simulate-and-Update procedure is then called to further improve $tmpR$ and return a new mapping relation $NewR$. If $NewR$ is not empty, it will call itself again; otherwise, it indicates a wrong selection of input mapping. Moreover, this algorithm can be easily modified to find all feasible mapping relations as shown in Fig. 11.10.

11.6.3 Implementation Issues

11.6.3.1 Control of Random Vector Generation

By our experimental results, most of the runtime was consumed by the simulation phase for some test cases. The reason is that too many random vectors generated for simulation are useless to improve current mapping relation. Thus it will incur a large amount of iterations on the simulating and updating steps in this phase. Instead of generating random vectors without using any criterion, we propose a simple heuristic to control the generation of two adjacent random vectors. Let v_1 be the first random vector. The second vector v_2 is generated by randomly complementing $n/2$ inputs in v_1, where n is the number of input variables of target functions. We expect it can evenly distribute the random vectors in the Boolean space and so that it can quickly converge to find a feasible mapping relation. Our experimental result shows the runtime can be greatly reduced for some benchmarking circuits.

11.6.3.2 Reduction of Simulation Time

Our matching algorithm can be easily extended to deal with Boolean functions with multiple outputs. While matching two target functions with multiple outputs, we can reduce the simulation time by utilizing the mapping relation found so far. Clearly, the more the unique mapping groups we find, the less the number of outputs with indistinguishable inputs is. So, rather than simulating the whole Boolean network, simulating the subnetwork involving these outputs and their transitive fanin nodes is enough. Our experimental result reveals that our matching algorithm can reduce the runtime significantly as it approaches the end of matching process.

11.6.3.3 Analysis of Space Complexity and Runtime

During the simulation process, we need to store the simulation vectors for all nodes in the Boolean network. Let the number of inputs and number of nodes in the Boolean network be I and N, respectively. The memory space used by our matching algorithm is $M \times I \times N$ words (4 bytes), where M is an adjustable parameter, i.e., the number of sets used in each simulation round. The smaller M means that it can reduce the storage space and simulation time in a simulation round. On the contrary, it may need more simulation rounds to improve the mapping relation. Besides, it may stop the simulation phase early and thus enter into the recursive matching phase. If there exist many large non-unique mapping groups, then the runtime will increase significantly because it may incur a large amount of simulation and SAT verification on infeasible mapping relations.

11.7 Experimental Results

The proposed S&S-based Boolean matching algorithm had been implemented into ABC system [23] on Linux platform with dual Intel Xeon 3.0 GHz CPU's. To demonstrate the efficiency of our algorithm, MCNC and LGSyn benchmarking sets were tested. For each tested circuit, we randomly permuted its input variables to generate a new circuit for being matched. In addition, to make our experimental results more convincible, we restructured this new circuit by executing a simple script file including some synthesis commands offered by ABC. Two sets of experiments were conducted to test our matching algorithm.

The first experiment was conducted to search all feasible mapping relations on 112 circuits with input number ranging from 4 to 257. The experimental results showed that three circuits C6288, i3, and o64 cannot be solved within 5000 s. To dissect these circuits, we found that one of the two mapping relations of C6288 cannot be verified by SAT technique while the other two circuits have a large amount of feasible mapping relations because they own a great many *$\mathcal{G}$-symmetries* [16]. If only to search one feasible mapping, the execution times for i3 and o64 are 0.49 s and 8.67 s, respectively. The experimental results are summarized in Table 11.2 w.r.t. the circuit input size. In this table, the first two columns show input ranges of benchmarking circuits and number of circuits in different input ranges, respectively. The third column labeled **#Solved** shows the number of circuits solved by our matching algorithm. The next three columns **Min**, **Avg**, and **Max** give the minimum, average, and maximum runtime for the solved circuits, respectively. It shows that our algorithm is very efficient for the circuits with moderate to large input sets.

Table 11.2 Boolean matching results for threshold 1000

			CPU time (s)		
#Input	**#Circuit**	**#Solved**	**Min**	**Avg**	**Max**
4∼10	31	31	0.00	0.04	0.30
11∼20	21	21	0.01	0.55	8.04
21∼30	14	14	0.03	0.21	1.29
31∼40	10	9	0.07	1.16	4.79
41∼50	8	8	0.22	2.94	5.83
51∼257	28	26	0.14	2.57	17.56

For those solved circuits, we also compared the effectiveness of three phases. Table 11.3 shows the comparison results. The rows labeled **#Circuit** and **#Inc** show the number of circuits that have been matched successfully and the number of increased matched circuits in each individual step, respectively. In the first (initialization) phase, we compared the effect of incrementally applying different functional properties. The columns named as **Unate**, **+Symm**, and **+SVS** show the results of only using functional unateness, adding *E-symmetry* and *NE-symmetry*, and adding *SV-symmetry*, respectively. Obviously the more functional properties are used, the more circuits can be solved in the first phase. It shows that 71 and

Table 11.3 Comparison on the effects of three phases

	Functional property (1)				
	Unate	**+Symm**	**+SVS**	**+Sim. (2)**	**+Rec. (3)**
#Circuit	19	49	71	94	109
#Inc	19	30	22	23	15
ratio (%)	17.4	45.0	65.1	86.2	100

94 circuits can be solved after the first phase and the second (simulation) phase, respectively. All the remaining circuits can be solved by the third (recursion) phase.

Table 11.4 shows the experimental results for the circuits with input size greater than 50. The first three columns labeled **Circuit**, **#I**, and **#O** show the circuit name, number of input variables, and number of outputs in this benchmarking circuit, respectively. The next two columns **O** and **S** are the numbers of mapping relations found by our matching algorithm without using and using functional properties. It should be noted that the solutions induced by *NE-symmetry* is not taken into account

Table 11.4 Benchmarking results for circuits with input number > 50

			#Sol		CPU time (s)		
Circuit	**#I**	**#O**	**O**	**S**	**Orig**	**Unate**	**+Symm**
apex3	54	50	1	1	0.10	0.10	0.38
apex5	117	88	144	1	7.11	2.96	0.68
apex6	135	99	2	1	1.86	0.42	0.33
C2670	233	140	–	2	*	*	7.96
C5315	178	123	4	1	6.31	2.86	3.29
C7552⋆	207	108	–	1	*	*	14.56
C880	60	26	8	1	0.28	0.20	0.25
dalu	75	16	2	1	1.20	3.36	5.47
des	256	245	1	1	10.21	0.25	2.33
e64	65	65	1	1	0.01	0.79	0.32
ex4p	128	28	–	4096	*	*	6.08
example2	85	66	1	1	0.05	0.02	0.23
frg2	143	139	1	1	0.45	0.10	0.72
i10	257	224	48	2	25.63	15.16	17.56
i2	201	1	–	1	*	*	1.02
i4	192	6	–	1	*	*	0.22
i5	133	66	1	1	0.18	0.03	0.35
i6	138	67	1	1	0.50	0.02	0.14
i7	199	67	1	1	0.82	0.04	0.19
i8	133	81	1	1	0.57	0.06	0.40
i9	88	63	1	1	0.18	0.03	0.16
pair	173	137	1	1	0.84	0.64	2.44
rot	135	107	72	1	3.79	1.69	1.25
x1	51	35	2	1	0.17	0.13	0.14
x3	135	99	2	1	2.05	0.28	0.32
x4	94	71	2	1	0.62	0.37	0.15
Total					> 25063	> 25029	66.94
Avg					> 964	> 963	2.57

–: unknown *: CPU time > 5000 s ⋆: memory explosion

on the numbers shown in the **S** column. Moreover, the numbers greater than one indicate that these benchmarking circuits own *𝒢-symmetry*. The last three columns named as **Orig**, **Unate**, and **+Symm** compare the execution times of without using functional property, using only functional unateness, and adding functional symmetries, respectively. The result shows that there are 5 out of 26 circuits cannot be solved by our matching algorithm without using full functional property within 5000 s. The reason why this situation occurs is that these circuits have a great many *NE-symmetries*. However, it can resolve all cases if functional symmetries are utilized to reduce the searching space. The average runtime of using full functional property is 2.57 s. It clearly reveals that our matching algorithm is indeed effective and efficient for solving the Boolean matching problem. In this experiment, the BDD's of these circuits were also built for comparison with AIGs. It shows the circuit C7752 had the memory explosion problem while constructing BDD without using dynamic ordering.

In order to test our matching algorithm on very large Boolean networks, the second experiment was conducted to test ISCAS89 benchmarking circuits. Since these circuits are sequential, the *comb* command in ABC was executed to transform them into combinational circuits. Table 11.5 shows the partial experimental results. For each circuit, the columns **#N, #BDD, Symmetry**, and **#Sol** show the number of nodes in the Boolean network (AIG), the number of nodes in the constructed BDD, *NE-symmetry*, and the number of feasible mappings. The CPU times of finding the first feasible mapping, finding all feasible mapping relations, and performing SAT verification of two target circuits are shown in the **First**, **All**, and **SAT** columns, respectively. The experimental results show that our algorithm cannot find all feasible mappings for those circuits with a great many *NE-symmetries* unless we detect them in advance. It also reveals that only searching the first feasible mapping relation without using symmetry is faster than the one using symmetry in some cases. The reason is that it takes too much time on detecting symmetries for the circuits. The result shows only a very small amount of runtime was consumed by SAT

Table 11.5 Benchmarking results for s-series circuits

							CPU Time (s)						
Circuit	**#I**	**#O**	**#N**	**#BDD**	**Symmetry**	**#Sol**	**Orig**		**Unate**		**+Symm**		
							First	**All**	**First**	**All**	**First**	**All**	**SAT**
s4863 †	153	120	3324	56691	1(8),1(9)	$4 \cdot 8! \cdot 9!$	*	*	2.6	*	1.9	1.9	0.0
s3384	264	209	2720	882	22(2)	2^{22}	4.8	*	2.1	*	4.0	4.0	0.0
s5378	199	213	2850	⋆	4(2),1(5),1(7)	$(2!)^4 \cdot 5! \cdot 7!$	1.3	*	3.4	*	2.4	2.4	0.0
s6669 †	322	294	4978	22957	32(2), 1(17)	$4 \cdot 2^{32} \cdot 17!$	6.3	*	2.8	*	4.1	50.5	46.5
s9234.1	247	250	4023	4545	–	1	3.4	3.4	5.8	5.8	7.8	7.8	0.0
s38584.1	1464	1730	26702	22232	1(3),1(9)	$3! \cdot 9!$	76.3	*	210.1	*	457.8	457.8	0.0
s38417	1664	1742	23308	55832099	2(2),1(3)	$(2!)^2 \cdot 3!$	91.8	*	324.6	*	998.5	998.6	0.1
Total							183.9		551.4		1476.5	1523.0	46.6
Avg							30.7		78.8		210.9	217.6	6.7
Ratio							0.15		0.37		1.00	1.03	0.03

†: circuits own *𝒢-symmetry*. -: no symmetry $m(n)$:m *NE-symmetric* sets with n inputs.
*: CPU time > 5000 s ⋆: memory explosion

verification for all circuits except the s6669 circuit. Besides, the AIG size was far less than the BDD size in many tested circuits and the s5378 circuit had the memory explosion problem. In summary, our S&S-based Boolean matching algorithm can be easily adjusted to fulfill different requirements for large Boolean networks.

11.8 Chapter Summary

In this chapter, we have presented a P-equivalent Boolean matching algorithm based on S&S approach. Signatures exploiting functional unateness and symmetries were applied to reduce the searching space quickly. Three types of input vectors were generated for simulation and their simulated results were checked to distinguish the input variables of two target functions. Our matching algorithm can find not only one feasible mapping solution but also all mapping solutions. We have implemented the matching algorithm and tested it on a set of large benchmarking circuits. The experimental results reveal that our matching is indeed effective and efficient to solve the Boolean matching problem for large-scale Boolean networks.

References

1. Mishchenko, A., Chatterjee S., Brayton, R., En N.: Improvements to combinational equivalence checking. In: Proceedings of the International Conference on Computer-Aided Design, pp. 836–843 San Jose, CA, USA (2006)
2. Plaza, S., Chang, K., Markov, I., Bertacco, V.: Node mergers in the presence of don't cares. In: Proceedings of the Asia and South Pacific Design Automation Conference, pp. 414–419 Yokohama, Japan (2007)
3. Zhu, Qi, Kitchen, N., Kuehlmann, A., Sangiovanni-Vincentelli, A.: SAT sweeping with local observability don't-cares. In: Proceedings of the Design Automation Conference, pp. 229–234 San Francisco, CA, USA (2006)
4. Mishchenko, A., Zhang, J.S., Sinha, S., Burch, J.R., Brayton, R., Chrzanowska-Jeske, M.: Using simulation and satisfiability to compute flexibilities in Boolean networks. IEEE Transaction on Computer-Aided-Design of Integrated Circuits and Systems, **25**(5) 743–755 (2006)
5. Zhang, J.S., Mishchenko, A., Brayton, R., Chrzanowska-Jeske, M.: Symmetry detection for large Boolean functions using circuit representation, simulation, and satisfiability. In: Proceedings of the Design Automation Conference, pp. 510–515 San Francisco, CA, USA (2006)
6. Benini, L., De Micheli, G.: A survey of Boolean matching techniques for library binding. ACM Transactions on Design Automation of Electronic Systems, **2**(3), 193–226 (1997)
7. Afshin Abdollahi: Signature based Boolean matching in the presence of don't cares. In: Proceedings of the Design Automation Conference, pp. 642–647 Anaheim, CA, USA (2008)
8. Agosta, G., Bruschi, F., Pelosi, G., Sciuto, D.: A unified approach to canonical form-based Boolean matching. In: Proceedings of the Design Automation Conference, pp. 841–846 San Diego, CA, USA (2007)
9. Abdollahi, A., Pedram, M.: A new canonical form for fast Boolean matching in logic synthesis and verification. In: Proceedings of the Design Automation Conference, pp. 379–384 San Diego, CA, USA (2005)
10. Debnath, D., Sasao, T.: Efficient computation of canonical form for Boolean matching in large libraries. In: Proceedings of the Asia and South Pacific Design Automation Conference, pp. 591–596 Yokohama, Japan (2004)

11. Safarpour, S., Veneris, A., Baeckler, G., Yuan, R.: Efficient SAT-based Boolean matching for FPGA technology mapping. In: Proceedings of the Design Automation Conference, pp. 466–471 San Francisco, CA, USA (2006)
12. Wang, K.H., Chan, C.M.: Incremental learning approach and SAT model for Boolean matching with don't cares. In: Proceedings of the International Conference on Computer-Aided Design of Integrated Circuits and Systems, pp. 234–239 San Jose, CA, USA (2007)
13. Wei, Z., Chai, D., Kuehlmann, A., Newton A.R.: Fast Boolean matching with don't cares," In: Proceedings of the International Symposium on Quality Electronic Design, pp. 346–351 Santa Clara, CA, USA (2006)
14. Bryant, R.: Graph based algorithm for Boolean function manipulation. IEEE Transactions on Computers, **C-35**(8), pp. 667–691 (1986)
15. Kuehlmann, A., Paruthi, V., Krohm, F., Ranai, M.K.: Robust Boolean reasoning for equivalence checking and functional property verification. IEEE Transactions on Computer-Aided Design of Integrated Circuits and Systems, **21**, 377–1394 (2002)
16. Mohnke, J., Molitor, P., Malik, S.: Limits of using signatures for permutation independent Boolean comparison. In: Proceedings of the Asia and South Pacific Design Automation Conference, pp. 459–464 Makuhari, Japan (1995)
17. Wang, K.H.: Exploiting k-distance signature for Boolean matching and G -symmetry detection. In: Proceedings of the Design Automation Conference, pp. 516–521 San Francisco, CA, USA (2006)
18. Marques-Silva, J., Sakallah, K.A.: GRASP: a search algorithm for propositional satisfiability. In: IEEE Transactions on Computer-Aided Design of Integrated Circuits and Systems, **48**(5), pp. 506–521 (1999)
19. Moskewicz, M.W., Madigan, C.F., Zhao, Y., Zhang, L., Malik, S.: Chaff: engineering an efficient SAT solver. In: Proceedings of the Design Automation Conference, pp. 530–535 Las Vegas, NV, USA (2001)
20. Eén, N., Sörensson, N.: The MiniSat page. http://minisat.se/ Cited 15 Jan 2008
21. Marques-Silva, J.P., Sakallah, K.A.: Boolean satisfiability in electronic design automation. In: Proceedings of the Design Automation Conference, pp. 675–680 Los Angeles, CA, USA (2000)
22. Mishchenko, A., Chatterjee, S., Jiang, R., Brayton, R.: FRAIGS: a unifying representation for logic synthesis and verification. In: ERL Technical Report, EECS Department, UC Berkeley (2005)
23. ABC: A System for Sequential Synthesis and Verification, by Berkeley Logic Synthesis and Verification Group, http://www.eecs.berkeley.edu/~alanmi/abc/ Cited 15 Jan 2008

Chapter 12
Logic Difference Optimization for Incremental Synthesis

Smita Krishnaswamy, Haoxing Ren, Nilesh Modi, and Ruchir Puri

Abstract During the IC design process, functional specifications are often modified late in the design cycle, often after placement and routing are completed. However, designers are left either to manually process such modifications by hand or to restart the design process from scratch–a very costly option. In order to address this issue, we present DeltaSyn, a tool and methodology for generating a highly optimized logic difference between a modified high-level specification and an implemented design. DeltaSyn has the ability to locate similar logic in the original design which can be reused to realize the modified specification through several analysis techniques that are applied in sequence. The first phase employs fast functional and structural analysis techniques to identify equivalent signals between the original and the modified circuits. The second phase uses a novel topologically-guided dynamic matching algorithm to locate reusable portions of logic close to the primary outputs. The third phase utilizes functional hashing to locate similar chunks of logic throughout the remainder of the circuit. Experiments on industrial designs show that, together, these techniques successfully implement incremental changes while preserving an average of 97% of the pre-existing logic. Unlike previous approaches, bit-parallel simulation and dynamic programming enable fast performance and scalability. A typical design of around 10K gates is processed and verified in about 200 s or less.

12.1 Introduction and Background

As the IC industry matures, it becomes common for existing designs to be modified incrementally. Since redesigning logic involves high expenditure of design effort and time, previous designs must be maximally re-utilized whenever possible.

S. Krishnaswamy (✉)
IBM TJ Watson Research Center, Yorktown Heights, NY
e-mail: skrishn@us.ibm.com

This work is based on an earlier work: DeltaSyn: an efficient logic difference optimizer for ECO synthesis, in Proceedings of the 2009 international Conference on Computer-Aided Design, ISBN:978-1-60558-800-1 (2009) © ACM, 2009. DOI= http://doi.acm.org/10.1145/1687399.1687546

S.P. Khatri, K. Gulati (eds.), *Advanced Techniques in Logic Synthesis, Optimizations and Applications*, DOI 10.1007/978-1-4419-7518-8_12,

Designers have noted that, in existing flows, even a small change in the specification can lead to large changes in the implementation [7]. More generally, the need for CAD methodologies to be less sequential in nature and allow for transformations that are "incremental and heterogeneous" has been recognized by leaders in industry [6].

Recent advances in incremental physical synthesis [2, 22], placement [17], routing [25], timing analysis [22], and verification [4] have made incremental tools practical. However, logic synthesis remains a bottleneck in incremental design for several reasons. First, it is difficult to process incremental changes in the design manually since logic optimizations can render intermediate signals unrecognizable. Second, the inherent randomness in optimization choices makes the design process unstable, i.e., a slight modification of the specification can lead to a different implementation. Therefore, a general incremental synthesis methodology that is able to quickly derive a small set of changes in logic to handle incremental updates is necessary.

Prior work on incremental synthesis tends to focus on small *engineering change orders* (ECOs). Such methods primarily fall into two categories. The first category consists of purely functional techniques which attempt to isolate point changes and perform in-place rectification. Such techniques can be unscalable [12, 21] due to the use of complex BDD manipulation and laborious analysis. Further, they may simply fail to identify multiple point changes and changes that cannot be easily isolated as originating at specific points in a circuit due to logic restructuring. The second category of methods is heavily reliant on structural correspondences [3, 19] and can result in large difference models that disrupt much of the existing design when such correspondences are unavailable.

In this chapter, we present DeltaSyn, a method to produce a synthesized delta or the *logic difference* between an RTL-level *modified specification* and an original implemented design. As illustrated in Fig. 12.1, DeltaSyn combines both functional

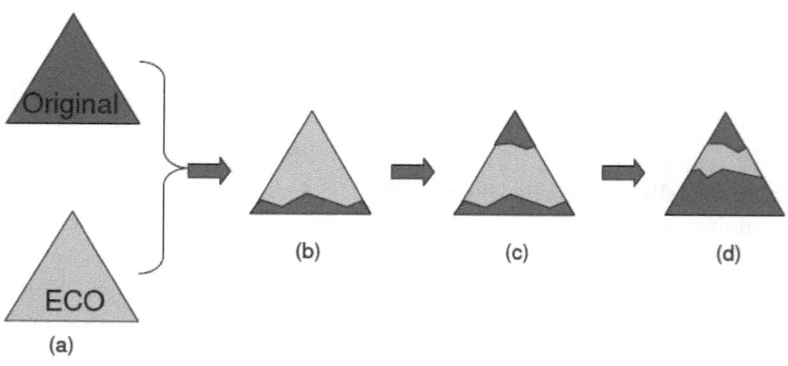

Fig. 12.1 The main phases of DeltaSyn: (**a**) the original design and the modified specification are given as inputs to DeltaSyn, (**b**) functional and structural equivalences forming the input-side boundary of the changes are identified, (**c**) matching subcircuits which form the output-side boundary of the changes are located and verified, (**d**) further reductions are identified through functional hashing

and structural analysis to minimize the logic difference. As a pre-processing step, we compile the modified specification into a preliminary technology-independent gate-level netlist with little optimization. Phase I finds structurally and functionally equivalent gates to determine the *input-side boundary* of the logic difference. Phase II uses a novel topologically guided functional technique that finds matching subcircuits starting from primary outputs and progressing upstream to determine the *output-side boundary* of the change. Phase III finds further logic for reuse through a novel functional hashing technique. DeltaSyn allows designers to avoid most design steps including much of logic synthesis, technology mapping, placement, routing, buffering, and other back-end optimizations on the unchanged logic.

The main features of our method include:

- An efficient multi-phase flow that integrates fast functional and structural techniques to reduce the logic difference through the identification of input- and output-side boundaries of the change.
- A novel dynamic algorithm that finds matching subcircuits between the modified specification and implemented design to significantly decrease the logic difference.
- A functional hashing technique to enable wider use of matching.

A key advantage of our approach is that, unlike traditional ECO methodologies, we make no assumptions about the type or extent of the changes in logic. The remainder of the chapter is organized as follows. Section 12.2 describes previous work in incremental synthesis. Section 12.3 describes the overall flow of DeltaSyn. Sections 12.3.1 and 12.3.2 describe our equivalence-finding and subcircuit-matching phases of logic difference reduction, while Section 12.3.3 presents the functional-hashing phase of difference reduction. Section 12.4 presents empirical results and analysis. Section 12.5 concludes the chapter.

12.2 Previous Work

Recently, the focus of incremental design has been on changes to routing or placement [2, 11, 18]. However, there have been several papers dealing specifically with logic ECO. Authors of [3, 19] present techniques that depend on structural correspondences. They find topologically corresponding nets in the design. Then, gates driving these nets are replaced by the correct gate type. While this type of analysis is generally fast, it can lead to many changes to the design since such structural correspondences are hard to find in designs that undergo many transformations.

In contrast, the method from [12] does not analyze topology. Instead, it uses a BDD-based functional decomposition technique to identify sets of candidate signals that are able to correct the outputs of the circuit to achieve the ECO. The authors rewrite functions of each output $O(X)$ in terms of internal signals t_1, t_2 to see whether there are functions that can be inserted at t_1, t_2 to realize a new function O'. In other words, they solve the Boolean equation $O(X, t_1, t_2) = O'$ for t_1 and t_2 and check for consistency. This method does not scale well due to the memory required for a BDD-based implementation of this technique.

More recently, Ling et al. [13] present a maximum satisfiability (MAX-SAT) formulation similar to that of [18] for logic rectification in FPGA-based designs. Rectification refers to corrections in response to missing or wrong connections in the design. They find the maximum number of clauses that can be satisfied between a miter that compares the original implementation and the modified specification. Then, gates corresponding to unsatisfied clauses are modified to correct the logic. They report that approximately 10% of the netlist is disrupted for five or fewer errors. For more significant ECO changes, MAX-SAT can produce numerous unsatisfied clauses since it depends on the existence of functional equivalences. Further, this method does not directly show how to correct the circuit. Deriving the correction itself can be a difficult problem – one that is circumvented by our method.

12.3 DeltaSyn

In this section, we describe the incremental logic synthesis problem and our solution techniques. First, we define terms that are used through the remainder of the chapter.

Definition 12.1 The *original model* is the original synthesized, placed, routed, and optimized design.

Definition 12.2 The *modified specification* is the modified RTL-level specification, i.e., the change order.

Definition 12.3 The *difference model* is a circuit representing the changes to the original model required to implement the modified specification. The set of gates in the difference model represent new gates added to the original model. Wires represent connections among these gates. The primary inputs and primary outputs of the difference model are annotated by connections to existing gates and pins in the original model.

Given the original model, the modified specification, and a list of corresponding primary outputs and latches, the objective of incremental synthesis is to derive a *difference model* that is minimal in the number of gates. We choose the minimal number of gates as our metric because the general procedure by which incremental synthesis occurs in the industry motivates the need to preserve as many gates as possible. Typically, when late-stage changes occur, the masks are already set for most metal layers. The changes are realized by rewiring spare gates in the top metal layer. Additionally, incremental placement and routing tools can optimize wire length and other physical concerns.

The new specification, generally written in an RTL-level hardware description language (such as VHDL), is compiled into a technology-independent form called the *modified model.* This step is relatively fast because the majority of the design time is spent in physical design including physical synthesis, routing, and analysis [Osler, P. Personal Communication (2009)] (see Fig. 12.17).

The circuits in Fig. 12.2 are used to broadly illustrate the three phases of our difference optimization. By inspection, it is clear that $f3$ and $f7$ are the only

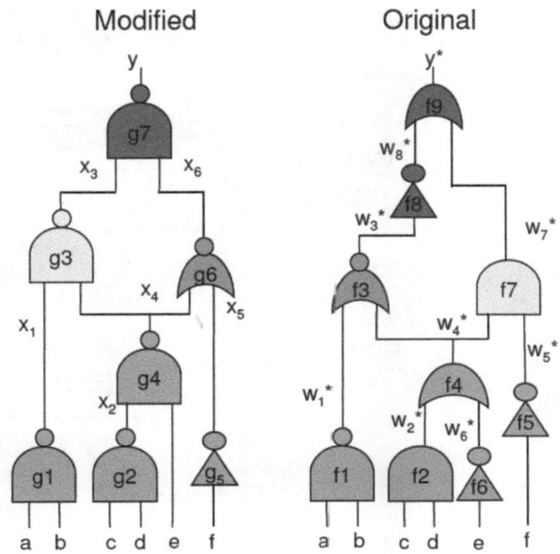

Fig. 12.2 Sample circuits to illustrate our method

differences between the two circuits. Although some of the local logic has undergone equivalent transformations (gates $g4$ and $g7$ have been modified through the application of DeMorgan's law), most of the circuit retains its global structure. The nearby logic in the original model being unrecognizable despite the actual change being small is typical of such examples.

DeltaSyn recognizes and isolates such changes as follows: Our first phase involves structural and functional equivalence checking. For the given example, the equivalences $x_1 \equiv w_1^*$, $x_4 \equiv w_4^*$, and $x_5 \equiv w_5^*$ are identified by these techniques. Our second phase is geared toward finding matching subcircuits from the primary outputs. Through a careful process of subcircuit enumeration and Boolean matching, the subcircuits consisting of $\{f8, f9\}$ from the original model and $\{g7\}$ are matched under the intermediate input mapping $\{(x_3, w_3^*), (x_6, w_7^*)\}$. This phase leaves $g3$ and $g6$ as the logic difference . However, in the third phase $f3$ and $g6$ are recognized as subcircuits performing the same functionality, therefore $f3$ can simply be rewired to realize $g6$. Therefore, the third phase leaves $g3$ as the optimized logic difference. The remainder of this section explains the algorithms involved in these steps.

12.3.1 Phase I: Equivalence-Based Reduction

Phase I is illustrated in Fig. 12.3. Starting with the given list of corresponding primary inputs and latches, DeltaSyn builds a new correspondence list L between matched signals in the original and modified models. Matches are found both structurally and functionally. Candidates for functional equivalence are identified by comparing simulation responses and verified using Boolean satisfiability (SAT).

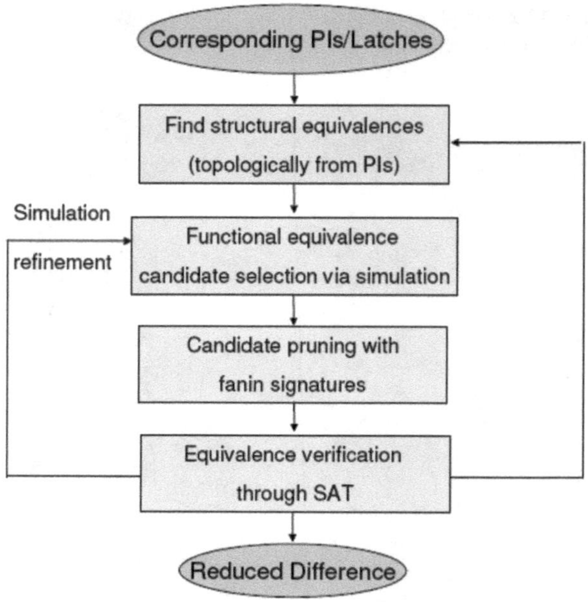

Fig. 12.3 Logic difference reduction through equivalence checking

Structural equivalences are found inductively, starting with corresponding primary inputs and latch outputs. All gates g, g' whose input signals correspond, and whose functions are identical, are added to the correspondence list. The correspondence list can keep track of all pairwise correspondences (in the case of one-to-many correspondences that can occur with redundancy removal). This process is then repeated until no further gate-outputs are found to structurally correspond with each other.

Example 12.1 In Fig. 12.4 the initial correspondence list is $L = \{(a, a^*)(b, b^*)(c, c^*)(d, d^*)(e, e^*)(f, f^*)\}$. Since both the inputs to the gate with output x are in L, we examine gate x^* in the original model. Since this gate is of the same type as x, (x, x^*) can be added to L.

After the structural correspondences are exhausted, the combinational logic is simulated in order to generate candidate functional equivalences. The simulation proceeds by first assigning common random input vectors to signal pairs in L. Signals with the same output response on thousands of input vectors (simulated in a bit-parallel fashion) are considered candidates for equivalence, as in [10, 15]. These candidates are further pruned by comparing a pre-computed *fanin signature* for each of these candidate signals. A fanin signature has a bit position representing each PI and latch in the design. This bit is set if the PI or latch in question is in the transitive fanin cone of the signal and unset otherwise. Fanin signatures for all internal signals can be pre-computed in one topological traversal of the circuit.

Example 12.2 In Figure 12.4, the same set of four random vectors are assigned to corresponding input and internal signals. The output responses to each of the inputs

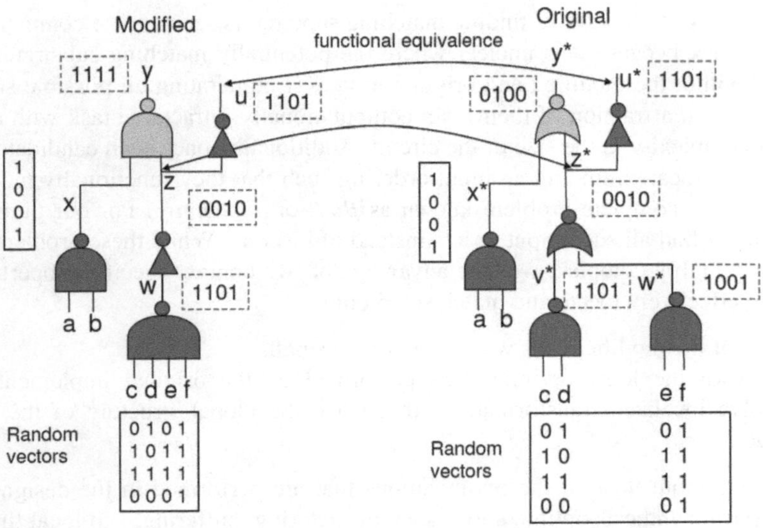

Fig. 12.4 Identifying structural and functional equivalences

are listed horizontally. The simulations suggest that (z, z^*), (u, u^*), (w, v^*) are candidate equivalences. However, the fanin list of $v*$ contains PIs c, d but the list for w contains c, d, e, f. Therefore, these signals are not equivalent.

Equivalences for the remaining candidates are verified using SAT. We construct miters between candidate signals by connecting the corresponding primary inputs together and check for satisfiability. UNSAT assignments can be used to update simulation vectors.

Note that it is not necessary for all intermediate signals to be matched. For instance, if two non-identical signals are merged due to local observability don't cares (ODCs) as in [27], then downstream equivalences will be detected after the point at which the difference between the signals becomes unobservable. After functional equivalences are found, all of the gates driving the signals in L can be deleted from the difference model.

12.3.2 Phase II: Matching-Based Reduction

Phase II of DeltaSyn finds subcircuits that are functionally equivalent under some permutation of intermediate signals. Since incremental synthesis is intended to be used for small changes in large netlists, there are large areas of logic that are identifiably unchanged once the dependence on the changed logic is removed. In other words, once the output-side boundary of the change is disconnected, the remaining logic should be equivalent under an appropriate association (connection) of internal signals (as illustrated in Fig. 12.2).

At the outset, the task of finding matching subcircuits seems to be computationally complex because it is unclear where the potentially matching subcircuits are located within the modified and original models. Enumerating all possible subcircuits (or even a fraction of them) is a computationally intractable task with exponential complexity in the size of the circuit. Additionally, once such candidate subcircuits are located, finding an input ordering such that they functionally match is itself an NP-complete problem known as *Boolean matching* . For our purposes, we actually find all such input orders instead of just one. While these problems are generally highly complex, we take advantage of two context-specific properties in order to effectively locate and match subcircuits:

1. Most of the modifications we encounter are small.
2. Many of the logic optimizations performed on the original implementation involve localized transformations that leave the global structure of the logic intact.

In fact, about 90% of the optimizations that are performed in the design flow are physical synthesis optimizations such as factoring, buffering, and local timing-driven expansions [9, 20, 22, 24]. While redundancy removal can be a non-local change, equivalent signals between the two circuits (despite redundancy removal) can be recognized by techniques in Phase I. Since we expect the change in logic to be small, regions of the circuit farther from the input-side boundaries are more likely to match. Therefore, we enumerate subcircuits starting from corresponding primary outputs in order to find upstream matches. Due to the second property, we are able to utilize local subcircuit enumeration. The subcircuits we enumerate are limited by a width of 10 or fewer inputs, thereby improving scalability. However, after each subcircuit pair is matched, the algorithm is recursively invoked on the corresponding inputs of the match.

Figure 12.5 illustrates the main steps of subcircuit identification and matching. Candidate subcircuits are generated by expanding two corresponding outputs along their fanin cones. For each candidate subcircuit pair, we find input symmetry classes, and one input order under which the two circuits are equivalent (if such an order exists). From this order, we are able to enumerate all input orders under which the circuits are equivalent. For each such order, the algorithm is called recursively on the corresponding inputs of the two subcircuits.

12.3.2.1 Subcircuit Enumeration

For the purposes of our matching algorithm we define a subcircuit as follows:

Definition 12.4 A subcircuit C consists of the logic cone between one output O, and a set of inputs $\{i_1, i_2, \ldots, i_n\}$.

Pairs of subcircuits, one from the original model and one from the modified model, are enumerated in tandem. Figure 12.6 illustrates the subcircuit enumeration algorithm. Each subcircuit in the pair starts as a single gate and expands to incorporate the drivers of its inputs. For instance, in Fig. 12.6, the subcircuit initially

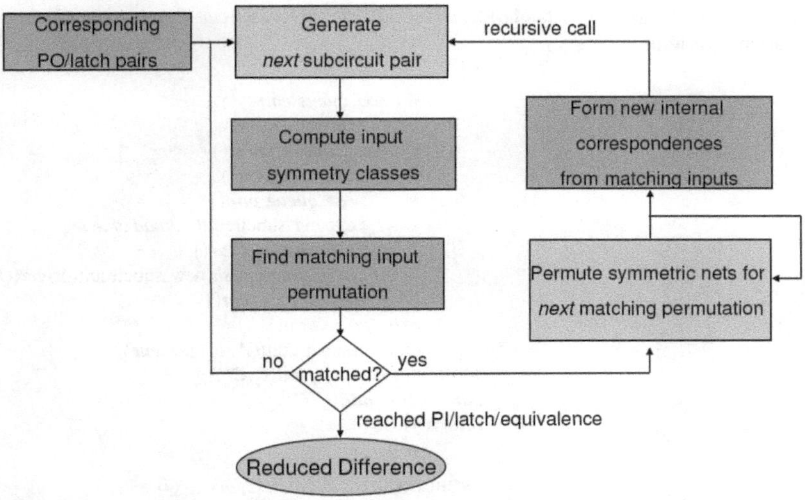

Fig. 12.5 Difference reduction through subcircuit matching

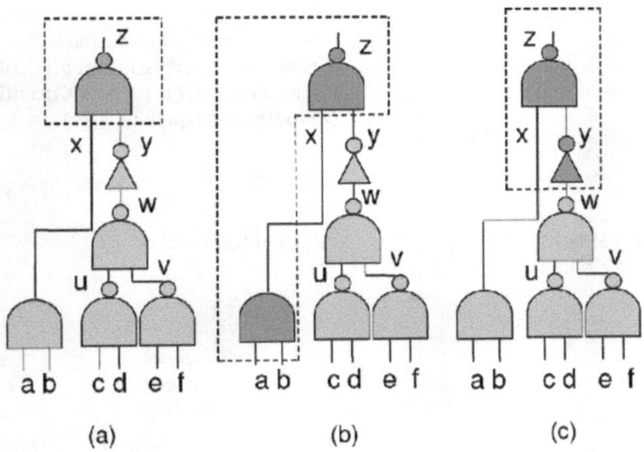

Fig. 12.6 Candidate subcircuit enumeration

contains only the gate driving primary output z and then expands in both the x- and y-directions. The expansion in a particular direction is stopped when the input being expanded is (a) a primary input, (b) a latch, (c) a previously identified equivalence, (d) a previously matched signal, (e) the added gate increases the subcircuit width beyond the maximum allowed width, or (f) the signal being expanded has other fanouts outside the subcircuit (signals with multiple fanouts can only be at the output of a subcircuit).

Pseudocode for subcircuit expansion and enumeration are shown in Fig. 12.7. A *subcirc_enum* structure (shown in Fig. 12.8) is instantiated for pairs of nets N, N^* where N is from the modified model and N^* is from the original model (starting

Fig. 12.7 Subcircuit pair enumeration algorithm

```
bool NEXT_SUBCIRCUIT_PAIR(subcircuit C, subcircuit C*)
{
        do
                if (mod_queue.empty())
                        return FALSE
                C = mod_queue.front()
                if(orig_queue.empty())
                        eco_queue.pop()
                        expand_subcircuit(C, mod_queue)
                        orig_queue.clear()
                        orig_queue.push(new_subcircuit(driver(O*)))
                C* = orig_queue.front()
                orig_queue.pop()
                expand_subcircuit(C*, orig_queue)
        while(pair_history.find(C,C*))
        pair_history.add(C,C*)
        return TRUE
}

EXPAND_SUBCIRCUIT(subcircuit C, queue Q)
{
        for(all inputs i ∈ C)
                if(has_outside_fanouts(i))continue
                g = get_driver(i)
                if(is_PI(g)|| is_latch(g)continue
                if(is_equivalent(g)|| is_matched(g))continue
                if(num_inputs((C ⋃ g) > MAX))continue
                Q.push(new_subcircuit(C ⋃ g))
}
```

Fig. 12.8 The data structure for enumerated subcircuits

```
STRUCT subcirc_enum
   {
          N
          N*
          mod_queue
          orig_queue
          pair_history

          next_subcircuit_pair(C, C*)
          expand_subcircuit( subcircuit C, queue Q)
   }
```

from corresponding primary outputs). The *next_subcircuit_pair* method fills in the variables C and C^*. First, the *orig_queue* is popped. If the *orig_queue* is empty, then all the possible subcircuits in the original model have already been enumerated for a particular subcircuit C in the modified model. In this case, a new modified subcircuit is found by popping the *mod_queue*. If a particular pair of subcircuits has already been seen (and recorded in the *pair_history*) then the next pair is generated. If the *mod_queue* is also empty, then all possible pairs of subcircuits have already been enumerated for the pair of nets (N, N^*) and the process terminates.

12.3.2.2 Subcircuit Matching

For two candidate subcircuits (C, C^*) realizing the Boolean functions $F(i_1, i_2, \ldots, i_n)$ and $F^*(i_1*, i_2*, \ldots, i_n*)$, respectively, our goal is to find *all* of the permutations of the inputs of F^* such that $F = F^*$. Note that this is not necessary for most uses of Boolean matching (such as technology mapping). We elaborate on this process below.

Definition 12.5 A *matching permutation* $\rho_{(F^*,F)}$ of a function $F^*(i_1, i_2, \ldots, i_n)$ with respect to a function F is a permutation of its inputs such that

$$F^*(\rho_{(F^*,F)}(i_1), \rho_{(F^*,F)}(i_2), \ldots, \rho_{(F^*,F)}(i_n)) = F$$

Definition 12.6 Two inputs i_x and i_y of a function F are said to be *symmetric* with respect to each other if

$$F(i_1, i_2, \ldots \mathbf{i_x}, \ldots \mathbf{i_y} \ldots i_n) = F(i_1, i_2, \ldots \mathbf{i_y}, \ldots \mathbf{i_x} \ldots i_n)$$

Definition 12.7 Given a function F and a partition of its inputs into symmetry classes

$$sym\_F = \{sym\_F[1], sym\_F[2], \ldots, sym\_F[n]\},$$

a *symmetric permutation* τ_F on the inputs of F is a composition of permutations on each symmetry class $\tau_F = \tau_{sym\_F[1]} \circ \tau_{sym\_F[2]} \circ \ldots \circ \tau_{sym\_F[n]}$. Each constituent permutation $\tau_{sym\_F[i]}$ leaves all variables not in $sym\_F[i]$ fixed.

We now state and prove the main property that allows us to derive all matching permutations.

Theorem 12.1 *Given a matching permutation $\rho_{(F^*,F)}$, all other matching permutations $\pi_{(F^*,F)}$ can be derived by composing a symmetric permutation τ with $\rho_{(F^*,F)}$, that is, for some symmetric permutation τ:*

$$\pi_{(F^*,F)} = \rho_{(F^*,F)} \circ \tau$$

Proof Assume there exists a matching permutation $\pi_{(F^*,F)}$ that cannot be derived by composing a symmetric permutation with $\rho_{(F,F^*)}$. Then, there is a permutation ϕ which permutes a set of non-symmetric variables S' such that $\rho_{(F,F^*)} \circ \phi = \pi_{(F^*,F)}$. However, by definition of symmetry

$$F^*(\rho_{(F,F^*)}(\phi(i_1)), \rho_{(F,F^*)}(\phi(i_2)), \rho_{(F,F^*)}(\phi(i_3)) \ldots)$$
$$\neq F^*(\rho_{(F,F^*)}(i_1), \rho_{(F,F^*)}(i_2), \rho_{(F,F^*)}(i_3))$$

By transitivity

$$F^*(\rho_{(F,F^*)}(\phi(i_1)), \rho_{(F,F^*)}(\phi(i_2)), \rho_{(F,F^*)}(\phi(i_3))\ldots) \neq F.$$

Therefore, $\pi_{(F^*,F)}$ cannot be a matching permutation . For the other side, suppose ϕ is any symmetric permutation of F^* then by definition of symmetry

$$F^*(\phi(i_1), \phi(i_2), \phi(i_3)\ldots) = F^*(i_1, i_2, i_3\ldots)$$

and by definition of matching permutation:

$$F^*(\rho_{(F,F^*)}(\phi(i_1)), \rho_{(F,F^*)}(\phi(i_2)), \rho_{(F,F^*)}(\phi(i_3))\ldots)$$
$$= F^*(\rho_{(F,F^*)}(i_1), \rho_{(F,F^*)}(i_2), \phi_{(F,F^*)}(i_3)) = F$$

Therefore, $\rho \circ \phi$ is also a matching permutation of F^* with respect to F. □

Theorem 12.1 suggests that all matching permutations can be derived in these steps:

1. Computing the set of input symmetry classes for each Boolean function, i.e., for a function F we compute $sym\_F = \{sym\_F[1], sym\_F[2], \ldots, sym\_F[n]\}$ where classes form a partition of the inputs of F and each input is contained in one of the classes of sym_F.
2. Deriving one matching permutation through the use of a Boolean matching method.
3. Permuting symmetric variables within the matching permutation derived in step 2.

To obtain the set of symmetry classes for a Boolean function F we recompute the truth table bitset after swapping pairs of inputs. This method has complexity $O(n^2)$ for a circuit of width n, and this method is illustrated in Fig. 12.9.

We derive a matching permutation of F^* or determine that one does not exist through the algorithm shown in Fig. 12.11. In the pseudocode, instead of specifying

```
COMPUTE_SYM_CLASSES(function F)
{
        unclassified[] = all_inputs(F)
        do
                curr = unclassified[0]
                for(i < |unclassified|)
                        F' = swap(F, curr, unclassified[i])
                        if(F' == F)
                                new_class.add(unclassified[i])
                        new_class.add(curr)
                        sym_F.add(new_class)
                        unclassified[] = all_inputs(F) -symmetry_classes
        while(!unclassified.empty())
        return sym_F
}
```

Fig. 12.9 Computing symmetry classes

permutations $\rho_{F^*,F}$, we directly specify the ordering on the variables in F^* that is induced by ρ when F is ordered in what we call a *symmetry class order*, i.e., F with symmetric variables adjacent to each other, as shown below:

$$F(sym\_F[1][1], sym\_F[1][2], \ldots, sym\_F[1][n], sym\_F[2][1], \\ sym\_F[2][2], \ldots, sym\_F[2][n], \ldots)$$

The $reorder(F, sym\_F)$ function in the pseudocode is used to recompute the functions F according to the order suggested by *sym_F* (and similarly with F^*). The overall function is explained below:

1. First, we check whether number of inputs in both the functions is the same.
2. Next, we check the sizes and number of symmetry classes. If the symmetry classes all have unique sizes, then the classes are considered *resolved*.
3. If the symmetry classes of F and F^* are resolved, they can be associated with each according to class size and immediately checked for equivalence.
4. If the symmetry classes do not have distinctive sizes, we use a simplified form of the method from [1], denoted by the function *matching_cofactor_order* in Fig. 12.10. Here, cofactors are computed for representative members of each unresolved symmetry class, and the minterm counts of the nth-order cofactors are used to associate the classes of F with those of F^*. This determines a permutation of the variables of F^* up to symmetry classes.

```
bool COMPUTE_MATCHING_PERM_ORDER(function F, function F*)
{
        if(|inputs(F)|! = |inputs(F*)|)
                return UNMATCHED
        sym_F = compute_sym_classes(F)
        sym_F* = compute_sym_classes(F*)
        sort_by_size(sym_F)
        sort_by_size(sym_F*)
        if(|sym_F|! = |sym_F*|)
                return UNMATCHED
        for(0 <= i < |sym_F|)
                if(|sym_F[i]|! = |sym_F*[i]|)
                        return UNMATCHED
        if(resolved(sym_F*))
                reorder(F*, sym_F*)
                reorder(F, sym_F)
                if(F* == F) return MATCHED
                else return UNMATCHED
        if(matching_cofactor_order(F, sym_F, F*, sym_F*))
                return MATCHED
        else
                return UNMATCHED

}
```

Fig. 12.10 Compute a matching permutation order

Fig. 12.11 Enumerating matching input orders

```
NEXT_MATCHING_PERM_ORDER(sym_classes sym_F*, function F*)
{
        index = -1
        for(0 <= i < |sym_F*|)
                if(next_permutation(sym_F*[i]))
                        index = i
                        break
        if(index == -1)
                return NULL
        for(0 <= j < i)
                next_permutation(sym_F*[j])
        reorder(sym_F*, F)
}
```

The remaining matching permutations are derived by enumerating symmetric permutations as shown in Fig. 12.11. The *next_permutation* function enumerates permutations of individual symmetry classes. Then all possible combinations of symmetry class permutations are composed with each other.

The different input orders induced by matching permutations define different associations of intermediate signals between the subcircuit from the original model C^* and that of the modified model C. Figure 12.12 illustrates that although two subcircuits can be equivalent under different input orders, the "correct" order leads to larger matches upstream.

Note that the discussion in this section can be applied to finding all matching permutations under negation-permutation-negation (NPN) equivalence, by simply negating the inputs appropriately at the outset as in [1]. This involves choosing the polarity of each input variable that maximizes its cofactor minterm count and using that polarity in deriving matching permutations. In other words, for a subcircuit C realizing function F, if $|F(i_0 = 0, \ldots)| > |F(i_0 = 1, \ldots)|$ then input i_0 is used in its negated form and the remainder of the analysis follows as discussed above. In practice, this helps in increasing design reuse by ignoring intermediate negations.

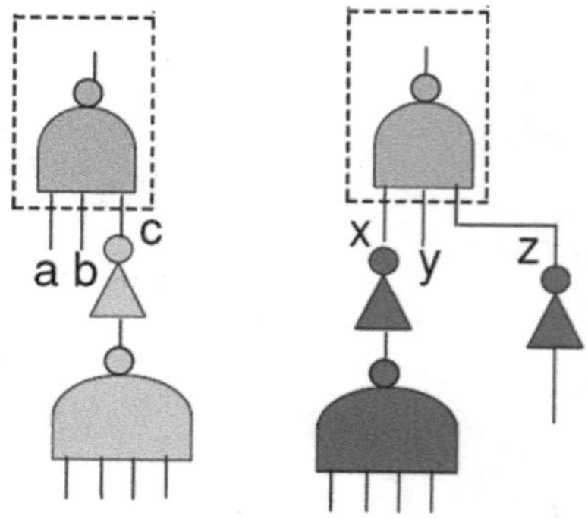

Fig. 12.12 Although the single-gate subcircuits in the boxes have completely symmetric inputs, the input ordering (c, b, a) leads to a larger upstream match than (a, b, c)

12.3.2.3 Subcircuit Covering

In this section, we describe a recursive covering algorithm which derives a set of subcircuits or *cover* of maximal size.

Definition 12.8 A *subcover* for two corresponding nets (N, N^*) is a set of connected matching subcircuit pairs that drive N and N^*.

Different subcircuit matches at nets (N, N^*) can lead to different subcovers as shown in Fig. 12.13. Once the subcircuit D of the original and D^* is generated through subcircuit enumeration algorithm of Fig. 12.8, the algorithm of Fig. 12.10 finds an input ordering under which they are functionally equivalent. Figure 12.13 shows the initial ordering where inputs (0, 1) of are associated with inputs (0, 1) of D^*. The subcover induced by this ordering is simply $\{(D, D^*)\}$, leaving the logic difference $\{A, B, C\}$. However, an alternate input ordering–derived by swapping the two symmetric inputs of D^*–yields a larger cover.

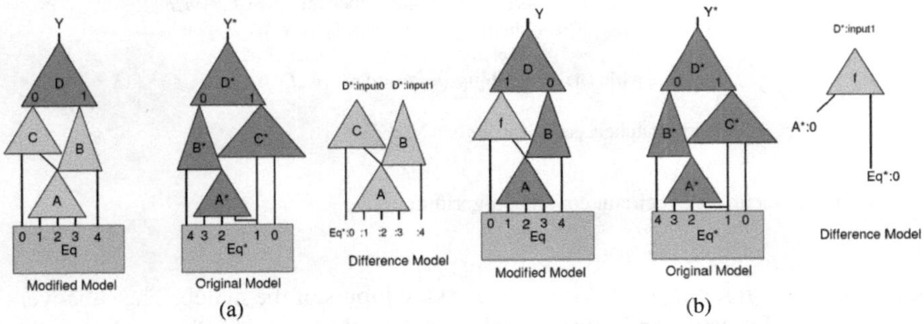

Fig. 12.13 Snapshots of subcircuit covering: (**a**) Subcover induced by input ordering $D^*(0, 1)$ on original model and resulting difference (**b**) Subcover induced by input ordering $D^*(1, 0)$, and the resulting (smaller) logic difference

Since $D(1, 0) = D^*(0, 1)$, the covering algorithm is invoked on the pairs of corresponding inputs of D and D^*. The subcircuits (B, B^*) are eventually found and matched. The inputs of B, B^* are then called for recursive cover computation. One of the inputs of B is an identified functional equivalence (from phase 1) so this branch of recursion is terminated. The recursive call on the other branch leads to the match (A, A^*) at which point this recursive branch also terminates due to the fact that all of the inputs of A are previously identified equivalences. The resultant logic difference simply consists of $\{C\}$. Note that this subcover requires a reconnection of the output of A to C which is reflected in the difference model.

Figure 12.14 shows the algorithm to compute the *optimal subcover* . The algorithm starts by enumerating all subcircuit pairs (see Fig. 12.6) and matching permutations (see Fig. 12.11) under which two subcircuits are equivalent. The function is recursively invoked at the input of each matching mapping in order to extend the cover upstream in logic. For each match C, C^* with input correspondence $\{(i_1, j_1), (i_2, j_2), (i_3, j_3), \ldots\}$ (defined by the matching permutation), the best induced subcover is computed by combining best subcovers

```
COMPUTE_COVER(net N, net N*)
{
        subcirc_enum N_enum
        while(N_enum.next_subcircuit_pair(C, C*)){
                F* = compute_truth_table(C*)
                F = compute_truth_table(C)
                sym_F = compute_symm_classes(F)
                sym_F* = compute_symm_classes(F*)
                if(!compute_matching_perm_order(F, F*))
                        continue
                do{
                        for(0 <= i < |sym_F|)
                        for(0 <= j < |sym_F[i]|)
                                if(is_PI_latch_matched(sym_F[i][j]))
                                        continue
                                if(is_PI_latch_matched(sym_F*[i][j]))
                                        continue
                                compute_cover(sym_F[i][j], sym_F*[i][j])
                        this_match = combine_subcovers(sym_F, sym_F*)
                        if(|this_match| > |opt_match(N, N*)|)
                                opt_match(N, N*) = this_match
                }while(next_matching_perm_order(sym_F*))
        }
        mark_matched_gates(opt_match(N, N*))
}
```

Fig. 12.14 The recursive subcircuit covering algorithm

$opt\_match(i_1, j_1), opt\_match(i_2, j_2) \ldots$ at the inputs of the match. The subcovers are corrected for any conflicting matches during the process of combining. For example, if a net in the modified model has higher fanout than a net in the original model then different subcovers may correspond the modified net with different original nets. When such conflicts occur, the correspondence that results in the larger subcover is retained.

In this process, we search for matches starting with topologically corresponding primary outputs, and further topological correspondences emerge from the matching processes. Since topologically connected gates are likely to be placed close to each other during physical design, many of the timing characteristics of the original implementation are preserved in reused logic. After a subcircuit cover is found, the outputs of subcircuit pairs are added to the correspondence list L and all the covered gates in the modified model are removed from the difference model.

12.3.3 Phase III: Functional Hashing-Based Reduction

In the previous section, we used a topologically-guided method to match regions of the circuit which can be reused starting from the primary outputs. However, it is possible to search for reusable logic at a finer level of granularity. Often, different logic functions have subfunctions in common. Further, certain logic optimizations, such as rewiring, can cause global changes in connectivity while still maintaining logic similarity. For instance, if two output pins were swapped, the method of Phase-II would fail because it searches based on topological connectivity. To address this issue, we present a method that searches for similar logic throughout the circuit and not just in topologically relevant.

This method proceeds by traversing all of the nets in the unmatched portions of the original and modified designs and hashing subcircuit functions of limited size at the fanin cones of the nets. The functions within the original design that hash to the same key as the modified design are candidates for reuse. These candidates are then verified for Boolean matching using the method of Fig. 12.10 and then the matches are dynamically extended using the recursive subcircuit covering algorithm of Fig. 12.16. In other words, the hashing enables us to restart the subcircuit coveri ng algorithm from new, promising locations. Additionally, reusable modules, such as adders or priority muxes, which may be available in the logic, can be appropriated for use in realizing the changed specification.

In previous literature, functional hashing has been used in logic *rewriting* [14] to hash 4-input cuts such that cuts realizing identical functionality can be replaced by each other. However, the representative member of the corresponding NPN equivalence class is simply referenced from an exhaustive list. It is only possible to exhaustively list 4-input functions, as classified by [16] as the number of Boolean functions of five or more gets prohibitively large. Authors of [5] use another method of functional hashing, where they derive a signature for 3- and 4-input cuts. However, that method does not scale to larger circuits either.

Here, we propose an efficient key for the hash function based on easy-to-compute functional characteristics such that likely Boolean matches are placed in the same hash bucket. These functional characteristics include a subset of what is computed in order to assess a full Boolean match.

Definition 12.9 Given a Boolean function $F(i_1, i_2, \ldots, i_n)$, the *matching key* $K(F, k)$ is the $(k+3)$-*tuple*,

$$< N, S, F_0, F_1, \ldots, F_k >$$

where

- N is the number of input symmetry classes in the function.
- S is a sequence containing the sizes of the symmetry classes in sorted order.
- F_0 is the minterm count of the function realized by the subcircuit.
- F_j for any $0 \leq j < k$ is a sequence containing the kth order positive cofactor or negative minterm counts (whichever is greater) in sorted order.

Note that if $k = n$, then $K(F, k)$ completely specifies the function K. However, in practice, one rarely needs more than $k = 2$ to sufficiently differentiate most functions. Since we do not need complete differentiation through hashing, we use $k = 1$. This observation has been corroborated by results in [1] where it is reported that $k = 2$ is enough to determine the existence of a Boolean match between two functions in most cases. The hash values simply consist of the boundaries of the subcircuit in question.

Figure 12.15 shows the overall matching algorithm using functional hashing . After the remainder of the original circuit is functionally hashed, the modified circuit is traversed and the largest matches starting at each net are found and stored in the map *BestMatch*. At this point the largest possible matches are known, and we essentially have an optimization version of the *set cover* problem, i.e., we want a cover of the largest number of gates in a circuit. Set cover is a well-known NP-complete problem, whose best-known approximation algorithms simply pick the largest covers greedily [8]. We follow the same approach in choosing a cover. We note that this finer granularity of gate preservation will enable farther-reaching changes to be incorporated into the incremental synthesis flow especially as the synthesis of larger and larger blocks begins to be automated.

```
FIND_FUNCTIONAL_MATCHES(C_orig, C_eco){
{
        foreach(unmatched net N* ∈ C_orig){
                while(N * _enum.next_subcircuit(C*)){
                        F* = compute_truth_table(C*)
                        K(F*, k) = compute_hash_key(F*)
                        H[K(F*, k)] = C
                }
        }
        foreach(unmatched net N ∈ C_eco){
                while(N_enum.next_subcircuit(C)){
                        F = compute_truth_table(C)
                        K(F, k) = compute_hash_key(F)
                        for(0 <= i < |H[K(F, k)]|){
                                C_cov(N) = compute_cover(N,H[K(F, k)][i])
                                if(|C_cov(N)| > |BestMatch[N]|){
                                        BestMatch[N] = C_cov
                                }
                        }
                }
        }
}
```

Fig. 12.15 Functional matching algorithm

12.4 Empirical Validation

We empirically validate our algorithms on actual ECOs, i.e., modifications to the VHDL specifications, performed in IBM server designs. Our experiments are conducted on AMD Opteron 242, 1.6 GHZ processors with 12 GB RAM. Our code is written in C++ and compiled with GCC on a GNU linux operating system. For our experimental setup, we initially compiled the modified VHDL into a technology independent netlist with some fast pre-processing optimizations [20] that took 0.01% of the design time. The result, along with the original mapped/placed/routed design, was analyzed by DeltaSyn to derive a logic difference. Results of this experiment are shown in Table 12.1. The logic difference is compared with the difference derived by the *cone-trace system*, which is used in industry. The cone-trace system copies the entire fanin cone of any mismatching primary output to the difference model and resynthesizes the cone completely. Table 12.1 shows an average improvement of 82% between the results of DeltaSyn and those of the cone-trace system. The entries with difference size 0 represent changes that were achieved simply by reconnecting nets.

Table 12.2 shows results on larger changes. These may be categorized as *incremental synthesis* benchmarks rather than traditional ECO benchmarks. On such cases, we measured the results of all three of the phases, and noted that the addition of a third phase offers an extra 8% reduction in delta size through the reuse of common subfunctions in logic. Note that the reduction numbers only reflect the results of DeltaSyn and not pre-processing optimizations.

Table 12.1 DeltaSyn statistics on IBM ECO benchmarks

Design	No. gates	Runtime CPU (s)	Cone size	Diff. model size	% Diff. reduced	% Design preserved
ibm1	3271	35.51	342	17	95.03	99.48
ibm2	2892	47.40	1590	266	83.27	90.80
ibm3	6624	192.40	98	1	98.98	99.98
ibm4	20773	20.32	774	4	99.48	99.98
ibm5	2681	10.01	1574	177	88.75	100.00
ibm6	1771	4.99	318	152	52.20	91.42
ibm7	3228	180.00	69	0	100.00	100.00
ibm8	5218	9.01	22	13	40.91	99.75
ibm9	532	38.34	77	20	74.03	96.24
ibm10	11512	0.40	1910	429	77.54	96.27
ibm11	6650	211.02	825	126	84.73	98.11
ibm12	611	0.23	47	0	100.00	100.00
ibm13	1517	6.82	21	6	71.43	99.60
Avg.					82.03	97.31

Table 12.2 DeltaSyn statistics on IBM incremental synthesis benchmarks. Compares two-phase difference reduction with three-phase difference reduction

Design	No. gates	Cone size	Diff. model size	New diff. model size	2-Phase runtime CPU(s)	3-Phase runtime CPU(s)	% 2-Phase diff. reduced	% 3-Phase diff. reduced
ibm14	7439	841	149	34	82.61	242.87	82.28	95.95
ibm15	4848	1679	447	169	24.77	29.27	73.38	89.93
ibm16	12681	4439	1310	584	179.13	474.86	70.49	86.84
ibm17	4556	510	12	9	23.93	22.58	97.65	98.23
ibm18	8711	1547	177	121	3.71	23.42	88.55	92.17
ibm19	3200	304	89	80	0.73	21.61	70.72	73.68
ibm20	5224	58	13	12	28.86	36.22	7.58	79.31
ibm21	6548	1910	429	261	190.82	266.69	77.54	86.33
ibm22	547	77	20	13	0.26	0.73	74.03	83.11
ibm23	8784	1299	249	174	13.93	85.74	80.83	86.61
Avg.							79.31	87.22

While the lack of standard benchmarks in this field makes it hard to directly compare to previous work, it should be noted that DeltaSyn is able to derive a small difference model for benchmarks that are significantly larger than previous work [3, 12]. DeltaSyn processes all benchmarks in 211 or fewer seconds. The more (global) structural similarity that exists between the modified model and the original model, the faster DeltaSyn performs. For instance, ibm12 is analyzed in less than 1 s because similarities between the implemented circuit and the modified model allow for the algorithm in Fig. 12.14 to stop subcircuit enumeration (i.e., stop searching for potential matches) and issue recursive calls frequently. Any fast logic optimizations that bring the modified model structurally closer to the original model can, therefore, be employed to improve results. Figure 12.16 shows the relative percentages of difference model size reduction achieved by our three phases. The first phase reduces the logic difference by about 50%. The second phase offers an additional 30% difference reduction. The third phase offers an additional 8% of reduction on average.

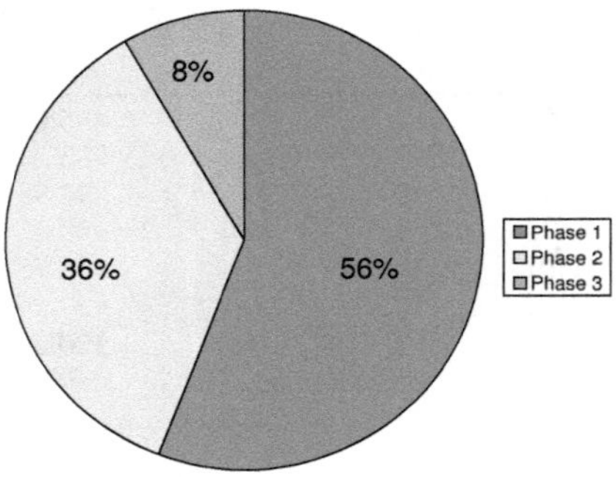

Fig. 12.16 Difference model reduction through phases I, II, III of DeltaSyn

Table 12.1 shows that our difference model disturbs only 3% of logic on average, which is important for preserving the design effort. Figure 12.17 gives a breakdown of the time spent in various parts of the design flow. This is derived from an

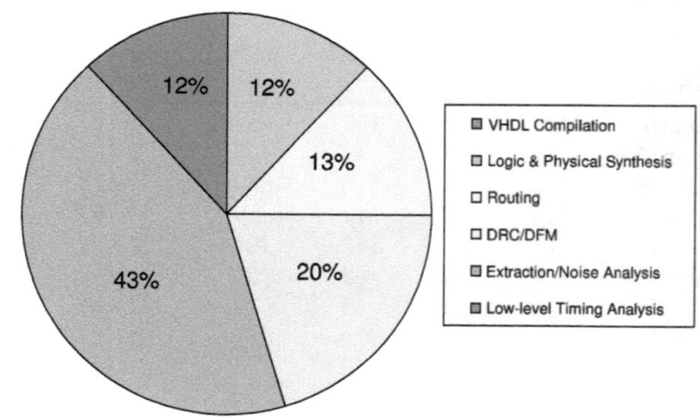

Fig. 12.17 Percentage of time spent in various parts of the design flow [Osler, P, Personal Communication (2009)]. The VHDL compilation step is too small to be visible

average of 44 circuits that were through the complete design flow [Osler, P, Personal Communication (2009)]. The first point to note in this figure is that the only step that DeltaSyn repeats is the VHDL compilation step which takes 0.01% of the entire design flow (not visible on the pie chart). Despite some additional overhead, DeltaSyn allows designers to essentially skip the rest of the process on the unperturbed parts of the design. To demonstrate this, we have embedded DeltaSyn into the PDSRTL physical synthesis and design system [22] which performs incremental placement and optimization only on gates in the difference model (leaving all other gates and latches fixed). Table 12.3 indicates that the runtime decreases drastically for all available benchmarks. In addition, the total slack generally improves or stays close to the same. In the case of ibm2, the fanout of a particular gate in the logic difference increased drastically and disturbed timing. We confirmed that the electrical correction step returns the slack to its original value.

Figure 12.18 shows an example of incremental placement enabled by DeltaSyn. The original model and the final model (with the difference model stitched in) look very similar while the entirely replaced modified model appears significantly different. Preserving the placement generally has the effect of preserving wire routes and also maintaining slack. In summary, DeltaSyn is able to automatically identify changes which leave a large portion of the design unperturbed through the design flow.

Table 12.3 PDSRTL [22] runtime and slack comparison between incremental design and complete redesign

Runtime (s)			% Runtime	% Slack
Design	Entire Design	Difference	Decrease	Increase
ibm1	23040	823	96.43	27.79
ibm2	3240	1414.13	56.35	−20.83
ibm3	10800	1567	85.49	21.95
ibm4	50400	2146	95.74	9.36
ibm5	22680	1315	94.20	99.02
ibm6	2160	665	69.21	−2.97
ibm7	2160	748	65.38	69.72
Avg.			80.40	29.15

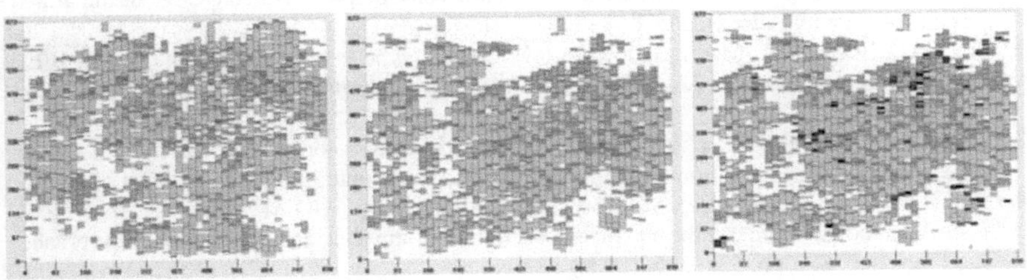

(a) (b) (c)

Fig. 12.18 Placement illustration of (**a**) the modified model placed from scratch, (**b**) the original model, and (**c**) incremental placement on the difference model stitched. Blue indicates deleted gates, red indicates newly added gates

12.5 Chapter Summary

In this chapter, we presented DeltaSyn, a method that analyzes an original and a modified design to maximize the design reuse and design preservation. DeltaSyn uses three phases of analysis in order to find redundant and usable subcircuits in logic. These phases use a variety of techniques such as functional equivalence checking, recursive topologically guided Boolean matching, and functional hashing. Results show that DeltaSyn reduces the logic difference by an average of 88% as compared to previous methods. Further, typical specification changes were processed by reusing an 97% of existing logic, on average. Future work involves extensions to handle changes in sequential logic.

References

1. Abdollahi, A., Pedram, M.: Symmetry detection and Boolean matching utilizing a signature based canonical form of Boolean functions. IEEE Transactions on Computer-Aided Design of Integrated Circuits and Systems **27**(6), 1128–1137 (2009)
2. Alpert, C., Chu, C., Villarrubia, P.: The coming of age of physical synthesis. In: Proceedings of the International Conference on Computer-Aided Design, San Jose, CA, pp. 246–249. (2007)
3. Brand, D., Drumm, A., Kundu, S., Narain, P.: Incremental synthesis. In: Proceedings of the International Conference on Computer-Aided Design, San Jose, CA, pp. 14–18. (1994)
4. Chang, K.H., Papa, D.A., Markov, I.L., Bertacco, V.: Invers: An incremental verification system with circuit similarity metrics and error visualization. IEEE Design and Test Magazine **26**(2), 34–43 (2009)
5. Ganai, M., Kuehlmann, A.: On-the-fly compression of logical circuits. In: Proceedings of the International Workshop on Logic Synthesis, Dana Point, CA, (2000)
6. Goering, R.: CAD foundations must change. EETimes (2006)
7. Goering, R.: Xilinx ISE handles incremental changes. EETimes (2007)
8. Kleinberg, J., Tardos, E.: Algorithm Design. Addison Wesley (2005)
9. Kravets, V., Kudva, P.: Implicit enumeration of structural changes in circuit optimization. In: Proceedings of the Design Automation Conference, San Diego, CA, pp. 438–441. (2004)
10. Kuehlmann, A., Paruthi, V., Krohm, F., Ganai, M.: Robust Boolean reasoning for equivalence checking and functional property verification. IEEE Transactions on Computer-Aided Design of Integrated Circuits and Systems **21**(12), 1355–1394 (2002)
11. Li, Y.L., Li, J.Y., Chen, W.B.: An efficient tile-based eco router using routing graph reduction and enhanced global routing flow. IEEE Transactions on Computer-Aided Design of Integrated Circuits and Systems **26**(2), 345–358 (2007)
12. Lin, C.C., Chen, K.C., Marek-Sadowska, M.: Logic synthesis for engineering change. IEEE Transactions on Computer-Aided Design of Integrated Circuits and Systems **18**(3), 282–292 (1999)
13. Ling, A.C., Brown, S.D., Zhu, J., Safarpour, S.: Towards automated ECOs in FPGAs. In: Proceedings of the International Symposium on FPGAs, Monterey, CA, pp. 3–12. (2009)
14. Mishchenko, A., Chatterjee, S., Brayton, R.: Dag-aware AIG rewriting: A fresh look at combinational logic synthesis. In: Proceedings of the Design Automation Conference, San Francisco, CA, pp. 532–536. (2006)
15. Mishchenko, A., Chatterjee, S., Jiang, R., Brayton, R.: Fraigs: A unifying representation for logic synthesis and verification. ERL Technical Report, EECS Department, UC Berkeley, March 2005.
16. Muroga, S.: Logic Design and Switching Theory, John Wiley, New York (1979)
17. Osler, P.: Personal communication (2009)

18. Roy, J., Markov, I., Eco-system: Embracing the change in placement. IEEE Transactions on Computer-Aided Design of Integrated Circuits and Systems **26**(12), 2173–2185 (2007)
19. Safarpour, S., Mangassarian, H., Veneris, A.G., Liffiton, M.H., Safarpour, S., Mangassarian, H., Veneris, A.G., Liffiton, M.H.: Improved design debugging using maximum satisfiability. In: Proceedings of Formal Methods in Computer-Aided Design, Austin, TX, pp. 13–19. (2007)
20. Shinsha, T., Kubo, T., Sakataya, Y., Ishihara, K.: Incremental logic synthesis through gate logic structure identification. In: Proceedings of the Design Automation Conference, Las Vegas, NV, pp. 391–397. (1986)
21. Stok, L., Kung, D.S., Brand, D., Drumm, A.D., Sullivan, A.J., Reddy, L.N., Hieter, N., Geiger, D.J., Chao, H.H., Osler, P.J.: Booledozer: Logic synthesis for ASICs. IBM Journal of Research and Development **40**(4), 407–430 (1996)
22. Swamy, G., Rajamani, S., Lennard, C., Brayton, R.K.: Minimal logic re-synthesis for engineering change. In: Proceedings of the International Symposium on Circuits and Systems, Hong Kong, pp. 1596–1599. (1997)
23. Trevillyan, L., Kung, D., Puri, R., Reddy, L.N., Kazda, M.A.: An integrated environment for technology closure of deep-submicron IC designs. IEEE Design and Test Magazine **21**(1), 14–22 (2004)
24. Visweswariah, C., Ravindran, K., Kalafa, K.,Walker, S., Narayan, S.: First-order incremental block-based statistical timing analysis. In: Proceedings of the Design Automation Conference, San Diego, CA, pp. 331–336. (2004)
25. Werber, C., Rautenback, D., Szegedy, C.: Timing optimization by restructuring long combinatorial paths. In: Proceedings of the International Conference on Computer-Aided Design, San Jose, CA, pp. 536–543. (2007)
26. Xiang, H., Chao, K.Y., Wong, M.: An ECO routing algorithm for eliminating coupling capacitance violations. IEEE Transactions on Computer-Aided Design of Integrated Circuits and Systems **25**(9), 1754–1762 (2006)
27. Zhu, Q., Kitchen, N., Kuehlmann, A., Sangiovanni-Vincentelli, A.L.: SAT sweeping with local observability don't-cares. In: Proceedings of the International Conference on Computer-Aided Design, San Jose, CA, pp. 229–234. (2006)

Chapter 13
Large-Scale Boolean Matching

Hadi Katebi and Igor Markov

Abstract In this chapter, we propose a methodology for Boolean matching under permutations of inputs and outputs (PP-equivalence checking problem) – a key step in incremental logic design that identifies large sections of a netlist that are not affected by a change in specifications. When a design undergoes incremental changes, large parts of the circuit may remain unmodified. In these cases, the original and the slightly modified circuits share a number of functionally equivalent subcircuits. Finding and reutilizing the equivalent subcircuits reduce the amount of work in each design iteration and accelerate design closure. In this chapter, we present a combination of fast and effective techniques that can efficiently solve the PP-equivalence checking problem in practice. Our approach integrates graph-based, simulation-driven, and SAT-based techniques to make Boolean matching feasible for large circuits. We verify the validity of our approach on ITC'99 benchmarks. The experimental results confirm scalability of our techniques to circuits with hundreds and even thousands of inputs and outputs.

13.1 Introduction

Boolean matching is the problem of determining whether two Boolean functions are equivalent under the permutation and negation of inputs and outputs. This formulation is usually referred to as the *generalized* Boolean matching problem or PNPN-equivalence checking (PNPN stands for Permutation and Negation of outputs and Permutation and Negation of inputs); however, different variants of the problem have been introduced for different synthesis and verification applications. The matching problem that we discuss in this chapter is PP-equivalence checking: two Boolean functions are called PP-equivalent if they are equivalent under permutation of inputs and permutation of outputs. The simplest method to determine

H. Katebi (✉)
University of Michigan, Ann Arbor, MI, USA
e-mail: hadik@eecs.umich.edu; hadi.katebi@gmail.com

Based on Katebi, H.; Markov, I.L.; "Large-scale Boolean matching," Design, Automation & Test in Europe Conference & Exhibition (DATE), 2010, pp.771–776, 8–12 March 2010 © [2010] IEEE.

S.P. Khatri, K. Gulati (eds.), *Advanced Techniques in Logic Synthesis, Optimizations and Applications*, DOI 10.1007/978-1-4419-7518-8_13,

whether two n-input m-output Boolean functions are PP-equivalent is to explicitly enumerate all the $m!n!$ possible matches and perform tautology checking on each. However, this exhaustive search is computationally intractable.

PP-equivalence checking finds numerous applications in verification and logic synthesis. In many cases, an existing design is modified incrementally leaving a great portion of the design untouched. In these cases, large isomorphic subcircuits exist in original and slightly modified circuits [16]. Identifying such subcircuits and reutilizing them whenever possible saves designers a great amount of money and time. Due to the fact that modifications to the original circuit are introduced by changing certain specifications and the fact that even a slight change in specifications can lead to large changes in implementation [9], PP-equivalence checking helps designers identify isomorphic and other equivalent subcircuits.

Specifically, PP-equivalence checking can be used to find the minimal set of changes in logic, known as *logic difference*, between the original design and the design with modified specification. DeltaSyn [10] is a tool developed at IBM Research that identifies and reports this logic difference. The current version of DeltaSyn uses a relatively inefficient and unscalable Boolean matcher that only exploits the symmetry of inputs to prune the search space.

Incremental Sequential Equivalence Checking (SEC) is another application of PP-equivalence checking where isomorphic subcircuits can be used to create a number of highly likely candidate equivalent nodes [16]. The current implementation of incremental SEC tires to find isomorphic subgraphs by performing extended simulation and finding structural similarities. Although the Boolean approach presented in our chapter does not fully exploit the structural similarities between two circuits, we believe that our techniques combined with structural verification techniques create a much more powerful tool for detecting isomorphic subgraphs.

Motivated by the practical importance of PP-equivalence checking in many EDA applications, we develop fast and scalable Boolean matching algorithms and implement them in the ABC package – an established system for synthesis and verification of combinational and sequential logic circuits [12]. The collection of all these techniques creates a powerful Boolean matching module that can be integrated into Combinational Equivalence Checking (CEC) to enhance its functionality. To this end, CEC requires two designs whose primary I/Os match by name. Our work allows one to relax this requirement with the help of a Boolean matcher. We call the new command *Enhanced CEC* (*ECEC*). Figure 13.1 shows how our Boolean matcher is integrated with CEC.

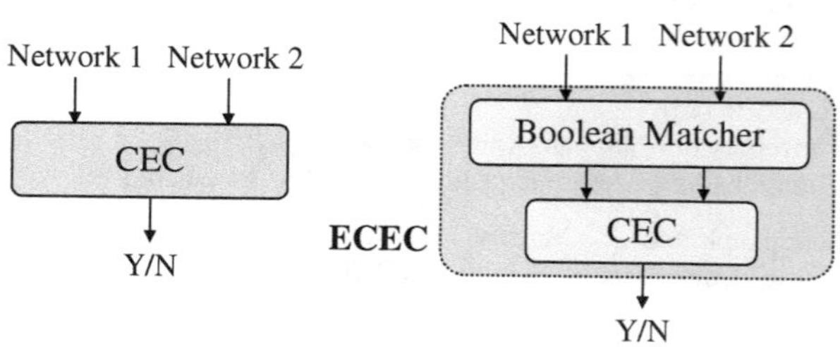

Fig. 13.1 The CEC (pre-existing) and ECEC (our) flows

In general, algorithms for Boolean matching fall into two major categories: signature-based and canonical form based. A signature is a property of an input or an output that is invariant under permutations and negation of inputs. The goal of signature-based matching is to prune the Boolean matching space by filtering out impossible I/O correspondences [1, 4]. On the other hand, in matching based on canonical forms, first canonical representations of two Boolean functions are computed and then compared against each other to find valid I/O matches [2, 3]. Here, our PP-equivalence checking method first prunes the search space using graph algorithms and simulation signatures, then it invokes SAT-solving until exact I/O matches are found (Fig. 13.2).

Fig. 13.2 Overview of our proposed Boolean matching approach

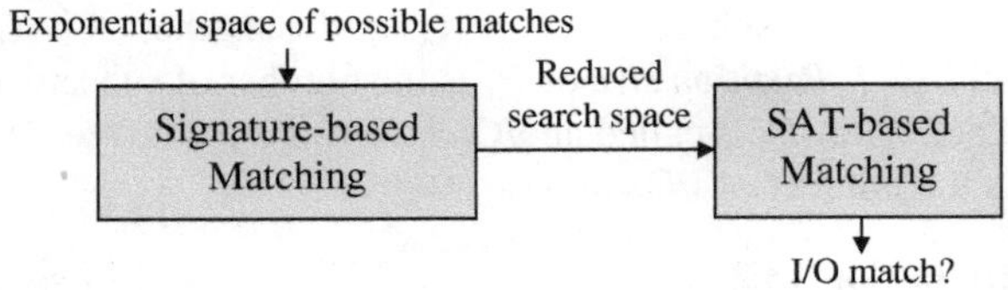

The key contributions of our work include:

1. *Analyzing functional dependency*. In a Boolean network with multiple outputs, some inputs affect only a fraction of the outputs, and different outputs are affected in different ways. Hence, by analyzing the functional dependency of outputs on inputs, we can distinguish the I/Os.
2. *Exploiting input observability and output controllability*. We use the observability of inputs and the controllability of outputs as (1) effective matching signatures and (2) ordering heuristics for our SAT-based matching.
3. *Building a SAT-tree*. When information about controllability, observability, and all simulation-based information are exhausted, we resort to SAT-solving and optimize the efficiency of SAT calls. This is accomplished through the concept of a SAT-tree, which is pruned in several ways.
4. *Pruning SAT-tree using SAT counterexamples*. In our SAT-based matching, the SAT-solver returns a counterexample whenever it finds an invalid match. The information in these counterexamples is then used to prune the SAT-tree.

The remainder of this chapter is organized as follows. Section 13.2 provides relevant background and discusses previous work on Boolean matching. Section 13.3 gives an overview of proposed signature-based techniques. Section 13.4 describes our SAT-based matching approach. Section 13.5 validates our method in experiments on available benchmarks, and Section 13.6 summarizes our work.

13.2 Background and Previous Work

In this section, we first review some common definitions and notation. Then, we explain the *And-Inverter* representation of Boolean networks and we compare its usage to that of conventional *Binary Decision Diagrams*. Next, we discuss *Boolean satisfiabilty* and explore its role in combinational equivalence checking. Relevant work in Boolean matching is reviewed at the end of this section.

13.2.1 Definitions and Notation

In the following definitions, an *input set* of a Boolean network N refers to the set of all the inputs of N. Similarly, an *output set* of N refers to the set of all the outputs of N. An *I/O set* is either an input set or an output set.

Definition 13.1 A *partition* $P_x = \{X_1, \ldots, X_k\}$ of an I/O set $X = \{x_1, \ldots, x_n\}$ is a collection of subsets $X_1, \ldots, X_k$ of X such that $\cup_{i=1}^{k} X_i = X$ and $X_i \cap X_j = \emptyset$ for all $i \neq j$. *Partition size* of P_x is the number of subsets in P_x and is denoted by $|P_x|$. Each X_i in P_x is called an I/O *cluster* of P_x. The *cardinality* of X_i, denoted by $|X_i|$, is the number of I/Os in X_i.

Definition 13.2 A partition $P_x = \{X_1, \ldots, X_k\}$ of set X is an *ordered partition* if the subsets $X_1, \ldots, X_k$ are *totally ordered*, i.e., for any two subsets X_i and X_j, it is known whether $X_i < X_j$ or $X_j < X_i$.

Definition 13.3 Two ordered partitions $P_x = \{X_1, \ldots, X_k\}$ of set X and $P_y = \{Y_1, \ldots, Y_k\}$ of set Y are *isomorphic* if and only if $|P_x| = |P_y| = k$ and $|X_i| = |Y_i|$ for all i, and *non-isomorphic* otherwise. Two isomorphic partitions are called *complete (discrete)* if and only if $|X_i| = |Y_i| = 1$ for all i.

Definition 13.4 The *positive cofactor* of function $f(x_1, .., x_n)$ with respect to variable x_i, denoted by f_{x_i}, is $f(x_1, .., x_i = 1, \ldots, x_n)$. Similarly, the *negative cofactor* of $f(x_1, .., x_n)$ with respect to variable x_i, denoted by $f_{x_i'}$, is $f(x_1, .., x_i = 0, \ldots, x_n)$.

Definition 13.5 A function $f(x_1, \ldots, x_n)$ is *positive unate* in variable x_i if and only if the negative cofactor of f with respect to x_i is covered by the positive cofactor of f with respect to x_i, i.e., $f_{x_i'} \subseteq f_{x_i}$. Likewise, f is *negative unate* in variable x_i if and only if $f_{x_i} \subseteq f_{x_i'}$. f is called *binate* in x_i if it is not unate in it.

13.2.2 And-Inverter Graphs (AIGs)

Recent tools for scalable logic synthesis, e.g., ABC, represent Boolean functions using the *And-Inverter Graph* (*AIG*) data structure. An AIG is a Boolean network composed of two-input AND gates and inverters. *Structural hashing* of an AIG is a transformation that reduces the AIG size by partially canonicalizing the AIG structure [13]. Representing a Boolean function in its AIG form is preferable to its *Binary Decision Diagram* (*BDD*) form mainly because AIGs result in smaller space complexity. Also, functional simulation can be performed much faster on AIGs, but AIGs are only locally canonical.

13.2.3 Boolean Satisfiability and Equivalence Checking

Boolean Satisfiability (*SAT*) is the problem of determining whether there exists a variable assignment to a Boolean formula that forces the entire formula evaluate to true; if such an assignment exists, the formula is said to be *satisfiable* and otherwise *unsatisfiable*. Pioneering techniques developed to solve the SAT problem were introduced by Davis, Putnam, Logemann, and Loveland in early 1960s. They are now referred to as the DPLL algorithm [6, 7]. Modern SAT-solvers, such as MiniSAT [8], have augmented DPLL search by adding efficient *conflict analysis*, *clause learning*, *back-jumping*, and *watched literals* to the basic concepts of DPLL.

SAT is studied in a variety of theoretical and practical contexts, including those arising in EDA. CEC is one of the main applications of SAT in EDA. If two single-output Boolean functions f and g are equivalent, then $f \bigoplus g$ must always evaluate to 0, and vice versa. Now, instead of simulating all input combinations, we take advantage of SAT solvers: if $f \bigoplus g$ is *unsatisfiable*, then $f \bigoplus g$ is zero for all input combinations and hence f and g are equivalent; and if $f \bigoplus g$ is *satisfiable*, then f and g are not equivalent and the satisfying assignment found by the SAT-solver is returned as a counterexample. $f \bigoplus g$ is called the *miter* of f and g [14]. If f and g have more than one output, say m outputs $f_1, \ldots, f_m$ and $g_1, \ldots, g_m$, $M_i = f_i \bigoplus g_i$ is first computed for all i and then $M_1 + \cdots + M_m$ is constructed as the miter of f and g. In our approach, instead of building one miter for the entire circuit and handing it off to the SAT solver, we try to find equivalent intermediate signals by simulation and use SAT to prove their equivalence. Counterexamples from SAT are used to refine simulation.

13.2.4 Previous Work

Research in Boolean matching started in the early 1980s with main focus on technology mapping (cell binding). A survey of Boolean matching techniques for library binding is given in [4]. Until recently, Boolean matching techniques scaled only to 10–20 inputs and one output [2, 5], which is sufficient for technology mapping, but not for applications considered in our work. In 2008, Abdollahi and Pedram presented algorithms based on canonical forms that can handle libraries with numerous cells limited to approximately 20 inputs [2]. Their approach uses generalized signatures (signatures of one or more variables) to find a canonicity-producing (CP) phase assignment and ordering for variables.

A DAC 2009 paper by Wang et al. [17] offers simulation-driven and SAT-based algorithms for checking P-equivalence that scale beyond the needs of technology mapping. Since our proposed techniques also use simulation and SAT to solve the PP-equivalence checking problem, we should articulate the similarities and the differences. First, we consider the more general problem of PP-equivalence checking where permutation of outputs (beside permutation of inputs) is allowed. In PP-equivalence, the construction of miters must be postponed until the outputs are matched, which seems difficult without matching the inputs. To address this challenge, we develop the concept of SAT-tree which is pruned to moderate the overall runtime of PP-matching. In addition to our SAT-based approach, we also use graph-based techniques in two different ways: to initially eliminate impossible I/O correspondences and to prune our SAT-tree. Furthermore, we have implemented three simulation types; two as signatures for outputs (type 1 and type 3) and one as a signature for inputs (type 2). While our type-2 simulation is loosely related to one of the techniques described in [17], the other two simulations are new. We additionally introduce effective heuristics that accelerate SAT-based matching.

13.3 Signature-Based Matching Techniques

We now formalize the PP-equivalence checking problem and outline our Boolean matching approach for two n-input m-output Boolean networks.

Definition 13.6 Consider two I/O sets X and Y of two Boolean networks N_1 and N_2 with their two isomorphic ordered partitions $P_x = \{X_1, \ldots, X_k\}$ and $P_y = \{Y_1, \ldots, Y_k\}$. A *cluster mapping* of X_i to Y_i, denoted by $X_i \mapsto Y_i$, is defined as the mapping of I/Os in X_i to all possible permutations of I/Os in Y_i. A *mapping* of X to Y with respect to P_x and P_y, denoted by $X \mapsto Y$, is defined as mapping of all same-index clusters of X and Y, i.e., $X_i \mapsto Y_i$ for all i. $X \mapsto Y$ is called a complete mapping if P_x and P_y are complete.

Given two input sets X and Y and two outputs sets Z and W of two Boolean networks N_1 and N_2, the goal of PP-equivalence checking is to find two complete mappings $X \mapsto Y$ and $Z \mapsto W$ such that those mappings make N_1 and N_2 behave functionally the same. In order to accomplish this, we first partition or *refine* these I/O sets based on some total ordering criteria. This so-called signature-based matching allows us to identify and eliminate impossible I/O matches. After this phase, we rely on SAT-solving to find the two complete mappings. Furthermore, Definition 13.6 implies the following lemma.

Lemma 13.1 *If at any point in the refinement process of two I/O sets X and Y, P_x and P_y become non-isomorphic, we conclude that N_1 and N_2 behave differently and we stop the Boolean matching process.*

As mentioned earlier, refinement at each step requires a total ordering criterion, tailored to the specific refinement technique used. Therefore, whenever we introduce a new matching technique, we also explain its ordering criterion. Furthermore, the following techniques are applied to the two input circuits one after another.

13.3.1 Computing I/O Support Variables

Definition 13.7 Input x is a *support variable* of output z and output z is a *support variable* of input x, if there exists an input vector V such that flipping the value of x in V flips the value of z.

Definition 13.8 The *support* of input (or output) x, denoted by $Supp(x)$, is the set of all the support variables of x. The *cardinality* of the support of x, denoted by $|Supp(x)|$, is the number of I/Os in $Supp(x)$. The *degree* of x, denoted by $D(x)$, is defined as the cardinality of its support.

The goal here is to find outputs that might be functionally affected by a particular input and inputs that might functionally affect a particular output. Here, we contrast *functionally* matching with *structurally* matching in the sense that two structurally different circuits with the same functionality should have the same I/O support. In general, the lack of structural dependency between an output and an input precludes a functional dependency, and the presence of a structural dependency most often indicates a functional dependency – this can usually be confirmed by random simulation and in rare cases require calling a SAT-solver [11].

Example 13.1 Consider a 4-bit adder with input set $X = \{C_{in}, A_0, \ldots, A_3, B_0, \ldots, B_3\}$ and output set $Z = \{S_0, \ldots, S_4\}$. The ripple-carry realization of this adder is shown in Fig. 13.3.

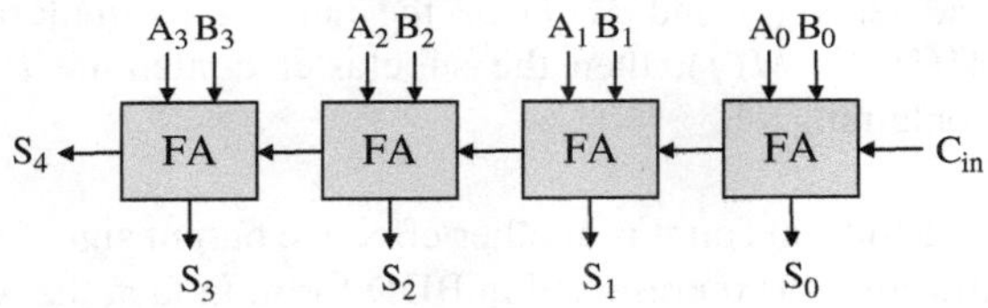

Fig. 13.3 A 4-bit ripple-carry adder

It is evident from the above circuit that A_0 can affect the values of $S_0, \ldots, S_4$ and A_1 can affect the value of $S_1, \ldots, S_4$. Hence, $Supp(A_0) = \{S_0, \ldots, S_4\}$ and $Supp(A_1) = \{S_1, \ldots, S_4\}$. Similarly, the value of S_0 is only affected by the value of A_0, B_0, and C_{in}. Hence, $Supp(S_0) = \{A_0, B_0, C_{in}\}$.

13.3.2 Initial refinement of I/O clusters

Lemma 13.2 *Two inputs (outputs) can match only if they have the same degree.*

Taking advantage of Lemma 13.2, we can initially refine the I/O sets by gathering all I/Os of the same degree in one subcluster and then sort the subclusters based on the following ordering criterion:

Ordering criterion 13.1 Let i and j be two inputs (outputs) with different degrees and assume that $D(i) < D(j)$. Then, the subcluster containing i precedes the subcluster containing j.

Example 13.2 Consider the 4-bit adder of Example 13.1. The degree of each input and output is given below:
$D(A_0) = D(B_0) = D(C_{in}) = 5$
$D(A_1) = D(B_1) = 4$
$D(A_2) = D(B_2) = 3$

$D(A_3) = D(B_3) = 2$
$D(S_0) = 3$
$D(S_1) = 5$
$D(S_2) = 7$
$D(S_3) = D(S_4) = 9$
The ordered partitions of the I/O sets of the 4-bit adder after initial refinement are
$P_x = \{\{A_3, B_3\}, \{A_2, B_2\}, \{A_1, B_1\}, \{A_0, B_0, C_{in}\}\}$
$P_z = \{\{S_0\}, \{S_1\}, \{S_2\}, \{S_3, S_4\}\}$

13.3.3 Refining Outputs by Minterm Count

Lemma 13.3 *Two outputs can match only if their Boolean functions have the same number of minterms.*

Ordering criterion 13.2 Let i and j be two outputs in the same output cluster and let $M(i)$ and $M(j)$ be the number of minterms of i and j, respectively. If $M(i) < M(j)$, then the subcluster containing i is smaller than the subcluster containing j.

Minterm count is another effective output signature which is only practical when the circuit is represented in BDD form. In fact, the widely adopted way to count the minterms of a Boolean network represented in AIG is to first convert it to a BDD, but this approach is limited in scalability [15].

13.3.4 Refining I/O by Unateness

Lemma 13.4 *Two outputs match only if they are unate in the same number of inputs. Similarly, two inputs match only if the same number of outputs is unate in them.*

Ordering criterion 13.3 Let i and j be two outputs in the same output cluster. Assume that $Unate(i)$ and $Unate(j)$ are the number of unate variables of i and j respectively, and let $Unate(i) < Unate(j)$. Then, the output subcluster containing i is smaller than the subcluster containing j. Similarly, let i and j be two inputs in one input cluster. Assume that $Unate(i)$ and $Unate(j)$ are the number of outputs that are unate in i and j, respectively, and let $Unate(i) < Unate(j)$. Then, the input subcluster containing i is smaller than the subcluster containing j.

Although unateness generates powerful signatures for Boolean matching, computing unateness in an AIG encounters the same limitation as was discussed for counting the number of minterms. Hence, refinement based on unateness is only practical for small Boolean networks.

13.3.5 Scalable I/O Refinement by Dependency Analysis

We mentioned earlier that the degree of each I/O is an effective signature for initial refinement of I/O sets. Here, we generalize this concept by not only considering the number of support variables but also carefully analyzing I/O dependencies.

Definition 13.9 Let x be an input (output) and let $Supp(x) = \{z_1, \ldots, z_k\}$. We define a sequence $S = (s_1, \ldots, s_k)$ of unsigned integers where each s_i is the index of the output (input) cluster that z_i belongs to. After sorting S, we call it *support signature* of x and we denote it by $Sign(x)$.

Lemma 13.5 *Two I/Os i and j in the same I/O cluster are distinguishable if $Sign(i) \neq Sign(j)$.*

Ordering criterion 13.4 Let i and j be two I/Os in the same I/O cluster. Assume that $Sign(i) < Sign(j)$ meaning that the support signature of i is lexicographically smaller than the support signature of j. Then, the subcluster containing i precedes the subcluster containing j.

Example 13.3 Consider a circuit with input set $X = \{x_1, x_2, x_3\}$ and output set $Z = \{z_1, z_2, z_3\}$ where $z_1 = \overline{x_1}$, $z_2 = x_1 \cdot x_2$ and $z_3 = \overline{x_2 \cdot x_3}$. The I/O supports of the circuit are $Supp(z_1) = \{x_1\}$, $Supp(z_2) = \{x_1, x_2\}$, $Supp(z_3) = \{x_2, x_3\}$ and $Supp(x_1) = \{z_1, z_2\}$, $Supp(x_2) = \{z_2, z_3\}$, $Supp(x_3) = \{z_3\}$. Since $D(z_1) = 1$ and $D(z_2) = D(z_3) = 2$, and $D(x_3) = 1$ and $D(x_1) = D(x_2) = 2$, we can initialize I/O clusters as follows: $P_z = \{\{z_1\}, \{z_2, z_3\}\}$, $P_x = \{\{x_3\}, \{x_1, x_2\}\}$. Now, we try refining based on support signatures. The signatures for z_2, z_3, x_1, and x_2 are $Sign(z_2) = (2, 2)$, $Sign(z_3) = (1, 2)$, $Sign(x_1) = (1, 2)$, $Sign(x_2) = (2, 2)$. Since $Sign(z_3) < Sign(z_2)$ and $Sign(x_1) < Sign(x_2)$, we can further partition $\{z_2, z_3\}$ and $\{x_1, x_2\}$, hence $P_z = \{\{z_1\}, \{z_3\}, \{z_2\}\}$ and $P_x = \{\{x_3\}, \{x_1\}, \{x_2\}\}$.

After each round of refinement based on I/O dependencies, we check if any I/O cluster is further partitioned. If a new partition is added, the algorithm performs another round of refinement. The procedure terminates when no new refinement occurs after a certain number of iterations.

13.3.6 Scalable I/O Refinement by Random Simulation

Functional simulation holds the promise to quickly prune away unpromising branches of search, but this seems to require a matching of outputs. Instead, we find pairs of input vectors that sensitize comparable functional properties of the two circuits. Let $V =< v_1, \ldots, v_n >$ be an input vector of Boolean network N. The result of simulating V on N is called the *output vector* of N under V and is denoted by $R_v =< r_1, \ldots, r_m >$.

Definition 13.10 Let N be a Boolean network with input set X and let $P_x = \{X_1, \ldots, X_k\}$ be an ordered partition of X. An input vector $V =< v_1, \ldots, v_n >$ is said to be *proper* if it assigns the same value (0 or 1) to all the inputs of N which are in the same input cluster, i.e., $v_i = v_j$ if $i, j \in X_l$ for some l. The input vectors consisting of all 0s or all 1s are the *trivial* proper input vectors.

Definition 13.11 Let $X = \{x_1, \ldots, x_n\}$ and $Y = \{y_1, \ldots, y_n\}$ be the input sets of two Boolean networks N_1 and N_2 and let $P_x = \{X_1, \ldots, X_k\}$ and $P_y = \{Y_1, \ldots, Y_k\}$ be two ordered partitions defined on them. Two proper random input vectors $V =< v_1, \ldots, v_n >$ and $U =< u_1, \ldots, u_n >$ of N_1 and N_2 are said to be *consistent* if, for all $1 \leq l \leq k$, $x_i \in X_l$ and $y_j \in Y_l$ imply that $v_i = u_j$.

Intuitively, two consistent random input vectors try to assign the same value to all potentially matchable inputs of the two Boolean networks. In the next three sections, we distinguish three types of simulation based on pairs of consistent random input vectors that help us sensitize certain functional properties of the two circuits. The flow of the I/O refinement by random simulation is shown in Fig. 13.4.

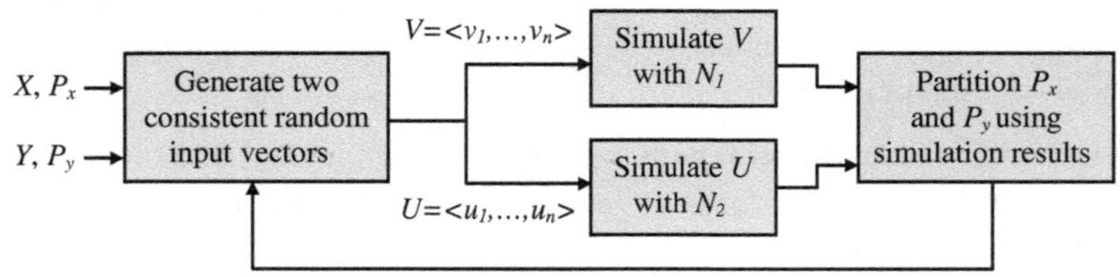

Fig. 13.4 Flow of the proposed I/O refinement by random simulation

13.3.6.1 Simulation Type 1

Lemma 13.6 *Let V be a proper random input vector and let $R_v =< r_1, \ldots, r_m >$ be the corresponding output vector under V. Two outputs i and j in one output cluster are distinguishable if $r_i \neq r_j$.*

The above lemma classifies outputs based on their values (0 or 1) using the following ordering criterion.

Ordering criterion 13.5 The output subcluster of all 0s precedes the output subcluster of all 1s.

13.3.6.2 Simulation Type 2

Definition 13.12 Let V be a proper random input vector and let $R_v =< r_1, \ldots, r_m >$ be the corresponding output vector under V. Let V' be another input vector created by flipping the value of input x in V and let $R_{v'} =< r'_1, \ldots, r'_m >$ be the corresponding output vector under V'. The *observability* of input x with respect to V denoted by $Obs(x)$ is defined as the number of flips in the outputs caused by V', i.e., the number of times $r_i \neq r'_i$.

Lemma 13.7 *Two inputs i and j in one input cluster are distinguishable if $Obs(i) \neq Obs(j)$.*

Ordering criterion 13.6 Let i and j be two inputs in one input cluster and let $Obs(i) < Obs(j)$. Then, the input subcluster containing i precedes the input subcluster containing j.

13.3.6.3 Simulation Type 3

Definition 13.13 Consider a proper random input vector V and its corresponding output vector $R_v =< r_1, \ldots, r_m >$. Let $V_1, \ldots, V_n$ be n input vectors where vector V_i is created by flipping the value of r_i in V. Also, let $R_{v_1} =< r_{1,1}, \ldots, r_{1,m} >, \ldots, R_{v_n} =< r_{n,1}, \ldots, r_{n,m} >$ be the corresponding output vectors under $V_1, \ldots, V_n$. The *controllability* of output z with respect to V denoted by $Ctrl(z)$ is defined as the number of times $r_i \neq r_{j,i}$, for all $1 \leq j \leq n$.

Lemma 13.8 *Two outputs i and j in one output cluster are distinguishable if $Ctrl(i) \neq Ctrl(j)$.*

Ordering criterion 13.7 Let i and j be two outputs in one output cluster and let $Ctrl(i) < Ctrl(j)$. Then, the output subcluster containing i precedes the output subcluster containing j.

Example 13.4 Take an 8-to-1 multiplexer with input set $X = \{a_0, \ldots, a_7, s_0, s_1, s_2\}$ and output z where $a_0, \ldots, a_7$ denote data signals and s_0, s_1, s_2 are control signals. Initially P_x has only one partition, namely X. Initial refinement and refinement by dependency analysis do not partition P_x, hence we try random simulation. Here, we can only use type 2 simulation since simulation of types 1 and 3 is for refining output clusters. First, we consider the trivial input vector V of all 0s. We flip one input in V at a time and we apply the resulting vectors to the multiplexer. Only flipping a_0 flips z; hence, $P_x = \{\{a_1, \ldots, a_7, s_0, s_1, s_2\}, \{a_0\}\}$. Then we try the trivial input vector V of all 1s. This time flipping a_7 flips z; hence, $P_x = \{\{a_1, \ldots, a_6, s_0, s_1, s_2\}, \{a_7\}, \{a_0\}\}$. Next, we put a_0 to 1 and all the other inputs to 0. Now flipping s_0, s_1, s_2 flips z, hence $P_x = \{\{a_1, \ldots, a_6\}, \{s_0, s_1, s_2\}, \{a_7\}, \{a_0\}\}$. If we continue partitioning based on the remaining proper input vectors no additional refinement will be gained.

After matching I/Os using random simulation, we check if any progress is achieved in refining I/O clusters. If a new cluster is added, the algorithm continues refining based on random simulation. The procedure terminates when no new refinement occurs in input or output subclusters after a certain number of iterations. Here, the number of iterations does not affect the correctness of the algorithm. However, too few iterations might diminish the impact of matching by random simulation, and excessive iterations offer no improvement. Our current implementation limits iterations to 200.

13.4 SAT-Based Search

The scalable methods we introduced so far typically reduce the number of possible matches from $n!m!$ to hundreds or less, often making exhaustive search (with SAT-based equivalence checking) practical. However, this phase of Boolean matching can be significantly improved, and the techniques we develop facilitate scaling to even larger instances.

13.4.1 SAT-Based Input Matching

The basic idea in our SAT-based matching approach is to build a tree data structure called *SAT-tree* that matches one input at a time from the remaining non-singleton input clusters. Subsequently, after an input is matched, all the outputs in its support which are not matched so far are also matched, one by one. In other words, we build a dual-purpose tree that repeatedly matches inputs and outputs until exact I/O matches are found. We take advantage of the following lemma to build our SAT-tree:

Lemma 13.9 *Assume that two Boolean networks N_1 and N_2 with input sets $X = \{x_1, \ldots, x_n\}$ and $Y = \{y_1, \ldots, y_n\}$ are functionally equivalent under two complete ordered partitions $P_x = \{X_1, \ldots, X_n\}$ and $P_y = \{Y_1, \ldots, Y_n\}$. Also, assume that $X_l = \{x_i\}$ and $Y_l = \{y_j\}$. Let N_1' be the positive (negative) cofactor of N_1 with respect to x_i and N_2' be the positive (negative) cofactor of N_2 with respect to y_j. N_1' and N_2' with input sets $X' = X - \{x_i\}$ and $Y' = Y - \{y_j\}$ behave functionally the same under two complete ordered partitions $P_{x'} = P_x - \{X_l\}$ and $P_{y'} = P_y - \{Y_l\}$.*

The inputs to the SAT-based matching algorithm are two ordered input partitions and two ordered output partitions. Here, we assume that some of the partitions are incomplete because if all partitions are complete, an exact match is already found. Without loss of generality, assume that in two ordered partitions $P_x = \{X_1, \ldots, X_k\}$ and $P_y = \{Y_1, \ldots, Y_k\}$ of sets X and Y, $X_1, \ldots, X_{l-1}$ and $Y_1, \ldots, Y_{l-1}$ are all singleton clusters, and $X_l, \ldots, X_k$ and $Y_l, \ldots, Y_k$ are non-singleton clusters. Repeatedly applying Lemma 13.9 allows us to create two new Boolean networks N_1' and N_2' by setting all the inputs in $X_l, \ldots, X_k$ and $Y_l, \ldots, Y_k$ to either constant 0 or constant 1. In other words, we shrink input sets X to $X' = X - \{x | x \in \{X_l, \ldots, X_k\}\}$ and input set Y to $Y' = Y - \{y | y \in \{Y_l, \ldots, Y_k\}\}$ such that X' and Y' only contain the inputs that have exact match in N_1 and N_2. Note that, by definition, the ordered partitions $P_x' = P_x - \{X_l, \ldots, X_k\}$ and $P_y' = P_y - \{Y_l, \ldots, Y_k\}$ are complete partitions of X' and Y'. According to Lemma 13.9, N_1' and N_2' must be functionally equivalent if N_1 and N_2 are equivalent. N_1' and N_2' are called the *Smallest Matching Subcircuits* (*SMS*) of N_1 and N_2.

After finding the SMS of N_1 and N_2, we try to expand X' and Y' back to X and Y by matching one input at a time. Let X_l and Y_l be the first two non-singleton input clusters of P_x and P_y and let $x_i \in X_l$. The goal here is to match x_i with one of the $|Y_l|$ inputs in Y_l. Assume that $y_j \in Y_l$, and we pick y_j as the first candidate to match x_i. Now, in order to reflect our matching decision, we partition X_l and Y_l to make $\{x_i\}$ and $\{y_j\}$ two singleton clusters; hence, X_l is partitioned to $X_{l,1} = \{x_i\}$ and $X_{l,2} = X_l - \{x_i\}$ and Y_l is partitioned to $Y_{l,1} = \{y_j\}$ and $Y_{l,2} = Y_l - \{y_j\}$. Complying with our previous notation, now $X_{l,2}, \ldots, X_k$ and $Y_{l,2}, \ldots, Y_k$ are the new non-singleton clusters. We then build two Boolean networks N_1'' and N_2'' from

N_1 and N_2 by setting all the inputs in non-singleton clusters to either constant 0 or constant 1, and we pass the miter of N_1'' and N_2'' to the SAT-solver. The SAT-solver may return either *satisfiable* or *unsatisfiable*. If the result is

- ***unsatisfiable***: N_1'' and N_2'' are functionally equivalent. In other words, x_i and y_j has been a valid match so far. Hence, first try to match the outputs in the supports of x_i and y_j (only the outputs that have not been matched so far) and then match the next two unmatched inputs in $X_{l,2}$ and $Y_{l,2}$.
- ***satisfiable***: N_1'' and N_2'' are not functionally equivalent. In other words, x_i cannot match y_j. Backtrack one level up and use the counterexample to prune the SAT-tree.

In a case where the SAT-solver times out, we terminate the matching process, and only report the I/Os matched by our signature-based techniques. Unlike early prototypes, our most optimized implementation does not experience this situation on the testcases used in our experiments.

13.4.2 Pruning Invalid Input Matches by SAT Counterexamples

Pruning the SAT-tree using counterexamples produced by SAT is a key step in our Boolean matching methodology. Continuing the scenario in Section 13.4.1, assume that the miter of N_1'' and N_2'' is *satisfiable*. Suppose that the SAT-solver returns an input vector $V =< v_1, \ldots, v_{l+1} >$ as the satisfying assignment. This input vector carries a crucial piece of information: the matching attempt before matching x_i and y_j was a successful match; otherwise we would have backtracked at the previous level and we would have never tried matching x_i and y_j. Thus, the input vector V sensitizes a path from x_i and y_j to the outputs of the miter.

According to Lemma 13.9, repeatedly computing negative and positive cofactors of N_1 and N_2 with respect to the values of $v_1, \ldots, v_l$ in V results in two new Boolean networks $\hat{N}_1$ and $\hat{N}_2$ that must be functionally equivalent under some ordered partition $P_x - \{X_1, \ldots, X_l\}$ and $P_y - \{Y_1, \ldots, Y_l\}$. In other words, $\hat{N}_1$ and $\hat{N}_2$ are two smaller Boolean networks that only contain the inputs of N_1 and N_2 that have not found exact match so far. Since $\hat{N}_1$ and $\hat{N}_2$ are computed with respect to the values of $v_1, \ldots, v_l$ in V and since V is a vector that sensitizes a path from x_i and y_j to the output of the miter, we conclude that there exists an output in $\hat{N}_1$ that is functionally dependent on x_i. The existence of such an output ensures that $D(x_i) > 0$. We can now apply our simple filtering signature from Lemma 13.2 to prune the SAT-tree. Specifically, $x_i \in X_l$ can match to $y_q \in Y_l$ $(q \neq j)$ only if $D(x_i) = D(y_q)$ in $\hat{N}_1$ and $\hat{N}_2$.

Example 13.5 Consider two 8-to-1 multiplexers with outputs z and z' and input sets $X = \{a_0, \ldots, a_7, s_0, s_1, s_2\}$ and $X' = \{a_0', \ldots, a_7', s_0', s_1', s_2'\}$. Refining X and X' based on the techniques explained in Section 13.3 would result in two ordered partitions $P_x = \{\{a_1, \ldots, a_6\}, \{s_0, s_1, s_2\}, \{a_7\}, \{a_0\}\}$ and $P_{x'} = \{\{a_1', \ldots, a_6'\}, \{s_0', s_1', s_2'\}, \{a_7'\}, \{a_0'\}\}$ (refer to Example 13.4). In order to find exact input matches, we build our SAT-tree and we first try matching s_2 and s'_0. The SAT-solver confirms the validity of this match. Then, s_1 matches s'_1 and s_0 matches s'_2. These two matches are also valid. So far, $P_x = \{\{a_1, \ldots, a_6\}, \{s_2\}, \{s_1\}, \{s_0\}, \{a_7\}, \{a_0\}\}$ and $P_{x'} = \{\{a'_1, \ldots, a'_6\}, \{s'_0\}, \{s'_1\}, \{s'_2\}, \{a'_7\}, \{a'_0\}\}$. Now, we look at the next non-singleton input cluster and we match a_1 and a'_1. Our SAT-solver specifies that matching a_1 and a'_1 do not form a valid match and it returns vector V with $s'_0 = s_2 = 0$, $s'_1 = s_1 = 0$, $s'_2 = s_0 = 1$, $a'_7 = a_7 = 0$, $a'_0 = a_0 = 0$, $a'_1 = a_1 = 1$ as a counterexample. In order to see why V is a counterexample of matching a_1 and a'_1, we look at the cofactors of the two multiplexers, c and c', where all the inputs in non-singleton clusters are set to 0: $c = a_0\bar{s}_2\bar{s}_1\bar{s}_0 + a_1\bar{s}_2\bar{s}_1 s_0 + a_7 s_2 s_1 s_0$ and $c' = a'_0\bar{s}'_0\bar{s}'_1\bar{s}'_2 + a'_1\bar{s}'_0\bar{s}'_1 s'_2 + a'_7 s'_0 s'_1 s'_2$. Applying V to c and c' would result in $c = 1$ and $c' = 0$. Since we know that a_1 does not match a'_1, we use the counterexample to prune the SAT-tree. Specifically, we compute cofactors of the two multiplexers, d and d', with respect to the values of matched inputs in V. So, $d = a_1\bar{s}_2\bar{s}_1 s_0$ and $d' = a'_4\bar{s}'_0\bar{s}'_1 s'_2$. In d and d', $D(a_1) = D(a'_4) = 1$. This means that a_1 can only match a'_4. In other words, we have pruned SAT search space by not matching a_1 to any of inputs a'_2, a'_3, a'_5, and a'_6. We continue matching inputs until we find valid matches.

13.4.3 SAT-Based Output Matching

Let Z and W be the output sets of two Boolean networks N_1 and N_2 and let $P_z = \{Z_1, \ldots, Z_k\}$ and $P_w = \{W_1, \ldots, W_k\}$ be two ordered partitions defined on them. Continuing the scenario in Section 13.4.1, assume that $z_i \in Z_l$ is a support variable of x_i, $w_j \in W_l$ is a support variable of y_j, and Z_l and W_l are two non-singleton output clusters of P_z and P_w. In order to verify if z_i and w_j match under current input correspondence, we add $z_i \bigoplus w_j$ to the current miter of N''_1 and N''_2 and we call SAT-solver once again. If SAT returns *unsatisfiable*, i.e., z_i matches w_j, we continue matching the remaining unmatched outputs in the support of x_i and y_j. If the result is *satisfiable*, we once again use the counterexample returned by SAT to prune the search space.

Example 13.6 Consider two circuits N_1 and N_2 with input sets $X = \{x_0, \ldots, x_3\}$ and $Y = \{y_0, \ldots, y_3\}$, and output sets $Z = \{z_0, z_1\}$ and $W = \{w_0, w_1\}$ where $z_0 = x_0 \cdot x_1 \cdot \overline{x_2} \cdot \overline{x_3}$, $z_1 = \overline{x_0} \cdot \overline{x_1} \cdot x_2 \cdot x_3$, $w_0 = \overline{y_0} \cdot \overline{y_1} \cdot y_2 \cdot y_3$, and $w_1 = y_0 \cdot y_1 \cdot \overline{y_2} \cdot \overline{y_3}$. For these two circuits, signature-based matching (discussed in Section 13.3) cannot distinguish any I/Os. Hence, we resort to SAT-solving. Assume that SAT search starts by matching x_0 to y_0. Since $\{z_0, z_1\} \in Supp(x_0)$ and $\{w_0, w_1\} \in Supp(y_0)$, the outputs of the circuits must be matched next. Among all valid matches, our SAT-solver can match z_0 to w_1 and z_1 to w_0. For the remaining space of the unmatched inputs, our SAT-solver can validly match x_1 to y_1, x_2 to y_3, and x_3 to y_2, and finish the search.

13.4.4 Pruning Invalid Output Matches by SAT Counterexamples

When output $z_i \in Z_l$ does not match output $w_j \in W_l$, the counterexample returned by SAT is a vector V that makes $z_i = 1$ and $w_j = 0$ or vice versa. This means that z_i matches output $w_q \in W_l$ $(q \neq j)$ only if $z_i = w_q$ under V. This simple fact allows us to drastically prune our SAT-tree whenever an invalid output match occurs.

13.4.5 Pruning Invalid I/O Matches Using Support Signatures

We demonstrated in Section 13.3.5 that support signatures of inputs and outputs can be used to refine I/O subclusters of a Boolean network. In this section, we show that support signatures can also be used in our SAT-tree to eliminate impossible I/O correspondences.

Lemma 13.10 *Suppose that N_1 and N_2 are two Boolean networks and $x_i \in X_l$ and $y_j \in Y_l$ are two unmatched inputs of N_1 and N_2. Then, x_i can match y_j only if $Sign(x_i) = Sign(y_j)$. Likewise, suppose that $z_i \in Z_l$ and $w_j \in W_l$ are two unmatched outputs of N_1 and N_2. Then, z_i matches w_j only if $Sign(z_i) = Sign(w_j)$.*

As indicated in Section 13.4.1, matching two I/Os during SAT search introduces new singleton cells. These new cells might change the support signature of the remaining unmatched I/Os (the ones in the supports of the recently matched inputs or outputs). According to Lemma 13.10, this change in the support signatures might preclude some I/Os from matching. Taking advantage of this lemma, we can prune the unpromising branches of the SAT-tree in the remaining space of matches.

13.4.6 Pruning Invalid Input Matches Using Symmetries

Our SAT-tree can exploit the symmetries of inputs to prune (1) impossible output matches and (2) symmetric portions of the search space. Since computing the input symmetries of a Boolean network is expensive, the techniques explained in this section may in some cases hamper the matching process.

Definition 13.14 Let $X = \{x_1, \ldots, x_n\}$ be an input set of a Boolean network N. Let $x_i \sim x_j$ (read x_i is *symmetric* to x_j) if and only if the functionality of N stays invariant under an exchange of x_i and x_j. This defines an equivalence relation on set X, i.e., $\sim$ partitions X into a number of *symmetry classes* where each symmetry class contains all the inputs that are symmetric to each other. The partition resulting from $\sim$ is called the *symmetry partition* of X.

For multi-output functions, symmetries of inputs are reported independently for each output. In other words, each output defines its own symmetry partition on inputs. Complying with the notion of symmetry in Definition 13.14, for a multi-output function, x_i is called symmetric to x_j if (1) x_i and x_j have the same output support, i.e., $Supp(x_i) = Supp(x_j)$ and (2) x_i and x_j are symmetric in all the outputs in their support, i.e., $x_i \sim x_j$ for all outputs in $Supp(x_i)$ (or equivalently $Supp(x_j)$).

Symmetries of inputs can serve as a signature for matching outputs in our SAT-based search. The following lemma explains the role of symmetries in detecting invalid output matches.

Lemma 13.11 *Output $z_i \in Z_l$ (from N_1) matches output $w_j \in W_l$ (from N_2) only if symmetry partition of z_i is isomorphic to the symmetry partition of w_j for at least one ordering of symmetry classes.*

Input symmetries can also be used to prune symmetric parts of the search space during SAT-based exploration. Specifically, assume that the miter of N_1'' and N_2'' from Section 13.4.1 is *satisfiable*, i.e., x_i does not match y_j. Based on the notion of input symmetries, if x_i does not match y_j, neither can it match another input in Y_l that is symmetric to y_j. In other words, x_i cannot match $y_q \in Y_l$, if y_j and y_q are symmetric.

In practice, the use of symmetries in Boolean matching encounters two major limitations: (1) finding symmetries of a large Boolean network usually takes a significant amount of time and, (2) in a case where a Boolean network does not have much symmetry, a considerable amount of time can be wasted.

13.4.7 A Heuristic for Matching Candidates

In order to reduce the branching factor of our SAT-tree, we first match I/Os of smaller I/O clusters. Also, within one I/O cluster, we exploit the observability of the inputs and the controllability of the outputs, to make more accurate guesses in our SAT-based matching approach. Heuristically, the probability that two I/Os match is higher when their observability/controllability are similar. We observed that, in many designs, the observability of control signals is higher than that of data signals. Therefore, we first match control signals. This simple heuristic can greatly improve the runtime – experiments indicate that once control signals are matched, data signals can be matched quickly.

13.5 Empirical Validation

We have implemented the proposed approach in ABC and we have experimentally evaluated its performance on a 2.67GHz Intel Xeon CPU running Windows Vista. Table 13.1 and Table 13.2 show the runtime of our algorithms on ITC'99 benchmarks for P-equivalence and PP-equivalence checking problems, respectively. In these two tables, #I is the number of inputs, #O is the number of outputs and |AIG| is the number of nodes in the AIG of each circuit. The last four columns demonstrate the initialization time (computing I/O support variables, initially refining I/O cluster and refining based on I/O dependencies), simulation time, SAT time, and overall time for each testcase. These runtimes are all shown in bold. In addition to the reported runtimes, (I%) and (I%,O%) respectively show the percentage of inputs and I/Os that are matched after each step. Note that, in these experiments, we did not perform refinement using minterm counts and unateness, and we did not account for input symmetries to prune our SAT-tree because these techniques appear less scalable than the ones reported in Tables 13.1 and 13.2. Also note that for each testcase we generated 20 new circuits each falling into one of the two following categories: (1) permuting inputs for verifying P-equivalence (2) permuting both inputs and outputs for verifying PP-equivalence. The results given in Tables 13.1 and 13.2 are the average results over all the generated testcases for each category. Furthermore, the AIGs of the new circuits are reconstructed using ABC's combinational synthesis commands to ensure that the new circuits are structurally different from the original ones.

Table 13.1 P-equivalence runtime (s) and percentage of matched inputs for ITC'99 benchmarks

Circuit	#I	#O	|AIG|	Initialization	Simulation	SAT	Overall
b01	6	7	48	**0.30** (66%)	**0** (100%)	**0** (100%)	**0.30**
b02	4	5	28	**0.28** (50%)	**0** (100%)	**0** (100%)	**0.28**
b03	33	34	157	**0.36** (97%)	**0** (97%)	**0.04** (100%)	**0.40**
b04	76	74	727	**0.41** (64%)	**0.04** (100%)	**0** (100%)	**0.45**
b05	34	70	998	**0.52** (84%)	**0.02** (100%)	**0** (100%)	**0.54**
b06	10	15	55	**0.37** (80%)	**0** (100%)	**0** (100%)	**0.37**
b07	49	57	441	**0.41** (67%)	**0.01** (100%)	**0** (100%)	**0.43**
b08	29	25	175	**0.36** (90%)	**0** (100%)	**0** (100%)	**0.36**
b09	28	29	170	**0.40** (100%)	**0** (100%)	**0** (100%)	**0.40**
b10	27	23	196	**0.34** (85%)	**0** (100%)	**0** (100%)	**0.34**
b11	37	37	764	**0.40** (95%)	**0.01** (100%)	**0** (100%)	**0.41**
b12	125	127	1072	**0.38** (60%)	**0.25** (100%)	**0** (100%)	**0.63**
b13	62	63	353	**0.38** (71%)	**0.01** (100%)	**0** (100%)	**0.39**
b14	276	299	10067	**6.89** (73%)	**3.29** (100%)	**0** (100%)	**10.18**
b15	484	519	8887	**14.26** (57%)	**5.82** (100%)	**0** (100%)	**20.08**
b17	1451	1512	32290	**246** (63%)	**46.14** (99%)	**1.41** (100%)	**294**
b18	3357	3343	74900	**2840** (69%)	**51.6** (99%)	**2.96** (100%)	**2895**
b20	521	512	20195	**52.8** (83%)	**2.23** (100%)	**0.01** (100%)	**55**
b21	521	512	20540	**52.8** (83%)	**2.30** (100%)	**0.01** (100%)	**55**
b22	766	757	29920	**150** (82%)	**3.85** (100%)	**0.32** (100%)	**154**

In the ITC'99 benchmark suite, 18 circuits out of 20 have less than a thousand I/Os. Checking P-equivalence and PP-equivalence for 12 out of these 18 circuits takes less than a second. There is only one circuit (b12) for which our software cannot match I/Os in 5000 s. The reason is that, for b12, 1033 out of 7750 input pairs (13%) are symmetric and since our implementation does not yet account for symmetries, our SAT-tree repeatedly searches symmetric branches that do not yield valid I/O matches. For b20, b21, and b22 and for b17 and b18 with more than a thousand I/Os, computing functional dependency is the bottleneck of the overall matching runtime. Note that checking PP-equivalence for b18 results in a very large SAT-tree that cannot be resolved within 5000 s, although our refinement techniques

Table 13.2 PP-equivalence runtime (s) and percentage of matched I/Os for ITC'99 benchmarks

Circuit	#I	#O	|AIG|	Initialization	Simulation	SAT	Overall
b01	6	7	48	**0.37** (50%, 43%)	**0** (83%, 85%)	**0.02** (100%, 100%)	**0.39**
b02	4	5	28	**0.28** (50%, 60%)	**0** (100%, 100%)	**0** (100%, 100%)	**0.28**
b03	33	34	157	**0.38** (48%, 38%)	**0.01** (54%, 47%)	**0.43** (100%, 100%)	**0.82**
b04	76	74	727	**0.37** (16%, 13%)	**0.1** (100%, 100%)	**0** (100%, 100%)	**0.47**
b05	34	70	998	**0.51** (34%, 24%)	**0.03** (54%, 47%)	**0.33** (100%, 100%)	**0.87**
b06	10	15	55	**0.39** (30%, 47%)	**0** (50%, 53%)	**0.04** (100%, 100%)	**0.43**
b07	49	57	441	**0.43** (67%, 70%)	**0.03** (94%, 95%)	**0.19** (100%, 100%)	**0.65**
b08	29	25	175	**0.41** (27%, 36%)	**0.12** (100%, 100%)	**0** (100%, 100%)	**0.53**
b09	28	29	170	**0.41** (46%, 48%)	**0.01** (46%, 48%)	**0.20** (100%, 100%)	**0.62**
b10	27	23	196	**0.37** (88%, 95%)	**0** (100%, 100%)	**0** (100%, 100%)	**0.37**
b11	37	37	764	**0.41** (65%, 65%)	**0** (100%, 100%)	**0.02** (100%, 100%)	**0.43**
b12	125	127	1072	**0.38** (21%, 25%)	**1.05** (41%, 41%)	—	—
b13	62	63	353	**0.35** (43%, 50%)	**0.05** (97%, 97%)	**0.14** (100%, 100%)	**0.54**
b14	276	299	10067	**7.99** (72%, 58%)	**3.89** (89%, 90%)	**27** (100%, 100%)	**38.8**
b15	484	519	8887	**16.40** (62%, 67%)	**45.6** (94%94, %)	**6.30** (100%, 100%)	**68.3**
b17	1451	1512	32290	**249** (62%, 65%)	**229** (94%, 94%)	**148** (100%, 100%)	**626**
b18	3357	3343	74900	**2862** (65%, 63%)	**530** (93%, 93%)	—	—
b20	521	512	20195	**53.3** (70%, 51%)	**13.82** (89%, 89%)	**146** (100%, 100%)	**213**
b21	521	512	20540	**53.3** (70%, 51%)	**11.70** (89%, 89%)	**159** (100%, 100%)	**225**
b22	766	757	29920	**151** (70%, 50%)	**26.28** (88%, 88%)	**473** (100%, 100%)	**650**

— indicates runtime > 5000 s.

before invoking SAT find exact matches for 3123 out of 3357 inputs (93%) and 3111 out of 3343 outputs (93%).

The results in Tables 13.1 and 13.2 assume that target circuits are equivalent. In contrast, Table 13.3 considers cases where input circuits produce different output values on at least some inputs. For this set of experiments, we constructed 20 inequivalent circuits for each testcase, using one of the following rules:

1. *Wrong signals*: outputs of two random gates were swapped.
2. *Wrong polarity*: an inverter was randomly added or removed.
3. *Wrong gate*: functionality of one random gate was altered.

In Table 13.3, columns Init, Sim, and SAT demonstrate the number of testcases (out of 20) for which our algorithms were able to prove inequivalence during initialization, simulation, and SAT search phases, respectively. Also, column Time (depicted in bold) shows the average runtime of our matcher for the P-equivalence and PP-equivalence problems. According to the results, our matcher resorts to SAT-solving in 45% of the testcases which suggests that many of our instances are not particularly easy. Moreover, calling SAT is due to the fact that our mismatched instances were all generated with minimal changes to the original circuits. Note that, even in the case of a slight mismatch, our signature-based techniques alone could effectively discover inequivalence for 55% of testcases. Furthermore, comparing the results in Tables 13.2 and 13.3, PP-equivalence checking is up to four times faster when mismatch exists. For instance, the inequivalence of b12 was confirmed by our matcher in less than 5s, even when SAT-solving was invoked. The reason is that in the case

Table 13.3 P-equivalence and PP-equivalence runtime (s) for ITC'99 benchmarks when mismatch exists

				P-equivalence				PP-equivalence			
Circuit	#I	#O	|AIG|	Init	Sim	SAT	Time	Init	Sim	SAT	Time
b01	6	7	48	4	2	14	**0.30**	1	13	6	**0.49**
b02	4	5	28	0	10	10	**0.27**	2	12	6	**0.33**
b03	33	34	157	9	0	11	**0.35**	10	7	3	**0.45**
b04	76	74	727	8	2	10	**0.42**	13	4	3	**0.39**
b05	34	70	998	7	0	13	**0.53**	6	10	4	**0.70**
b06	10	15	55	3	3	14	**0.31**	14	5	1	**0.46**
b07	49	57	441	10	0	10	**0.43**	15	1	4	**0.71**
b08	29	25	175	9	2	9	**0.36**	12	6	2	**0.46**
b09	28	29	170	4	1	15	**0.40**	10	4	6	**0.45**
b10	27	23	196	10	5	5	**0.33**	11	3	6	**0.31**
b11	37	37	764	5	0	15	**0.40**	10	2	8	**0.53**
b12	125	127	1072	6	10	4	**0.45**	10	8	2	**3.5**
b13	62	63	353	6	9	5	**0.38**	7	7	6	**0.55**
b14	276	299	10067	3	0	17	**9.89**	10	3	7	**10.65**
b15	484	519	8887	4	2	14	**20.03**	8	4	8	**38.2**
b17	1451	1512	32290	11	0	9	**260**	3	7	10	**373**
b18	3357	3343	74900	2	0	18	**2864**	0	9	11	—[a]
b20	521	512	20195	7	0	13	**54**	1	4	15	**75.4**
b21	521	512	20540	2	0	18	**54**	5	11	4	**59.4**
b22	766	757	29920	7	1	12	**154**	0	4	16	**181**

— indicates runtime > 5000 s.

[a]The average runtime excluding instances requiring SAT was 2902 s.

of a mismatch, our SAT-tree usually encounters invalid I/O matches early in the tree, which results in a vast pruning in the space of invalid matches.

In order to compare our work to that in [17], we have tested our algorithms on circuits from [17] that have more than 150 inputs. Results are listed in Table 13.4 (the overall runtime of our algorithms is shown in bold). For the results reported from [17], Orig, Unate, and +Symm, respectively, show the runtime when no functional property is used, only functional unatness is used and, both unateness and symmetries are used. These three runtimes are also shown in bold. Note that experiments reported in [17] used 3GHz Intel CPUs, while our runs were on a 2.67 GHz Intel CPU. To make the numerical comparisons entirely fair, our runtimes would need to be multiplied by 0.89. However, we omit this step, since our raw runtimes are already superior in many cases. According to Table 13.4, our matching algorithm times out in 5000 s on C2670, i2, and i4. This is again due to the symmetries that are present in the inputs of these circuits. Note that the approach in [17] cannot solve these three circuits without symmetry search, either. For some other circuits, such as C7552, our approach verifies P-equivalence in less than 10 s but the approach in [17] cannot find a match without invoking symmetry finder. It is also evident from the results that checking P-equivalence for very large circuits, such as s38584 and s38417, is 3.5–11 times slower when symmetry finding and unateness calculations are performed during Boolean matching. This confirms our intuition that

Table 13.4 P-equivalence runtime (s) compared to runtime (s) from [17]

			P-equivalence Runtime (sec.)				CPU Time (sec.) in [17]		
Circuit	#I	#O	Init	Sim	SAT	Overall	Orig	+Unate	+Sym
C2670	233	140	0.14	1.18	—	—	—	—	**7.96**
C5315	178	123	0.33	0.11	0.06	**0.5**	**6.31**	**2.86**	**3.29**
C7552	207	108	0.51	3.76	4.83	**9.10**	—	—	**14.56**
des	256	245	0.38	0.07	0	**0.45**	**10.21**	**0.25**	**2.33**
i10	257	224	0.43	1.03	1.23	**2.69**	**25.63**	**15.16**	**17.56**
i2	201	1	0.34	0.28	—	—	—	—	**1.02**
i4	192	6	0.31	0.27	—	—	—	—	**0.22**
i7	199	67	0.36	0.18	0	**0.54**	**0.82**	**0.04**	**0.19**
pair	173	137	0.32	0.14	0	**0.46**	**0.84**	**0.64**	**2.44**
s3384	226	209	0.10	0.25	0.47	**0.82**	**4.79**	**2.14**	**4.02**
s5378	199	213	0.11	0.53	0.63	**1.27**	**1.31**	**3.38**	**2.42**
s9234	247	250	3.11	0.53	2.85	**6.49**	**3.41**	**5.84**	**7.82**
s38584	1464	1730	58	1.66	1.54	**61**	**76**	**210**	**458**
s38417	1664	1742	50	9.46	30.9	**90**	**91**	**324**	**999**

— indicates runtime > 5000 s.

symmetry and unateness are not essential to Boolean matching in many practical cases, although they may occasionally be beneficial.

13.6 Chapter Summary

In this chapter, we proposed techniques for solving large-scale PP-equivalence checking problem. Our approach integrates graph-based, simulation-driven, and SAT-based techniques to efficiently solve the problem. Graph-based techniques limit dependencies between inputs and outputs and are particularly useful with word-level arithmetic circuits. Simulation quickly discovers inputs on which inequivalent circuits differ. Equivalences are confirmed by invoking SAT, and these invocations are combined with branching on possible matches. Empirical validation of our approach on available benchmarks confirms its scalability to circuits with thousands of inputs and outputs. Future advances in Boolean matching, as well as many existing techniques, can also be incorporated into our framework to improve its scalability.

References

1. Abdollahi, A.: Signature based Boolean matching in the presence of don't cares. In: DAC '08: Proceedings of the 45th annual Design Automation Conference, pp. 642–647. ACM, New York, NY (2008). DOI: http://doi.acm.org/10.1145/1391469.1391635
2. Abdollahi, A., Pedram, M.: A new canonical form for fast Boolean matching in logic synthesis and verification. In: DAC '05: Proceedings of the 42nd annual Design Automation Conference, pp. 379–384. ACM, New York, NY (2005). DOI: http://doi.acm.org/10.1145/1065579.1065681
3. Agosta, G., Bruschi, F., Pelosi, G., Sciuto, D.: A unified approach to canonical form-based Boolean matching. In: DAC '07: Proceedings of the 44th annual

Design Automation Conference, pp. 841–846. ACM, New York, NY (2007). DOI: http://doi.acm.org/10.1145/1278480.1278689
4. Benini, L., Micheli, G.D.: A survey of Boolean matching techniques for library binding. ACM Transactions on Design Automation of Electronic Systems **2**, 193–226 (1997)
5. Chai, D., Kuehlmann, A.: Building a better Boolean matcher and symmetry detector. In: DATE '06: Proceedings of the conference on Design, automation and test in Europe, pp. 1079–1084. European Design and Automation Association, 3001 Leuven, Belgium, Belgium (2006)
6. Davis, M., Logemann, G., Loveland, D.: A machine program for theorem-proving. Communications of the ACM **5**(7), 394–397 (1962). DOI: http://doi.acm.org/10.1145/368273.368557
7. Davis, M., Putnam, H.: A computing procedure for quantification theory. Journal of the ACM **7**(3), 201–215 (1960). DOI: http://doi.acm.org/10.1145/321033.321034
8. Eén, N., Sörensson, N.: An extensible SAT-solver. In: E. Giunchiglia, A. Tacchella (eds.) SAT, Lecture Notes in Computer Science, vol. 2919, pp. 502–518. Springer, Heidelberg (2003)
9. Goering, R.: Xilinx ISE handles incremental changes. http://www.eetimes.com/showArticle.jhtml?articleID=196901122 (2009)
10. Krishnaswamy, S., Ren, H., Modi, N., Puri, R.: Deltasyn: An efficient logic difference optimizer for ECO synthesis. In: ICCAD '09: Proceedings of the 2009 International Conference on Computer-Aided Design, pp. 789–796. ACM, New York, NY (2009). DOI: http://doi.acm.org/10.1145/1687399.1687546
11. Lee, C.C., Jiang, J.H.R., Huang, C.Y.R., Mishchenko, A.: Scalable exploration of functional dependency by interpolation and incremental SAT solving. In: ICCAD '07: Proceedings of the 2007 IEEE/ACM International Conference on Computer-aided design, pp. 227–233. IEEE Press, Piscataway, NJ (2007)
12. Mishchenko, A.: Logic synthesis and verification group. ABC: A system for sequential synthesis and verification, release 70930. http://www.eecs.berkeley.edu/~alanmi/abc/ (2008)
13. Mishchenko, A., Chatterjee, S., Brayton, R.: FRAIGs: A unifying representation for logic synthesis and verification. Technical report, UC Berkeley (2005)
14. Mishchenko, A., Chatterjee, S., Brayton, R., Een, N.: Improvements to combinational equivalence checking. In: ICCAD '06: Proceedings of the 2006 IEEE/ACM international conference on Computer-aided design, pp. 836–843. ACM, New York, NY (2006). DOI: http://doi.acm.org/10.1145/1233501.1233679
15. Nocco, S., Stefano, Q.: A probabilistic and approximated approach to circuit-based formal verification. Journal of Satisfiability, Boolean Modeling and Computation **5**, 111–132. http://jsat.ewi.tudelft.nl/content/volume5/JSAT5_5_Nocco.pdf (2008)
16. Ray, S., Mishchenko, A., Brayton, R.: Incremental sequential equivalence checking and subgraph isomorphism. In: Proceedings of the International Workshop on Logic Synthesis, pp. 37–42. (2009)
17. Wang, K.H., Chan, C.M., Liu, J.C.: Simulation and SAT-based Boolean matching for large Boolean networks. In: DAC '09: Proceedings of the 46th Annual Design Automation Conference, pp. 396–401. ACM, New York, NY (2009). DOI: http://doi.acm.org/10.1145/1629911.1630016

Part IV
Logic Optimization

The first chapter in logic optimization enhances common subexpression elimination using novel logic transformations. Approximate SPFDs for sequential circuits are investigated in the second chapter. The third chapter provides details of Boolean relation determinization using quantification and interpolation techniques while the last chapter presents logic minimization using a window-based multi-node optimization technique.

Chapter 14
Algebraic Techniques to Enhance Common Sub-expression Extraction for Polynomial System Synthesis

Sivaram Gopalakrishnan and Priyank Kalla

Abstract Datapath designs that perform polynomial computations over bit-vectors are found in many practical applications, such as in Digital Signal Processing, communication, multi-media, and other embedded systems. With the growing market for such applications, advancements in synthesis and optimization techniques for polynomial datapaths are desirable. Common sub-expression extraction (CSE) serves as a useful optimization technique in the synthesis of such polynomial systems. However, CSE has limited potential for optimization when many common sub-expressions are not exposed in the given symbolic representation. Given a suitable set of transformations (or decompositions) that expose many common sub-expressions, subsequent application of CSE can offer a higher degree of optimization. This chapter proposes algebraic (algorithmic) techniques to perform such transformations and presents a methodology for their integration with CSE. Experimental results show that designs synthesized using our integrated approach are significantly more area-efficient than those synthesized using contemporary techniques.

14.1 Introduction

High-level descriptions of arithmetic datapaths that perform *polynomial computations* over bit-vectors are found in many practical applications, such as in Digital Signal Processing (DSP) for multi-media applications and embedded systems. These polynomial designs are initially specified using behavioral or Register-Transfer-Level (RTL) descriptions, which are subsequently synthesized into hardware using high-level and logic synthesis tools [23]. With the widespread use of

---

S. Gopalakrishnan (✉)
Synopsys Inc., Hillsboro, Oregon, USA
e-mail: sivaram.gopalakrishnan@synopsys.com

Based on Gopalakrishnan, S.; Kalla, P.; "Algebraic techniques to enhance common sub-expression elimination for polynomial system synthesis," Design, Automation & Test in Europe Conference & Exhibition, 2009. DATE '09, pp.1452–1457, 20–24 April 2009

S.P. Khatri, K. Gulati (eds.), *Advanced Techniques in Logic Synthesis, Optimizations and Applications*, DOI 10.1007/978-1-4419-7518-8_14,

such designs, there is a growing need to develop more sophisticated synthesis and optimization techniques for polynomial datapaths at high-level/RTL.

The general area of high-level synthesis has seen extensive research over the years. Various algorithmic techniques have been devised, and CAD tools have been developed that are quite adept at capturing hardware description language (HDL) models and mapping them into control/data-flow graphs (CDFGs), performing scheduling, resource allocation and sharing, binding, retiming, etc. [7]. However, these tools lack the mathematical wherewithal to perform sophisticated algebraic manipulation for arithmetic datapath-intensive designs. Such designs implement a sequence of ADD, MULT type of algebraic computations over bit-vectors; they are generally modeled at RTL or behavioral-level as *systems of multi-variate polynomials of finite degree* [19, 22]. Hence, there has been increasing interest in exploring the use of algebraic manipulation of polynomial expressions, for RTL synthesis of arithmetic datapaths. Several techniques such as Horner decomposition, factoring with common sub-expression extraction [13], term-rewriting [1] have been proposed. Symbolic computer algebra [10, 19, 22] has also been employed for polynomial datapath optimization. While these methods are useful as stand-alone techniques, they exhibit limited potential for optimization as explained below.

Typically, in a system of polynomials representing an arithmetic datapath, there are many common sub-expressions. In such systems, common sub-expression extraction (CSE) serves as a useful optimization technique, where isomorphic patterns in an arithmetic expression tree are identified, extracted, and merged. This prevents the cost of implementing multiple copies of the same expression. However, CSE has a limited potential for optimization if the common expressions are not exposed in the given symbolic representation. Hence, application of a "suitable set of transformations" (or decompositions) of the given polynomial representation to expose more common sub-expressions offers a higher potential for optimization by CSE. The objective of this chapter is to develop *algorithmic and algebraic* techniques to perform such transformations, to present a methodology for their integration with CSE, and to achieve a higher degree of optimization.

14.1.1 Motivation

Consider the various decompositions for a system of polynomials P_1, P_2, and P_3, implemented with variables x, y, and z, as shown in Table 14.1. The direct implementation of this system will require 17 multipliers and 4 adders. To reduce the size of the implementation, a Horner-form decomposition may be used. This implementation requires the use of 15 multipliers and 4 adders. However, a more sophisticated factoring method employing kernel/co-kernel extraction with CSE [13, 14] can further reduce the size of the implementation, using 12 multipliers and 4 adders. Now, consider the proposed decomposition of the system, also shown in the table. This implementation requires only 8 multipliers and 1 adder. Clearly, this is an efficient implementation of the polynomial system. This decomposition achieves a high degree of optimization by analyzing common sub-expressions across multiple

Table 14.1 Various decompositions for a polynomial system

Original system	Horner-form decomposition
$P_1 = x^2 + 6xy + 9y^2$;	$P_1 = x(x + 6y) + 9y^2$;
$P_2 = 4xy^2 + 12y^3$;	$P_2 = 4xy^2 + 12y^3$;
$P_3 = 2x^2z + 6xyz$;	$P_3 = x(2xz + 6yz)$;
Factorization + CSE	Proposed decompsoition
$P_1 = x(x + 6y) + 9y^2$;	$d_1 = x + 3y$; $P_1 = {d_1}^2$;
$P_2 = y^2(4x + 12y)$;	$P_2 = 4y^2d_1$;
$P_3 = xz(2x + 6y)$;	$P_3 = 2xzd_1$;

polynomials. This is not a trivial task and is not achieved by any earlier manipulation techniques [13, 14]. Note that d_1 is a good building block (common sub-expression) for these system of equations. Identifying and factoring out such building blocks across multiple polynomial datapaths can yield area-efficient hardware implementations.

14.1.2 Contributions

In this chapter, we develop techniques to transform the given system of polynomials by employing certain algebraic manipulations. These transformations have the potential to expose more common terms among the polynomials. These terms can be easily identified by the CSE routines and can be used as good "building blocks" for the design. Our expression manipulations are based on the following algebraic concepts:

- Canonical representation of polynomial functions over finite integer rings of the type Z_{2^m} [4]
- Square-free factorization
- Common coefficient extraction
- Factoring with kernel/co-kernel computation
- Algebraic division

We show how the above-mentioned algebraic methods are developed and employed in a synergistic fashion. These methods form the foundation of an integrated CSE technique for area-efficient implementations of the polynomial system.

14.1.3 Paper Organization

The next section presents the previous work in the area of polynomial datapath synthesis. Section 14.3 describes some preliminary concepts related to polynomial functions and their algebraic manipulations. Section 14.4 describes the optimization methods developed in this chapter. Section 14.5 presents our overall integrated approach. The experimental results are presented in Section 14.6. Finally, Section 14.7 concludes the chapter.

14.2 Previous Work

Contemporary high-level synthesis tools are quite adept in extracting control/data-flow graphs (CDFGs) from the given RTL descriptions and also in performing scheduling, resource-sharing, retiming, and control synthesis. However, they are limited in their capability to employ sophisticated algebraic manipulations to reduce the cost of the implementation. For this reason, there has been increasing interest in exploring the use of algebraic methods for RTL synthesis of arithmetic datapaths.

In [20, 21], the authors derive new polynomial models of complex computational blocks by the way of polynomial approximation for efficient synthesis. In [19], symbolic computer algebra tools are used to search for a decomposition of a given polynomial according to available components in a design library, using a Buchberger-variant algorithm [2, 3] for Gröbner bases. Other algebraic transforms have also been explored for efficient hardware synthesis: factoring with common sub-expression elimination [13], exploiting the structure of arithmetic circuits [24], term re-writing [1], etc. Similar algebraic transforms are also applied in the area of code optimization. These include reducing the height of the operator trees [18], loop expansion, induction variable elimination. A good review of these approaches can be found in [8].

Taylor Expansion Diagrams (TEDs) [5] have also been used for data-flow transformations in [9]. In this technique, the arithmetic expression is represented as a TED. Given an objective (design constraint), a sequence of decomposition cuts are applied to the TED that transforms it to an optimized data-flow graph. Modulo arithmetic has also been applied for polynomial optimization/decomposition of arithmetic datapaths in [10, 11]. By accounting for the bit-vector size of the computation, the systems are modeled as polynomial functions over finite integer rings. Datapath optimization is subsequently performed by exploiting the number theoretic properties of such rings, along with computational commutative algebra concepts.

14.2.1 Kernel/Co-kernel Extraction

Polynomial systems can be manipulated by extracting common expressions by using the kernel/co-kernel factoring. The work of [13] integrates factoring using kernel/co-kernel extraction with CSE. However, this approach has its limitations. Let us understand the general methodology of this approach before describing its limitations. The following terminologies are mostly referred from [13].

A **literal** is a variable or a constant. A **cube** is a product of variables raised to a non-negative integer power, with an associated sign. For example, $+acb$, $-5cde$, $-7a^2bd^3$ are cubes. A sum of product (SOP) is said to be cube-free if no cube (except "1") divides all the cubes of the SOP. For a polynomial P and a cube c, the expression P/c is a **kernel** if it is cube-free and has at least two terms. For example, when $P = 4abc - 3a^2b^2c$, the expression $P/abc = 4 - 3ab$ is a **kernel**. The cube that is used to obtain the kernel is the **co-kernel** (abc). This approach has two major limitations:

Coefficient Factoring: Numeric coefficients are treated as literals, not numbers. For example, consider a polynomial $P = 5x^2 + 10y^3 + 15pq$. According to this approach, coefficients {5, 10, 15} are also treated as literals like variables $\{x, y, p, q\}$. Since it does not use algebraic division, it cannot determine the following decomposition: $P = 5(x^2 + 2y^3 + 3pq)$.

Symbolic Methods: Polynomials are factored without regard to their algebraic properties. Consider a polynomial $P = x^2 + 2xy + y^2$, which can actually be transformed as $(x + y)^2$. Such a decomposition is also not identified by this kernel/co-kernel factoring approach. The reason for the inability to perform such a decomposition is due to the lack of symbolic computer algebra manipulation.

This chapter develops certain algebraic techniques that address these limitations. These techniques, along with kernel/co-kernel factoring, can be seamlessly integrated with CSE to provide an additional degree of optimization. With this integration, we seek to extend the optimization potential offered by the conventional methods.

14.3 Preliminary Concepts

This section will review some fundamental concepts of factorization and polynomial function manipulation, mostly referred from [4, 6].

14.3.1 Polynomial Functions and Their Canonical Representations

A bit-vector of size m represents integer values reduced modulo 2^m. Therefore, polynomial datapaths can be considered as polynomial functions over finite integer rings of the form Z_{2^m}. Moreover, polynomial datapaths often implement bit-vector arithmetic with operands of different bit-widths. Let $x_1, \ldots, x_d$ represent the bit-vector variables, where each bit-vector has bit-width $n_1, \ldots, n_d$. Let f be the bit-vector output of the datapath, with m as its bit-width. Then the bit-vector polynomial can be considered as a function $f : Z_{2^{n_1}} \times Z_{2^{n_2}} \times \cdots Z_{2^{n_d}} \rightarrow Z_{2^m}$.

A function f from $Z_{2^{n_1}} \times Z_{2^{n_2}} \times \cdots \times Z_{2^{n_d}} \rightarrow Z_{2^m}$ is said to be a polynomial function if it is represented by a polynomial $F \in Z[x_1, x_2, \ldots, x_d]$; i.e., $f(x_1, x_2, \ldots, x_d) \equiv F(x_1, x_2, \ldots, x_d)$ for all $x_i \in Z_{2^{n_i}}$, $i = 1, 2, \ldots, d$ and $\equiv$ denotes congruence (mod 2^m).

Let $f : Z_{2^1} \times Z_{2^2} \rightarrow Z_{2^3}$ be a function defined as: $f(0, 0) = 1$, $f(0, 1) = 3$, $f(0, 2) = 5$, $f(0, 3) = 7$, $f(1, 0) = 1$, $f(1, 1) = 4$, $f(1, 2) = 1$, $f(1, 3) = 0$. Then, f is a polynomial function representable by $F = 1 + 2y + xy^2$, since $f(x, y) \equiv F(x, y) \bmod 2^3$ for $x = 0, 1$ and $y = 0, 1, 2, 3$.

Polynomial functions implemented over specific bit-vector sizes can be represented in a unique canonical form. According to [4, 10], any polynomial representation F for a function f, from $Z_{2^{n_1}} \times Z_{2^{n_2}} \times \cdots Z_{2^{n_d}}$ to Z_{2^m}, can be uniquely represented as a sum-of-product of falling factorial terms:

$$F = \Sigma_{\mathbf{k}} c_{\mathbf{k}} \mathbf{Y}_{\mathbf{k}} \tag{14.1}$$

where,

- $\mathbf{k} =< k_1, \ldots, k_d >$ for each $k_i = 0, 1, \ldots, \mu_i - 1$;
- $\mu_i = \min(2^{n_i}, \lambda)$, for each $i = 1, \ldots, d$;
- λ is the least integer such that 2^m divides $\lambda!$;
- $c_{\mathbf{k}} \in Z$ such that $0 \leq c_{\mathbf{k}} < \frac{2^m}{gcd(2^m, \prod_{i=1}^{d} k_i!)}$;

In (14.1), $\mathbf{Y_k}$ is represented as

$$\begin{aligned} \mathbf{Y_k}(\mathbf{x}) &= \prod_{i=1}^{d} Y_{k_i}(x_i) \\ &= Y_{k_1}(x_1) \cdot Y_{k_2}(x_2) \cdots Y_{k_d}(x_d) \end{aligned} \tag{14.2}$$

where $Y_k(x)$ is a falling factorial defined as follows:

Definition 14.1 Falling factorials of degree k are defined according to:

- $Y_0(x) = 1$
- $Y_1(x) = x$
- $Y_2(x) = x(x - 1)$

  $\vdots$
- $Y_k(x) = (x - k + 1) \cdot Y_{k-1}(x)$

Intuitively, this suggests that while having a canonical form representation as in (14.1), it is possible to find common $Y_{k_i}(x_i)$ terms.

For example, consider the following polynomials implementing a 16-bit datapath, i.e., as polynomial functions over $f : Z_{2^{16}} \times Z_{2^{16}} \rightarrow Z_{2^{16}}$:

$$F = 4x^2y^2 - 4x^2y - 4xy^2 + 4xy + 5z^2x - 5zx \tag{14.3}$$
$$G = 7x^2z^2 - 7x^2z - 7xz^2 + 7zx + 3y^2x - 3yx \tag{14.4}$$

Using the canonical form representation, we get

$$F = 4Y_2(x)Y_2(y) + 5Y_2(z)Y_1(x) \tag{14.5}$$
$$G = 7Y_2(x)Y_2(z) + 3Y_2(y)Y_1(x) \tag{14.6}$$

Such a representation exposes many common terms in $Y_{k_i}(x_i)$. These terms may subsequently serve as a good basis for common sub-expression extraction.

For a detailed description of the above canonical form representation, the canonical reduction operations, and their impact on hardware implementation costs for polynomial datapaths, the reader is referred to [10].

14.3.2 Factorization

Definition 14.2 Square-free polynomial Let F be a field or an integral domain Z. A polynomial u in $F[x]$ is a square-free polynomial if there is no polynomial v in $F[x]$ with $\deg(v, x) > 0$, such that $v^2|u$.

Although the definition is expressed in terms of a squared factor, it implies that the polynomial does not have a factor of the form v^n with $n \geq 2$.

Example 14.1 The polynomial $u_1 = x^2 + 3x + 2 = (x + 1)(x + 2)$ is square-free. However, $u_2 = x^4 + 7x^3 + 18x^2 + 20x + 8 = (x + 1)(x + 2)^2$ is not square-free, as v^2 (where $v = x + 2$) divides u_2.

Definition 14.3 Square-free factorization A polynomial u in $F[x]$ has a unique factorization

$$u = cs_1{s_2}^2 \cdots {s_m}^m \tag{14.7}$$

where c is in F and each s_i is monic and square-free with $gcd(s_i, s_j) = 1$ for $i \neq j$. This unique factorization in (14.7) is called square-free factorization of u.

Example 14.2 The polynomial $u = 2x^7 - 2x^6 + 24x^5 - 24x^4 + 96x^3 - 96x^2 + 128x - 128$ has a square-free factorization $2(x - 1)(x^2 + 4)^3$ where $c = 2$, $s_1 = x - 1$, $s_2 = 1$, and $s_3 = x^2 + 4$. Note that a square-free factorization may not contain all the powers given in (14.7).

A square-free factorization only involves the square-free factors of a polynomial and leaves the deeper structure that involves the irreducible factors intact.

Example 14.3 Using square-free factorization

$$x^6 - 9x^4 + 24x^2 - 16 = (x^2 - 1)(x^2 - 4)^2 \tag{14.8}$$

both factors are reducible. This suggests that even after obtaining square-free polynomials, there is a potential for additional factorization. In other words, consider 14.8, where $(x^2 - 1)$ can be further factored as $(x + 1)(x - 1)$ and $(x^2 - 4)^2$ can be factored as $((x + 2)(x - 2))^2$.

14.4 Optimization Methods

The limitations of contemporary techniques come from their narrow approach to factorization, relying on single types of factorization, instead of the myriad of optimization techniques available. We propose an integrated approach, to polynomial optimization, to overcome these limitations. This section describes the various optimization techniques that are developed/employed in this chapter.

14.4.1 Common Coefficient Extraction

The presence of many coefficient multiplications in polynomial systems increases the area-cost of the hardware implementation. Moreover, existing coefficient factoring techniques [13] are inefficient in their algebraic manipulation capabilities. Therefore, it is our focus to develop a coefficient factoring technique that employs efficient algebraic manipulations and as a result reduces the number of coefficient multiplications in the given system.

Consider the following polynomial $P_1 = 8x + 16y + 24z$. When coefficient extraction is performed over P_1, it results in three possible transformations, given as follows:

$$P_1 = 2(4x + 8y + 12z) \quad (14.9)$$

$$P_1 = 4(2x + 4y + 6z) \quad (14.10)$$

$$P_1 = 8(x + 2y + 3z) \quad (14.11)$$

From these three transformations, (14.11) extracts the highest common term in P_1. This results in the best transformation (reduced set of operations). A method to determine the highest common coefficient is the greatest common divisor (GCD) computation. Therefore, in this approach, GCD computations are employed to perform common coefficient extraction (CCE) for a system of polynomials. The pseudocode to perform CCE is shown in Algorithm 6.

Algorithm 6 Common Coefficient Extraction (CCE)

CCE($a_1, \cdots, a_n$)
/* $(a_1, \cdots, a_n)$ = Coefficients of the given polynomial;*/
for every pair (a_i, a_j) in n **do**
    Compute GCD(a_i, a_j);
    Ignore GCDs = "1";
    if GCD(a_i, a_j) $< a_i$ and GCD(a_i, a_j) $< a_j$ **then**
        Ignore the GCDs;
    end if
end for
Order the GCDs in decreasing order;
while GCD list is non-empty **do**
    Perform the extraction using that order
    Store the linear/non-linear blocks created as a result of extraction
    Remove GCDs corresponding to extracted terms and update the GCD list
end while

Let us illustrate the operation of the CCE routine. Consider the polynomial P_1 computed as

$$P_1 = 8x + 16y + 24z + 15a + 30b + 11 \quad (14.12)$$

The input to CCE is the coefficients of the given polynomial that are involved in coefficient multiplications. In other words, if there is a coefficient addition in the polynomial, it is not considered while performing CCE. For example in (14.12), only the coefficients $\{8, 16, 24, 15, 30\}$ are considered and 11 is ignored. The reason is because there is no benefit in extracting this coefficient and a direct implementation is the cheapest in terms of area-cost.

The algorithm then begins by computing the GCDs for every pair-wise combination of the coefficients in the input set. Computing pair-wise GCDs of the coefficients:

$$
\begin{aligned}
GCD(8, 16) &= 8 \\
GCD(8, 24) &= 8 \\
&\vdots \\
GCD(15, 30) &= 15
\end{aligned}
\tag{14.13}
$$

we get the following set $\{8, 8, 1, 2, 8, 1, 2, 3, 6, 15\}$. However, only a subset is generated by ignoring "GCDs = 1" and "GCDs $(a_i, a_j) < a_i$ and a_j." The reason for ignoring these GCDs is that we only want to extract the highest common coefficients that would result in a reduced cost. For example, the GCD(24, 30) $= 6$. However, extracting 6 does not reduce the cost of the sub-expression $24z + 30b$ in (14.12), as $6(4z + 5b)$ requires more coefficient multipliers.

Applying the above concepts, the final subset is $\{8, 15\}$. This set is then arranged in the decreasing order to get $\{15, 8\}$. The first element is "15." On performing the extraction using coefficient "15," the following decomposition is realized:

$$P_1 = 8x + 16y + 24z + 15(a + 2b) \tag{14.14}$$

This creates a smaller polynomial $(a + 2b)$. It should be noted that this is a linear polynomial. This polynomial is stored and the extraction continues until the GCD list is empty. After CCE, the polynomial decomposition obtained is

$$P_1 = 8(x + 2y + 3z) + 15(a + 2b) \tag{14.15}$$

Two linear blocks $(a + 2b)$ and $(x + 2y + 3z)$ are finally obtained. The motivation behind storing these polynomials is that they can serve as potentially good building blocks in the subsequent optimization methods.

14.4.2 Common Cube Extraction

Common cubes, that consist of products of variables, also need to be extracted from the given polynomial representation. The kernel/co-kernel extraction technique from [13] is quite efficient for this purpose. Therefore, we employ this approach to perform the common cube extraction. Note that the cube extraction technique of [13] also considers coefficients as variables. We do not allow the technique of [13] to treat coefficients as variables – as we employ CCE for coefficient extraction. We employ this technique of [13] for extracting cubes composed only of variables.

Consider the following system of polynomials:

$$\begin{aligned} P_1 &= x^2y + xyz \\ P_2 &= ab^2c^3 + b^2c^2x \\ P_3 &= axz + x^2z^2b \end{aligned} \tag{14.16}$$

A kernel/co-kernel cube extraction results in the following representation. (Here, c_k is the co-kernel cube and k is the kernel.)

$$\begin{aligned} P_1 &= (xy)_{ck}(x + z)_k \\ P_2 &= (b^2c^2)_{ck}(ac + x)_k \\ P_3 &= (xz)_{ck}(a + xzb)_k \end{aligned} \tag{14.17}$$

Note that this procedure (which we call Cube_Ex()) exposes both cubes and kernels as potential (common) building blocks, which CSE can further identify and extract.

14.4.3 Algebraic Division

This method can potentially lead to a high degree of optimization. The problem essentially lies in identifying a good divisor, which can lead to an efficient decomposition. Given a polynomial $a(x)$, and a set of divisors $(b_i(x))$, $\forall i$ we can perform the division $a(x)/b_i(x)$ and determine if the resulting transformation is optimized for hardware implementation.

Using common coefficient extraction and cube extraction , a large number of linear blocks, that are simpler than the original polynomial, are exposed. These linear blocks can subsequently be used for performing algebraic division. For our overall synthesis approach, we consider only the exposed "linear expressions" as algebraic divisors. The motivation behind using the exposed "linear" blocks for division is that

- Linear blocks cannot be decomposed any further, implying that they have to be certainly implemented.
- They also serve as good building blocks in terms of (cheaper) hardware implementation.

For example, using cube extraction the given system in Table 14.1 is transformed to

$$P_1 = x(x+6y) + 9y^2 \quad or \quad P_1 = x^2 + y(6x+9y)$$
$$P_2 = 4y^2(x+3y)$$
$$P_3 = 2xz(x+3y) \tag{14.18}$$

The following linear blocks are now exposed: $\{(x+6y), (6x+9y), (x+3y)\}$. Using these blocks as divisors, we divide P_1, P_2, and P_3. $(x+3y)$ serves as a good building-block because it divides all the three polynomials as

$$P_1 = (x+3y)^2$$
$$P_2 = 4y^2(x+3y)$$
$$P_3 = 2xz(x+3y) \tag{14.19}$$

Such a transformation to (14.19) is possible only through algebraic division. None of the other expression manipulation techniques can identify this transformation.

14.5 Integrated Approach

The overall approach to polynomial system synthesis is presented in this section. We show how we integrate the algebraic methods presented previously with common sub-expression elimination. The pseudocode for the overall integrated approach is presented in Algorithm 7.

The algorithm operates as follows:

- The given system of polynomials is initially stored in a list of arrays. Each element in the list represents a polynomial. The elements in the array for each list represent the transformed representations of the polynomial. Figure 14.1a shows the polynomial data structure representing the system of polynomials in its expanded form, canonical form (can), and square-free factored form (sqf).
- The algorithm begins by computing the canonical forms and the square-free factored forms, for all the polynomials in the given system. At this stage, the polynomial data structure looks like in Fig. 14.1a.
- Then, the best-cost implementation among these representations is chosen and stored as P_{initial}. The cost is stored as C_{initial}. We estimate the cost using the number of adders and multipliers required to implement the polynomial.
- Common coefficient extraction (CCE) and common cube extraction (Cub_Ex) are subsequently performed. The linear/non-linear polynomials obtained from these extractions are stored/updated. Also, the resulting transformations for each polynomial are updated in the polynomial data structure. At this stage, the data structure looks like in Fig. 14.1b. To elaborate further, in this figure, $\{P_1, P_{1a}, P_{1b}, P_{1c}\}$ are various representations of P_1 (as a result of CCE and Cub_Ex), and so on.
- Using the linear blocks, algebraic division is performed and the polynomial data-structure is further populated, with multiple representations.

Algorithm 7 Approach to Polynomial System Synthesis

/*Given: $(P_1, P_2, \cdots, P_n)$ = Polys $(P_i's)$ representing the system; Each P_i is a list to store multiple representations of P_i;*/
Poly_Synth($P_1, P_2, \cdots, P_n$)
/*Initial set of Polynomials, P_{orig}*/
$P_{orig} = \langle P_1, \cdots, P_n \rangle$;
P_{can} = Canonize(P_{orig});
P_{sqf} = Sqr_free(P_{orig});
Initial_cost $C_{initial} = min\_cost(P_{orig}, P_{can}, P_{sqf})$;
/*The polynomial with cost $C_{initial}$ is $P_{initial}$*/
$CCE(P_{initial})$; Update resulting linear/non-linear polynomials;
/* P_{CCE} = Polynomial representation after CCE();*/ Update $P_i's$;
Cube_Ex($P_i's$); Update resulting linear/non-linear polynomials;
/* P_{CCE_Cube} = Polynomial representation after Cube_Ex();*/ Update $P_i's$;
Linear polynomials exposed are $lin\_poly = \langle l_1, \cdots, l_k \rangle$
for every l_j in lin_poly **do**
    ALG_DIV($P_i's, l_j$);
    Update $P_i's$ and $l_j's$;
end for
for every *combination of* $P_i's$ (P_{comb}) representing P_{orig} **do**
    $Cost$ = CSE(P_{comb});
    if $(Cost < C_{initial})$ **then**
        $C_{initial} = Cost$;
        $P_{final} = P_{comb}$;
    end if
end for
return P_{final};

- The entire polynomial system can be represented using a list of polynomials, where each element in the list is some representation for each polynomial. For example, $\{P_1, P_{2a}, P_{3b}\}$ is one possible list that represents the entire system (refer Fig. 14.1b). The various lists that represent the entire system are given by

$$\begin{aligned} &\{(P_1, P_2, P_3), \ (P_1, P_2, P_{3a}), \ (P_1, P_2, P_{3b}), \\ &\qquad \vdots \\ &(P_{1a}, P_{2b}, P_3), (P_{1a}, P_{2b}, P_{3a}), (P_{1a}, P_{2b}, P_{3b}), \\ &\qquad \vdots \\ &(P_{1c}, P_{2b}, P_3), (P_{1c}, P_{2b}, P_{3a}), (P_{1c}, P_{2b}, P_{3b})\} \end{aligned} \tag{14.20}$$

- Finally, we can pick the decomposition with the least estimated cost. For example, Fig. 14.1c shows that the least-cost implementation of the system is identified as:

$$P_{final} = (P_{1a}, P_{2b}, P_{3a}) \tag{14.21}$$

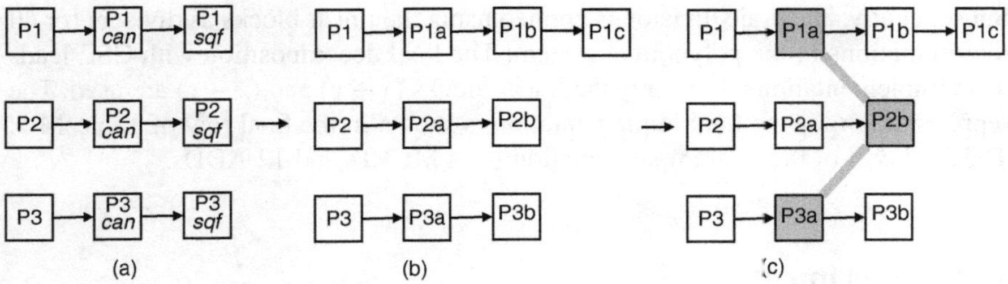

Fig. 14.1 Polynomial system representations

The working of Algorithm 7 is explained with the polynomial system presented in Table 14.2.

Table 14.2 Illustration of algorithm 7

Original system
$P_1 = 13x^2 + 26xy + 13y^2 + 7x - 7y + 11;$
$P_2 = 15x^2 - 30xy + 15y^2 + 11x + 11y + 9;$
$P_3 = 5x^3y^2 - 5x^3y - 15x^2y^2 + 15x^2y + 10xy^2 - 10xy + 3z^2;$
$P_4 = 3x^2y^2 - 3x^2y - 3xy^2 + 3xy + z + 1;$
After canonization and CCE
$P_1 = 13(x^2 + 2xy + y^2) + 7(x - y) + 11;$
$P_2 = 15(x^2 - 2xy + y^2) + 11(x + y) + 9;$
$P_3 = 5x(x-1)(x-2)y(y-1) + 3z^2;$
$P_4 = 3x(x-1)y(y-1) + z + 1;$
After cube extraction
$P_1 = 13(x(x + 2y) + y^2) + 7(x - y) + 11;$
$P_2 = 15(x(x - 2y) + y^2) + 11(x + y) + 9;$
$P_3 = 5x(x-1)(x-2)y(y-1) + 3z^2;$
$P_4 = 3x(x-1)y(y-1) + z + 1;$
Final decomposition
$d_1 = x + y;\ d_2 = x - y;\ d_3 = x(x-1)y(y-1)$
$P_1 = 13(d_1{}^2) + 7d_2 + 11;\ P_2 = 15(d_2{}^2) + 11d_1 + 9;$
$P_3 = 5d_3(x-2) + 3z^2;\ P_4 = 3d_3 + z + 1;$

Initially, canonical reduction and square-free factorization are performed. In this example, this technique does not result in any decomposition for square-free factorization. For P_3 and P_4, there is a low-cost canonical representation.

We then compute the initial cost of the polynomial by using only CSE. In the original system, there are no common sub-expressions. The total cost of the original system is estimated as 51 MULTs and 21 ADDs. Then CCE is performed, resulting in the transformation, as shown in the Table 14.2.

The linear polynomials obtained are $(x - y)$ and $(x + y)$. The non-linear polynomials are $(x^2 + 2xy + y^2)$ and $(x^2 - 2xy + y^2)$. After performing common cube extraction (Cube_Ex()), the additional linear blocks added are $(x+2y)$ and $(x-2y)$.

Subsequently, algebraic division is applied using the linear blocks as divisors for all representations of the polynomial system. The final decomposition with CSE leads to an implementation where only the linear blocks $(x+y)$ and $(x-y)$ are used. The representation for the final implementation is shown in the final row of Table 14.2. The total cost of the final implementation is 14 MULTs and 12 ADDs.

14.6 Experiments

The datapath computations are provided as a polynomial system, operating over specific input/output bit-vector sizes. All algebraic manipulations are implemented in Maple [15]; however, for Horner-form decomposition and factorization, we used the routines available in MATLAB [17]. For common sub-expression elimination, we use the JuanCSE tool available at [14]. Based on the given decomposition (for each polynomial in the system), the individual blocks are generated using the Synopsys Design Compiler [23]. These units are subsequently used to implement the entire system.

The experiments are performed on a variety of DSP benchmarks. The results are presented in Table 14.3. The first column lists the polynomial systems used for the experiments. The first five benchmarks are Savitzky-Golay filters. These filters are widely used in image-processing applications. The next benchmark is a polynomial system implementing quadratic filters from [16]. The next benchmark is from [12], used in automotive applications. The final benchmark is a multi-variate cosine wavelet used in graphics application from [13]. In the second column, we list the design characteristics: number of variables (bit-vectors), the order (highest degree), and the output bit-vector size (m). Column 3 lists the number of polynomials representing the entire system. Columns 4 and 5 list the implementation area and delay, respectively, of the polynomial system implemented using Factorization + common sub-expression elimination. Columns 6 and 7 list the implementation area and delay of the polynomial system, implemented using our proposed method. Columns 8 and 9 list the improvement in the implementation area and delay using our polynomial decomposition technique, respectively. Considering all the benchmarks, we show

Table 14.3 Comparison of proposed method with factorization/CSE

			Factorization/CSE		Proposed method		Improvement	
Systems	Var/Deg/m	# polys	Area	Delay	Area	Delay	Area %	Delay %
SG_3X2	2/2/16	9	204805	186.6	102386	146.8	50	21.3
SG_4X2	2/2/16	16	449063	211.7	197599	262.8	55.9	−24.1
SG_4X3	2/3/16	16	690208	282.3	557252	328.5	19.2	−16.3
SG_5X2	2/2/16	25	570384	205.6	271729	234.2	52.3	−13.9
SG_5X3	2/3/16	25	1365774	238.1	614955	287.4	54.9	−20.7
Quad	2/2/16	2	36405	118.4	30556	129.7	16	−9.5
Mibench	3/2/8	2	20359	64.8	8433	67.2	58.6	−3.7
MVCS	2/3/16	1	31040	119.1	22214	157.8	28.4	−32

an average improvement in the actual implementation area of approximately 42%. However, this area optimization does come at a cost of higher delay.

14.7 Conclusions

This chapter presents a synthesis approach for arithmetic datapaths implemented using a system of polynomial functions. We develop algebraic techniques that efficiently factor coefficients and cubes from the polynomial system, resulting in the generation of linear blocks. Using these blocks as divisors, we perform algebraic division, resulting in a decomposition of the polynomial system. Our decomposition exposes more common terms which can be identified by CSE, leading to a more efficient implementation. Experimental results demonstrate significant area savings using our approach as compared against contemporary datapath synthesis techniques. As part of future work, as datapath designs consume a lot of power, we would like to investigate the use of algebraic transformations in low-power synthesis of arithmetic datapaths.

References

1. Arvind, Shen, X.: Using term rewriting systems to design and verify processors. IEEE Micro **19**(2), 36–46 (1998)
2. Buchberger, B.: Ein Algorithmus zum Auffinden der Basiselemente des Restklassenringes nach einem nulldimensionalen Polynomideal. Ph.D. thesis, Philosophiesche Fakultat an der Leopold-Franzens-Universitat, Austria (1965)
3. Buchberger, B.: A theoretical basis for reduction of polynomials to canonical forms. ACM SIG-SAM Bulletin **10**(3), 19–29 (1976)
4. Chen, Z.: On polynomial functions from $Z_{n1} \times Z_{n2} \times \cdots \times Z_{nr}$ to Z_m. Discrete Mathematics **162**(1–3), 67–76 (1996)
5. Ciesielski, M., Kalla, P., Askar, S.: Taylor expansion diagrams: A canonical representation for verification of dataflow designs. IEEE Transactions on Computers **55**(9), 1188–1201 (2006)
6. Cohen, J.: Computer Algebra and Symbolic Computation. A. K. Peters, Wellesley, MA (2003)
7. DeMicheli, G.: Synthesis and Optimization of Digital Circuits. McGraw-Hill, New York, NY (1994)
8. DeMicheli, G., Sami, M.: Hardware/Software Co-Design. Kluwer, Norwell, MA (1996)
9. Gomez-Prado, D., Ciesielski, M., Guillot, J., Boutillon, E.: Optimizing data flow graphs to minimize hardware implementation. In: Proceedings of Design Automation and Test in Europe, pp. 117–122. Nice, France (2009)
10. Gopalakrishnan, S., Kalla, P.: Optimization of polynomial datapaths using finite ring algebra. ACM Transactions on Design Automation of Electronic System **12**(4), 49 (2007)
11. Gopalakrishnan, S., Kalla, P., Meredith, B., Enescu, F.: Finding linear building-blocks for RTL synthesis of polynomial datapaths with fixed-size bit-vectors. In: Proceedings of the International Conference on Computer Aided Design, pp. 143–148. San Jose, CA (2007)
12. Guthaus, M.R., Ringenberg, J.S., Ernst, D., Austin, T.M., Mudge, T., Brown, R.B.: Mibench: A free, commercially representative embedded benchmark suite. In: IEEE 4th Annual Workshop on Workload Characterization. Austin, TX (2001)
13. Hosangadi, A., Fallah, F., Kastner, R.: Optimizing polynomial expressions by algebraic factorization and common subexpression elimination. IEEE Transactions on Computer-Aided Design of Integrated Circuits and Systems **25**(10), 2012–2022 (2006)

14. JuanCSE: Extensible, programmable and reconfigurable embedded systems group. http://express.ece.ucsb.edu/suif/cse.html
15. Maple. http://www.maplesoft.com
16. Mathews, V.J., Sicuranza, G.L.: Polynomial Signal Processing. Wiley-Interscience, New York (2000)
17. MATLAB/Simulink. http://www.mathworks.com/products/simulink
18. Nicolau, A., Potasman, R.: Incremental tree-height reduction for high-level synthesis. In: Proceedings of the Design Automation Conference. San Francisco, CA (1991)
19. Peymandoust, A., DeMicheli, G.: Application of symbolic computer algebra in high-level data-flow synthesis. IEEE Transactions on CAD **22**(9), 1154–11656 (2003)
20. Smith, J., DeMicheli, G.: Polynomial methods for component matching and verification. In: Proceedings of the International Conference on Computer-Aided Design (ICCAD). San Jose, CA (1998)
21. Smith, J., DeMicheli, G.: Polynomial methods for allocating complex components. In: Proceedings of Design, Automation, and Test in Europe. Munich, Germany (1999)
22. Smith, J., DeMicheli, G.: Polynomial circuit models for component matching in high-level synthesis. IEEE Transactions on VLSI **9**(6), 783–800 (2001)
23. Synopsys: Synopsys Design Compiler and DesignWare library. htpp://www.synopsys.com
24. Verma, A.K., Ienne, P.: Improved use of the Carry-save representation for the synthesis of complex arithmetic circuits. In: Proceedings of the International Conference on Computer Aided Design. San Jose, CA (2004)

Chapter 15
Automated Logic Restructuring with *a*SPFDs

Yu-Shen Yang, Subarna Sinha, Andreas Veneris, Robert Brayton, and Duncan Smith

Abstract This chapter presents a comprehensive methodology to automate logic restructuring in combinational and sequential circuits. This technique algorithmically constructs the required transformation by utilizing a functional flexibility representation called *Set of Pairs of Function to be Distinguished* (SPFD). SPFDs can express more functional flexibility than the traditional don't cares and have proved to provide additional degrees of flexibility during logic synthesis [21, 27]. Computing SPFDs may suffer from memory or runtime problems [16]. Therefore, a simulation-based approach to approximate SPFDs is presented to alleviate those issues. The result is called *Approximate SPFDs* (*a*SPFDs). *a*SPFDs approximate the information contained in SPFDs using the results of test vector simulation. With the use of *a*SPFDs as a guideline, the algorithm searches for the necessary nets to construct the required function. Experimental results indicate that the proposed methodology can successfully restructure locations where a previous approach that uses a dictionary model [1] as the underlying transformation template fails.

15.1 Introduction

During the chip design cycle, small structural transformations in logic netlists are often required to accommodate different goals. For example, the designer needs to rectify designs that fail functional verification at locations identified by a debugging program [6, 23]. In the case of engineering changes (EC) [13], a logic netlist is modified to reflect specification changes at a higher level of abstraction. Logic transformations are also important during rewiring-based post-synthesis performance

---

Y.-S. Yang (✉)
University of Toronto, Toronto, ON, Canada
e-mail: terry.yang@utoronto.ca

Based on Yang, Y.-S.; Sinha, S.; Veneris, A.; Brayton, R.K.; Smith, D.; "Sequential logic rectifications with approximate SPFDs," Design, Automation & Test in Europe Conference & Exhibition, 2009, pp. 1698–1703, 20–24 April 2009

S.P. Khatri, K. Gulati (eds.), *Advanced Techniques in Logic Synthesis, Optimizations and Applications*, DOI 10.1007/978-1-4419-7518-8_15,

optimization [10, 24] where designs are optimized at particular internal locations to meet specification constraints.

Although these problems can be resolved by another round of full logic synthesis, directly modifying logic netlists is usually preferable in order to preserve any engineering effort that has been invested. Hence, logic restructuring has significant merits when compared to re-synthesis. Today, most of these incremental logic changes are implemented manually. The engineer examines the netlist to determine what changes need to be made and how they can affect the remainder of the design.

One simple ad hoc logic restructuring technique modifies the netlist by using permissible transformations from a *dictionary model* [1], which contains a set of simple modifications, such as single wire additions or removals. This technique is mostly adapted for design error correction [6, 18] and has been used in design rewiring as well [24]. A predetermined dictionary model, although effective at times, may not be adequate when complex transformations are required. It has been shown that a dictionary-model based design error correction tool can only successfully rectify 10–30% of cases [28]. Complex modifications perturb the functionality of the design in ways that simple dictionary-driven transformations may not be able to address. Therefore, automated logic transformation tools that can address these problems effectively are desirable.

The work presented in this chapter aims to develop a comprehensive methodology to automate logic restructuring in combinational and sequential circuits. It first presents a simulation-based technique to approximate SPFDs, or simply *a*SPFDs. Using aSPFDs can keep the process memory and runtime efficient while taking advantage of most of the benefits of SPFDs. Then, an *a*SPFD-based logic restructuring methodology is presented. It uses *a*SPFDs as a guideline to algorithmically restructure the functionality of an internal node in a design. Two searching algorithms, an SAT-based algorithm and a greedy algorithm, are proposed to find nets required for restructuring the transformation. The SAT-based algorithm selects minimal numbers of wires, while the greedy algorithm returns sub-optimal results with a shorter runtime.

Extensive experiments confirm the theory of the proposed technique and show that *a*SPFDs provide an effective alternative to dictionary-based transformations. It returns modifications where dictionary-based restructuring fails, increasing the impact of tools for debugging, rewiring, EC, etc. For combinational circuits, the proposed approach can identify five times more valid transformations than a dictionary-based one. Experiments also demonstrate the feasibility of using *a*SPFDs to restructure sequential designs. Although this method bases its results on a small sample of the input test vector space, empirical results show that more than 90% of the first solution returned by the method passes formal validation.

The remainder of this chapter is structured as follows. Section 15.2 summarizes previous work in logic restructuring, as well as the basic concept of SPFDs. Section 15.3 defines *a*SPFDs and the procedures used to generate *a*SPFDs. Section 15.4 presents the transformation algorithms utilizing *a*SPFDs. Experimental results are given in Section 15.5, followed by the conclusion in Section 15.6.

15.2 Background

This section reviews previous work on logic restructuring and summarizes the concept of SPFDs.

15.2.1 Prior Work on Logic Restructuring

Most research done on logic restructuring deals with combinational designs. In [26], the authors insert circuitry before and after the original design so that the functionality of the resulting network complies with the required specifications. The main disadvantage of this approach is that the additional circuitry may be too large and can dramatically change the performance of the design.

Redundancy addition and removal (RAR) [5, 10] is a post-synthesis logic optimization technique. It optimizes designs through the iterative addition and removal of redundant wires. All logic restructuring operations performed by RAR techniques are limited to single wire additions and removals. There is little success in trying to add and remove multiple wires simultaneously due to a large search space and complicated computation [4].

In [24], the authors view logic optimization from a logic debugging angle. It introduces a design error into the design, identifies locations for correction with a debug algorithm, and rectifies those locations with a dictionary model [1]. This method has been shown to exploit the complete solution space and offers great flexibility in optimizing a design and achieving larger performance gains. The technique presented in this chapter adapts the same viewpoint to logic restructuring.

Two recent approaches [3, 14] are similar to the one presented in this chapter. They construct the truth table of the new function at the location that requires restructuring and synthesize the new function based on the table. However, both approaches provide few descriptions on the application on sequential designs.

15.2.2 Sets of Pairs of Functions to Be Distinguished

Set of Pairs of Function to be Distinguished (SPFD) is a representation that provides a powerful formalism to express the functional flexibility of nodes in a multilevel circuit. The concept of SPFDs was first proposed by Yamashita et al. [27] for applications in FPGA synthesis and has been used in many applications for logic synthesis and optimization [7, 21, 22].

Formally, an SPFD

$$R = \{(g_{1a}, g_{1b}), (g_{2a}, g_{2b}), \cdots, (g_{na}, g_{nb})\} \tag{15.1}$$

denotes a set of pairs of functions that must be *distinguished*. That is, for each pair $(g_{ia}, g_{ib}) \in R$, the output of the node (wire) associated with R must have different values between a minterm of g_{ia} and a minterm of g_{ib}.

In [21], an SPFD is represented as a graph, $G = (V, E)$, where

$$
\begin{aligned}
V &= \{m_k \mid m_k \in g_{ij}, 1 \leq i \leq n, j = \{a, b\}\} \\
E &= \{(m_i, m_j) \mid \{(m_i \in g_{pa})\text{and}(m_j \in g_{pb})\} \\
&\quad \text{or}\{(m_i \in g_{pb})\text{and}(m_j \in g_{pa})\}, \\
&\quad 1 \leq p \leq n\}
\end{aligned} \tag{15.2}
$$

This graphical representation makes it possible to visualize SPFDs and can explain the concept of SPFDs more intuitively. Figure 15.1 depicts the graph representation of the SPFD, $R = \{(ab, \overline{a}b), (\overline{a}\overline{b}, a\overline{b})\}$. The graph contains four vertices that represent minterms {00, 01, 10, 11} in terms of $\{a, b\}$. Two edges are added for $(ab, \overline{a}b)$ and $(\overline{a}\overline{b}, a\overline{b})$. The edge is referred to an *SPFD edge*.

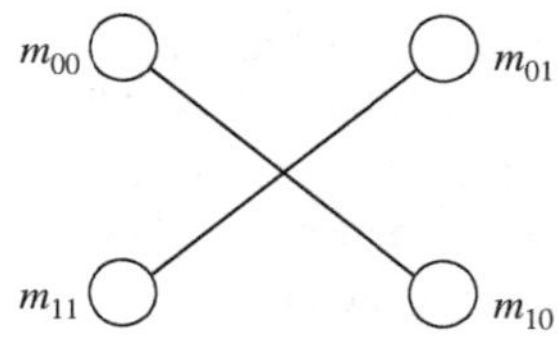

Fig. 15.1 The graphical representation of SPFD $R = \{(ab, \overline{a}b), (\overline{a}\overline{b}, a\overline{b})\}$

SPFDs of a node or wire can be derived in a multitude of ways, depending on their application during logic synthesis. For instance, SPFDs can be computed from the primary outputs in reverse topology order [21, 27]. An SPFD of a node represents the minterm pairs in which the function of the node must evaluate to different values. In rewiring applications, the SPFD of a wire, (η_a, η_b), can denote the minimum set of edges in the SPFD of η_b that can only be distinguished by η_a (but none of the remaining fanins of η_b) [21]. In all these methods, the SPFD of a node complies with Property 15.1, which indicates that the ability of a node to distinguish minterm pairs cannot be better than the ability of all of its fanins. This property is the key to performing logic restructuring with SPFDs in this work.

Property 15.1 Given a node η_k whose fanins are $\{\eta_1, \eta_2, \cdots, \eta_n\}$, the SPFD of η_k is the subset of the union of the SPFDs of its fanin nodes [20].

Finally, a function of a node can be synthesized from its SPFD into a two-level AND-OR network with an automated approach by Cong et al. [7].

15.3 Approximating SPFDs

SPFDs are traditionally implemented with BDDs or with SAT. However, each approach has its own disadvantage. Computing BDDs of some types of circuits (e.g., multipliers) may not be memory efficient [2]. The SAT-based approach alleviates the memory issue with BDDs, but it can be computationally intensive to obtain all the minterm pairs that need to be distinguished [16].

Intuitively, the runtime and memory overhead of the aforementioned approaches can be reduced if fewer minterms are captured by the formulation. Hence, this section presents a simulation-based approach to "approximate" SPFDs to reduce the information that needs to be processed. The main idea behind *a*SPFDs is that they only consider a subset of minterms that are important to the problem. Although *a*SPFDs are based on a small set of the complete input space, experiments show that *a*SPFDs include enough information to construct valid transformations.

To determine a good selection of minterms, logic restructuring can be effectively viewed as a pair of "error/correction" operations [24]. In this context, the required transformation simply corrects an erroneous netlist to a new specification. From this point of view, it is constructive to see that test vectors used for diagnosis are a good means of determining minterms required to construct *a*SPFDs for logic restructuring. This is because test vectors can be thought of as the description of the erroneous behavior and minterms explored by test vectors are more critical than others. Since an *a*SPFD of a node stores less information than its respected SPFD, it is inherently less expensive to represent, manipulate, and compute.

Due to the loss of information, the transformation is guaranteed to be valid only under the input space exercised by the given set of input test vectors only. The transformations may fail to correct designs respected to the complete input space. Hence, the design has to undergo verification after restructuring to guarantee its correctness. However, in some cases, such as in rewiring, a full blown verification may not be required, but a faster proof method can be used [11, 12, 24].

The next two sections present the procedures used to compute *a*SPFDs using a test vector set for nodes in combinational and sequential circuits, respectively.

15.3.1 Computing aSPFDs for Combinational Circuits

Consider two circuits, C_e and C_c, with the same number of the primary inputs and primary outputs. Let $\mathcal{V} = \{v_1, \cdots, v_q\}$ be a set of vectors. For combinational circuits, each $v_i \in \mathcal{V}$ is a single vector, while for sequential circuits, each v_i is a sequence of input vectors. Let η_{err} be the node in C_e where the correction is required, such that C_e is functionally equivalent to C_c after restructuring. Node η_{err} can be identified using diagnosis [6, 23] or formal synthesis [13] techniques and is referred to as a *transformation node* in the remaining discussion.

Let $f'_{\eta_{err}}$ denote the new function of η_{err}. As discussed earlier, the *a*SPFD of η_{err} should contain the pairs of primary input minterms that $f'_{\eta_{err}}$ needs to distinguish. To identify those pairs, the correct values of η_{err} under the test vectors $\mathcal{V}$ are first identified. Those values are what $f'_{\eta_{err}}$ should evaluate for $\mathcal{V}$ after restructuring is implemented. Such a set of values is referred to as *expected trace*, denoted as E_T. Finally, $on(n)(off(n))$ denotes the set of minterms that n is equal to 1(0).

After the expected trace of η_{err} is calculated, the procedure uses the trace to construct *a*SPFDs of η_{err}. In practice, $\mathcal{V}$ includes vectors that detect errors ($\mathcal{V}_e$), as well as ones that do not ($\mathcal{V}_c$). Both types of vectors can provide useful information about the required transformation.

The procedure to generate the *a*SPFD of the transformation node in C_e w.r.t. $\mathcal{V} = \{\mathcal{V}_e \cup \mathcal{V}_c\}$ is as follows. First, C_e is simulated with the input vector $\mathcal{V}$. Let $V_c(\eta_{err})$ and $V_e(\eta_{err})$ denote the value of η_{err} when C_e is simulated with $\mathcal{V}_c$ and $\mathcal{V}_e$. To rectify the design, $f'_{\eta_{err}}$ has to evaluate to the complemented values of $V_e(\eta_{err})$. That is, the expected trace of η_{err}, denoted by $E_T^{\eta_{err}}$, is $\{\overline{V_e(\eta_{err})}, V_c(\eta_{err})\}$ for vectors $\{\mathcal{V}_e, \mathcal{V}_c\}$. Finally, The *a*SPFD of η_{err} states that minterms in $on(E_T^{\eta_{err}})$ have to be distinguished from minterms in $off(E_T^{\eta_{err}})$.

Example 15.1 Figure 15.2(a) depicts a circuit; its truth table is shown in Figure 15.2(b). Let the wire $e \rightarrow z$ (the dotted line) be the target to be removed. After the removal of $e \rightarrow z$, an erroneous circuit is created where the new z, labeled z_{mod}, becomes AND($\overline{d}$, f). The value of z_{mod} is shown in the eighth column of the truth table.

Suppose the design is simulated with test vectors $\mathcal{V} = \{001, 100, 101, 110, 111\}$. The discrepancy is observed when the vector 110 is applied. Therefore, $\mathcal{V}_e = \{110\}$ and $\mathcal{V}_c = \{001, 100, 101, 111\}$. Let z_{mod} be the transformation node. $V_e(z_{mod}) = \{1\}$ and $V_c(z_{mod}) = \{0, 1, 0, 0\}$. Hence, the expected trace of z_{mod} consists of the complemented values of $V_e(z_{mod})$ and $V_c(z_{mod})$, as shown in the final column of Fig. 15.2(b). Finally, the *a*SPFD of z_{mod} w.r.t. $\mathcal{V}$ is generated according to E_T and contains four edges, as shown in Fig. 15.2(a). The dotted vertices indicate that the labeled minterm is a don't care w.r.t. $\mathcal{V}$. For comparison, the SPFD of z_{mod} is shown in Fig. 15.3(b). One can see that information included in *a*SPFD of z_{mod} is much less than what the SPFD representation includes. The minterms that are not encountered during the simulation are considered don't cares in *a*SPFDs. For instance, the minterm pair, (110, 000), does not need to be distinguished in the *a*SPFD of z_{mod} because the vector 000 is not simulated.

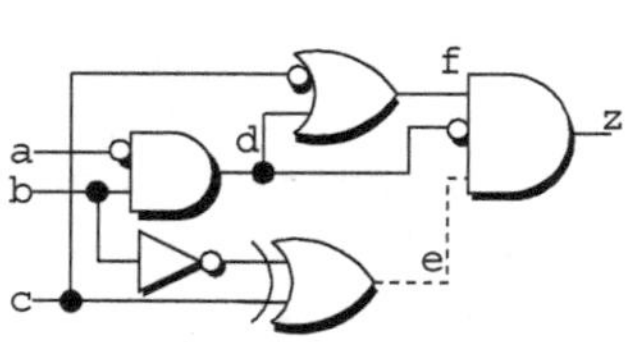

(a)

a	b	c	d	e	f	z	z_{mod}	E_T
0	0	0	0	1	1	1	1	-
0	0	1	0	0	0	0	0	0
0	1	0	1	0	1	0	0	-
0	1	1	1	1	1	0	0	-
1	0	0	0	1	1	1	1	1
1	0	1	0	0	0	0	0	0
1	1	0	0	0	1	0	1	0
1	1	1	0	1	0	0	0	0

(b)

Fig. 15.2 The circuit for Examples 15.1 and 15.5. (**a**) Circuit, (**b**) Truth table

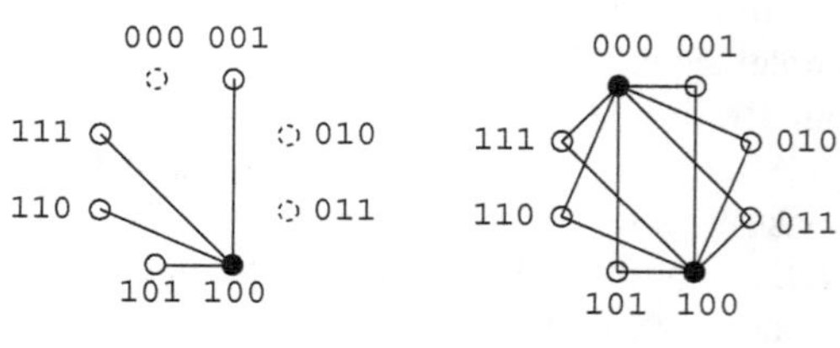

Fig. 15.3 SPFD and *a*SPFDs of z_{mod} in Fig. 15.2. (**a**) *a*SPFD of z_{mod}, (**b**) SPFD of z_{mod}

15.3.2 Computing aSPFDs for Sequential Circuits

The procedure of building *a*SPFDs of nodes in sequential circuits is more complicated than the procedure for combinational circuits due to the state elements. Each node in the circuit depends not only on the present value of the primary inputs but also on values applied to them in previous timeframes. This time dependency characteristic of sequential designs prohibits the application of the procedure of generating *a*SPFDs presented in Section 15.3.1 directly to sequential designs. First of all, the expected trace of the transformation node, η_{err}, is not simply the complemented values of the node under the erroneous vector sequences. Because it is not known in which timeframe the error condition is excited, complementing values in all timeframes risks the introduction of more errors. Moreover, when modeling sequential designs in the ILA representation, the value of nets in the circuit at T_i for some input vector sequences is a function of the initial state input and the sequence of the primary input vectors up to and including cycle T_i. Hence, the input space of *a*SPFD of a node is different in each timeframe.

To simplify the complexity of the problem, the proposed procedure constructs one *a*SPFD over the input space $\{\mathcal{S} \cup \mathcal{X}\}$ that integrates information stored in the *a*SPFD in each timeframe. It first determines the values of the state elements in each timeframe for the given set of input vectors that should take place after the transformation. Then, a partially specified truth table of the new function at η_{err}, in terms of the primary input and the current states, can be generated. The *a*SPFD of η_{err} over the input space $\{\mathcal{S} \cup \mathcal{X}\}$ is constructed based on the truth table. The complete procedure is summarized below:

Step 1. Extract the expected trace E_T of η_{err} for an input vector sequence v. Given the expected output response $(\mathcal{Y})$ under v, a satisfiability instance, $\Phi = \prod_{i=0}^{k} \Phi^i_{C_e}(v^i, \mathcal{Y}^i, \eta^i_{err})$, is constructed. Each $\Phi^i_{C_e}$ represents a copy of C_e at timeframe i, where η^i_{err} is disconnected from its fanins and treated as a primary input. The original primary inputs and the primary outputs of C^i_e are constrained with v^i and $\mathcal{Y}^i$, respectively. The SAT solver assigns values to $\{\eta^0_{err}, \cdots, \eta^k_{err}\}$ to make C_e comply with the expected responses. These values are the desired value of η_{err} for v.

Step 2. Simulate C_e with v at the primary inputs and E_T at η_{err} to determine state values in each timeframe. Those state values are what should be expected after the transformation is applied. Subsequently, a partial specified truth table (in terms of $\{\mathcal{X} \cup \mathcal{S}\}$) of $f'_{\eta_{err}}$ in C_e can be constructed.

Step 3. The *a*SPFD of η_{err} contains an edge for each minterm pair in $\{on(\eta) \times off(\eta)\}$ according to the partially specified truth table.

Example 15.2 Figure 15.4(a) depicts a sequential circuit unrolled for three cycles under the simulation of a single input vector sequence. Assume the correct response at o^1 should be 0 and net p is the transformation node. To determine the expected trace of p, p's are made into primary inputs, as shown in Fig. 15.4(b). A SAT instance is constructed from the modified circuit with the input and output

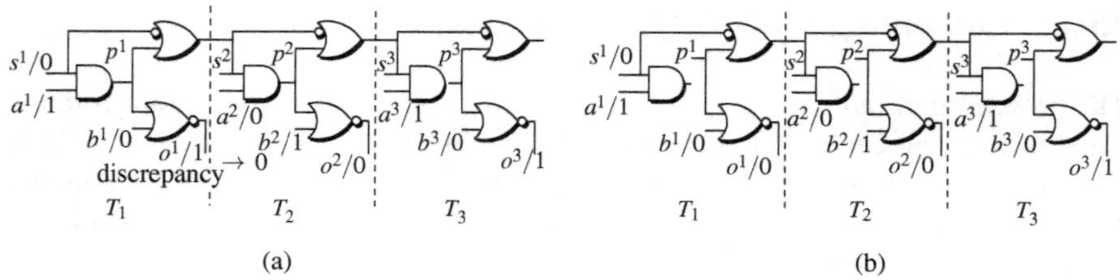

Fig. 15.4 The circuit for Example 15.2. (**a**) Unrolled circuit (**b**) Unrolled circuit with p's as primary inputs

constraints. Given the instance to a SAT solver, 110 is returned as a valid expected trace for p. Next, simulating C_e with the input vector and the expected value of p, $s^2 = 1$, and $s^3 = 1$ are obtained. Then, the partially specified truth table of p states that p evaluates to 1 under minterms (in terms of $\{a, b, s\}$) $\{100, 011\}$ and to 0 under $\{101\}$. Therefore, the *a*SPFD of p contains two edges: (100, 101) and (011, 101).

A special case needs to be considered in Step 1. For any two timeframes, T_i and T_j, of the same test vector, if the values of the primary inputs and the states at these two timeframes are the same, the value of η^i_{err} must equal the value of η^j_{err}. Hence, additional clauses are added to Φ to ensure that the values of η_{err} are consistent when such conditions occur.

Example 15.3 With respect to Example 15.2, another possible assignment to (p^1, p^2, p^3) is 100. However, in this case, the values of $\{a, b, s\}$ at T_1 and T_3 are both 100, while p^1 and p^3 have opposite values. Consequently, this is not a valid expected trace. To prevent this assignment returned by the SAT solver, the clauses

$$(s^1 + s^3 + r) \cdot (\overline{s^1} + \overline{s^3} + r) \cdot (\overline{r} + \overline{p^1} + p^3) \cdot (\overline{r} + p^1 + \overline{p^3})$$

are added to the SAT instance. The new variable, r, equals 1, if s^1 equals s^3. When that happens, the last two clauses ensure that p^1 and p^3 have the same value.

15.3.3 Optimizing aSPFDs with Don't Cares

The procedure of *a*SPFDs generation described above does not take into account all external don't cares in the design. Identifying don't cares for η_{err} can further reduce the size of *a*SPFDs, since all SPFD edges connected to a don't care can be removed from the *a*SPFDs. Consequently, the constraints of qualified solutions for restructuring are relaxed.

There are two types of combinational don't cares: Satisfiability Don't Cares (SDCs) and Observability Don't Cares (ODCs). Since *a*SPFDs of nodes in designs are built over the minterms explored by test vectors, only ODCs need to be considered. ODCs are minterm conditions where the value of the node has no effect on the behavior of the design. Hence, ODCs of η_{err} can only be found under $\mathcal{V}_c$.

Minterms encountered under the simulation of $\mathcal{V}_e$ cannot be ODCs, because, otherwise, no erroneous behavior can be observed at the primary outputs. ODCs can be easily identified by simulating the circuit with $\mathcal{V}_c$ and complement of the original simulation value at η_{err}. If no discrepancy is observed at the primary outputs, the respected minterm is an ODC.

Similarly, combinational ODCs of node η in a sequential design are assignments to the primary inputs and current states such that a value change at η is not observed at the primary outputs or at the next states. However, combinational ODCs of sequential designs may be found in erroneous vector sequences. This is because the sequential design behaves correctly until the error is excited. Having this in mind, the following procedures can be added after Step 2 in Section 15.3.2 to obtain combinational ODCs.

Step 2a. Let E_{T1} denote the expected trace obtained in Step 2 in Section 15.3.2 and $\hat{S}$ denote the values of states in each timeframe. Another expected trace E_{T2} can be obtained by solving the SAT instance Φ again with additional constraints that (a) force $\hat{S}$ on all state variables, and (b) block E_{T1} from being selected as a solution again. Consequently, the new expected trace consists of different values at the transformation node such that the same state transition is maintained.

Step 2b. Let E_{T2} be the second expected trace and T_i be the timeframe where E_{T1} and E_{T2} have different values. It can be concluded that the minterm at T_i is a combinational don't care, since, at T_i, the values of the primary outputs and the next states remain the same, regardless of the value of η.

Step 2c. Repeat this procedure until no new expected trace can be found.

Example 15.4 In Example 15.2, an expected trace, $E_T = 110$, has been obtained, and the state value, $\{s^1, s^2, s^3\}$, is $\{0, 1, 1\}$. To obtain another expected trace at p, additional clauses, $(\overline{s^1})(s^2)(s^3)$, are added to ensure states $\{s^1, s^2, s^3\}$ to have the value $\{0, 1, 1\}$. Another clause, $(\overline{p^1} + \overline{p^2} + p^3)$, is added to prevent the SAT solver to assign $\{1, 1, 0\}$ to $\{p^1, p^2, p^3\}$ again. In this example, another expected trace of p, $E_{T2} = 010$, can be obtained. The values of E_T and E_{T2} are different at T_1, which implies that the minterm 100 in terms of $\{a, b, s\}$ is a don't care. Hence, the *a*SPFD of p can be reduced to contain only one edge, (011, 101).

15.3.3.1 Conflicts in Multiple Expected Traces

In the case of sequential circuits, there can exist multiple expected traces for the given input sequences and the expected output responses. The procedure described earlier for obtaining ODCs in sequential circuits identifies expected traces with the same state transitions. To further explore equivalent states, one can obtain a trace with different state transitions. This can be done by adding additional constraints to block $\hat{S}$ assigned to state variables and solving the SAT instance again. However, these additional traces may assign conflict logic values to the transformation node for the same minterms.

Let E_{T1} and E_{T2} represent two expected traces of the same node for the same test vector sequence. Assume a conflict occurs for minterm m (in terms of the primary input and the current state) between the assignment to E_{T1} at cycle T_i and the assignment to E_{T2} at cycle T_j. In this instance, one of the two cases below is true:

- *Case 1:* The output responses and the next states at cycle T_i for E_{T1} and T_j for E_{T2} are the same. This implies that the value of the transformation node under m does not affect the behavior of the design. Hence, m is a *combinational ODC*.
- *Case 2:* The next states are different. This can happen when the circuit has multiple state transition paths with the same initial transitions. Figure 15.5 shows an example of this scenario. Let η be the transformation node. The graph depicts a state transition diagram for a single-output design. The state transition depends on the value of η; the value of the output is indicated inside the state. Assume a test vector makes the design start at S_0. It takes at least three cycles to differentiate the transition $Path_a$ and $Path_b$, since the value of the primary output is not changed until the design is in S_4. Since the proposed analysis is bounded by the length of the input vector sequences, it may not process enough cycles to differentiate these different paths. Hence, multiple assignments at the transformation node can be valid within the bounded cycle range and, consequently, cause conflicts. In the example, if the circuit is only unrolled for two cycles, both paths ($S_0 \rightarrow S_1$ and $S_0 \rightarrow S_2$) would seem to be the same from the observation of the primary outputs. It implies that η can have either logic 0 and logic 1 in S_0. Since the algorithm does not have enough information to distinguish the correct assignment, minterm m in this case is considered to be a don't care as well. This issue can be resolved if vector sequences that are long enough are used instead.

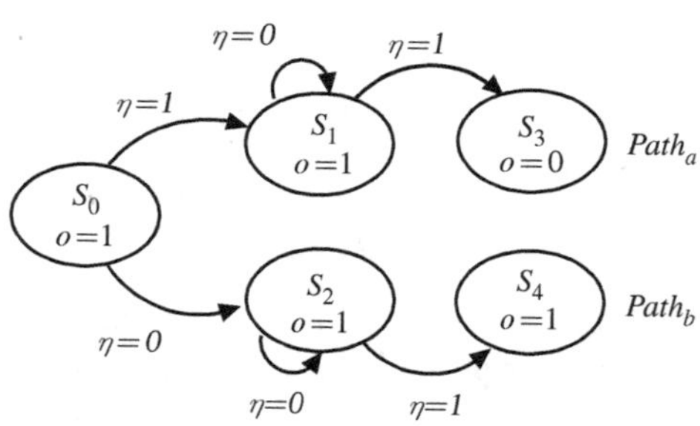

Fig. 15.5 State transition

Algorithm 8 Transformation using *a*SPFDs

C_e := Erroneous circuit
$\mathcal{V}$:= A set of input vectors
η_{err} := Transformation node
TRANSFORMATION_WITH_ASPFD(C_e, $\mathcal{V}$. η_{err}) {
    Compute *a*SPFDs of η_{err}
    $E \leftarrow (m_i, m_j) \in R^{appx}_{\eta_{err}} | (m_i, m_j)$ cannot be distinguished by any fanin of η_{err}
    Let $\mathcal{N} := \{\eta_k \mid \eta_k \in C_e$ and $\eta_k \notin \{TFO(\eta_{err}) \cup \eta_{err}\}$
    $Cover \leftarrow$SELECTCOVER($\mathcal{N}$)
    Re-implementing η_{err} with the original fanins and the nodes in $Cover$
}

15.4 Logic Transformations with *a*SPFDs

In this section, the procedure to systematically perform logic restructuring with *a*SPFDs is presented. The proposed restructuring procedure uses *a*SPFDs to seek transformations at the transformation node, η_{err}. The transformations are constructed with one or more additional fanins.

The procedure is summarized in Algorithm 8. The basic idea is to find a set of nets such that every minterm pair of the *a*SPFD of the new transformation implemented at η_{err} is distinguished by at least one of the nets, as stated in Property 15.1. Hence, the procedure starts by constructing the *a*SPFD of η_{err}, $R^{appx}_{\eta_{err}}$. To minimize the distortion that may be caused by the rectification, the original fanins are kept for restructuring. In other words, it is sufficient that *a*SPFDs of additional fanins only need to distinguish edges in $R^{appx}_{\eta_{err}}$ that cannot be distinguished by any original fanins. Those undistinguished edges are referred to as *uncovered edges*. A function is said to *cover* an SPFD edge if it can distinguish the respected minterm pair. Let $TFO(\eta_{err})$ denote the transitive fanout of n_{err}. The function SELECTCOVER is used to select a set of nodes ($Cover$) from nodes not in $TFO(\eta_{err})$ such that each uncovered edge is distinguished by at least one node in $Cover$. The function SELECTCOVER is further discussed in the next sections. Finally, a new two-level AND-OR network is constructed at η_{err} using the nodes in $Cover$ as additional fanins as discussed in Section 15.2.2.

Example 15.5 Returning to Example 15.1, the *a*SPFD of z_{mod} is shown in Fig. 15.6(a) and the partial truth table of remaining nodes is shown in Fig. 15.6(b). Figure 15.6(a) shows that the edge (110, 100) (the dotted line) is the only SPFD edge that is not distinguished by the fanin of z_{mod}, $\{f, d\}$. Hence, the additional fanins required for restructuring at z_{mod} must distinguish this edge. According to the truth table, this edge can be distinguished by b. As a result, b is used as the additional fanin for restructuring z_{mod}. Since the minterm 100 is the only minterm in the onset of new z_{mod} w.r.t. $\mathcal{V}$, it implies $b = 0$, $d = 0$ and $f = 1$. Therefore, the new function of z_{mod} is AND($\overline{b}, \overline{d}, \overline{f}$), as shown in Fig. 15.6(c).

Two approaches to find the set, $Cover$, are presented in the following sections: a SAT-based approach that finds the minimal number of fanin wires and a greedy approach that exchanges optimality for performance.

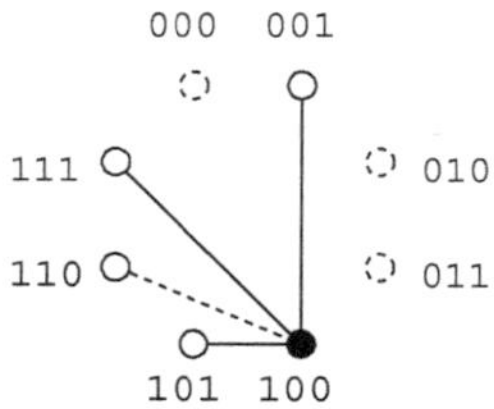

(a)

a	b	c	d	e	f
0	0	1	0	0	0
1	0	0	0	1	1
1	0	1	0	0	0
1	1	0	0	0	1
1	1	1	0	1	0

(b)

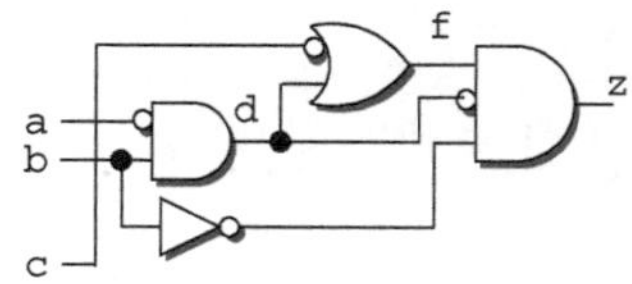

(c)

Fig. 15.6 aSPFD of z_{mod} and the partially specified truth table of nodes in Fig. 15.2 and modified circuit. (**a**) aSPFD of z_{mod}, (**b**) partial truth table, (**c**) Modified circuit

15.4.1 SAT-Based Searching Algorithm

The search problem in Algorithm 8 is formulated as an instance of Boolean satisfiability. Recall that the algorithm looks for a set of nodes outside $TFO(\eta_{err})$ such that those nodes can distinguish SPFD edges of $R_{\eta_{err}}^{appx}$ that cannot be distinguished by any fanins of η_{err}.

Construction of the SAT instance is fairly straightforward. Each uncovered SPFD edge in the aSPFD of η_{err} has a list of nodes that can distinguish the edge. The SAT solver selects a node from each list such that at least one node in the list of each uncovered SPFD edge is selected. The set, $Cover$, consists of these selected nodes. The formulation of the SAT instance Φ is as follows. Each node η_k is associated with a variable w_k. Node η_k is added to the set $Cover$ if w_k is assigned a logic value 1. The instance contains two components: $\Phi_C(\mathcal{W})$ and $\Phi_B(\mathcal{W}, \mathcal{P})$, where $\mathcal{W} = \{w_1, w_2, \cdots\}$ is the set of variables that are associated with nodes in the circuit and $\mathcal{P}$ is a set of new variables introduced.

- *Covering clauses* ($\Phi_C(\mathcal{W})$): A covering clause lists the candidate nodes for an uncovered edge. A satisfied covering clause indicates that the associated edge is covered. One covering clause, c_j, is constructed for each uncovered edge e_j in the aSPFD of η_{err}. Let $\mathcal{D}_j$ be the candidate nodes which can cover edge e_j. Clause c_j contains w_k if η_k in $\mathcal{D}_j$ covers e_j; that is, $c_j = \bigvee_{\eta_k \in \mathcal{D}_j} w_k$. Hence, this clause is satisfied if one of the included candidate nodes is selected.
- *Blocking clauses* ($\Phi_B(\mathcal{W}, \mathcal{P})$): Blocking clauses define the condition where a candidate node η_k should not be considered as a solution. They help to prune the solution space and prevent spending time on unnecessary searches. For each node $\eta_k \notin \{TFO(\eta_{err}) \cup \eta_{err}\}$, according to Property 15.1, η_k does not distinguish more edges if all of its fanins are selected already. Hence, for each candidate node η_k, a new variable, p_k, is introduced; p_k is assigned a logic value 1 if all of the fanins of η_k are selected, and 0 otherwise. Consequently, w_k is assigned a logic value 0 (i.e., η_k is not considered for the solution) when p_k has a logic value 1. The blocking clause for node $\eta_k = f(\eta_1, \eta_2, \cdots, \eta_m)$, where η_i, $1 \le i \le m$, is a fanin of η_k, is as follows: $(\bigvee_{i=1}^{m} \overline{w_i} + p_k) \cdot \bigwedge_{i=1}^{m}(w_i + \overline{p_k}) \cdot (\overline{p_k} + \overline{w_k})$.

Given a satisfying assignment for Φ, a node η_k is added to the set $Cover$ if $w_k = 1$. The covering clauses ensure that $Cover$ can cover all the edges in the *a*SPFD of η_{err}. Blocking clauses reduce the possibility of the same set of edges being covered by multiple nodes in $Cover$. If the function derived by the satisfying assignment from the set $Cover = \{w_1, w_2, \cdots, w_n\}$ fails formal verification, then $(\overline{w_1} + \overline{w_2} + \cdots + \overline{w_n})$ is added as an additional blocking clause to Φ and the SAT solver is invoked again to find another solution.

Note that in the above formulation because there are no constraints on the number of nodes that should be selected to cover the edges, the solution returned by the solver may not be optimal. In order to obtain the optimal solution, in experiments, SAT instances are solved with a pseudo-Boolean constraint SAT solver [8] that returns a solution with the smallest number of nodes. The use of a pseudo-Boolean solver is not mandatory and any DPLL-based SAT solvers [15, 17] can be used instead. One way to achieve this is to encode the countercircuitry from [23] to count the number of selected nodes. Then, by enumerating values $N = 1, 2, \ldots$, the constraint enforces that no more than N variables can be set to a logic value 1 simultaneously or Φ becomes unsatisfiable. Constraining the number N in this manner, any DPLL-based SAT solver can find the minimum size of $Cover$.

15.4.2 Greedy Searching Algorithm

Although the SAT-based formulation can return the minimum set of fanins to resynthesize η_{err}, experiments show that, at times, it may require excessive runtime. To improve the runtime performance, a greedy approach to search solutions is proposed as follows:

Step 1. Let E be the set of SPFD edges in the *a*SPFD of η_{err} that needs to be covered. For each edge $e \in E$, let N_e be the set of nodes $\eta \notin \{TFO(\eta_{err}) \cup \eta_{err}\}$ which can distinguish the edge. Sort $e \in E$ in descending order by the cardinality of N_e

Step 2. Select the edge, e_{min}, with the smallest cardinality of $N_{e_{min}}$. This step ensures that the edge that can be covered with the least number of candidates is targeted first

Step 3. Select η_k from $N_{e_{min}}$ such that η_k covers the largest set of edges in E and add η_k to $Cover$

Step 4. Remove edges that can be covered by η_k from E. If E is not empty, go back to Step 1 to select more nodes

The solutions identified by the greedy approach may contain more wires than the minimum set. However, experiments indicate that the greedy approach can achieve results of a similar quality to the SAT-based approach in a more computationally efficient manner.

15.5 Experimental Results

The proposed logic restructuring methodology using *a*SPFDs is evaluated in this section. ISCAS'85 and ISCAS'89 benchmarks are used. The diagnosis algorithm from [23] is used to identify the restructuring locations and Minisat [9] is the underlying SAT solver. The restructuring potential of the *a*SPFD-based algorithms is compared with that of a logic correction tool from [25] which uses the dictionary model of [1]. Both methodologies are compared against the results of a formal method, called *error equation* [6]. This method answers with certainty whether there exists a modification that corrects a design at a location. Experiments are conducted on a Core 2 Duo 2.4 GHz processor with 4 GB of memory while the runtime is reported in seconds.

Table 15.1 summarizes the characteristics of benchmarks used in this experiment. Combinational benchmarks are listed in the first four columns, while sequential benchmarks are shown in the last four columns. The table includes the number of primary inputs, the number of flip-flops, and the total number of gates in each column, respectively.

Table 15.1 Characteristics of benchmarks

Combinational				Sequential			
Circ.	# PI	# FF	# Gates	Circ.	# PI	# FF	# Gates
c1355	41	0	621	s510	19	6	256
c1908	33	0	940	s713	35	19	482
c2670	157	0	1416	s953	16	29	476
c3540	50	0	174	s1196	14	18	588
c5315	178	0	2610	s1238	14	18	567
c7552	207	0	3829	s1488	8	6	697

In this work, performance of the proposed methodology is evaluated with the ability to correct errors in logic netlists. Three different complexities of modifications are injected in the original benchmark. The locations and the types of modifications are randomly selected. Simple complexity modifications (suffix "s") involve the addition or deletion of a single wire, replacement of a fanin with another node, and a gate-type replacement. Moderate modifications (suffix "m") on a gate include multiple aforementioned changes on a single gate. The final type of modification complexity, complex (suffix "c"), injects multiple simple complexity modifications on a gate and those in the fanout-free fanin cone of the gate.

For each of the above types, five testcases are generated from each benchmark. The proposed algorithm is set to find, at most, 10 transformations for each location identified first by the diagnosis algorithm. Functional verification is carried out at the end to check whether the 10 transformations are valid solutions.

15.5.1 Logic Restructuring of Combinational Designs

The first set of experiments evaluates the proposed methodology for a single location in combinational circuits. Experimental results are summarized in Table 15.2. In this

Table 15.2 Combinational logic transformation results for various complexities of modifications

Circ.	Error loc.	Error equat. (%)	Dict. model (%)	*a*SPFD(%)	Avg time (s)	Avg # wires (greedy)	Min # wires (SAT)	Avg # corr/ loc.	% verified First	% verified All
c1355_s	5.3	100	19	81	3.5	1.7	1.7	8.3	100	46
c1908_s	18.0	84	13	84	18.9	1.4	1.4	8.1	90	62
c2670_s	9.2	98	11	82	21.9	2.4	2.2	6.2	100	75
c3540_s	7.2	100	28	86	9.3	1.1	1.1	4.5	100	66
c5315_s	6.4	100	25	100	7.6	1.9	–	5.4	89	77
c7552_s	11.8	88	19	50	25.7	1.7	–	3.1	88	54
c1355_m	2.7	100	13	100	32.0	2.1	2.0	7.0	100	52
c1908_m	5.8	100	3	83	11.0	2.5	2.5	5.6	100	68
c2670_m	5.2	96	4	60	95.4	3.4	2.9	9.4	100	60
c3540_m	3.2	100	25	100	54.2	1.6	1.6	6.1	84	78
c5315_m	9.6	94	2	100	46.7	2.9	–	5.7	100	77
c7552_m	8.8	100	9	91	39.2	1.9	–	6.9	100	79
c1355_c	3.7	96	0	73	38.4	2.9	2.9	3.3	100	40
c1908_c	15.8	47	41	70	19.0	1.4	1.3	7.2	100	88
c2670_c	12.4	98	31	62	33.2	1.7	1.7	4.7	100	76
c3540_c	3.0	100	7	67	122.4	3.6	3.4	3.8	100	33
c5315_c	6.4	97	16	100	20.0	2.7	–	9.1	100	79
c7552_c	20.6	64	20	50	23.7	1.9	–	3.5	91	43
Average	8.6	93	16	80	29.2	2.0	–	6.4	96	67

experiment, circuits are simulated with a set of 1000 input vectors that consists of a set of vectors with high stuck-at fault coverage and random-generated vectors.

The first column lists the benchmarks and the types of modifications inserted as described earlier. The second column has the average number of locations returned by the diagnosis program for the five experiments. The percentage of those locations where the error equation approach proves the existence of a solution is shown in the third column. The next two columns show the percentage of locations (out of those in the second column) for which the dictionary approach and the proposed *a*SPFD approach can successfully find a valid solution. A valid solution is one in which the restructured circuit passes verification. The sixth column contains the average runtime, including the runtime of verification, to find all 10 transformations using greedy heuristics.

Taking `c1908_s` as an example, there are, on average, 18 locations returned by the diagnosis program. The error equation check returns that 15 (84% of 18) out of those locations can be fixed by re-synthesizing the function of the location. The dictionary approach successfully identifies two locations (13% of 15) while the *a*SPFD approach can restructure 13 locations (84% of 15). This shows that the proposed approach is seven times more effective than the dictionary approach. Overall, the proposed methodology outperforms the dictionary approach in all cases and achieves greater improvement when the modification is complicated.

The quality of the transformations, in terms of the wires involved as well as some algorithm performance metrics, are summarized in column 7–11 of Table 15.2. Here, only cases where a valid solution is identified by the proposed algorithm are considered. The seventh and the eighth columns list the average number of additional wires returned by the greedy algorithm and by the SAT-based searching algorithm, respectively. As shown in the table, the greedy heuristic performs well compared to the SAT-based approach. Because the SAT-based approach may run into runtime problems as the number of new wires increases, it times out ("-") after 300 s if it does not return with a solution.

As mentioned earlier, the algorithm is set to find, at most, 10 transformations for each location. The ninth column shows the average number of transformations identified for each location. It shows that, for all cases, more than one transformation can be identified. This is a desirable characteristic, since engineers can have more options to select the best fit for the application. The final two columns show the percentage of transformations that pass verification. The first column of these two columns only considers the first identified transformation, while the second column has the percentage of all 10 transformations that pass verification. One can observe that the vast majority of first-returned transformations pass verification.

Next, the performance of restructuring with various numbers of test vectors is investigated. Four sizes are used: 250, 500, 1000, and 2000 test vectors. The results are depicted in Fig. 15.7. Figure 15.7(a) shows the percentage of the locations where the proposed algorithm can identify a valid transformation. As shown, the success rate increases as the size of input vectors increases for each error complexity group. This is expected, since more vectors provide more information for *a*SPFDs. The chance that the algorithm incorrectly characterizes a minterm as a don't care is also reduced.

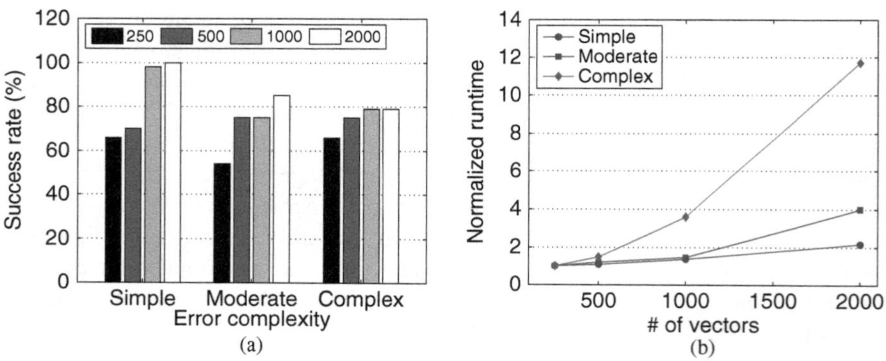

Fig. 15.7 Performance of restructuring with variable vector sizes for combinational designs. (**a**) Solution success rate (**b**) Runtime profiling

Although using a larger vector set can improve the success rate of the restructuring, it comes with the penalty that more computational resources are required to tackle the problem. The average runtime is plotted in Fig. 15.7(b) and normalized by comparing it to the runtime of the case with 250 vectors. Each line represents one error complexity type. Taking `Complex` as an example, the runtime is 12 times longer when the vector size is increased from 250 to 2000. Note that there is a significant increase when the size of the vector set increases from 1000 to 2000.

Since the success rate of cases when 1000 vectors are used is close to the success rate of those with 2000 vectors (Fig. 15.7a), this suggests that, for those testcases, 1000 input vectors can be a good size to have a balance between the resolution of solutions and the runtime performance.

15.5.2 Logic Restructuring of Sequential Designs

The second set of the experiments evaluates the performance of logic restructuring with *a*SPFDs in sequential designs. The vector set for sequential circuits contains 500 random input vector sequences with a length of 10 cycles. To verify the correctness of transformations, a bounded sequential equivalent checker [19] is used. This tool verifies the resulting design against the reference within a finite number of cycles, which is set to 20 cycles in our experiment.

The performance of the proposed methodology is recorded in Table 15.3. The benchmarks and the type of the modification inserted are listed in the first column. The second column presents the average number of locations for transformations reported by the diagnosis program while the percentage of these locations that are proven to be correctable by error equation is recorded in the third column. The percentage of locations in which the proposed methodology finds a valid transformation is reported in the fourth column, followed by runtime.

Table 15.3 Sequential logic transformation results for various complexities of modifications

Circ.	Error loc.	Error equat. (%)	*a*SPFD(%)	Avg. time (s)	Avg # wires	Avg # corr/ loc.	% verified First	All	% unique
s510_s	2.4	100	75	384	0.3	1.8	100	92	100
s713_s	5.0	72	0	325	–	–	–	–	–
s953_s	1.8	100	33	223	1.0	3.3	100	37	70
s1196_s	1.8	100	56	237	2.0	5.0	83	92	64
s1238_s	1.6	100	38	781	1.1	5.0	100	100	55
s1488_s	2.8	86	43	258	1.7	5.0	83	46	68
s510_m	2.0	100	90	68	0.3	4.2	100	38	99
s713_m	2.8	43	36	689	0.6	1.4	100	41	60
s953_m	1.6	63	40	105	1.2	1.2	100	100	100
s1196_m	1.2	83	66	27	1.8	2.6	100	72	83
s1238_m	2.6	85	72	218	2.2	4.3	100	76	47
s1488_m	3.4	100	0	83	–	–	–	–	–
s510_c	1.6	100	38	166	0.5	1.5	100	92	100
s713_c	3.4	71	47	1124	1.0	1.0	100	100	75
s953_c	2.2	73	0	122	–	–	–	–	–
s1196_c	2.0	50	20	588	0.5	2.3	50	32	100
s1238_c	1.2	100	14	328	0	–	100	–	100
s1488_c	1.8	71	30	98	1.7	1.5	33	27	100
Average	2.1	90	39	236	1.0	3.1	92	68	82

Note that the error equation approach in [6] is developed for combinational circuits. Hence, here the sequential circuits are converted into combinational ones by treating the states as pseudo-primary inputs/outputs. In this way, the number of locations reported by the error equation approach is the lower bound of the locations that are correctable, since it constrains that the design after restructuring has to be combinationally functional equivalent to the reference design. This constraint discards any solutions that utilize equivalent states. Overall, the proposed approach can restructure 39% of the locations. The reason why the algorithm fails to correct some of the locations is because the input vector sequences do not provide enough information to generate a good *a*SPFD. This occurs when the algorithm characterizes a minterm as a don't care when this minterm is not exercised by the input vector sequences, or when conflict values are required for this minterm, as discussed in Section 15.3.3. Consequently, the resulting transformation does not distinguish all the necessary minterm pairs that are required to correct the design.

The sixth and the seventh columns report the average number of additional wires used in the transformations and the average number of transformations per location, respectively. Note, because the transformation at some locations only needs to be re-synthesized with the existing fanin nets without any additional wires, cases such as `s510_s` use less than one additional wire on average. The next two columns show the percentage of cases where the first transformation passes verification, and the percentage of 10 transformations that pass verification. Similar to the combinational circuits, there is a high percentage of the first transformations that passes verification if the proposed methodology returns a valid transformation. This indicates that *a*SPFD is a good metric to prune out invalid solutions.

As the result shown in the last column, the valid solutions are further checked for whether or not they are unique to the sequential *a*SPFD-based algorithm. That is, the modified design is not combinationally functional equivalent to the reference design; otherwise, such restructuring can be identified by the combinational approach. If two designs are not combinationally equivalent, it means that the transformation changes the state assignments as well. Consequently, these transformations will be pruned out by the combinational approach. Overall, 82% of the valid transformations is uniquely identified by the sequential approach, restructuring method.

Finally, the impact of test vector sizes on the performance of the presented methodology is studied. Here, the number of test vector sequences is set to 100, 200, 500, and 700. These test sequences have a length of 10 cycles and are randomly generated. The success rate and the normalized runtime are shown in Fig. 15.8(a) and Fig. 15.8(b), respectively. One can see that the behavior observed earlier for the combinational cases is also observed here. The success rate of the restructuring decreases as the number of the test sequences decreases. Among the different error complexities, the benchmarks with complex errors are affected most. This is because a complex error can be excited in various ways and requires more test sequences to fully characterize the erroneous behavior. As a result, the algorithm needs more vector sequences to construct an accurate transformation. Moreover, Fig. 15.8(b)

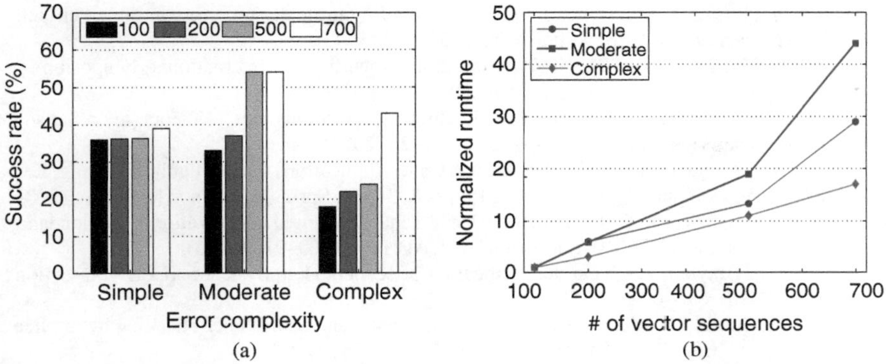

Fig. 15.8 Performance of restructuring with variable vector sizes for sequential designs. (**a**) Solution success rate, (**b**) Runtime profiling

shows a significant reduction of the runtime with the decrease of the number of vector sequences.

15.6 Summary

In this chapter, a simulation-based procedure to approximate SPFDs, namely *a*SPFDs, is first presented. An *a*SPFD is an approximation of the original SPFD, as it only contains information that is explored by the simulation vectors. Next, an *a*SPFD-based logic restructuring algorithm for both combinational and sequential designs is presented. This technique can be used for a wide range of applications, such as logic optimization, debugging, and applying engineer changes. Experiments demonstrate that *a*SPFDs provide a powerful approach to restructuring a logic design to a new set of specifications. This approach is able to construct required logic transformations algorithmically and restructure designs at a location where other methods fail.

References

1. Abadir, M.S., Ferguson, J., Kirkland, T.E.: Logic verification via test generation. IEEE Transactions on CAD **7**, 138–148 (1988)
2. Bryant, R.E.: On the complexity of VLSI implementations and graph representations of Boolean functions with application to integer multiplication. IEEE Transactions on Computer **40**(2), 205–213 (1991)
3. Chang, K.H., Markov, I.L., Bertacco, V.: Fixing design errors with counterexamples and resynthesis. IEEE Transactions on CAD **27**(1), 184–188 (2008)
4. Chang, S.C., Cheng, D.I.: Efficient Boolean division and substitution using redundancy addition and removing. IEEE Transactions on CAD **18**(8), 1096–1106 (1999)
5. Chang, S.C., Marek-Sadowska, M., Cheng, K.T.: Perturb and simplify: Multi-level Boolean network optimizer. IEEE Transactions on CAD **15**(12), 1494–1504 (1996)

6. Chung, P.Y., Hajj, I.N.: Diagnosis and correction of multiple design errors in digital circuits. IEEE Transactions on VLSI Systems **5**(2), 233–237 (1997)
7. Cong, J., Lin, J.Y., Long, W.: SPFD-based global rewiring. In: International Symposium on FPGAs, pp. 77–84. Monterey, CA, USA (2002)
8. Eén, N., Sörensson, N.: Translating pseudo-Boolean constraints into SAT. Journal on Satisfiability, Boolean Modeling and Computation **2**, 1–26 (2006)
9. In: E. Giunchiglia, A. Tacchella (eds.) Theory and Applications of Satisfiability Testing, Lecture Notes in Computer Science, vol. 2919, pp. 333–336. Springer Berlin / Heidelberg (2004)
10. Entrena, L., Cheng, K.T.: Combinational and sequential logic optimization by redundancy addition and removal. IEEE Transactions on CAD **14**(7), 909–916 (1995)
11. Jiang, J.H., Brayton, R.K.: On the verification of sequential equivalence. IEEE Transactions on CAD **22**(6), 686–697 (2003)
12. Kunz, W., Stoffel, D., Menon, P.R.: Logic optimization and equivalence checking by implication analysis. IEEE Transactions on CAD **16**(3), 266–281 (1997)
13. Lin, C.C., Chen, K.C., Marek-Sadowska, M.: Logic synthesis for engineering change. IEEE Transactions on CAD **18**(3), 282–292 (1999)
14. Ling, A.C., Brown, S.D., Zhu, J., Saparpour, S.: Towards automated ECOs in FPGAs. In: International symposium on field programmable gate arrays, pp. 3–12. Monterey, CA, USA (2009)
15. Marques-Silva, J.P., Sakallah, K.A.: GRASP: A new search algorithm for satisfiability. IEEE Transactions on Computer **48**(5), 506–521 (1999)
16. Mishchenko, A., Chatterjee, S., Brayton, R.K., Eén, N.: Improvements to combinational equivalence checking. In: Proceedings of International Conference on CAD, pp. 836–843. San Jose, CA, USA (2006)
17. Moskewicz, M.W., Madigan, C.F., Malik, S.: Chaff: Engineering an efficient SAT solver. In: Design Automation Conference, pp. 530–535. Las Vegas, NV, USA (2001)
18. Nayak, D., Walker, D.M.H.: Simulation-based design error diagnosis and correction in combinational digital circuits. In: VLSI Test Symposium, pp. 70–78. San Diego, CA, USA (1999)
19. Safarpour, S., Fey, G., Veneris, A., Drechsler, R.: Utilizing don't care states in SAT-based bounded sequential problems. In: Great Lakes VLSI Symposium. Chicago, IL, USA (2005)
20. Sinha, S.: SPFDs: A new approach to flexibility in logic synthesis. Ph.D. thesis, University of California, Berkeley (2002)
21. Sinha, S., Brayton, R.K.: Implementation and use of SPFDs in optimizing Boolean networks. In: Proceedings of International Conference on CAD, pp. 103–110. San Jose, CA, USA (1998)
22. Sinha, S., Kuehlmann, A., Brayton, R.K.: Sequential SPFDs. In: Proceedings of International Conference on CAD, pp. 84–90. San Jose, CA, USA (2001)
23. Smith, A., Veneris, A., Ali, M.F., Viglas, A.: Fault diagnosis and logic debugging using Boolean satisfiability. IEEE Transactions on CAD **24**(10), 1606–1621 (2005)
24. Veneris, A., Abadir, M.S.: Design rewiring using ATPG. IEEE Transactions on CAD **21**(12), 1469–1479 (2002)
25. Veneris, A., Hajj, I.N.: Design error diagnosis and correction via test vector simulation. IEEE Transactions on CAD **18**(12), 1803–1816 (1999)
26. Watanabe, Y., Brayton, R.K.: Incremental synthesis for engineering changes. In: International Conference on Computer Design, pp. 40–43. Cambridge, MA, USA (1991)
27. Yamashita, S., H.Sawada, Nagoya, A.: A new method to express functional permissibilities for LUT based FPGAs and its applications. In: Proceedings of International Conference on CAD, pp. 254–261. San Jose, CA, USA (1996)
28. Yang, Y.S., Sinha, S., Veneris, A., Brayton, R.K.: Automating logic rectification by approximate SPFDs. In: Proceedings of ASP Design Automation Conference, pp. 402–407. Yokohama, Japan (2007)

Chapter 16
Extracting Functions from Boolean Relations Using SAT and Interpolation

Jie-Hong Roland Jiang, Hsuan-Po Lin, and Wei-Lun Hung

Abstract Boolean relations are an important tool in system synthesis and verification to characterize solutions to a set of Boolean constraints. For hardware realization of a system specified by a Boolean relation, a deterministic function often has to be extracted from the relation. Prior methods, however, are unlikely to handle large problem instances. From the scalability standpoint this chapter demonstrates how interpolation can be exploited to extend determinization capacity. A comparative study is performed on several proposed computation techniques, specifically, expansion- versus composition-based quantification and interpolation- versus cofactoring-based function extraction. Experimental results show that Boolean relations with thousands of variables can be effectively determinized and the extracted functional implementations are of reasonable quality.

16.1 Introduction

Relations are a powerful tool to represent mappings. Admitting one-to-many mappings, they are strictly more generic than functions. Taking the Boolean mappings $\{(x_1, x_2) \in \mathbb{B}^2\} \rightarrow \{(y_1, y_2) \in \mathbb{B}^2\}$ of Fig. 16.1 as an example, we can express the (one-to-one) mapping of Fig. 16.1a using Boolean functions $f_1 = x_1 x_2$ and $f_2 = \neg x_1 \neg x_2$ for outputs y_1 and y_2, respectively. On the other hand, there is no similar functional representation for the mapping of Fig. 16.1b due to the one-to-many mapping under $(x_1, x_2) = (0, 1)$. However, this mapping can be specified by the relation (with characteristic function) $R = \neg x_1 \neg x_2 \neg y_1 y_2 \vee \neg x_1 x_2 \neg y_1 \neg y_2 \vee \neg x_1 x_2 y_1 y_2 \vee x_1 \neg x_2 \neg y_1 \neg y_2 \vee x_1 x_2 y_1 \neg y_2$.

Owing to their generality, relations can be exploited to specify the permissible behavior of a design. For instance, the behavior of a system can be specified using

J.-H.R. Jiang (✉)
National Taiwan University, Taipei, Taiwan
e-mail: jhjiang@cc.ee.ntu.edu.tw

This work is based on an earlier work: Interpolating functions from large Boolean relations, in Proceedings of the 2009 International Conference on Computer-Aided Design, ISBN:978-1-60558-800-1 (2009) © ACM, 2009. DOI= http://doi.acm.org/10.1145/1687399.1687544

S.P. Khatri, K. Gulati (eds.), *Advanced Techniques in Logic Synthesis, Optimizations and Applications*, DOI 10.1007/978-1-4419-7518-8_16,

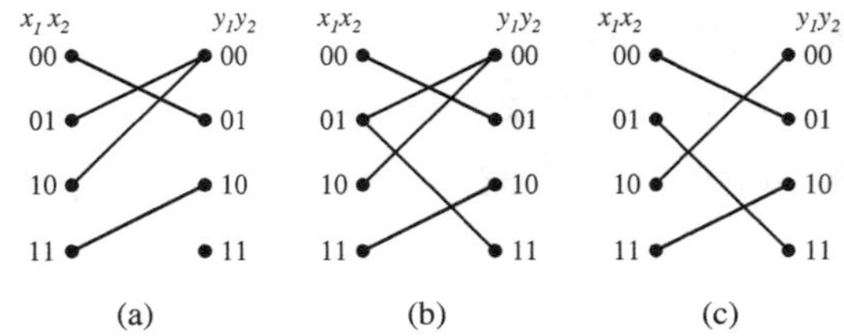

Fig. 16.1 Boolean mappings

relations as constraints over its input stimuli, state transitions, and output responses. Moreover, the flexibility of a circuit can be naturally characterized by a relation. In fact, relations subsume the conventional notion of don't cares. To see it, assume Fig. 16.1b to be a relaxed permissible mapping of Fig. 16.1a. That is, under input $(x_1, x_2) = (0, 1)$ the output (y_1, y_2) can be $(1, 1)$ in addition to $(0, 0)$. This flexibility is not expressible using the conventional don't care representation, and it can be useful in circuit optimization. By trimming off the output choice $(0, 0)$ under input $(0, 1)$, the resulting mapping in Fig. 16.1c has new output functions $f_1 = x_2$ and $f_2 = \neg x_1$, simpler than those of Fig. 16.1a.

In fact circuit flexibility can be abundant in a given multi-level netlist structure. To illustrate, Fig. 16.2 shows an example revealing the origin of flexibility and its potential use for circuit minimization. The circuit structure of Fig. 16.2a induces the mapping relation shown by the solid edges in Fig. 16.2c. Observe that output (z_1, z_2) equals $(0, 0)$ when (y_1, y_2) is $(0, 0)$ or $(1, 1)$. Consequently, as indicated by the dashed edge in Fig. 16.2c, (y_1, y_2) can be $(1, 1)$ as well in addition to the original $(0, 0)$ when $(x_1, x_2) = (0, 1)$. Hence the circuit can be simplified to Fig. 16.2b by choosing $(y_1, y_2) = (1, 1)$ under $(x_1, x_2) = (0, 1)$.

Compared with relations, functions, though restrictive, are often closer to physical realization due to their deterministic nature. Therefore conversions between relations and functions are usually indispensable. To name two examples, in reachability analysis, the transition functions of a state transition system are often converted to a transition relation to abstract away input variables; in circuit synthesis, optimal

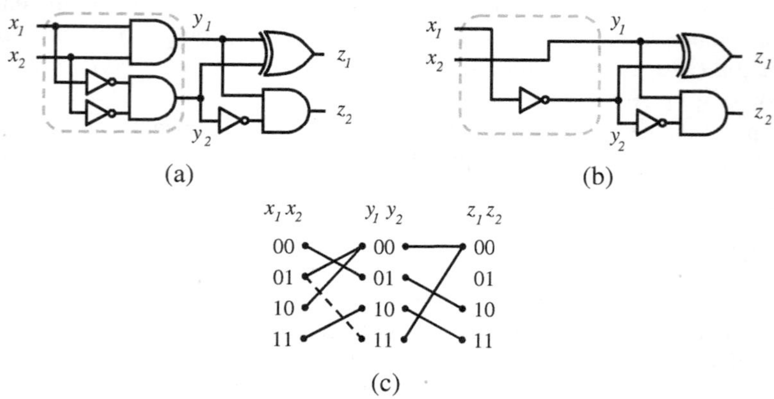

Fig. 16.2 Exploring the flexibilities described by a relation for optimization

functions can be extracted from a relation representing some specification or permissible behavior.

Whereas converting functions to relations is straightforward, converting relations to functions involves much complication. Among many possibilities, relation-to-function conversion in its two extremes can be *range-preserving*, where all possible output responses under every input stimulus[1] are produced (with the help of parametric variables), and can be *deterministically reducing*, where only one output response under every input stimulus is produced (without the need of parametric variables). Both extremes have important applications in system synthesis and verification. The former is particularly useful in verification. As the constraints specified by a relation are preserved, the conversion helps create a testbench to generate simulation stimuli [10, 22, 23] mimicking the constrained system environment. The later on the other hand is particularly useful in synthesis [4]. As the synthesized components are typically much compact than those with range preserved [3], it is attractive for generating the final implementation. This chapter is concerned with the latter, in particular, determinizing a relation and extracting functional implementation in the Boolean domain.

State-of-the-art methods, such as [1, 20], are based on decision diagrams. As BDDs are not memory efficient in representing large Boolean functions, these methods have intrinsic limitations and are not scalable to large problem instances. There has been growing demands for scalable determinization of large Boolean relations. Synthesis from specifications shows one such example [3]. The quest for scalable determinization methods remains.

From the scalability standpoint, we seek reasonable representation of large Boolean functions and, in particular, use *and-inverter graphs* (AIGs) (see, e.g., [18]) as the underlying data structure. Due to their simple and multi-level nature, AIGs are often much compact and are closer to final logic implementation than the two-level sum-of-products (SOP) form. Moreover they provide a convenient interface with SAT solvers in terms of conversion to CNFs and representation of interpolants [15]. Therefore, unlike previous efforts on relation solving, our objective is to convert a large relation to a set of functions with reasonable quality. Similar attempts were pursued recently in other efforts of scalable logic synthesis, e.g., [11, 12, 14, 17].

Our main exploration includes (1) exploiting *interpolation* [6] for Boolean relation determinization and function extraction, (2) studying expansion- and substitution-based quantifications with reuse, and (3) showing support minimization in the interpolation framework. A comparative empirical study is performed on various computation schemes. Experimental results suggest that interpolation is essential to scalable relation determinization and function extraction. Boolean relations with thousands of variables can be determinized effectively and the extracted functions are typically of reasonable quality compared with their respective reference models.

---

[1]In some occasions input stimuli are unimportant and their correspondences with output responses need not be preserved.

16.2 Previous Work

Brayton and Somenzi [4] were among the first to observe the utility of Boolean relations in logic minimization. Boolean relations are useful not only in characterizing circuit flexibilities [16, 21] but also in characterizing solutions to design specifications/constraints [3]. There were intensive efforts which have been proposed in recent decades on relation determinization. These efforts can be categorized into exact and approximate methods. The former approaches focus on finding the optimum functional implementation [5, 8]. The later methods adopt heuristic strategy to optimize the functions implementing the given relation [1, 7, 13, 20]. Such optimization objectives, for instance, can be in terms of two-level logic minimization under the SOP representation [5, 7, 8, 13, 20] or in term of some polynomial functions over BDD sizes [1].

An exact approach to minimizing SOP function expressions for a given Boolean relation was proposed [5]. This method is similar to the Quine–McCluskey procedure, building all prime cubes of permissible functions and solving a binate covering problem to derive minimal SOP representation. Extending [5], the work [8] represents constraints of the binate covering problem using multiple BDDs. Although these methods are able to find the exact minimal function representation, they are only applicable to small instances.

In [7], heuristic SOP minimization for a given Boolean relation is achieved using the automatic test pattern generation (ATPG) technique in testing. There are other BDD-based approaches [1, 13, 20]. In [13], the covering problem of SOP minimization was formulated using BDDs. The method [20] adopted multiple-valued decision diagrams (MDDs) to represent a multiple-valued relation and applied EXPRESSO-like procedure [19] for SOP representation. Another recursive approach was proposed in [1], where split and projection operations were applied iteratively to find an optimal BDD representation for a given Boolean relation.

All of the above prior methods focused on the optimization of functional implementation without considering the scalability issue. In contrast, our approach focuses on scalability with reasonable optimization quality.

16.3 Preliminaries

As a notational convention, substituting function g for variable v in function f is denoted as $f[v/g]$.

16.3.1 Boolean Relation

A relation $R \subseteq X \times Y$ can be equivalently represented as a characteristic function $R : X \times Y \to \mathbb{B}$ such that $(a, b) \in R$ for $a \in X, b \in Y$ if and only if $R(a, b) = 1$.

Definition 16.1 A relation $R : X \times Y \to \mathbb{B}$ is *total* (in X) if $\forall a \in X, \exists b \in Y. R(a, b) = 1$. Otherwise, R is *partial*.

Unless otherwise said we shall assume that a relation $R \subseteq X \times Y$ is total in X.

Definition 16.2 Given a partial relation $R : X \times Y \to \mathbb{B}$, an (input) assignment $a \in X$ is *undefined* if no (output) assignment $b \in Y$ makes $R(a, b) = 1$.

Figure 16.3 shows examples of total and partial relation. The relation of Fig. 16.3a is total because every input element maps to some output element(s); that of Fig. 16.3b is partial because the mapping under $(x_1, x_2) = (1, 0)$ is undefined.

Fig. 16.3 Examples of total and partial relation

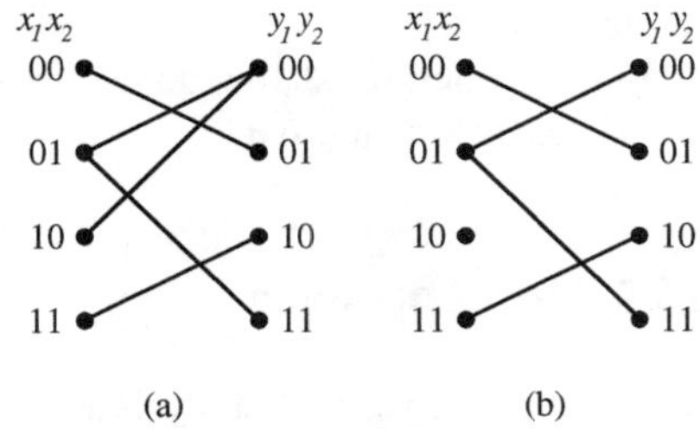

We assume that X is the *input space* $\mathbb{B}^n$ spanned by *input variables* $\mathbf{x} = (x_1, \ldots, x_n)$ and Y is the *output space* $\mathbb{B}^m$ spanned by *output variables* $\mathbf{y} = (y_1, \ldots, y_m)$.

Given a Boolean relation $R : \mathbb{B}^n \times \mathbb{B}^m \to \mathbb{B}$ with input variables $\mathbf{x} = (x_1, \ldots, x_n)$ and output variables $\mathbf{y} = (y_1, \ldots, y_m)$, we seek a *functional implementation* $\mathbf{f} = (f_1, \ldots, f_m)$ with $f_i : \mathbb{B}^n \to \mathbb{B}$ such that

$$D = \bigwedge_{i=1}^{m} (y_i \equiv f_i(\mathbf{x}))$$

is contained by R, i.e., the implication $D \Rightarrow R$ holds. Equivalently, the relation after substituting f_i for y_i

$$R[y_1/f_1, \ldots, y_m/f_m]$$

equals constant 1.

Note that the above relation D is a *deterministic* relation, i.e.,

$$\forall a \in X, \forall b, b' \in Y.((D(a, b) \wedge D(a, b')) \Rightarrow (b = b')).$$

Therefore seeking a functional implementation of a total relation can be considered as determinizing the relation. On the other hand, any deterministic total relation has a unique functional implementation.

16.3.2 Satisfiability and Interpolation

The reader is referred to prior work [11] for a brief introduction to SAT solving and circuit-to-CNF conversion, which are essential to our development. To introduce terminology and convention for later use, we restate the following theorem:

Theorem 16.1 (Craig Interpolation Theorem) *[6] Given two Boolean formulas ϕ_A and ϕ_B, with $\phi_A \wedge \phi_B$ unsatisfiable, then there exists a Boolean formula ψ_A referring only to the common variables of ϕ_A and ϕ_B such that $\phi_A \Rightarrow \psi_A$ and $\psi_A \wedge \phi_B$ is unsatisfiable.*

The Boolean formula ψ_A is referred to as the *interpolant* of ϕ_A with respect to ϕ_B. Modern SAT solvers can be extended to construct interpolants from resolution refutations [15].

In the sequel, we shall assume that Boolean relations, functions, and interpolants are represented using AIGs.

16.4 Our Approach

In this section, we first study function extraction from single-output relations and generalize the computation to multiple-output relations. In addition, we consider the special case of extracting functions from a deterministic relation. Finally, methods for function simplification are discussed.

16.4.1 Single-Output Relation

We consider first the functional implementation of a single-output relation $R(\mathbf{x}, y)$ with y the only output variable.

16.4.1.1 Total Relation

Proposition 16.1 *A relation $R(\mathbf{x}, y)$ is total if and only if the conjunction of $\neg R(\mathbf{x}, 0)$ and $\neg R(\mathbf{x}, 1)$ is unsatisfiable.*

Theorem 16.2 *Given a single-output total relation $R(\mathbf{x}, y)$, the interpolant ψ_A of the refutation of*

$$\neg R(\mathbf{x}, 0) \wedge \neg R(\mathbf{x}, 1) \tag{16.1}$$

with $\phi_A = \neg R(\mathbf{x}, 0)$ and $\phi_B = \neg R(\mathbf{x}, 1)$ corresponds to a functional implementation of R.

Proof Since R is total, Formula (16.1) is unsatisfiable by Proposition 16.1. That is, the set $\{a \in X \mid R(a, 0) = 0 \text{ and } R(a, 1) = 0\}$ is empty. Hence ϕ_A (respectively ϕ_B) characterizes the set $S_A = \{a \in X \mid R(a, 1) = 1 \text{ and } R(a, 0) = 0\}$ (respectively $S_B = \{a \in X \mid R(a, 0) = 1 \text{ and } R(a, 1) = 0\}$). As $\phi_A \Rightarrow \psi_A$ and $\psi_A \Rightarrow \neg\phi_B$, the interpolant ψ_A maps every element of S_A to 1, every element of S_B to 0, and every other element to either 0 or 1. Let D be $(y \equiv \psi_A)$. Then $D \Rightarrow R$.

Therefore interpolation can be seen as a way to exploit flexibility for function derivation without explicitly computing don't cares.

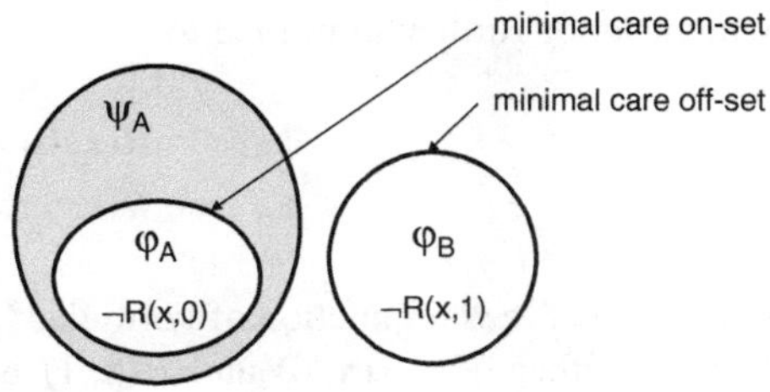

Fig. 16.4 Construct of function extraction for single-output total relation

Figure 16.4 shows the main construct of Theorem 16.2. By assuming $R(\mathbf{x}, y)$ is a total relation, $\neg R(\mathbf{x}, 0)$ and $\neg R(\mathbf{x}, 1)$ corresponds to the minimal care on-set and minimal care off-set of the function for output variable y, respectively. Hence, the interpolant ψ_A, which is implied by $\neg R(\mathbf{x}, 0)$ and implies $R(\mathbf{x}, 1)$, is a legitimate functional implementation of y.

Corollary 16.1 *Given a single-output total relation R, both $R(\mathbf{x}, 1)$ and $\neg R(\mathbf{x}, 0)$ are legitimate functional implementation of R.*

Proof Let $\phi_A = \neg R(\mathbf{x}, 0)$ and $\phi_B = \neg R(\mathbf{x}, 1)$. Because $\phi_A \Rightarrow R(\mathbf{x}, 1)$ and $R(\mathbf{x}, 1) \Rightarrow \neg \phi_B$, $R(\mathbf{x}, 1)$ is a legitimate interpolant. Similarly, $\neg R(\mathbf{x}, 0)$ is a legitimate interpolant, too.

In fact, the cofactored relations $R(\mathbf{x}, 1)$ and $\neg R(\mathbf{x}, 0)$ are the largest (weakest) and smallest (strongest) interpolants, respectively, in terms of solution spaces. Therefore to derive a functional implementation of a single-output total relation, interpolation is unnecessary. However, practical experience suggests that functional implementations obtained through interpolation are often much simpler in AIG representation.

16.4.1.2 Partial Relation

Note that Theorem 16.2 works only for *total* relations because *partial* relations make Formula (16.1) satisfiable. To handle partial relations, we treat undefined input assignments as don't care conditions (this treatment is legitimate provided that the undefined input assignments can never be activated) and define complete totalization as follows.

Definition 16.3 Given a single-output partial relation R, its *complete totalization* is the new relation

$$T(\mathbf{x}, y) = R(\mathbf{x}, y) \vee \forall y. \neg R(\mathbf{x}, y) \tag{16.2}$$

Note that $T = R$ if and only if R is total.

Accordingly Theorem 16.2 is applicable to a totalized relation T with

$$\phi_A = \neg T(\mathbf{x}, 0) \text{ and} \tag{16.3}$$

$$\phi_B = \neg T(\mathbf{x}, 1) \tag{16.4}$$

which can be further simplified to

$$\phi_A = \neg R(\mathbf{x}, 0) \wedge R(\mathbf{x}, 1) \text{ and} \tag{16.5}$$

$$\phi_B = \neg R(\mathbf{x}, 1) \wedge R(\mathbf{x}, 0) \tag{16.6}$$

Observe that the conjunction of Formulas (16.5) and (16.6) is trivially unsatisfiable. Further, either of $\neg R(\mathbf{x}, 0)$ and $R(\mathbf{x}, 1)$ is a legitimate interpolant. Therefore, as long as the undefined input assignments of R are never activated, the interpolant is a legitimate functional implementation of R. (This fact will play a role in the development of Section 16.4.2.2.)

Given a (partial or total) relation R with y being the only output variable, in the sequel we let $FI(y, R)$ denote a functional implementation of y with respect to R. Among many possibilities, $FI(y, R)$ can be derived through the interpolation of Formulas (16.5) and (16.6).

16.4.2 Multiple Output Relation

We now turn attention to the functional implementation of a multiple output relation $R(\mathbf{x}, y_1, \ldots, y_m)$ with $m > 1$. In essence, we intend to reduce the problem so as to apply the previous determinization of single-output relations.

A determinization procedure contains two phases: The first phase reduces the number of output variables; the second phase extracts functional implementation. We study two determinization procedures with different ways of reducing the number of output variables. One is through existential quantification; the other is through substitution.

16.4.2.1 Determinization via Expansion Reduction

As a notational convention, we let $R^{(i)}$ denote $\exists y_m, \ldots, y_i.R$ for $1 \leq i \leq m$. Through standard existential quantification by formula expansion, i.e., $\exists x.\varphi = \varphi[x/0] \vee \varphi[x/1]$ for some formula φ and variable x, one can reduce a multiple output relation R to a single-output relation $R^{(2)}$.

In the first phase, $R^{(i)}$ can be computed iteratively as follows:

$$\begin{aligned} R^{(m)} &= \exists y_m.R \\ &\vdots \\ R^{(i)} &= \exists y_i.R^{(i+1)} \\ &\vdots \\ R^{(2)} &= \exists y_2.R^{(3)} \end{aligned}$$

for $i = m - 1, \ldots, 2$.

In the second phase, functional implementations of all output variables can be obtained through the following iterative calculation:

$$f_1 = FI(y_1, R^{\langle 2\rangle})$$
$$\vdots$$
$$f_i = FI(y_i, R^{\langle i+1\rangle}[y_1/f_1, \ldots, y_{i-1}/f_{i-1}])$$
$$\vdots$$
$$f_m = FI(y_m, R[y_1/f_1, \ldots, y_{m-1}/f_{m-1}])$$

for $i = 2, \ldots, m-1$.

The above procedure is similar to prior work (see, e.g., [20]) with some subtle differences: First, the quantification results of the first phase are reused in the second-phase computation. It reduces the number of quantifications from $O(m^2)$ to $O(m)$. Second, interpolation is the key element in the computation and AIGs are the underlying data structure.

16.4.2.2 Determinization via Substitution Reduction

Alternatively the solution to the determinization of a single-output relation can be generalized as follows. Each time we treat all except one of the output variables as the input variables. Thereby we see a single-output relation rather than a multiple output relation. For example, let y_m be the only output variable and treat $y_1, \ldots, y_{m-1}$ be additional input variables. In the enlarged input space (spanned by $y_1, \ldots, y_{m-1}$ as well as $\mathbf{x}$), however, R may not be total even though it is total in the original input space X. Let $f'_m = FI(y_m, R)$, obtained through interpolation mentioned in Section 16.4.1.2. Note that since f'_m depends not only on $\mathbf{x}$ but also on $y_1, \ldots, y_{m-1}$, it is not readily a functional implementation of y_m.

In the first phase, the number of output variables can be iteratively reduced through the following procedure:

$$f'_m = FI(y_m, R)$$
$$R^{\langle m\rangle} = R[y_m/f'_m]$$
$$\vdots$$
$$f'_i = FI(y_i, R^{\langle i+1\rangle})$$
$$R^{\langle i\rangle} = R^{\langle i+1\rangle}[y_i/f'_i]$$
$$\vdots$$
$$f'_2 = FI(y_2, R^{\langle 3\rangle})$$
$$R^{\langle 2\rangle} = R^{\langle 3\rangle}[y_2/f'_2]$$

for $i = m-1, \ldots, 2$.

In the second phase, the functional implementations can be obtained through the following iterative calculation:

$$f_1 = FI(y_1, R^{\langle 2\rangle})$$
$$\vdots$$
$$f_i = FI(y_i, R^{\langle i+1\rangle}[y_1/f_1, \ldots, y_{i-1}/f_{i-1}])$$
$$\vdots$$
$$f_m = FI(y_m, R[y_1/f_1, \ldots, y_{m-1}/f_{m-1}])$$

for $i = 2, \ldots, m-1$.

The following fact can be shown.

Lemma 16.1 *[9] Given a relation R and* $f'_m = FI(y_m, R)$*, the equality* $R[y_m/f'_m] = \exists y_m.R$ *holds.*

It may be surprising, at first glance, that any $f'_m = FI(y_m, R)$ results in the same $R[y_m/f'_m]$. This fact is true, however, and a detailed exposition can be found in the work [9]. By induction on $i = m, \ldots, 2$ using Lemma 16.1, one can further claim that $R^{\langle i\rangle} = R^{(i)}$.

Note that the above computation implicitly relies on the don't care assumption of partial relations. This assumption is indeed legitimate because the don't cares for deriving f'_i can never be activated when substituting f'_i for y_i in $R^{\langle i+1\rangle}$.

Comparing $R^{(i)}$ of Section 16.4.2.1 and $R^{\langle i\rangle}$ of Section 16.4.2.2, one may notice that the AIG of $R^{(i)}$ is in general wider in width but shallower in depth, and, in contrast, that of $R^{\langle i\rangle}$ narrower but deeper.

As an implementation technicality, relations $R^{(i)}$ (similarly $R^{\langle i\rangle}$) can be stored in the same AIG manger. So structurally equivalent nodes are hashed together, and logic sharing is possible among relations $R^{(i)}$ (similarly $R^{\langle i\rangle}$).

16.4.3 Deterministic Relation

We consider the special case of extracting functions from a deterministic relation.

Lemma 16.2 *Given a deterministic relation* $D(\mathbf{x}, \mathbf{y})$ *total in the input space X with*

$$D = \bigwedge_{i=1}^{m} (y_i \equiv f_i), \tag{16.7}$$

let

$$\phi_A = D(\mathbf{x}, y_1, \ldots, y_{i-1}, 1, y_{i+1}, \ldots, y_m) \text{ and} \tag{16.8}$$
$$\phi_B = D(\mathbf{x}, y'_1, \ldots, y'_{i-1}, 0, y'_{i+1}, \ldots, y'_m) \tag{16.9}$$

where y and y' are independent variables. Then the interpolant of ϕ_A with respect to ϕ_B is functionally equivalent to f_i.

Proof Since D is deterministic and total in X, for every $a \in X$ there exists a unique $b \in Y$ such that $D(a, b) = 1$. It follows that the formulas

$$\exists y_1, \ldots, y_{i-1}, y_{i+1}, \ldots, y_m.D[y_i/0] \tag{16.10}$$

and

$$\exists y_1, \ldots, y_{i-1}, y_{i+1}, \ldots, y_m.D[y_i/1] \tag{16.11}$$

must induce a partition on the input space X, and thus the interpolant of ϕ_A with respect to ϕ_B must logically the equivalent to Formula (16.11), which is unique.

Back to the computation of Section 16.4.2.2, let

$$D = R \wedge \bigwedge_i (y_i \equiv f_i') \tag{16.12}$$

Since the relation D is deterministic, the computation of Lemma 16.2 can be applied to compute f_i. The strengths of this new second-phase computation are twofold: First, no substitution is needed, in contrast to the second-phase computation of Section 16.4.2.2. Hence the formula sizes of ϕ_A and ϕ_B in interpolant computation do not grow, unlike the previous second-phase computation. As interpolant sizes are more or less proportional to the formula sizes of ϕ_A and ϕ_B, this approach is particularly desirable. Second, only functions f_i', but not relations $R^{\langle i \rangle}$, are needed in the computation. Since the formula sizes of $R^{\langle i \rangle}$ are typically much larger than those of f_i', this approach saves memory by discharging $R^{\langle i \rangle}$.

16.4.4 Function Simplification

16.4.4.1 Support Minimization

The following lemma can be exploited in reducing the support variables of a functional implementation.

Lemma 16.3 *For two Boolean formulas ϕ_A and ϕ_B with an unsatisfiable conjunction, there exists an interpolant without referring to variable x_i if and only if the conjunction of $\exists x_i.\phi_A$ and ϕ_B (equivalently the conjunction of ϕ_A and $\exists x_i.\phi_B$) is unsatisfiable.*

Proof ($\Longleftarrow$) Assume the conjunction of $\exists x_i.\phi_A$ and ϕ_B (similarly ϕ_A and $\exists x_i.\phi_B$) is unsatisfiable. Since $\phi_A \Rightarrow \exists x_i.\phi_A$ (similarly $\phi_B \Rightarrow \exists x_i.\phi_B$), the conjunction of ϕ_A and ϕ_B is unsatisfiable as well. Also by the common-variable property of Theorem 16.1, the existence condition holds.

($\Longrightarrow$) Observe that $\exists x_i.\phi_A$ (respectively $\exists x_i.\phi_B$) is the tightest x_i-independent formula that is implied by ϕ_A (respectively ϕ_B). The existence of an interpolant of ϕ_A with respect to ϕ_B without referring to x_i infers the unsatisfiability of the conjunction of $\exists x_i.\phi_A$ and ϕ_B as well as that of ϕ_A and $\exists x_i.\phi_B$.

By the lemma, we can possibly knock out some variables from an interpolant.

Note that, in Lemma 16.3, it suffices to quantify x_i over ϕ_A or ϕ_B even though it is okay to quantify on both. In practice, quantification on just one formula results in smaller interpolants because the unsatisfiability is easier to be shown in SAT solving.

16.4.4.2 Determinization Scheduling

Practical experience suggests that interpolant sizes are typically proportional to the formula sizes of ϕ_A and ϕ_B. Assuming formulas ϕ'_A and ϕ'_B are logically equivalent to ϕ_A and ϕ_B, respectively, and are larger in sizes, we may expect the interpolant of ϕ'_A with respect to ϕ'_B is typically larger than that of ϕ_A with respect to ϕ_B. Therefore, it is usually beneficial to keep interpolants small during the iterative determinization process.

To keep interpolant sizes small throughout determinization, two approaches can be pursued. One is to apply logic synthesis to minimize interpolants and relations. It avoids AIG sizes being amplified greatly in the iterative computation process. The other is to choose good ordering for determinizing output variables, especially for the second-phase computation. In essence, there is a trade-off between minimizing complex functions first (to have more flexibility for minimization) and minimizing simple functions first (to have less AIG size increase).

Note that, by reusing the quantification results of the first phase, the determinization procedures of Sections 16.4.2.1 and 16.4.2.2 only require a linear number, $O(m)$, of quantifications. The determinization order of the second phase, however, is constrained by that of the first phase. On the other hand, there can be more freedom in choosing determinization order for not reusing quantification results. In this case, a quadratic number, $O(m^2)$, of quantifications may be needed.

16.5 Experimental Results

The proposed methods were implemented in the ABC package [2]; the experiments were conducted on a Linux machine with Xeon 3.4 GHz CPU and 6 GB RAM.

To prepare Boolean relations, we constructed the transition relations of circuits taken from ISCAS and ITC benchmark suites. Different amounts of don't cares were inserted to the transition relations to introduce nondeterminism. We intended to retrieve a circuit's transition functions in the following experiments.

The original circuits[2] were minimized with the ABC command dc2, and so were the AIGs produced during determinization and function extraction. The profile of the original circuits (after the removal of primary output functions and after dc2 synthesis) is shown in Table 16.1, where "(n, m)" denotes the pair of input- and output-variable sizes of the transition relation, "#n" denotes the number of AIG nodes, "#l" AIG logic levels, and "#v" the summation of support variables of all transition functions.

Table 16.1 Profile of original benchmark circuits

		Orig		
Circuit	(n, m)	#n	#l	#v
s5378	(214, 179)	624	12	1570
s9234.1	(247, 211)	1337	25	3065
s13207	(700, 669)	1979	23	3836
s15850	(611, 597)	2648	36	15788
s35932	(1763, 1728)	8820	12	7099
s38584	(1464, 1452)	9664	26	19239
b10	(28, 17)	167	11	159
b11	(38, 31)	482	21	416
b12	(126, 121)	953	16	1639
b13	(63, 53)	231	10	383

We first study the usefulness of interpolation in contrast to cofactoring, which can be considered as a way of deriving special interpolants as mentioned in Section 16.4.1.1. In the experiment, a circuit was determinized via expansion reduction, where the functional implementations extracted in the second phase were derived differently using interpolation and cofactoring to compare. Taking circuit b11 as a typical example, Fig. 16.5 contrasts the difference between the two techniques. As can be seen, by cofactoring, the function sizes grow almost exponentially during the iterative computation; by interpolation, the function sizes remain under control. In fact, derived by cofactoring, say, $R^{(i+1)}$ with $y_i = 1$, function f_i has almost the same size as $R^{(i+1)}$ unless command dc2 can effectively minimize f_i. However, dc2 is unlikely to be helpful for large f_i as justified by experiments.

Our another experiment studies the connection between interpolant sizes and formula sizes of ϕ_A and ϕ_B. We, however, omit the detailed statistics and simply conclude that, for logically equivalent sets of ϕ_A and ϕ_B formulas, their corresponding interpolant sizes tend to be linearly proportional to the sizes of these formulas.

Below we compare different determinization methods, including BDD-based computation, that via expansion reduction (Section 16.4.2.1), denoted XP, that via substitution reduction (Section 16.4.2.2), denoted ST, and that via constructing deterministic relation (Section 16.4.3), denoted SD. Dynamic variable reordering and BDD minimization are applied in BDD-based computation.

[2]Since a circuit's primary output functions are immaterial to our evaluation, we are concerned only about the sub-circuit responsible for transition functions.

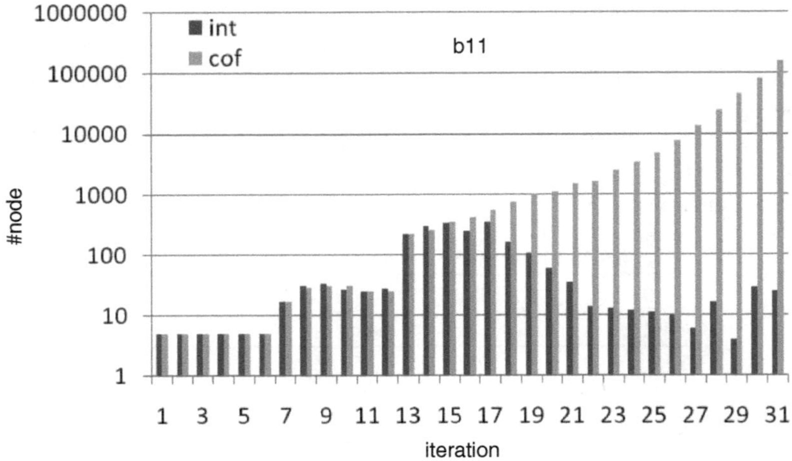

Fig. 16.5 Circuit b11 determinized by expansion reduction with function extraction by interpolation versus cofactoring in the second phase

Table 16.2 shows the results of function extraction from relations without don't care insertion. BDD-based computation is not scalable as expected. There are five circuits whose transition relations cannot be built compactly using BDDs under 500K nodes and the computation cannot finish either within 30 h CPU time or within the memory limitation. Ratio 1 and Ratio 2 are normalized with respect to the data of the original circuits of Table 16.1, whereas Ratio 3 is normalized with respect to the BDD-based derivation. Ratio 1 covers all the ten circuits, whereas Ratio 2 and Ratio 3 cover only the five circuits that BDD-based derivation can finish.

By Ratio 1, we observe that the derived functions (without further postprocessing to minimize) are about three times larger in nodes, four times larger in logic levels, and 9% smaller in support sizes. To be shown in Table 16.4, with postprocessing, the derived functions can be substantially simplified and are comparable to the original sizes. By Ratio 2, we see that even BDD-based derivation may increase circuit sizes by 70% while logic levels are reduced by 11%. By Ratio 3, we see that the results of the SAT-based methods are about 40% larger in nodes and two times larger in logic levels.

We examine closely the case of circuit s5378. Figures 16.6, 16.7, and 16.8 show the relation and function sizes at every iterations of XP, ST, and SD computations, respectively. For XP and ST, the size of function f_i and the sizes of relations in every step of the first- and second-phase computations are shown. For SD, the sizes of f_i' and f_i are plotted. Note that in these figures the first-phase (second-phase) iterations proceed forwardly (backwardly). Hence the relation sizes decrease in the first phase and those increase in the second phase, whereas the functions extracted remain small.

Table 16.3 shows the results of function extraction from relations with don't cares inserted. For a circuit with r registers, $\lceil r \cdot 10\% \rceil$ random cubes (conjunction of

Table 16.2 Function extraction from relations – without don't care insertion

Circuit	BDD				XP				ST				SD			
	#n	#l	#v	Time	#n	#l	#v	Time	#n	#l	#v	Time	#n	#l	#v	Time
s5378	783	10	1561	286.4	1412	25	1561	49.6	1328	23	1561	58.5	1625	34	1561	32.4
s9234.1	—	—	—	—	7837	59	2764	158.6	8015	45	2765	282.4	8637	45	2766	100.7
s13207	—	—	—	—	5772	140	3554	769.3	6625	223	3554	949.5	6642	109	3554	247.2
s15850	—	—	—	—	42622	188	13348	2700.0	42902	153	13318	3029.6	41014	357	13329	404.7
s35932	—	—	—	—	7280	10	6843	4178.5	7310	10	6843	3982.7	7293	12	6843	2039.2
s38584	—	—	—	—	22589	277	17678	5772.8	22691	387	17676	8481.0	17018	178	17676	2438.1
b10	200	10	152	0.1	197	8	152	0.9	231	14	152	1.7	234	14	152	1.0
b11	1301	18	394	0.9	1504	57	394	5.1	1759	55	394	14.9	1959	53	394	8.0
b12	1663	14	1574	56.7	2166	25	1574	24.0	2368	35	1575	78.8	2662	33	1576	38.6
b13	240	10	349	3.1	224	10	349	2.2	222	11	349	3.5	222	11	349	2.7
Ratio 1					3.40	4.16	0.91		3.47	4.98	0.91		3.24	4.41	0.91	
Ratio 2	1.70	0.89	0.97		2.24	1.79	0.97		2.40	1.97	0.97		2.53	1.90	0.97	
Ratio 3	1.00	1.00	1.00		1.31	2.02	1.00		1.41	2.23	1.00		1.48	2.15	1.00	

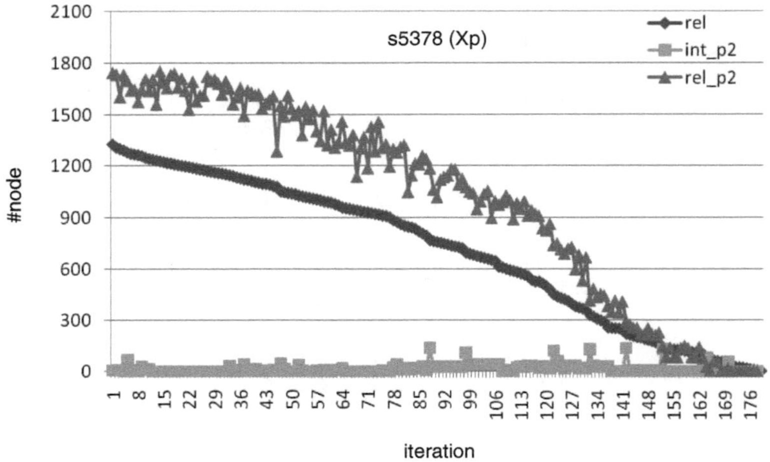

Fig. 16.6 Relation and function sizes of circuit s5378 under XP computation

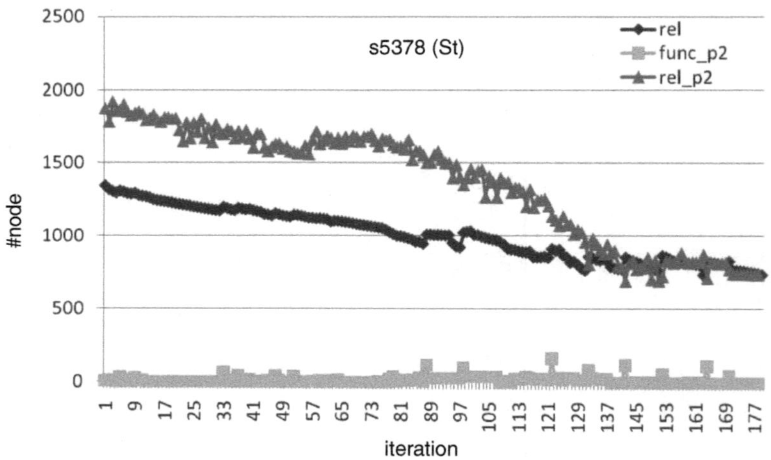

Fig. 16.7 Relation and function sizes of circuit s5378 under ST computation

literals of input and state variables) are created. Each cube represents some don't cares for a randomly selected set of functions. Presumably the more the don't cares are inserted, the simpler the transition functions are extracted. In practice, however, such simplification[3] is not guaranteed (even in BDD-based computation). The reasons can be twofold: First, the simplification achieves only local optimums. Second,

---

[3] Notice that, unlike BDD-based computation, our methods do not explicitly perform don't care based minimization on the extracted transition functions. The don't care choices are made implicitly by SAT solving for interpolation.

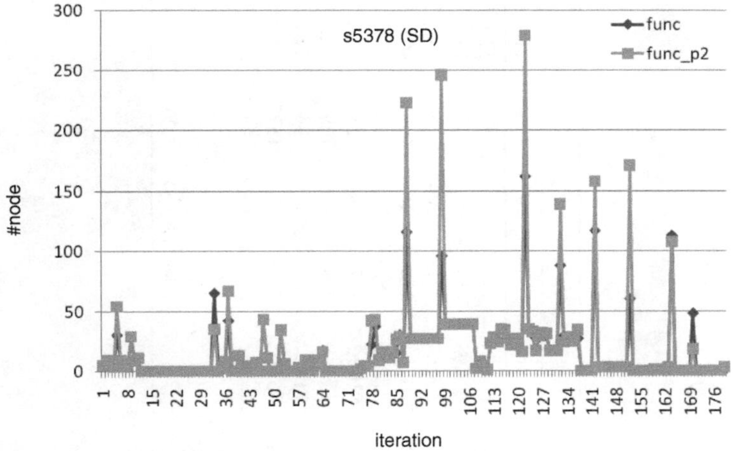

Fig. 16.8 Function sizes of circuit s5378 under SD computation

relations with don't cares inserted become more sophisticated and affect interpolant derivation. Nevertheless the results of Tables 16.2 and 16.3 are comparable.

Taking circuit b07 for case study, Fig. 16.9 shows the results of function extraction by XP for relations under different amounts of don't care insertion. The original transition relation is augmented with a certain number of don't care cubes. The AIG sizes of augmented relations and their extracted functions are shown with the light and dark bars, respectively. The light bars increase monotonically with respect to the insertion of don't cares, while the dark bars oscillate. To screen out the effects of relation increase on interpolant derivation, the normalized ratios are plotted as the line chart whose indices are on the right-hand side. The derived functions indeed tend to get simplified with the increasing amount of don't cares.

The above experiments of XP, ST, and SD used only light synthesis operations in minimizing the extracted functions. Nonetheless, it is possible to greatly simplify these functions with heavier synthesis operations. To justify such possibilities, we applied ABC command `collapse` once followed by `dc2` twice as postprocessing. Table 16.4 shows the statistics of extracted functions by XP for relations without don't care insertion. (Similar results were observed for ST and SD, and for cases with don't care insertion.) The postprocessing results of the original functions and extracted functions are shown. This postprocessing time is listed in the last column. Operation `collapse` failed on circuit `s15850`, and the two ratios shown excludes the data of `s15850`. As can be seen, the postprocessing makes the extracted functions comparable to the original ones. Since the postprocessing time is short, our method combined with some powerful synthesis operations can effectively extract simple functions from large relations, where pure BDD-based computation fails. Our method can be used as a way of bypassing the BDD memory explosion problem.

Table 16.3 Function extraction from relations – with don't care insertion

	BDD				XP				ST				SD			
Circuit	#n	#l	#v	Time	#n	#l	#v	Time	#n	#l	#v	Time	#n	#l	#v	Time
s5378	769	11	1561	200.2	1332	25	1561	49.05	1196	27	1561	60.99	1919	42	1561	32.74
s9234.1	—	—	—	—	7696	55	2765	166.74	8739	64	2764	325.98	11613	99	2927	120.37
s13207	—	—	—	—	5818	202	3554	897.86	6882	228	3554	1062.15	6218	204	3554	287.47
s15850	—	—	—	—	40078	136	13309	2596.94	42097	164	13318	3012.36	41240	212	14276	467.95
s35932	—	—	—	—	7360	25	6843	4811.1	7300	10	6843	7168.53	8823	19	8756	2775.32
s38584	—	—	—	—	23726	331	17676	5476.67	21595	285	17676	8160.28	17708	281	18556	2591.71
b10	199	9	152	0.1	193	8	152	0.99	217	9	152	1.62	239	8	168	1.13
b11	1221	20	394	0.9	1562	52	394	5.5	1638	46	394	15.19	1896	54	394	8.54
b12	1619	15	1574	452.5	2261	23	1574	26.98	2081	24	1574	86.96	2458	25	1575	44.19
b13	243	11	349	1.6	229	12	349	2.45	236	10	349	4.2	232	10	349	2.9
Ratio 1					3.35	4.53	0.91		3.42	4.52	0.91		3.43	4.97	0.98	
Ratio 2	1.65	0.94	0.97		2.27	1.71	0.97		2.18	1.66	0.97		2.74	1.99	0.97	
Ratio 3	1.00	1.00	1.00		1.38	1.82	1.00		1.33	1.76	1.00		1.66	2.11	1.00	

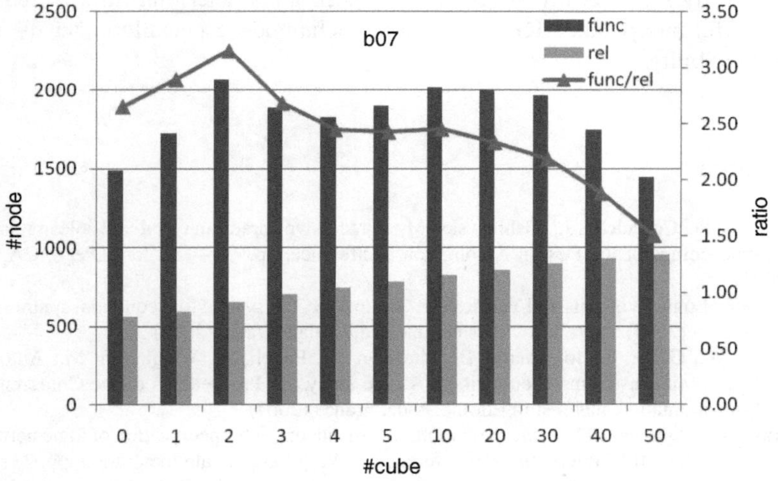

Fig. 16.9 Connection between relations and extracted functions of circuit b07 under different amounts of don't care insertion

Table 16.4 Function extraction from relations – effect of collapse minimization

	Orig						XP						
Circuit	#n	#l	#v	#n_c	#l_c	#v_c	#n	#l	#v	#n_c	#l_c	#v_c	Time_c
s5378	624	12	1570	772	11	1561	1412	25	1561	760	11	1561	0.4
s9234.1	1337	25	3065	2791	25	2764	7837	59	2764	2751	22	2764	6.4
s13207	1979	23	3836	2700	20	3554	5772	140	3554	2709	20	3554	5.2
s15850*	2648	36	15788	—	—	—	42622	188	13348	—	—	—	—
s35932	8820	12	7099	7825	9	6843	7280	10	6843	7857	9	6843	39.9
s38584	9664	26	19239	12071	23	17676	22589	277	17678	12132	21	17676	36.5
b10	167	11	159	195	10	152	197	8	152	195	10	152	0.0
b11	482	21	416	1187	21	394	1504	57	394	1226	21	394	0.2
b12	953	16	1639	1556	17	1574	2166	25	1574	1645	16	1574	0.4
b13	231	10	383	237	9	349	224	10	349	237	9	349	0.1
Ratio 1	1.00	1.00	1.00	1.21	0.93	0.93	2.02	3.92	0.93	1.22	0.89	0.93	
Ratio 2				1.00	1.00	1.00				1.01	0.96	1.00	

16.6 Chapter Summary

In this chapter, we have shown that Boolean relations with thousands of variables can be determinized inexpensively using interpolation. The extracted functions from a relation are of reasonable sizes. With such extended capacity, we would anticipate real-world applications, which might in turn enable constraint-based synthesis and verification, synthesis from specifications, and other areas that require solving large Boolean relations.

As we just presented a first step, there remain some obstacles to overcome. In particular, the unpredictability of interpolation prevents relation determinization from being robustly scalable. Moreover, we may need good determinization scheduling and powerful interpolant/AIG minimization techniques, especially under the presence of flexibility.

References

1. Baneres, D., Cortadella, J., Kishinevsky, M.: A recursive paradigm to solve Boolean relations. In: Proceedings of the Design Automation Conference, pp. 416–421. San Diego, CA, USA (2004)
2. Berkeley Logic Synthesis and Verification Group: ABC: A system for sequential synthesis and verification (2005). http://www.eecs.berkeley.edu/~alanmi/abc/ (2009)
3. Bloem, R., Galler, S., Jobstmann, B., Piterman, N., Pnueli, A., Weiglhofer, M.: Automatic hardware synthesis from specifications: A case study. In: Proceedings of the Conference on Design Automation and Test in Europe. Nice, France (2007)
4. Brayton, R., Somenzi, F.: Boolean relations and the incomplete specification of logic networks. In: Proceedings IFIP International Conference on Very Large Scale Integration, pp. 231–240. Munich, Germany (1989)
5. Brayton, R., Somenzi, F.: An exact minimizer for Boolean relations. In: Proceedings of the International Conference on Computer-Aided Design, pp. 316–319. San Jose, CA, USA (1989)
6. Craig, W.: Linear reasoning: A new form of the Herbrand-Gentzen theorem. Journal of Symbolic Logic **22**(3), 250–268 (1957)
7. Ghosh, A., Devadas, S., Newton, A.: Heuristic minimization of Boolean relations using testing techniques. In: Proceedings of the International Conference on Computer Design. Cambridge, MA, USA (1990)
8. Jeong, S., Somenzi, F.: A new algorithm for the binate covering problem and its application to the minimization of Boolean relations. In: Proceedings of the International Conference on Computer-Aided Design, pp. 417–420. San Jose, CA, USA (1992)
9. Jiang, J.H.R.: Quantifier elimination via functional composition. In: Proceedings of the International Conference on Computer Aided Verification, pp. 383–397. Grenoble, France (2009)
10. Kukula, J., Shiple, T.: Building circuits from relations. In: Proceedings of the International Conference on Computer Aided Verification, pp. 113–123. Chicago, IL, USA (2000)
11. Lee, C.C., Jiang, J.H.R., Huang, C.Y., Mishchenko, A.: Scalable exploration of functional dependency by interpolation and incremental SAT solving. In: Proceedings of the International Conference on Computer-Aided Design, pp. 227–233. San Jose, CA, USA (2007)
12. Lee, R.R., Jiang, J.H.R., Hung, W.L.: Bi-decomposing large Boolean functions via interpolation and satisfiability solving. In: Proceedings of the Design Automation Conference, pp. 636–641. Anaheim, CA, USA (2008)
13. Lin, B., Somenzi, F.: Minimization of symbolic relations. In: Proceedings of the International Conference on Computer-Aided Design, pp. 88–91. San Jose, CA, USA (1990)
14. Lin, H.P., Jiang, J.H.R., Lee, R.R.: To SAT or not to SAT: Ashenhurst decomposition in a large scale. In: Proceedings of the International Conference on Computer-Aided Design, pp. 32–37. San Jose, CA, USA (2008)
15. McMillan, K.L.: Interpolation and SAT-based model checking. In: Proceedings of the International Conference on Computer Aided Verification, pp. 1–13. Boulder, CO, USA (2003)
16. Mishchenko, A., Brayton, R.K.: Simplification of non-deterministic multi-valued networks. In: Proceedings of the International Conference on Computer-Aided Design, pp. 557–562. San Jose, CA, USA (2002)
17. Mishchenko, A., Brayton, R.K., Jiang, J.H.R., Jang, S.: Scalable don't care based logic optimization and resynthesis. In: Proceedings of the International Symposium on Field-Programmable Gate Arrays, pp. 151–160. Monterey, CA, USA (2009)

18. Mishchenko, A., Chatterjee, S., Jiang, J.H.R., Brayton, R.K.: FRAIGs: A unifying representation for logic synthesis and verification. Technical report, ERL Technical Report, UC Berkeley (2005)
19. Rudell, R., Sangiovanni-Vincentelli, A.: Multiple-valued minimization for PLA optimization. IEEE Transactions on Computer-Aided Design of Integrated Circuits and Systems **CAD-6**(5), 727–750 (1987)
20. Watanabe, Y., Brayton, R.: Heuristic minimization of multi-valued relations. IEEE Transactions on Computer-Aided Design of Integrated Circuits and Systems **12**(10), 1458–1472 (1993)
21. Wurth, B., Wehn, N.: Efficient calculation of Boolean relations for multi-level logic optimization. In: Proceedings of the European Design and Test Conference, pp. 630–634. Paris, France (1994)
22. Yuan, J., Albin, K., Aziz, A., Pixley, C.: Constraint synthesis for environment modeling in functional verification. In: Proceedings of the Design Automation Conference, pp. 296–299. Anaheim, CA, USA (2003)
23. Yuan, J., Shultz, K., Pixley, C., Miller, H., Aziz, A.: Modeling design constraints and biasing in simulation using BDDs. In: Proceedings of the International. Conference on Computer-Aided Design, pp. 584–589. San Jose, CA, USA (1999)

Chapter 17
A Robust Window-Based Multi-node Minimization Technique Using Boolean Relations

Jeff L. Cobb, Kanupriya Gulati, and Sunil P. Khatri

Abstract In this chapter, a scalable dual-node technology-independent logic optimization technique is presented. This technique scales well and can minimize both small designs and large designs typical of industrial circuits. It is experimentally demonstrated that this technique produces minimized technology-independent networks that are on average 12% smaller than networks produced by single-node minimization techniques.

17.1 Introduction

The optimization of industrial multi-level Boolean networks is traditionally performed using algebraic techniques. The main reason for this is that Boolean techniques such as don't care-based optimization, though more powerful, do not scale well with design size. Don't cares are calculated for a single node, and they specify the flexibility for implementing the node function. These don't cares (for a node) are computed using a combination of Satisfiability Don't Cares (SDCs), Observability Don't Cares (ODCs), or External Don't Cares (XDCs). These are described further in [12].

ODCs [13, 20] of a node are a powerful representation of the node's flexibility. However, the minimization of a node with respect to its ODCs can potentially change the ODCs at other nodes in the circuit, resulting in a need to recompute ODCs for all circuit nodes. A subset of ODCs, termed as Compatible Observability Don't Cares (CODCs) [20], was formulated to remove this limitation. By definition, if a node n is minimized with respect to its CODCs, then the CODCs of all other circuit nodes are still valid (and therefore do not need to be recomputed). However, in the CODC computation, the order of selecting nodes during the CODC

K. Gulati (✉)
Intel Corporation, Hillsboro, OR, USA
e-mail: kanupriya.gulati@intel.com

This work is based on an earlier work: Robust window-based multi-node technology-independent logic minimization, in Proceedings of the 19th ACM Great Lakes Symposium on VLSI, ISBN:978-1-60558-522-2 (2009) © ACM, 2009. DOI= http://doi.acm.org/10.1145/1531542.1531623

S.P. Khatri, K. Gulati (eds.), *Advanced Techniques in Logic Synthesis, Optimizations and Applications*, DOI 10.1007/978-1-4419-7518-8_17,

computation becomes important. The maximum flexibility that can be obtained at the fanin node i of a node n is a function of the CODCs of the fanins computed prior to i. In both the ODC and CODC approaches, network optimization is performed on one node at a time.

As significant improvement (in terms of optimization power) over don't care-based techniques can be obtained by considering *multiple nodes at once*, the formulation of such an optimization results in a Boolean relation [6], which implicitly represents the flexibility available in optimizing the nodes *simultaneously*. The flexibility inherent in multi-node optimization cannot be expressed using functions. Table 17.1 represents a Boolean relation, which, for a single input vector {10}, can express more than one *allowed* output vector, {00,11}. On the other hand, no Boolean function can represent the fact that both vectors {00,11} are allowed at the outputs, for the output {10}.

Table 17.1 Example of a Boolean relation

Inputs	Outputs
00	00
01	01
10	{00,11}
11	10

The superiority of a multi-node optimization approach (using Boolean relations) over don't cares has been pointed out in [5, 27]. The reason for this superior optimization flexibility is that in the computation of a node's don't cares, the functions of all the other nodes are *not* allowed to change. This restriction does not apply to the multi-node optimization approach (using Boolean relations) since they allow the *simultaneous* modification of all nodes being targeted. However, this superior optimization flexibility has a price. The multi-node optimization approach requires that a Boolean relation be solved, which is typically a highly time and memory-intensive operation. As a result, not much attention has been devoted to these approaches, although there have been theoretical works which have suggested the superiority of this technique over don't care-based approaches [27]. However, there has been no robust, scalable approach which demonstrates the applicability of multi-node optimization techniques to large designs.

The power of a multi-node optimization approach can be illustrated by way of a small example [10]. Consider the network η shown in Fig. 17.1a, where node V_1's output f implements the Boolean function $x \cdot y$ and node V_2's output g implements $x + y$. Given a network η with primary outputs Z, the ODC of a node y is given by

$$ODC(y) = \prod_{z_i \in PO(\eta)} \overline{\left(\frac{\partial z_i}{\partial y}\right)}$$

where

$$\frac{\partial z_i}{\partial y} = z_i|_y \oplus z_i|_{\overline{y}}$$

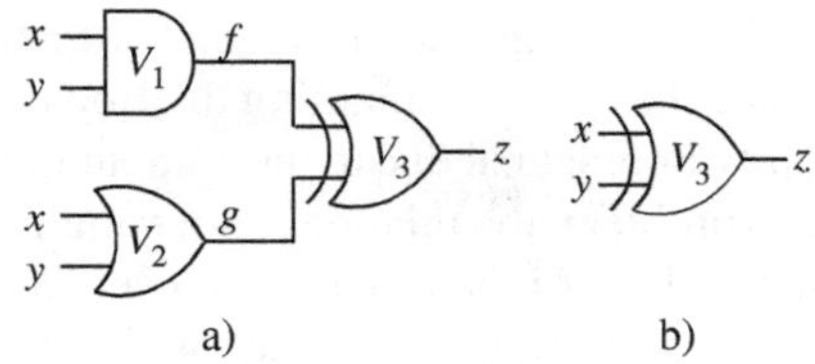

Fig. 17.1 Network η before and after optimization

Using this equation for the network of Fig. 17.1, the ODCs are computed to be $ODC(V_1) = ODC(V_2) =$ NULL. As a result, no optimization is possible using ODCs. However, one can observe that z is equivalent to $x \oplus y$ as shown in Fig. 17.1b. This optimization can *only* be obtained when V_1 and V_2 are optimized simultaneously. The Boolean relation resulting from such an optimization can express this flexibility. After minimizing this Boolean relation, nodes V_1 and V_2 can be deleted from the network without compromising the network's functionality.

This chapter is organized as follows. The problem addressed is formally defined in Section 17.2 and relevant previous work is described in Section 17.3. Section 17.4 provides the necessary background knowledge and terminology used in this chapter, as well as a description of the data structures used. In Section 17.5, the algorithms used to select node pairs and to compute the Boolean relation expressing the flexibility of a two-node optimization are given. The windowing method is also described in this section, as well as the early quantification technique is employed when computing the relation.

Section 17.6 presents the experimental results for the algorithms described in Section 17.5. Section 17.6.2 shows the approach used to choose the parameters that control the node selection algorithm. Section 17.6.3 reports the results obtained when the proposed method is compared with the *mfsw* approach of [17]. Section 17.6.4 presents results of many variations on the original approach, as well as a timing and quantification analysis.

Section 17.7 summarizes the chapter and discusses the implications of the experimental results. Ideas for future improvements are also presented in this section.

17.2 Problem Definition

This chapter addresses the problem of creating a scalable dual-node optimization technique which can handle the large designs typical of industrial circuits. The approach presented uses Reduced Ordered Binary Decision Diagrams (ROBDDs) [7, 24] to represent the Boolean relations expressing the flexibility. The terms ROBDD and BDD are used interchangeably in this chapter. ROBDDs provide an efficient mechanism to compute and store the Boolean relation. Many of the steps that occur frequently in the dual-node optimization technique, including quantification and complementation, can be performed extremely efficiently using this data structure.

Large designs can have tens of thousands of nodes in the network, which would make the task of computing the Boolean relation that represents the dual-node flexibility impractical due to the computational time and/or memory required. To address this problem, the relation is not built in terms of the primary inputs and outputs of the network, but it is instead built using a subnetwork η' rooted around the nodes being targeted. Building the relations in terms of the primary inputs and outputs would be applicable to small networks. However, the ROBDDs representing the relation would be intractable for larger networks. Working with the subnetwork allows the resulting Boolean relation to be *significantly* smaller, which enables the approach to work on large networks.

Another feature of this approach, which allows it to scale elegantly, is that it uses *aggressive early quantification* [8] while computing the Boolean relation. During the computation, the size of the ROBDDs can blow up rapidly if the relevant intermediate variables are not quantified out. Performing operations on the large ROBDDs can be very expensive in time as well as memory, especially on large networks. To reduce the size of the ROBDDs, intermediate variables are quantified out during each step of the computation when possible.

Additionally, since the work presented in this optimizes two nodes at a time, the node pair must be carefully selected. Optimizing all node pairs in a network would result in a quadratic cost. By choosing only those node pairs which have a high likelihood of minimizing the network, the algorithm remains efficient for large designs.

Finally, the dual-node optimization approach results in a Boolean relation that encodes the flexibility in implementing the targeted nodes. To re-implement the targeted functions, this relation needs to be minimized. The technique used to do this comes from [1].

17.3 Previous Work

Some of the previous research efforts which are relevant to the technique and objective of this approach are discussed next. In [16], the authors describe a method to compute don't cares using overlapping subnetworks, computed using a varying window size. Their method does not optimize wires, but only the gates in a design, in contrast to the approach described in this chapter (which frequently removes wires in a circuit). Further, the technique of [16] uses [15] to optimize a single subnetwork. In [15], optimization is done by manipulating a cover of the subnetwork explicitly. The authors indicate that this requires large amounts of runtime for small networks. As a consequence, the technique of [16], in many examples, requires run-times which are dramatically larger than MIS [3].

The approach of [11] partitions the circuit into subnetworks, each of which is flattened and optimized using ESPRESSO [4]. The technique presented in this chapter uses a similar approach of circuit partitioning but with a relation-based optimization method in place of ESPRESSO and achieves a significantly lower literal count.

In [2], the CODC computation of [19] was shown to be dependent on the current implementation of a node, and an implementation-independent computation was proposed. In [18], the authors perform CODC computation on overlapping subnetworks and demonstrate a faster technique compared to the full CODC computation. They report achieving a good literal count reduction (within 10% of the *full_simplify* (FS) command of SIS [22]) with a faster runtime (25x faster than FS). The method presented in this chapter improves on these results due to the additional flexibility encoded in the dual-node optimization technique (using Boolean relations).

In [17], the authors present a Boolean Satisfiability (SAT)-based methodology for computing the ODC and SDC, termed as complete DC (CDC), for every node in a network. They also propose a windowing scheme to maintain robustness. This approach provides the best results in terms of both optimization ability and runtime among all the previous single-node approaches mentioned here.

While [17] explores the flexibility of exactly one node at a time, a much greater flexibility can be availed by optimizing multiple nodes of a network simultaneously. This is a relatively unexplored aspect of multi-level optimization. There are research efforts which recognize the power of such a technique [5, 23, 27], but none of these work on even medium-sized circuits. The survey described in [5] only points out the advantage of multi-node minimization over don't cares. The approach in [23] describes a BDD-based computation of SPFDs [28], which can encode the flexibility of more than one node, but it is limited to small circuits and shown not to be scalable to large designs.

In [27], an approach for computing the Boolean relation of a single subnetwork of the original network is described. However, no approach or intuition for selecting the subnetwork is discussed. The approach in this chapter, in contrast, uses an efficient method to find pairs of nodes to optimize together. This method effectively filters out pairs of nodes for which the expected flexibility is low. Also, the results reported in [27] are for very small circuits and incur extremely high runtimes. The implementation in this chapter is powerful and robust, resulting in the ability to optimize large networks extremely fast, with a high quality of results. Further, [27] does not use a relation minimizer, but instead it calls ESPRESSO in order to minimize the Boolean relation that represents the optimization flexibility. The authors do acknowledge this as a possible limitation in their paper. The work in this chapter uses BREL [1] to minimize the Boolean relation which is constructed for each pair of nodes being optimized simultaneously.

There are some earlier research efforts in the context of multi-node optimization (using Boolean relations to express the multi-node optimization flexibility), but the approach in this chapter is very different. A technique which calculates this Boolean relation in terms of primary inputs is presented in [10]. The work in this chapter computes this Boolean relation in terms of the 'primary input' variables *of the extracted subnetwork*, allowing the technique to handle large designs.

A technique to compute the maximal Boolean relation that represents the optimization flexibility for the nodes in an arbitrary subnetwork is presented in [9], which was improved by [21] to additionally compute approximate Boolean relations. However, they do not support their work with experimental results.

Techniques for minimizing a Boolean relation are reported in [1, 25, 26]. In [26] the authors represent a Boolean function as a multi-valued decision diagram and propose a heuristic to minimize it. The authors of [25] formulate the problem of minimizing a Boolean relation as a binate-covering problem. The more recent approach used in BREL [1] follows a recursive branch-and-bound heuristic for minimizing a given Boolean relation. This approach demonstrates better results and runtimes as compared to those reported in [25, 26]. Therefore the work in this chapter uses BREL [1] for minimizing the Boolean relation that is computed. The details of the BREL algorithm are described in the next section.

17.4 Preliminaries and Definitions

The goal of the approach presented is to reduce the size and complexity of a Boolean network at the technology-independent level. A Boolean network is defined as the following:

Definition 17.1 A Boolean network η is a directed acyclic graph (DAG) $G = (V, E)$ in which every node has a Boolean function f_i associated with it. Also, f_i has a corresponding Boolean variable y_i associated with it, such that $y_i \equiv f_i$.

There is a directed edge $e_{ij} \in E$ from y_i to y_j if f_j explicitly depends on y_i.

A node y_i is a *fanin* of a node y_j iff there exists a directed edge $e_{ij} \in E$. Node y_i is a *fanout* of y_j iff there exists a directed edge $e_{ji} \in E$. $FI(y)$ and $FO(y)$ represent the set of fanins and the set of fanouts of y, respectively. $FI(y)$ and $FO(y)$ are equivalently referred to as *immediate fanins* and *immediate fanouts*, respectively.

A node y_i is in the *transitive fanin* of a node y_j if there is a directed path from y_i to y_j. Node y_i is in the *transitive fanout* of node y_j if there is a directed path from y_j to y_i. The transitive fanin of a node y_i up to a k levels, $TFI(y_i, k)$, is the set of nodes $\{y_j\}$ such that there is a directed path of length less than or equal to k, between y_j and y_i. Similarly, the transitive fanout of a node $TFO(y_i, k)$ is the set of nodes $\{y_j\}$ such that there is a directed path of length less than or equal to k, between y_i and y_j.

The *transitive fanin frontier* of a node y_i at k levels, $TFI_{frontier}(y_i, k)$, is the set of nodes $\{y_j\}$ such that there is a directed path of length exactly equal to k, between y_j and y_i. The *transitive fanout frontier* of a node $TFO_{frontier}$ is the set of nodes $\{y_j\}$ such that there is a directed path of length exactly equal to k, between y_i and y_j.

These definitions are illustrated in Fig. 17.2. The gray nodes are the immediate fanins and fanouts of the node y. The white nodes represent nodes in the $TFI_{frontier}$ and $TFO_{frontier}$ of node y. The nodes of $TFI(y, 2)$ and $TFO(y, 2)$ are also shown, as well as the nodes in $TFI(y)$ and $TFO(y)$. These classifications are used extensively in Section 17.5 of this chapter.

Definition 17.2 The **consensus operator** or **universal quantification** of a function f with respect to a variable x_i is

$$\forall_{x_i} f = f_{x_i} \cdot f_{\overline{x_i}}$$

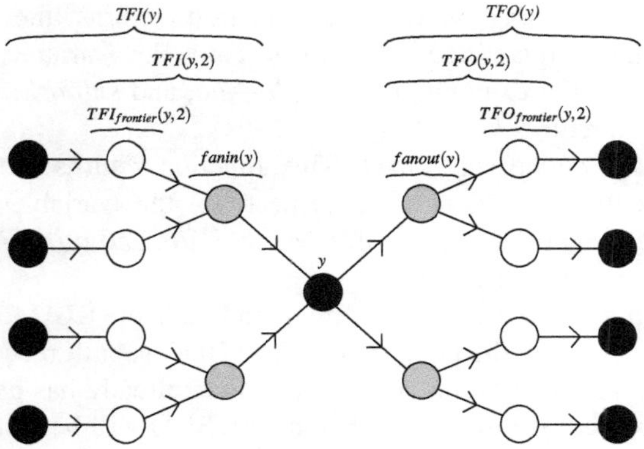

Fig. 17.2 Terminology for nodes in a boolean network

Definition 17.3 The **smoothing operator** or **existential quantification** of a function f with respect to a variable x_i is

$$\exists_{x_i} f = f_{x_i} + f_{\overline{x_i}}$$

A Boolean relation is used to represent the flexibility available in optimizing multiple nodes simultaneously. Related definitions are given next.

Definition 17.4 A **Boolean relation** $\mathcal{R}$ is a one-to-many multi-output Boolean mapping, $\mathcal{R} : B^n \rightarrow B^m$.

An output vector $y^l \in B^m$ is *allowed for* an input vector $x^k \in B^n$ iff $(x^k, y^l) \in \mathcal{R}$.

Definition 17.5 A **multi-output Boolean function** f is a mapping compatible with $\mathcal{R}$ if $f(x) \in \mathcal{R}, \forall_x \in B^n$. This is denoted by $f \prec \mathcal{R}$.

A Boolean relation $\mathcal{R}$ can be represented by its *characteristic function* $\Phi : B^n \times B^m \rightarrow B$ such that $\Phi(x^k, y^l) = 1$ iff $(x^k, y^l) \in \mathcal{R}$.

For a network η which implements the multi-output Boolean function $\boldsymbol{z} = f(\boldsymbol{x})$, the characteristic function is denoted by Φ^η, where

$$\Phi^\eta = \prod_{k=1}^{m} (z_i \overline{\oplus} f_{z_i}(\boldsymbol{x}))$$

where m is the number of outputs of η and $f_{z_i}(\boldsymbol{x})$ is the function of z_i in terms of $\boldsymbol{x}$.

Note that in the sequel a set of variables $\{a\}$ is represented as $\boldsymbol{a}$.

The following BDD operations are used in the work presented:

- *bdd_smooth(f, smoothing_vars)*: This function returns the BDD formula of f existentially quantified with respect to the variables in the array $smoothing\_vars$. For example, if $f = ab + \overline{a}c$, and $smoothing\_vars = [a]$, then the function returns $b + c$.
- *bdd_consensus(f, quantifying_vars)*: This function returns the BDD formula of f universally quantified with respect to the variables in the array $quantifying\_vars$. For example, if $f = ab + \overline{a}c$, and $quantifying\_vars = [a]$, then the function returns bc.
- *bdd_node_to_bdd(node,leaves)*: This function builds the BDD for $node$ in terms of the variables given in the table $leaves$. The BDD is built recursively from the BDD's of its immediate fanins. If a visited node already has a BDD, then this BDD will be reused; if it does not, then a new BDD will be constructed for the node.

17.4.1 BREL Boolean Relation Minimizer

Finding a set of multi-output functions that are compatible with a Boolean relation is a trivial task. For the relation in Table 17.1, arbitrarily choosing either $\{00\}$ or $\{11\}$ as outputs for the input vector $\{10\}$ would yield a compatible solution. However, this solution may not be minimal in terms of the literal count of the resulting functions. BREL uses a recursive algorithm to explore a wide range of solutions and chooses the best result based on a given cost function.

First, a quick initial solution is found. This is done by projecting the relation onto each output, and then minimizing the resulting incompletely specified function using the maximum flexibility provided by the relation. The constraints of the solution are passed on to the rest of the outputs to ensure that the final solution is compatible with the relation. Once this is done for all outputs, an initial cost for the solution is determined. However, this initial solution depends on the order that the outputs are minimized. In addition, it favors outputs minimized first, since they have the most flexibility, while the last outputs inherit little flexibility.

Next, a recursive algorithm is used to find an optimal solution. Each output is first minimized independently. If the resulting solution is compatible with the relation and has the lowest cost explored so far, then the solution is returned to the calling function. If the resulting solution is incompatible with the relation, then the relation R is split into two relations R_1 and R_2, which are compatible with R. This is done by selecting an incompatible input vertex x and an output y_i and defining R_1 and R_2 as

$$R_1 = R \cdot (\overline{x} + \overline{y_i})$$
$$R_2 = R \cdot (\overline{x} + y_i)$$

The algorithm is recursively called on R_1 and R_2, until either the cost is greater than the best cost previously explored or if the terminal case is reached where R is a function. In the end, the output of BREL is the minimum cost set of functions that are compatible with R.

17.5 Approach

In general, the exact computation of the Boolean relation expressing the optimization flexibility of multiple nodes is extremely memory intensive, even for small networks. This is one of the reasons why past research efforts in this area have been mostly theoretical in nature. The approach for simultaneous multi-node minimization of a multi-level network presented in this chapter has several salient features.

- The flexibility is computed for simultaneously optimizing a pair of nodes of the network at a time, using an ROBDD-based approach.
- Memory explosion is avoided by a *windowing* technique which first creates a subnetwork around the two nodes being optimized. This subnetwork has a user-controllable topological depth. The Boolean relation representing the flexibility for simultaneously optimizing the two nodes is built in terms of the primary inputs of the *subnetwork*. This keeps the sizes of the ROBDDs under control, and effectively allows the approach to scale robustly for large networks, with very good result quality.
- During the computation of the ROBDD of the characteristic function of the Boolean relation, memory utilization is aggressively controlled by performing careful early quantification.
- Further, instead of running this algorithm on all pairs of nodes, it is run on only those node pairs that are likely to yield good optimization opportunities. This is done *without* enumerating all node pairs.

Algorithm 1 describes the flow of the multi-level optimization methodology. The input is a Boolean network η, and the output is an optimized Boolean network η', which is functionally equivalent to η.

The algorithm begins by efficiently selecting pairs of nodes to optimize from the original multilevel network η. Given a pair of nodes (n_i, n_j) to optimize simultaneously, the algorithm then finds a subnetwork $\eta_{i,j}$ which is rooted around these nodes. The Boolean relation $\mathcal{R}$ representing the simultaneous flexibility of these two nodes is computed in terms of the primary inputs of the subnetwork $\eta_{i,j}$. Finally, the Boolean relation $\mathcal{R}$ is minimized using a relation minimizer (BREL [1] in the approach presented). The relation minimizer returns a multi-output function (in particular a 2 output function) f, such that f is compatible with $\mathcal{R}$ ($f \prec \mathcal{R}$). The optimized pair of nodes are then grafted back into η. At the end of the for loop, a *minimized* multi-level network η' is obtained.

The details of the steps of the algorithm are described next.

Algorithm 9 Boolean relation-based multi-node optimization

$L = select\_nodes(thresh, k_1, k_2, \alpha)$
for all $(n_i, n_j) \in L$ **do**
  $\eta_{i,j} = extract\_subnetwork(n_i, n_j, k_1)$
  $\mathcal{R}^Y(\boldsymbol{s}, \boldsymbol{y}) = build\_relation\_bdd(\eta_{i,j}, X, Z, S, Y)$
  $(n_i^{'}, n_j^{'}) = BREL(\mathcal{R}^Y(\boldsymbol{s}, \boldsymbol{y}))$
  Graft $(n_i^{'}, n_j^{'})$ *in* η
  Delete n_i *and* n_j from η
end for
Return $\eta^{'} = network\_sweep(\eta)$

17.5.1 Algorithm Details

17.5.1.1 Selecting Node Pairs

When selecting node pairs, it is important to find nodes that share common fanins and fanouts when the subnetwork is created. Not only will this make the subnetwork smaller, but it will also increase the likelihood that more flexibility will be found from the resulting relation.

To generate a list of all node pairs to minimize, $select\_nodes(thresh, k_1, k_2, \alpha)$ is called. This algorithm is shown in pseudocode in Algorithm 2 and graphically in Fig. 17.3.

Algorithm 10 Pseudocode of node selection algorithm

for all $(n_i) \in \eta$ **do**
  $\boldsymbol{m} \leftarrow TFI_{frontier}(n_i, k_1)$
  for all $m_l \in \boldsymbol{m}$ **do**
    $\boldsymbol{n} \leftarrow TFO(m_l, k_2)$
    for all $n_j \in \boldsymbol{n}$ **do**
      if $n_j \notin$ fanin or fanout of n_i **then**
        $\delta \leftarrow common\_pi(n_i, n_j, k_1)$
        $\epsilon \leftarrow common\_po(n_i, n_j, k_1)$
        if $\alpha \cdot \delta + (1 - \alpha) \cdot \epsilon \geq thresh$ **then**
          $L \leftarrow (n_i, n_j)$
        end if
      end if
    end for
  end for
end for
Return L

This function starts by selecting a node n_i in the network. To find a potential partner n_j for this node, $TFI_{frontier}(n_i, k_1)$ is called, which returns only the nodes $\boldsymbol{m}$ in the transitive fanin frontier of n_i which have a backward depth of exactly k_1 levels from n_i. This step is shown in Fig. 17.3a. For each of these nodes $m_l \in \boldsymbol{m}$, $TFO(m_l, k_2)$ is called, which returns nodes $\boldsymbol{n}$ in the transitive fanout of m_l that have a forward depth of up to k_2 levels from m_l. This gives all potential partners

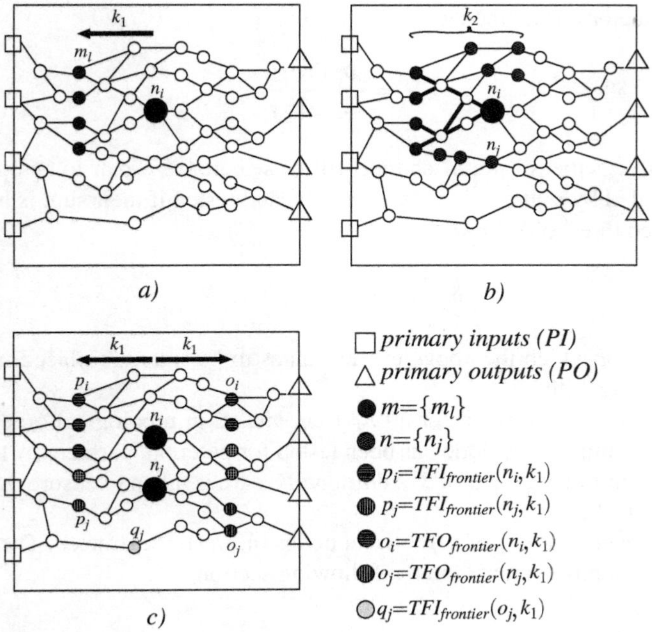

Fig. 17.3 Selection of node pairs

$\boldsymbol{n}$ and ensures that they will later share at least one common primary input in the subnetwork with n_i.

Note that nodes in the transitive fanin or fanout of n_i are not included in $\boldsymbol{n}$. The reason for this is explained in Section 17.5.1.3. Figure 17.3b shows the nodes (circles) that are included in $\boldsymbol{n}$, and the darkened edges show paths in the TFI of n_i where nodes will not be included in $\boldsymbol{n}$.

Next, each node $n_j \in \boldsymbol{n}$ is tested against n_i to measure their compatibility. The TFI and TFO (up to k_1 levels) of both nodes are considered when determining compatibility; however, only nodes at the frontier of these sets are used, shown in Fig. 17.3c. Node sets $\boldsymbol{p_i}$ and $\boldsymbol{p_j}$ are the $TFI_{frontier}$ sets for n_i and n_j, respectively. Node sets $\boldsymbol{o_i}$ and $\boldsymbol{o_j}$ are the TFO_{frontier} sets for n_i and n_j respectively. In addition, as will be explained in Section 17.5.1.2, the sets $\boldsymbol{q_i}$ and $\boldsymbol{q_j}$ are the TFI_{frontier} sets for $\boldsymbol{o_i}$ and $\boldsymbol{o_j}$, respectively. In Fig. 17.3c, the set of nodes $\boldsymbol{q_i}$ is empty because all nodes in $\boldsymbol{o_i}$ can be expressed completely in terms of nodes already in $\boldsymbol{p_i}$. These frontier nodes will later be the PIs and POs of the $k_1 \times k_1$ window around the nodes. The more of these nodes that n_i and n_j share in common, the fewer the number of nodes required in the subnetwork. The PI and PO compatibility of both nodes are calculated. The more these sets overlap, the more likely they are to be selected as a pair. The PI factor δ is defined as

$$\delta = \frac{|\boldsymbol{p_i} \cap \boldsymbol{p_j} \cap \boldsymbol{q_i} \cap \boldsymbol{q_j}|}{|\boldsymbol{p_i} \cup \boldsymbol{p_j} \cup \boldsymbol{q_i} \cup \boldsymbol{q_j}|}$$

and the PO factor ϵ is defined as

$$\epsilon = \frac{|\boldsymbol{o_i} \cap \boldsymbol{o_j}|}{|\boldsymbol{o_i} \cup \boldsymbol{o_j}|}$$

To determine whether or not n_i and n_j will be selected as a pair to be optimized, δ and ϵ are scaled by α and $1 - \alpha$, respectively, and tested if their sum is higher than a user-defined threshold $thresh$:

$$\alpha\delta + (1 - \alpha) \cdot \epsilon \geq thresh$$

All nodes n_j for which the above test evaluates to be $true$ are placed in the node pair list L, along with n_i.

These steps are performed for all $n_i \in \eta$, visited in topological order from the POs to the PIs, until every node has been tested for potential partners. A list L of all node pairs to optimize is returned. Additionally, care is taken to ensure that no pairs appear twice in L,

Next, a subnetwork $\eta_{i,j}$ of η, rooted at nodes (n_i, n_j), is extracted. The technique for this extraction is explained in the following section.

17.5.1.2 Building the Subnetwork

For each pair of nodes (n_i, n_j) found, subnetworks of η rooted at the nodes n_i and n_j are extracted by calling *extract_subnetwork*(n_i, n_j, k_1). This function constructs a subnetwork $\eta_{i,j}$ such that if node $m \in \{TFO(n_i, k_1) \cup TFO(n_j, k_1)\}$, then $m \in \eta_{i,j}$ and if node $p \in \{TFI(n_i, k_1) \cup TFI(n_j, k_1)\}$, then $p \in \eta_{i,j}$. Here k_1 is the same value used when calling *select_nodes*. The result of this step is illustrated in Fig. 17.4a as the shaded subnetwork.

Node $m \in \eta_{i,j}$ is designated as a primary input of $\eta_{i,j}$ if $\exists_{n \in FI(m)}, n \notin \eta_{i,j}$. Similarly, a node m is designated as a primary output of $\eta_{i,j}$ if $\exists_{n \in FO(m)}, n \notin \eta_{i,j}$. The set of primary inputs (outputs) of $\eta_{i,j}$ is referred to as X (Z).

Next the set of all nodes $m \in TFI(v, k_1)$ is collected, where v is a primary output of the subnetwork $\eta_{i,j}$. This step is illustrated in Fig. 17.4b. Let this set be called D. The nodes in the dotted and shaded region of Fig. 17.4b constitute the set D. These nodes are included in the subnetwork as well, by setting $\eta_{i,j} \leftarrow \eta_{i,j} \cup D$. Figure 17.4c zooms into the region of interest for the subsequent discussion.

Next, for each $d \in D$ a check is done to see if $FI(d)$ can be expressed completely in terms of the current nodes in $\eta_{i,j}$. This check is performed by recursively traversing the network topologically from d toward the primary inputs X^{global} of η. If this traversal visits a node in $\eta_{i,j}$, the traversal terminates and all nodes visited in this traversal are added to $\eta_{i,j}$. If the traversal visits a node in X^{global} instead, then the set of primary inputs of $\eta_{i,j}$ is augmented with d, i.e., X is updated as $X \leftarrow X \cup d$. This step is illustrated in Fig. 17.4d.

Nodes w and $r \in D$ could be considered as primary inputs to the subnetwork; however, all of their fanins can be expressed completely in terms of X. Thus, the

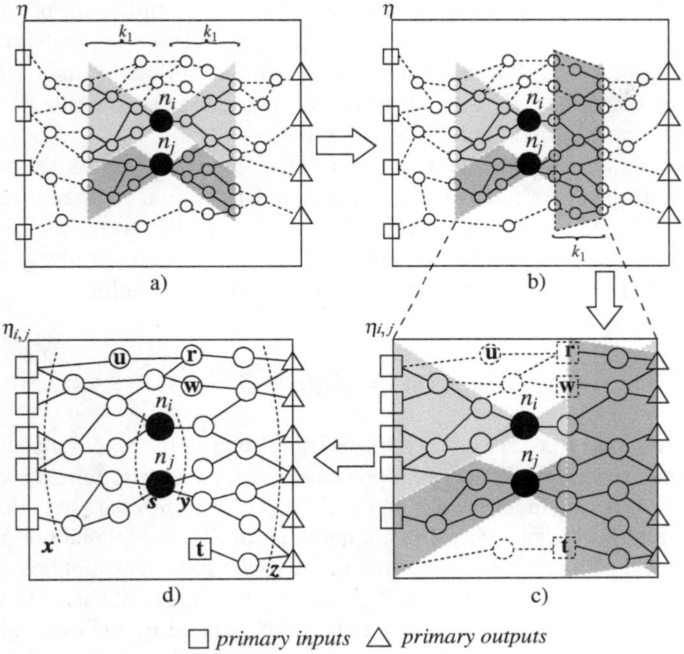

Fig. 17.4 Extraction of subnetwork

fanin of the node $w \in D$ and the fanin u of $r \in D$ are added to $\eta_{i,j}$. However, the fanin of node $t \in D$ cannot be expressed in terms of nodes in $\eta_{i,j}$, and so t is added to X. This check avoids the addition of unnecessary primary inputs for representing the subnetwork $\eta_{i,j}$. A larger number of primary input variables typically results in larger intermediate ROBDDs in the computation of the Boolean relation $\mathcal{R}$, and consequently more time needed for the computations.

Note that the size of each subcircuit $\eta_{i,j}$ is determined by the depth parameter k_1. Hence, by suitably choosing k_1, it can be guaranteed that the subcircuits are never too large, and the Boolean relation can be computed with low memory utilization, even for an extremely large network η. The final subnetwork $\eta_{i,j}$ is shown in Fig. 17.4d. This subnetwork is then used to create a Boolean relation which inherently represents the simultaneous flexibility of both n_i and n_j, as discussed in the following section.

17.5.1.3 Computing the Boolean Relation $\mathcal{R}^Y$

As mentioned previously, the exact computation of a Boolean relation expressing the flexibility in a medium to large design could be extremely memory intensive. Additionally, ROBDD-based computations are used for this relation. ROBDDs can, by nature, exhibit very irregular memory requirements, especially for medium to large designs. A goal of this approach is to develop a robust methodology for computing the Boolean relation. This is achieved by keeping a tight control on the sizes of the BDDs of the Boolean relation. Not only is this relation computed for a node pair (n_i, n_j) using a windowed subnetwork $\eta_{i,j}$ (thus ensuring that the ROBDDs are small) but also careful early quantification is performed to ensure that the ROBDD sizes stay tractable during the relation computation.

Consider a subnetwork $\eta_{i,j}$, its set of primary inputs X and its set of primary outputs Z. Let the set of nodes being simultaneously optimized be referred to as Y and their combined support be S. Note that S, Y, X and Z correspond to a set of nodes of $\eta_{i,j}$. Let the variables for these be $\boldsymbol{s}$, $\boldsymbol{y}$, $\boldsymbol{x}$ and $\boldsymbol{z}$, respectively, as shown in Fig. 17.4d. The characteristic function of the Boolean relation $\mathcal{R}$ is a mapping $B^{|S|} \times B^{|Y|} \rightarrow B$ s.t.

$$\mathcal{R}^Y(\boldsymbol{s}, \boldsymbol{y}) = \forall_{\boldsymbol{x}}[(\boldsymbol{s} = g_S(\boldsymbol{x})) \Rightarrow \forall_{\boldsymbol{z}}[(\boldsymbol{z} = Z_i^M(\boldsymbol{x}, \boldsymbol{y})) \Rightarrow \Phi(\boldsymbol{x}, \boldsymbol{z})]]$$

In this expression, $\Phi(\boldsymbol{x}, \boldsymbol{z})$ is the characteristic function of the circuit outputs $\boldsymbol{z} = f(\boldsymbol{x})$. The subexpression $Z^M(\boldsymbol{x}, \boldsymbol{y})$ represents the characteristic function of the circuit outputs expressed in terms of $\boldsymbol{x}$ and $\boldsymbol{y}$. Also, $g_S(\boldsymbol{x})$ is the characteristic function of the $\boldsymbol{s}$ variables in terms of $\boldsymbol{x}$. The computation of $\mathcal{R}^Y$ is explained intuitively as follows. For all primary input minterms $\boldsymbol{x}$, let $\boldsymbol{s}$ take on values dictated by $\boldsymbol{x}$ (i.e., $\boldsymbol{s} = g_S(\boldsymbol{x})$). If this is the case, then if $\boldsymbol{z}$ takes on the values dictated by $\boldsymbol{x}$ and the node values of $\boldsymbol{y}$, the values of $\boldsymbol{x}$ and $\boldsymbol{z}$ should be related by the original network functionality (i.e., $\Phi(\boldsymbol{x}, \boldsymbol{z})$).

One caveat of this computation is that the two nodes (n_1, n_2) for which the relation is calculated cannot be in each others' *TFI* or *TFO*. The reason for this is explained as follows. Suppose node n_1 is a fanin of node n_2. If that is true, then $n_1 \in \boldsymbol{s}$ and $n_1 \in \boldsymbol{y}$ simultaneously. If the relation is then minimized, BREL produces functions $\boldsymbol{y} = f(\boldsymbol{s})$ that are cyclic, with variables being on both sides of the equation. This could lead to feedback in the optimized circuit. For this reason, a node in the other node's *TFI* or *TFO* is not chosen in the node selection algorithm.

17.5.1.4 Quantification Scheduling

In the approach presented, the Boolean relation $\mathcal{R}^Y(\boldsymbol{s}, \boldsymbol{y})$ is computed using ROBDDs. In order to avoid a possible memory explosion problem, early quantification is performed as explained next.

The computation for $\mathcal{R}^Y(\boldsymbol{s}, \boldsymbol{y})$ is rewritten as

$$\mathcal{R}^Y(\boldsymbol{s}, \boldsymbol{y}) = \forall_{\boldsymbol{x}}\left[(\boldsymbol{s} = g_S(\boldsymbol{x})) \Rightarrow \forall_{\boldsymbol{z}}\left[\prod_i (z_i \overline{\oplus} Z_i^M(\boldsymbol{x}, \boldsymbol{y})) \Rightarrow \prod_i (z_i \overline{\oplus} Z_i(\boldsymbol{x}))\right]\right]$$

This expression can be rewritten as

$$\mathcal{R}^Y(\boldsymbol{s}, \boldsymbol{y}) = \forall_{\boldsymbol{x}}\left[(\boldsymbol{s} = g_S(\boldsymbol{x})) \Rightarrow \forall_{\boldsymbol{z}}\left[\prod_i [(z_i \overline{\oplus} Z_i^M(\boldsymbol{x}, \boldsymbol{y})) \Rightarrow (z_i \overline{\oplus} Z_i(\boldsymbol{x}))]\right]\right]$$

The first observation is that the quantification over $\boldsymbol{z}$ ($\forall_{\boldsymbol{z}}$) and the product term over i ($\prod_i$) can be swapped to obtain a new expression for $\mathcal{R}^Y(\boldsymbol{s}, \boldsymbol{y})$:

$$\mathcal{R}^Y(\boldsymbol{s}, \boldsymbol{y}) = \forall_{\boldsymbol{x}} \left[(\boldsymbol{s} = g_S(\boldsymbol{x})) \Rightarrow \prod_i [\forall_{\boldsymbol{z}}[(z_i \overline{\oplus} Z_i^M(\boldsymbol{x}, \boldsymbol{y})) \Rightarrow (z_i \overline{\oplus} Z_i(\boldsymbol{x}))]] \right]$$

This is correct because in general,

$$\forall_\omega(f \cdot g) = \forall_\omega(f) \cdot \forall_\omega(g)$$

Quantifying out the $\boldsymbol{z}$ variables earlier results in smaller intermediate ROBDDs for the expression to the right of the first implication. The computation can therefore be expressed as

$$\mathcal{R}^Y(\boldsymbol{s}, \boldsymbol{y}) = \forall_{\boldsymbol{x}}[(\boldsymbol{s} = g_S(\boldsymbol{x})) \Rightarrow P(\boldsymbol{x})] = \forall_{\boldsymbol{x}}[\overline{(\boldsymbol{s} = g_S(\boldsymbol{x}))} + P(\boldsymbol{x})]$$

where P(**x**) is the ROBDD obtained after applying the first observation.

$$P(\boldsymbol{x}) = \prod_i [\forall_{\boldsymbol{z}}[(z_i \overline{\oplus} Z_i^M(\boldsymbol{x}, \boldsymbol{y})) \Rightarrow (z_i \overline{\oplus} Z_i(\boldsymbol{x}))]]$$

In general, however,

$$\forall_\omega(f + g) \neq \forall_\omega(f) + \forall_\omega(g)$$

Let the common variables between f and g be ω^*. Let $\omega' = \omega \cap \omega^*$. Then,

$$\forall_\omega(f + g) = \forall_{\omega'}(\forall_{\omega \backslash \omega'}(f) + \forall_{\omega \backslash \omega'}(g))$$

The second observation is that $g_S(\boldsymbol{x})$ depends on a smaller subset ($\boldsymbol{x}'$) of the primary inputs ($\boldsymbol{x}$) of the network. Hence, $\mathcal{R}^Y(\boldsymbol{s}, \boldsymbol{y})$ can be computed as

$$\mathcal{R}^Y(\boldsymbol{s}, \boldsymbol{y}) = \forall_{\boldsymbol{x}'}[\forall_{\boldsymbol{x} \backslash \boldsymbol{x}'} \overline{(\boldsymbol{s} = g_S(\boldsymbol{x}))} + \forall_{\boldsymbol{x} \backslash \boldsymbol{x}'}(P(\boldsymbol{x}))]$$

which reduces to

$$\mathcal{R}^Y(\boldsymbol{s}, \boldsymbol{y}) = \forall_{\boldsymbol{x}'}[\overline{(\boldsymbol{s} = g_S(\boldsymbol{x}))} + \forall_{\boldsymbol{x} \backslash \boldsymbol{x}'}(P(\boldsymbol{x}))]$$

In practice, both observations are applied in tandem. First $g_S(\boldsymbol{x})$ is found, as well as the set $\boldsymbol{x}'$. Then, while computing P(**x**), $\boldsymbol{x} \backslash \boldsymbol{x}'$ is quantified out. The final computing step is

$$\mathcal{R}^Y(\boldsymbol{s}, \boldsymbol{y}) = \forall_{\boldsymbol{x}'}[\overline{(\boldsymbol{s} = g_S(\boldsymbol{x}))} + P'(\boldsymbol{x}')]$$

where $P^{'}(x^{'}) = \forall_{x \setminus x^{'}}(P(x))$. By implementing both these techniques, intermediate ROBDD never blows up in size. Without using the early quantification ideas, the ROBDD size is dramatically larger, hence the early quantification is key to the robustness and scalability of the approach. The final ROBDD representing $\mathcal{R}^{Y}(s, y)$ is returned to the calling function.

17.5.1.5 Endgame

Next, BREL is called to minimize $\mathcal{R}^{Y}(s, y)$. The output of BREL is a pair of completely specified functions for the nodes $n_i^{'}$ and $n_j^{'}$ such that these functions are compatible with $\mathcal{R}^{Y}(s, y)$ and the total cost of $n_i^{'}$ and $n_j^{'}$ is minimal. The new nodes $n_i^{'}$ and $n_j^{'}$ are grafted back into η and the original nodes n_i and n_j are deleted from η.

At the end of the *for* loop in Algorithm 1, when all node pairs have been processed by the relation-based minimization procedure, the *network_sweep* command of SIS [22] is run. This command quickly eliminates any constant-valued nodes in the network that may have been created during the minimization process. Finally, the *network_verify* command of SIS is run to check if the resulting network $\eta^{'}$ is functionally equivalent to the original network η.

17.6 Experimental Results

This section presents the experimental results for the algorithm described in the previous section. Section 17.6.1 describes the preprocessing steps used in all the experimental results. Section 17.6.2 shows the methodology used to determine the the parameters that control the node selection algorithm. Section 17.6.3 reports the results obtained when comparing with the *mfsw* approach of [17], which is the most powerful technique among single-node optimized approaches (in terms of runtime and quality of results). Section 17.6.4 discusses some variations on the original algorithm and their results.

The metric for quality that is used throughout this section is literal count. This is the sum of the number of literals for each node in the network. The fewer the number of literals in the network, the better the optimization technique. The literal counts shown in the results are all relative to another approach used for comparison. Runtimes are also reported in these results. For some experiments, absolute runtime is reported, and for others, runtime relative to another approach is reported.

The approach was implemented in SIS [22], a logic synthesis package written in C. The ROBDD package used was the CUDD package [24]. A sample of 15 medium and large circuits from the *mcnc91* and *itc99* benchmarks was used in the experiments. The experiments were performed on a Linux-based Dell Optiplex with a 2.6 GHz Core 2 Quad CPU with 4 GB of RAM.

17.6.1 Preprocessing Steps

Before any minimization is performed on the original network, two preprocessing steps are performed. The first is the *network_sweep* command of SIS, which eliminates constant-valued nodes as well as nodes which do not fanout anywhere.

The second step is running *sat_sweep* [14] on the network. This command uses a Boolean Satisfiability (SAT) checker to determine if two nodes u and v are functionally identical by calling SAT_CHECK($u \oplus v$). This checks if there is any input vector to u and v for which the outputs of u and v differ. If there is, then the nodes are functionally different and cannot be merged, and a new pair is selected. Otherwise, then the nodes are functionally equivalent and can be merged together. This algorithm quickly reduces the literal count of a circuit by removing redundancies. The result is obtained over and above what *sat_sweep* achieves. The results reported for the competing technique *mfsw* [17] were also preceded by a *sat_sweep* command. In other words, *sat_sweep* is run first. Then the additional improvements obtained by the proposed method are compared to those obtained by *mfsw*.

17.6.2 Parameter Selection

As described in Section 17.5.1, the node selection algorithm is based on four user-defined parameters, namely $thresh$, k_1, k_2, and α. Tuning these parameters can customize the trade-off between quality of results and runtime. In general, the longer the runtime, the better the quality of results. However, the runtime is heavily dependent on the number of nodes chosen. If changing a parameter increases the runtime, this is because either more node pairs were selected, or the processing time of a node pair is increased. Depending on the 'quality' of the additional pairs, the literal reduction could change as well. Because of this, optimal values need to be determined for all parameters as a first step.

The experiments in this section are conducted to find a 'golden' set of parameter values for the proposed approach. In these experiments, the ranges of values for each parameter are listed in Table 17.2. The nominal values of these parameters are also listed in this table.

Table 17.2 Initial values, final values, increments, and nominal values of the node selection parameters

Parameter	Low	High	Increment	Nominal
α	0	1.0	0.1	0.5
k_1	2	3	1	2
k_2	2	4	1	3
thresh	0	1.0	0.1	0.5

17.6.2.1 Selecting α

The first parameter to determine is α. This parameter determines the weight that PIs and POs of the subnetwork are given when selecting a node pair. The parameter α can range from 0, which considers only POs, to 1.0, which considers only PIs. The reason for determining α first is because it is the parameter least dependent on the others. Since $thresh$, k_1, and k_2 affect only the *number* of pairs selected and the window size, α can be chosen first.

Figure 17.5 shows α being swept from 0 to 1.0, while the other three parameters are held constant. The nominal values for $thresh$, k_1, and k_2 were chosen in the middle of their ranges at 0.5, 2, and 3, respectively. The left axis represents the ratio of literals obtained compared to that obtained after running *sat_sweep*, and the right axis represents the average runtime of the method used here. For each value of α, the average literal ratio and runtime are presented in Fig. 17.5, across all the benchmark examples.

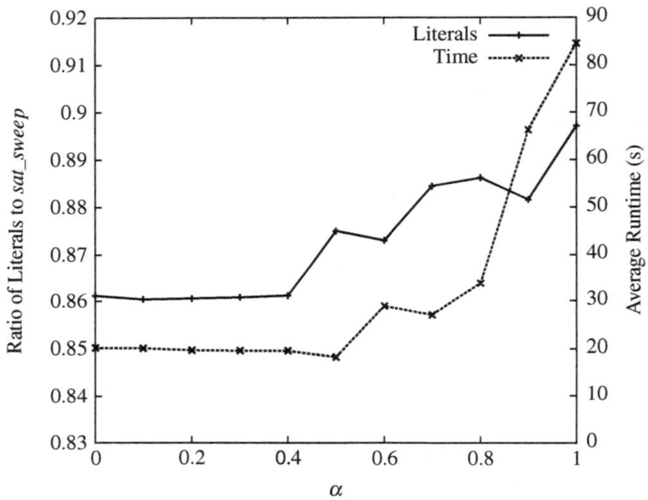

Fig. 17.5 Sweeping α with $thresh = 0.5$, $k_1 = 2$, $k_2 = 3$

The main observation is that in general, lower values of α yield lower runtimes as well as fewer literals. This means that PO compatibility is more important when choosing node pairs than PI compatibility. At the extremes, it is seen than an α value of 0 provides 40% more literal reduction in a quarter of the runtime than with an α value of 1.0. Since both runtime and literal count increase with α, one can infer that with the higher α values, more node pairs were chosen, but the pairs gave less literal count reduction. This shows that when two nodes are minimized together, more flexibility is obtained if they reconverge quickly than if they share a common variable support.

For $0 \leq \alpha < 0.5$, both runtime and literal count are relatively flat. Therefore, the value of 0.25 is chosen for α for the rest of the experiments. Although there is no empirical data to guide the selection of α for values less than 0.5, a value in the middle of the range is chosen. This is so that PI compatibility still contributes to the node selection, but not enough to degrade the results.

17.6.2.2 Selecting k_1 and k_2

The parameters k_1 and k_2 are determined next. The size of the window is determined by k_1, because the subnetwork created includes nodes k_1 levels back and k_1 levels forward from the nodes to be minimized, as shown in Fig. 17.4a. Therefore a larger value of k_1 means more nodes are included in the subnetwork. The parameter k_2 affects the number of nodes in $\boldsymbol{n}$, shown in Fig. 17.3b, which are tested against the first node n_i for compatibility. A larger value of k_2 means that more partners are tested for n_i.

Figure 17.6 shows the literal ratio and average runtime for different values of (k_1, k_2). The first observation is that the point $(3, 2)$ has a distinctly higher literal ratio than the other points. This is because node n_j is selected by going back three topological levels from n_i but then forward only two levels from there. This precludes any node n_j that is on the same level as n_i from being selected. By comparing points $(3, 2)$ and $(3, 3)$ in Fig. 17.6, is clear that these nodes account for a large portion of the gains in literal ratio.

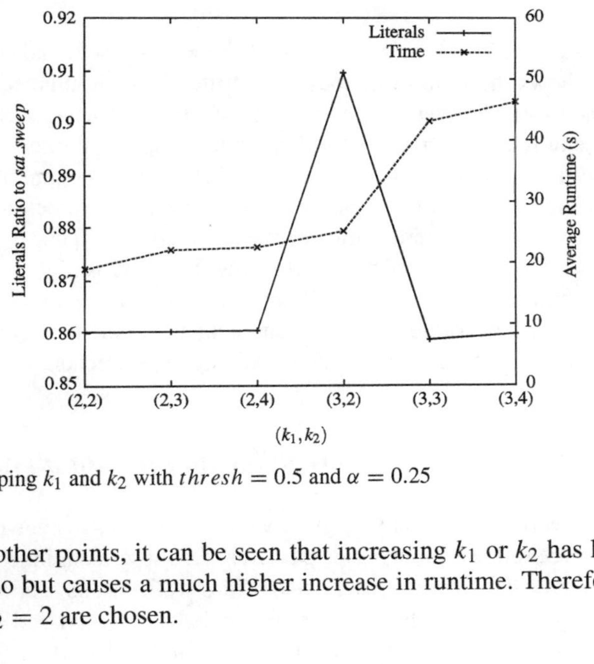

Fig. 17.6 Sweeping k_1 and k_2 with $thresh = 0.5$ and $\alpha = 0.25$

From the other points, it can be seen that increasing k_1 or k_2 has little effect on the literal ratio but causes a much higher increase in runtime. Therefore the values $k_1 = 2$ and $k_2 = 2$ are chosen.

17.6.2.3 Selecting *thresh*

The final parameter to determine is $thresh$. This parameter controls how 'compatible' two nodes must be for them to be selected as a pair. A high value of $thresh$ means that only node pairs with a high percentage of outputs and inputs in common are chosen for minimization. A low value of $thresh$ allows the nodes with fewer inputs and outputs in common to be minimized as well.

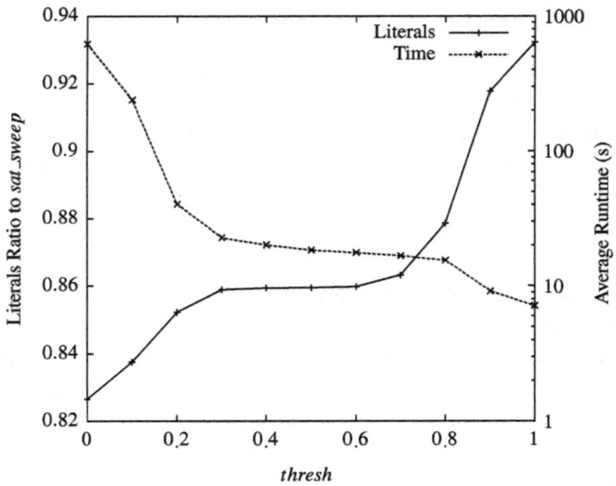

Fig. 17.7 Sweeping $thresh$ with $k_1 = 2$, $k_2 = 2$, and $\alpha = 0.25$

Figure 17.7 shows $thresh$ being swept with $k_1 = 2$, $k_2 = 2$, and $\alpha = 0.25$. The left axis again shows the ratio of the number of literals using this technique to the literals from the original network after calling *sat_sweep*. The right axis shows the average runtime in seconds for the minimization to complete, plotted on a log scale. This shows that for $thresh \leq 0.2$, the runtime increases exponentially as $thresh$ decreases, and the literal ratio decreases linearly. For large values of $thresh$, the runtime decreases, but the literal ratio increases drastically. This is because very few nodes are selected for minimization when the threshold is high.

While either extreme can be chosen if speed or literal ratio alone were desired, selecting an intermediate value of $thresh$ can achieve a balance between the two. Therefore *thresh* $= 0.4$ is selected for the remaining experiments.

17.6.3 Comparison of the Proposed Technique with mfsw

As described in Section 17.3, many single-node techniques have been developed for minimizing Boolean networks using don't cares. Of these techniques, the method of [17], called *mfsw*, has the best results and scales well. In this section, the results from the method presented in this chapter are labeled as *relation* and are compared with the results from *mfsw*. The *mfsw* technique uses a SAT-based CDC method and a 2×2 window for creating subnetworks.

For the remaining experiments, the 'golden' values of the parameters as described in Section 17.6.2 are used. In particular, *thresh* $= 0.4$, $k_1 = 2$, $k_2 = 2$, and $\alpha = 0.25$.

For the results shown in Table 17.3, *sat_sweep* is run first, providing the starting point for both the *mfsw* technique and the method described in this chapter. The literal count after *sat_sweep* is shown in Column 2 of Table 17.3. The literal count and

Table 17.3 Results after *sat_sweep*

	Orig	*mfsw*		*relation*		Ratio			
Circuit	Lits	Lits	Time	Lits	Time	Lits	Time	Mem	% Gain
c1355	992	992	0.09	598	1.53	0.603	16.96	339	1
c1908	759	748	0.09	595	6.74	0.795	74.38	54939	0.41
c2670	1252	1197	0.11	901	4.04	0.753	36.76	1025	0.78
c5315	3062	2935	0.29	2372	12.37	0.808	42.65	2683	0.74
c7552	3796	3549	0.43	2990	14.30	0.842	33.25	3314	0.68
b15	15084	14894	1.78	14654	49.31	0.984	27.70	5594	0.64
b17	49096	48595	5.74	48047	228.57	0.989	39.82	6578	0.60
b20	22037	21816	2.56	21501	91.40	0.986	35.70	3489	0.26
b21	22552	22306	2.59	21933	92.08	0.983	35.55	3489	0.32
b22	33330	33001	3.97	32321	203.11	0.979	51.16	3519	0.28
s1494	1239	1177	0.13	1195	3.68	1.015	28.31	594	0.20
s5378	2327	2283	0.27	1993	7.36	0.873	27.26	3306	0.38
s13207	5052	4833	0.38	4259	27.35	0.881	71.96	1430	0.30
s15850	6624	6342	0.52	5519	26.10	0.870	50.19	1234	0.51
s38417	17531	17314	1.43	17158	68.68	0.991	48.02	2598	0.41
Average	1	0.974	–	0.859	–	0.882	38.62	–	0.45

runtime after running *mfsw* is reported in Column 3 and Column 4, respectively. The literal count and runtime after running *relation* is reported in Column 5 and Column 6, respectively. For these columns, the average literal count relative to *sat_sweep* is shown in the last row. Column 7 shows the ratio of literals in Column 5 to Column 3, and Column 8 shows the ratio of runtimes in Column 6 to Column 4. Column 9 reports the peak number of ROBDD nodes for *relation*, and Column 10 shows the percentage of node pairs selected by *relation* that actually reduce the number of literals in the network.

From Table 17.3 it is seen that after *sat_sweep*, the *relation* method reduces the literal count by approximately 12% over what *mfsw* achieves. This shows that minimizing two nodes simultaneously has significant benefits in terms of literal count over the don't care approach of *mfsw*. The memory requirements are also very low regardless of the size of the circuit, due to the aggressive quantification scheduling performed. This supports the claim that *relation* scales well and is a robust technique. Column 7 shows that the node selection method is quite efficient. On average, 45% of the node pairs chosen resulted in a reduction in the number of literals in the network.

In terms of runtime, *mfsw* is clearly more efficient than *relation*, which requires nearly 40× more runtime on average. However, the absolute time values must be taken into account. Column 6 shows that for most circuits runtimes are under 1 min, and the peak runtime is still under 4 min for the largest circuit. Compared to the time scale of the entire design process, which is measured in months or years, these times are therefore quite small. In addition, as discussed in Section 17.6.2, the parameters for selecting node pairs can be altered to decrease the runtime.

It should also be noted that increasing the window size of *mfsw* to a 10 × 10 window greatly increases the runtime of that method but reduces the literal count by less than 1%. This means that while *relation* does require more runtime, the minimization it performs cannot be matched by *mfsw* regardless of the time it is allowed to run.

17.6.4 Additional Experiments

Section 17.6.3 presented the gains of the relation-based minimization approach after running *sat_sweep*. In this section, a variety of other experiments are performed to further explore the relation-based technique.

17.6.4.1 Running *relation* After *mfsw*

For this experiment, *relation* is run on networks that have *already* been reduced by *sat_sweep* and *mfsw*. The purpose is to test how much *relation* can improve upon the minimization results of *mfsw*. Table 17.4 shows the results of this experiment. Column 2 (3) reports the literal count (runtime) of running *sat_sweep* followed *mfsw*. Columns 4 and 5 show the literal count and runtime (respectively) of running *relation* on the netlist obtained by *sat_sweep* followed by *mfsw*. The literal and runtime ratios are shown in Column 6 and Column 7, respectively.

Table 17.4 Results after *sat_sweep* and *mfsw*

	mfsw		$mfsw + relation$		Ratio		
Circuit	Lits	Time	Lits	Time	Lits	Time	Mem
c1355	992	0.09	600	1.53	0.605	16.994	336
c1908	748	0.09	588	2.86	0.786	31.759	12026
c2670	1197	0.11	906	4.21	0.757	38.307	742
c5315	2935	0.29	2298	11.57	0.783	39.912	1452
c7552	3549	0.43	2795	13.17	0.788	30.635	1842
b15	14894	1.78	14558	44.18	0.977	24.822	1262
b17	48595	5.74	47639	213.90	0.980	37.264	5648
b20	21816	2.56	21293	91.65	0.976	35.802	3490
b21	22306	2.59	21711	91.97	0.973	35.509	3489
b22	33001	3.97	32050	202.72	0.971	51.063	3511
s1494	1177	0.13	1142	3.29	0.970	25.319	673
s5378	2283	0.27	1972	6.95	0.864	25.738	5666
s13207	4833	0.38	4256	27.15	0.881	71.442	13121
s15850	6342	0.52	5331	23.92	0.841	46.004	1212
s38417	17314	1.43	16968	55.62	0.980	38.897	2009
Average	–	–	–	–	0.868	34.684	–

It is seen from Column 6 that running *relation* after *mfsw* can further reduce the literals by about 13%. Since *the window sizes of both methods were identical*, this improvement represents the benefits of two-node minimization over single-node minimization. In the specific case of circuit c1355, nearly 40% of the literals can be removed only through the node pair technique. Columns 6 and 7 demonstrate again that the memory utilization is very low, and the node selection method is effective.

17.6.4.2 Running *relation* Twice

In this experiment, after running *sat_sweep*, *relation* is run twice in succession on the same network. The purpose of this experiment is to determine if there are improvements that can be had by minimizing a network multiple times. Three separate experiments are tried. In the first, only nodes pairs that did not give any literal count reduction during the first run are minimized again. In the second, only node pairs that *did* give a reduction in literal count are minimized again. And finally, all node pairs were rerun regardless of whether they yielded a literal reduction or not during the first run.

The results for all three experiments showed a less than 1% improvement in literal count compared to the first run of *relation*. These experiments show that even though the network has changed significantly after the first run of *relation*, these changes have almost no impact on the ability of other nodes to be further minimized in a subsequent iteration. The same conclusion can be drawn for the *mfsw* method as well, which also yields almost no further reductions when run more than once on a network.

17.6.4.3 Minimizing Single Nodes

In this experiment, *sat_sweep* is run first, followed by *relation*. During the *relation* algorithm, some nodes get minimized while others do not, either because they were not selected in a node pair or because the algorithm did not reduce their literal count. For such nodes, an additional step of minimization was performed after running *relation*. After *relation*, these nodes are again minimized individually using ODCs. This is implemented using the same steps in Section 17.5, by creating a relation corresponding to the subnetwork, and then using BREL to minimize it. The only difference is that only one node is used. This experiment ensures that some type of minimization is attempted for *each* node in the network.

However, experimental results showed that this idea does not further reduce the literal count by more than 1%, across all the circuits. Almost all of the nodes which were subjected to single-node optimization techniques were those that were originally selected but did not reduce their literal count when minimized with other nodes in a pair. The conclusion that can be drawn is that if a node cannot be minimized with another node, then minimizing it alone does not yield any gains either.

17.6.4.4 Effects of Early Quantification

Section 17.5.1.4 discusses the methods for early quantification used for the approach presented. Figure 17.8 shows the effects of quantifying during different stages of the computation of $P(\boldsymbol{x})$ from $\mathcal{R}(Y)$. One node pair of the network *c*432 is being minimized in this example, and the number of nodes in the BDD during each iteration of the $P(\boldsymbol{x})$ computation is reported.

The plot Q_1 represents the incremental size of the relation BDD without any early quantification. Note that the BDD size is reported on a logarithmic scale. After

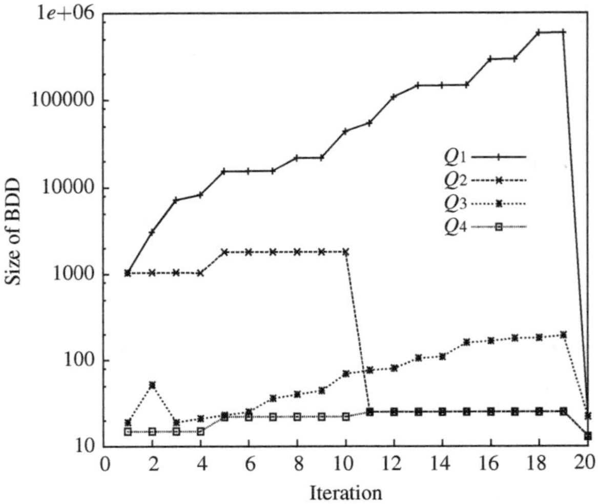

Fig. 17.8 Effects of quantification scheduling on BDD size

only 20 BDD calculations, the size of the BDD is almost one million nodes. Only at the end of the computation, when the $\boldsymbol{z}$ and $\boldsymbol{x}$ variables are quantified out, does the size of the BDD drop.

The plot Q_2 shows the BDD size when the output variables $\boldsymbol{z}$ are quantified out after each iteration of the computation. The number of nodes stays near 1000 until a particular output in $\boldsymbol{z}$ is quantified out, and then the BDD size drops to 25.

The plot Q_3 is the case when only the $\boldsymbol{x}\backslash\boldsymbol{x}'$ variables are quantified out after each iteration. The BDD size steadily climbs to over 100, until the final iteration, when the $\boldsymbol{x}$ variables are quantified out.

Q_4 shows the results when the quantification techniques Q_2 and Q_3 are applied in tandem. The size of the BDD never grows past 25 during *the entire computation.*

This example demonstrates that for a single-node pair, even when a windowing technique is employed, the BDD of $\mathcal{R}(Y)$can blow up in size unless both the early quantification techniques of Section 17.5.1.4 are employed.

17.7 Chapter Summary

In this chapter, a scalable dual-node technology-independent logic optimization technique was presented. This technique scales well and can minimize both small designs and large designs typical of industrial circuits.

The algorithm presented first selects which node pairs will be minimized. For each node pair, a subnetwork is created around the nodes. This windowing is done in order to make this approach feasible for large industrial circuits. Once the subnetwork is created, the Boolean relation, which represents the flexibility of the nodes, is computed. During this process, early quantification is performed. BREL is used to minimize the Boolean relation, and the new nodes replace the original nodes in the original circuit. This is done for all node pairs that were selected.

It is experimentally demonstrated that this technique produces minimized technology-independent networks that are on average 12% smaller than networks produced by a single-node minimization technique called *mfsw*. Although the runtimes of *mfsw* are significantly smaller than this approach, the runtime for any given circuit using this approach is never more than 4 min. In addition, the memory usage is very low and is independent of the circuit size.

Additionally, the approach can *further* reduce the literal count of networks that have already been minimized by *mfsw*, by 13%. This result shows how the increased flexibility from two-node minimization can simplify networks better than single-node techniques.

Some of the future work involves using a SAT-based approach for constructing the Boolean relation. An alternative SAT-based replacement for BREL can be implemented as well. Both of these have the potential to reduce runtimes of the technique. In addition, modifications to minimize three or more nodes simultaneously can be made to gain even more flexibility using the Boolean relation-based multi-output optimization technique.

References

1. Baneres, D., Cortadella, J., Kishinevsky, M.: A recursive paradigm to solve Boolean relations. In: Proceedings of the Design Automation Conference, pp. 416–421. San Diego, CA (2004)
2. Brayton, R.: Compatible output don't cares revisited. In: Proceedings of International Conference on Computer-Aided Design, pp. 618–623. San Jose, CA (2001)
3. Brayton, R., Rudell, R., Sangiovanni-Vincentelli, A., Wang, A.: MIS: A multiple-level logic optimization system. IEEE Transactions on CAD/ICAS **CAD-6**(6), 1062–1082 (1987)
4. Brayton, R.K., Hachtel, G.D., McMullen, C.T., Sangiovanni-Vincentelli, A.L.: Logic Minimization Algorithms for VLSI Synthesis. Kluwer, Norwell, MA (1984)
5. Brayton, R.K., Hachtel, G.D., Sangiovanni-Vincentelli, A.L.: Multilevel logic synthesis. In: Proceedings of IEEE, vol.78, pp. 264–300. (1990)
6. Brayton, R.K., Somenzi, F.: Boolean relations and the incomplete specification of logic networks. In: Proceedings of International Conference on VLSI. Cambridge, MA (1989)
7. Bryant, R.E.: Graph based algorithms for Boolean function representation. IEEE Transactions on Computers **C-35**, 677–690 (1986)
8. Burch, J.R., Clarke, E.M., Long, D.E.: Symbolic model checking with partitioned transition relations. In: Proceedings of International Conference on VLSI. Cambridge, MA (1991)
9. Cerny, E., Marin, M.A.: An approach to unified methodology of combinational switching circuits. In: Proceedings of IEEE Transactions on Computers, vol. 26, pp. 745–756. (1977)
10. Chen, K.C., Fujita, M.: Efficient sum-to-one subsets algorithm for logic optimization. In: Proceedings of Design Automation Conference, pp. 443–448. Anaheim, CA (1992)
11. Dey, S., Brglez, F., Kedem, G.: Circuit partitioning and resynthesis. In: Proceedings of Custom Integrated Circuits Conference, pp. 29.4/1–29.4/5. (1990)
12. Hassoun, S. (ed.): Logic Synthesis and Verification. Kluwer, Norwell, MA (2001)
13. Jiang, Y., Brayton, R.K.: Don't cares and multi-valued logic network minimization. In: Proceedings of International Conference on Computer-Aided Design. San Jose, CA (2000)
14. Kuehlmann, A.: Dynamic transition relation simplification for bounded property checking. In: Proceedings of International Conference on Computer-Aided Design, pp. 50–57. San Jose, CA (2004)

15. Limqueco, J.C., Muroga, S.: SYLON-REDUCE: An MOS network optimization algorithms using permissible functions. In: Proceedings of International Conference on Computer-Aided Design, pp. 282–285. Santa Clara, CA (1990)
16. Limqueco, J.C., Muroga, S.: Optimizing large networks by repeated local optimization using windowing scheme. In: IEEE International Symposium on Circuits and Systems, ISCAS, vol. 4, pp. 1993–1996. San Diego, CA (1992)
17. Mishchenko, A., Brayton, R.K.: SAT-based complete don't care computation for network optimization. In: Proceedings of Design, Automation and Test in Europe, pp. 412–417. Munich, Germany (2005)
18. Saluja, N., Khatri, S.P.: A robust algorithm for approximate compatible observability don't care (CODC) computation. In: Proceedings of Design Automation Conference, pp. 422–427. San Diego, CA (2004)
19. Savoj, H., Brayton, R., Touati, H.: Extracting local don't cares for network optimization. In: Proceedings of IEEE Transactions on Computer-Aided Design, pp. 514–517. Santa Clara, CA, USA (1991)
20. Savoj, H., Brayton, R.K.: The use of observability and external don't cares for the simplification of multi-level networks. In: Proceedings of Design Automation Conference, pp. 297–301. Orlando, Florida (1990)
21. Savoj, H., Brayton, R.K.: Observability relations for multi-output nodes. In: Proceedings of International Workshop on Logic Synthesis. Tahoe City, CA (1993)
22. Sentovich, E.M., Singh, K.J., Lavagno, L., Moon, C., Murgai, R., Saldanha, A., Savoj, H., Stephan, P.R., Brayton, R.K., Sangiovanni-Vincentelli, A.L.: SIS: A System for Sequential Circuit Synthesis. Technical Report, UCB/ERL M92/41, Electronics Research Lab, University of California, Berkeley, CA 94720 (1992)
23. Sinha, S., Brayton, R.K.: Implementation and use of SPFDs in optimizing Boolean networks. In: Proceedings of International Conference on Computer-Aided Design, pp. 103–110. San Jose, CA (1998)
24. Somenzi, F.: CUDD: CU decision diagram package. Accessed November 2007 [Online]. http://vlsi.colorado.edu/ fabio/CUDD/cudd.html
25. Somenzi, F., Brayton, R.K.: An exact minimizer for Boolean relations. In: Proceedings of International Conference on Computer-Aided Design, pp. 316–319. Santa Clara, CA (1989)
26. Watanabe, Y., Brayton, R.: Heuristic minimization of multi-valued relations. In: Proceedings of IEEE Transactions on Computer-Aided Design, vol.12, pp. 1458–1472. (1993)
27. Wurth, B., Wehn, N.: Efficient calculation of Boolean relations for multi-level logic optimization. In: Proceedings of European Design and Test Conference, pp. 630–634. Paris, France (1994)
28. Yamashita, S., Sawada, H., Nagoya, A.: A new method to express functional permissibilities for LUT based FPGAs and its applications. In: Proceedings of the International Conference on Computer-Aided Design, pp. 254–261. San Jose, CA (1996)

Part V
Applications to Specialized Design Scenarios

In this section, the first chapter describes the generation of arbitrary decimal probabilities from small sets of probabilities (or pairs of probabilities) through combinational logic. The next work presents a circuit reliability calculator which employs Boolean difference. A realizable combination logic circuit using sinusoidal signals, along with gates that can operate on such signals, is presented in the third chapter in the applications category. The last chapter investigates throughput reduction and synchronization failures introduced by existing GALS pausible clocking schemes for SoCs and NoCs and proposes an optimized scheme for more reliable GALS system design.

Chapter 18
Synthesizing Combinational Logic to Generate Probabilities: Theories and Algorithms

Weikang Qian, Marc D. Riedel, Kia Bazargan, and David J. Lilja

Abstract As CMOS devices are scaled down into the nanometer regime, concerns about reliability are mounting. Instead of viewing nano-scale characteristics as an impediment, technologies such as PCMOS exploit them as a source of randomness. The technology generates random numbers that are used in probabilistic algorithms. With the PCMOS approach, different voltage levels are used to generate different probability values. If many different probability values are required, this approach becomes prohibitively expensive. In this chapter, we demonstrate a novel technique for synthesizing logic that generates new probabilities from a given set of probabilities. We focus on synthesizing combinational logic to generate arbitrary *decimal* probabilities from a given set of input probabilities. We demonstrate how to generate arbitrary decimal probabilities from small sets – a single probability or a pair of probabilities – through combinational logic.

18.1 Introduction and Background

It can be argued that the entire success of the semiconductor industry has been predicated on a single, fundamental abstraction, namely, that digital computation consists of a deterministic sequence of zeros and ones. From the logic level up, the precise Boolean functionality of a circuit is prescribed; it is up to the physical layer to produce voltage values that can be interpreted as the exact logical values that are called for. This abstraction delivers all the benefits of the digital paradigm: precision, modularity, extensibility. And yet, as circuits are scaled down into the nanometer regime, delivering the physical circuits underpinning the abstraction is increasingly costly and challenging. Power consumption is a major concern [6].

W. Qian (✉)
University of Minnesota, Minneapolis, MN, USA
e-mail: qianx030@umn.edu

This work is based on an earlier work: The synthesis of combinational logic to generate probabilities, in Proceedings of the 2009 international Conference on Computer-Aided Design, ISBN:978-1-60558-800-1 (2009) © ACM, 2009. DOI= http://doi.acm.org/10.1145/1687399.1687470

S.P. Khatri, K. Gulati (eds.), *Advanced Techniques in Logic Synthesis, Optimizations and Applications*, DOI 10.1007/978-1-4419-7518-8_18,

Also, soft errors caused by ionizing radiation are a problem, particularly for circuits operating in harsh environments [1].

We advocate a novel view for digital computation: instead of transforming definite inputs into definite outputs – say, Boolean, integer, or real values into the same – we design circuits that transform probability values into probability values; so, conceptually, real-valued probabilities are both the inputs and the outputs. The circuits process random bit streams; these are digital, consisting of zeros and ones; they are processed by ordinary logic gates, such as AND and OR. The inputs and outputs are encoded through the statistical distribution of the signals instead of specific values. When cast in terms of probabilities, the computation is robust [10].

The topic of computing probabilistically dates back to von Neumann [9]. Many flavors of probabilistic design have been proposed for circuit-level constructs. For instance, [8] presents a design methodology based on Markov random fields, geared toward nanotechnology. Recent work on *probabilistic* CMOS (PCMOS) is a promising approach. Instead of viewing variable circuit characteristics as an impediment, PCMOS exploits them as a source of randomness. The technology generates random numbers that are used in probabilistic algorithms [3].

A PCMOS switch is an inverter with the input coupled to a noise source, as shown in Fig. 18.1. With the input V_{in} set to 0 volts, the output of the inverter has a certain probability p $(0 \leq p \leq 1)$ of being at logical one. Suppose that the probability density function of the noise voltage V is $f(V)$ and that the trip point of the inverter is $V_{dd}/2$, where V_{dd} is the supply voltage. Then, the probability that the output is one equals the probability that the input to the inverter is below $V_{dd}/2$, or

$$p = \int_{-\infty}^{V_{dd}/2} f(V)\,\mathrm{d}V,$$

which corresponds to the shaded area in Fig. 18.2. Thus, with a given noise distribution, p can be modulated by changing V_{dd}.

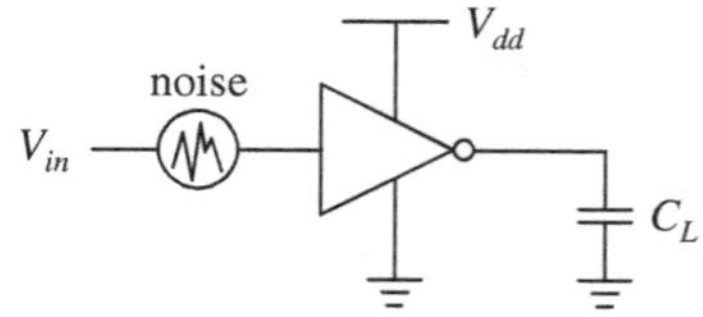

Fig. 18.1 A PCMOS switch. It consists of an inverter with its input coupled to a noise source

In [2], PCMOS switches are applied to form a probabilistic system-on-a-chip (PSOC) architecture that is used to execute probabilistic algorithms. In essence, the PSOC architecture consists of a host processor that executes the deterministic part of the algorithm, and a coprocessor built with PCMOS switches that executes the probabilistic part of the algorithm. The PCMOS switches in the coprocessor are configured to realize the set of probabilities needed by the algorithm. This approach achieves an energy-performance-product improvement over conventional architectures for some probabilistic algorithms.

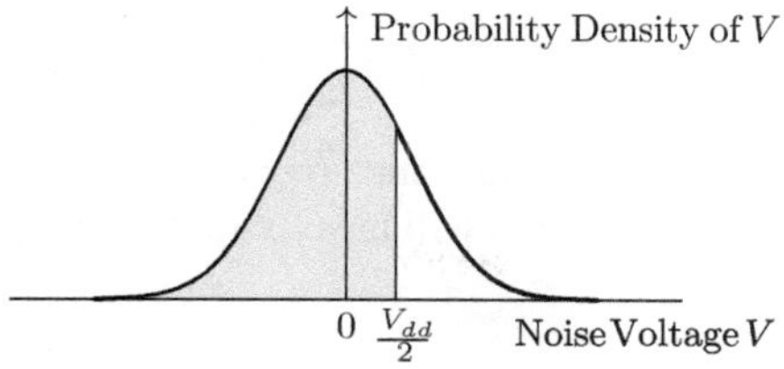

Fig. 18.2 Probability density function of the noise source. The probability that the output of the PCMOS switch is one equals the *shaded area* in the figure. Changing V_{dd} will change this probability

However, as is pointed out in [2], a serious problem must be overcome before PCMOS can become a viable design strategy for many applications: since the probability p for each PCMOS switch is controlled by a specific voltage level, different voltage levels are required to generate different probability values. For an application that requires many different probability values, many voltage regulators are required; this is costly in terms of area as well as energy.

In this chapter, we present a synthesis strategy to mitigate this issue: we describe a method for transforming probability values from a small set to many different probability values entirely through combinational logic. For what follows, when we say "with probability p," we mean "with a probability p of being at logical one." When we say "a circuit," we mean a combinational circuit built with logic gates.

Example 18.1 Suppose that we have a set of probabilities $S = \{0.4, 0.5\}$. As illustrated in Fig. 18.3, we can generate new probabilities from this set:

1. An inverter with an input x with probability 0.4 will have output z with probability 0.6 since for an inverter,

$$P(z = 1) = P(x = 0) = 1 - P(x = 1). \tag{18.1}$$

2. An AND gate with inputs x and y with *independent* probabilities 0.4 and 0.5, respectively, will have an output z with probability 0.2 since for an AND gate,

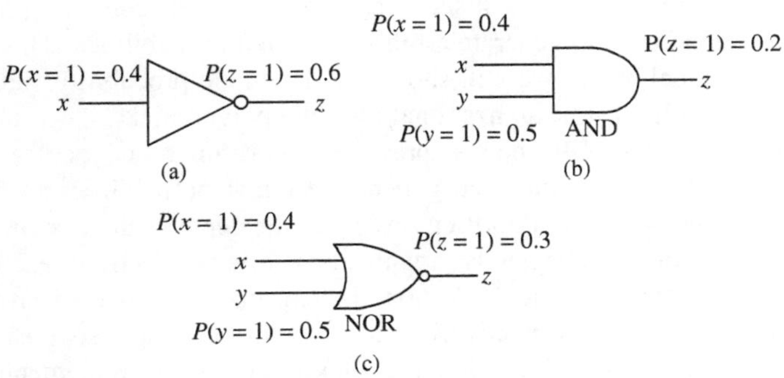

Fig. 18.3 An illustration of generating new probabilities from a given set of probabilities through logic. (**a**): An inverter implementing $p_z = 1 - p_x$. (**b**): An AND gate implementing $p_z = p_x \cdot p_y$. (**c**): A NOR gate implementing $p_z = (1 - p_x) \cdot (1 - p_y)$

$$P(z = 1) = P(x = 1, y = 1) = P(x = 1)P(y = 1). \tag{18.2}$$

3. A NOR gate with inputs x and y with *independent* probabilities 0.4 and 0.5, respectively, will have output z with probability 0.3 since for a NOR gate,

$$\begin{aligned} P(z = 1) &= P(x = 0, y = 0) = P(x = 0)P(y = 0) \\ &= (1 - P(x = 1))(1 - P(y = 1)). \end{aligned} \tag{18.3}$$

Thus, using only combinational logic, we can get the additional set of probabilities $\{0.2, 0.3, 0.6\}$. □

Motivated by this example, we consider the problem of how to synthesize combinational logic to generate a required probability q from a given set of probabilities $S = \{p_1, p_2, \ldots, p_n\}$. Specifically, we focus on synthesizing arbitrary decimal probabilities (i.e., q is a decimal number). We assume that the probabilities in a set S can be freely chosen and each element in S can be used as the input probability any number of times. (We say that the probability is *duplicable*.) The problem is to find a good set S such that, for an arbitrary decimal probability, we can construct a circuit to generate it.

As a result, in Section 18.3, we will show that there exist sets consisting of two elements that can be used to generate arbitrary decimal probabilities. In fact, in Section 18.3.1, we will first show that we can generate arbitrary decimal probabilities from the set $S = \{0.4, 0.5\}$. The proof is constructive: we will show a procedure for synthesizing logic that generates such probabilities. Next, in Section 18.3.2, we will show that we can generate arbitrary decimal probabilities from the set $S = \{0.5, 0.8\}$. We will show that with this set of input probabilities, for an output probability of n decimal digits, we can synthesize combinational logic with $2n$ inputs.

Further, in Section 18.4, we will show that there exist sets consisting of a single element that can be used to generate arbitrary decimal probabilities. This is essentially a mathematical result: we will show that the single probability value cannot be a rational value; it must be an irrational root of a polynomial.

In Section 18.5, we will show a practical algorithm based on fraction factorization to synthesize circuits that generate decimal probabilities from the set $S = \{0.4, 0.5\}$. The proposed algorithm optimizes the depth of the circuit.

The remainder of this chapter is organized as follows: Section 18.2 describes related work. Sections 18.3 and 18.4 show the existence of a pair of probabilities and of a single probability, respectively, that can be used as input sources to generate arbitrary decimal probabilities. Section 18.5 describes our implementation and presents algorithms for optimizing the resulting circuits. Section 18.6 demonstrates the effectiveness of the proposed algorithms. Finally, Section 18.7 summarizes this chapter.

18.2 Related Work

We point to three related pieces of research:

- In an early set of papers, Gill discussed the problem of generating a new set of probabilities from a given set of probabilities [4, 5]. He focused on synthesizing a sequential state machine to generate the required probabilities.
- In recent work, the proponents of PCMOS discussed the problem of synthesizing combinational logic to generate probability values [2]. These authors suggest a tree-based circuit. Their objective is to realize a set of required probabilities with minimal additional logic. This is positioned as future work; no details are given.
- Wilhelm and Bruck [11] proposed a general method for synthesizing *switching circuits* to achieve a desired probability. Their designs consist of relay switches that are open or closed with specified probabilities. They proposed an algorithm that generates circuits of optimal size for any binary fraction.

In contrast to Gill's work and Wilhelm and Bruck's work, we focus on combinational circuits built with logic gates. Our approach dovetails nicely with the circuit-level PCMOS constructs. It is complementary and orthogonal to the switch-based approach of Wilhelm and Bruck. Our scheme can generate arbitrary decimal probabilities, whereas the method of Wilhelm and Bruck only generates binary fractions.

18.3 Sets with Two Elements that Can Generate Arbitrary Decimal Probabilities

In this section, we will show two input probability sets that contain only two elements and can generate arbitrary decimal probabilities. The first one is the set $S = \{0.4, 0.5\}$ and the second one is the set $S = \{0.5, 0.8\}$.

18.3.1 Generating Decimal Probabilities from the Input Probability Set $S = \{0.4, 0.5\}$

We will first show that we can generate arbitrary decimal probabilities from the input probability set $S = \{0.4, 0.5\}$. Then, we will show an algorithm to synthesize circuits that generate arbitrary decimal probabilities from the set of input probabilities.

Theorem 18.1 *With circuits consisting of fanin-two AND gates and inverters, we can generate arbitrary decimal fractions as output probabilities from the input probability set $S = \{0.4, 0.5\}$.*

Proof First, we note that an inverter with a probabilistic input gives an output probability equal to one minus the input probability, as was shown in (18.1). An AND gate with two probabilistic inputs performs a multiplication on the two input probabilities, as was shown in (18.2). Thus, we need to prove that with the two operations

$1 - x$ and $x \cdot y$, we can generate arbitrary decimal fractions as output probabilities from the input probability set $S = \{0.4, 0.5\}$. We prove this statement by induction on the number of digits n after the decimal point.

Base case:

1. $n = 0$. It is trivial to generate 0 and 1.
2. $n = 1$. We can generate 0.1, 0.2, and 0.3 as follows:

$$0.1 = 0.4 \times 0.5 \times 0.5,$$
$$0.2 = 0.4 \times 0.5,$$
$$0.3 = (1 - 0.4) \times 0.5.$$

Since we can generate the decimal fractions 0.1, 0.2, 0.3, and 0.4, we can generate 0.6, 0.7, 0.8, and 0.9 with an extra $1 - x$ operation. Together with the given value 0.5, we can generate any decimal fraction with one digit after the decimal point.

Inductive step:

Assume that the statement holds for all $m \leq (n - 1)$. Consider an arbitrary decimal fraction z with n digits after the decimal point. Let $u = 10^n \cdot z$. Here u is an integer.

Consider the following four cases.

1. The case where $0 \leq z \leq 0.2$.
   a. The integer u is divisible by 2. Let $w = 5z$. Then $0 \leq w \leq 1$ and $w = (u/2) \cdot 10^{-n+1}$, having at most $(n - 1)$ digits after the decimal point. Thus, based on the induction hypothesis, we can generate w. It follows that z can also be generated as $z = 0.4 \times 0.5 \times w$.
   b. The integer u is not divisible by 2 and $0 \leq z \leq 0.1$. Let $w = 10z$. Then $0 \leq w \leq 1$ and $w = u \cdot 10^{-n+1}$, having at most $(n - 1)$ digits after the decimal point. Thus, based on the induction hypothesis, we can generate w. It follows that z can also be generated as $z = 0.4 \times 0.5 \times 0.5 \times w$.
   c. The integer u is not divisible by 2 and $0.1 < z \leq 0.2$. Let $w = 2 - 10z$. Then $0 \leq w < 1$ and $w = 2 - u \cdot 10^{-n+1}$, having at most $(n - 1)$ digits after the decimal point. Thus, based on the induction hypothesis, we can generate w. It follows that z can also be generated as $z = (1 - 0.5 \times w) \times 0.4 \times 0.5$.

2. The case where $0.2 < z \leq 0.4$.
   a. The integer u is divisible by 4. Let $w = 2.5z$. Then $0 < w \leq 1$ and $w = (u/4) \cdot 10^{-n+1}$, having at most $(n - 1)$ digits after the decimal point. Thus, based on the induction hypothesis, we can generate w. It follows that z can be generated as $z = 0.4 \times w$.
   b. The integer u is not divisible by 4 but is divisible by 2. Let $w = 2 - 5z$. Then $0 \leq w < 1$ and $w = 2 - (u/2) \cdot 10^{-n+1}$, having at most $(n - 1)$ digits after

the decimal point. Thus, based on the induction hypothesis, we can generate w. It follows that z can be generated as $z = (1 - 0.5 \times w) \times 0.4$.

c. The integer u is not divisible by 2 and $0.2 < u \leq 0.3$. Let $w = 10z - 2$. Then $0 < w \leq 1$ and $w = u \cdot 10^{-n+1} - 2$, having at most $(n - 1)$ digits after the decimal point. Thus, based on the induction hypothesis, we can generate w. It follows that z can also be generated as $z = (1 - (1 - 0.5 \times w) \times 0.5) \times 0.4$.

d. The integer u is not divisible by 2 and $0.3 < u \leq 0.4$. Let $w = 4 - 10z$. Then $0 \leq w < 1$ and $w = 4 - u \cdot 10^{-n+1}$, having at most $(n - 1)$ digits after the decimal point. Thus, based on the induction hypothesis, we can generate w. It follows that z can be generated as $z = (1 - 0.5 \times 0.5 \times w) \times 0.4$.

3. The case where $0.4 < z \leq 0.5$. Let $w = 1 - 2z$. Then $0 \leq w < 0.2$ and w falls into case 1. Thus, we can generate w. It follows that z can be generated as $z = 0.5 \times (1 - w)$.
4. The case where $0.5 < z \leq 1$. Let $w = 1 - z$. Then $0 \leq w < 0.5$ and w falls into one of the above three cases. Thus, we can generate w. It follows that z can be generated as $z = 1 - w$.

For all of the above cases, we proved that z can be generated with the two operations $1 - x$ and $x \cdot y$ on the input probability set $S = \{0.4, 0.5\}$. Thus, we proved the statement for all $m \leq n$. Thus, the statement holds for all integers n. □

Based on the proof above, we derive an algorithm to synthesize a circuit that generates an arbitrary decimal fraction output probability z from the input probability set $S = \{0.4, 0.5\}$. See Algorithm 11.

The function GetDigits(z) in Algorithm 11 returns the number of digits after the decimal point of z. The while loop continues until z has at most one digit after the decimal point. During the loop, it calls the function ReduceDigit(ckt, z), which synthesizes a partial circuit such that the number of digits after the decimal point of z is reduced, which corresponds to the inductive step in the proof. Finally, Algorithm 11 calls the function AddBaseCkt(ckt, z) to synthesize a circuit that realizes a number having at most one digit after the decimal point; this corresponds to the base case of the proof.

Algorithm 11 Synthesize a circuit consisting of AND gates and inverters that generates a required decimal fraction probability from the given probability set $S = \{0.4, 0.5\}$.

1: {Given an arbitrary decimal fraction $0 \leq z \leq 1$.}
2: Initialize ckt;
3: **while** GetDigits(z) > 1 **do**
4:     (ckt, z) $\Leftarrow$ ReduceDigit(ckt, z);
5: **end while**
6: AddBaseCkt(ckt, z); {Base case: z has at most one digit after the decimal point.}
7: Return ckt;

Algorithm 12 ReduceDigit(ckt, z)

```
1: {Given a partial circuit ckt and an arbitrary decimal fraction 0 ≤ z ≤ 1.}
2: n ⇐ GetDigits(z);
3: if z > 0.5 then {Case 4}
4:     z ⇐ 1 − z; AddInverter(ckt);
5: end if
6: if 0.4 < z ≤ 0.5 then {Case 3}
7:     z ⇐ z/0.5; AddAND(ckt, 0.5);
8:     z ⇐ 1 − z; AddInverter(ckt);
9: end if
10: if z ≤ 0.2 then {Case 1}
11:     z ⇐ z/0.4; AddAND(ckt, 0.4);
12:     z ⇐ z/0.5; AddAND(ckt, 0.5);
13:     if GetDigits(z) < n then
14:         go to END;
15:     end if
16:     if z > 0.5 then
17:         z ⇐ 1 − z; AddInverter(ckt);
18:     end if
19:     z = z/0.5; AddAND(ckt, 0.5);
20: else {Case 2: 0.2 < z ≤ 0.4}
21:     z ⇐ z/0.4; AddAND(ckt, 0.4);
22:     if GetDigits(z) < n then
23:         go to END;
24:     end if
25:     z ⇐ 1 − z; AddInverter(ckt);
26:     z ⇐ z/0.5; AddAND(ckt, 0.5);
27:     if GetDigits(z) < n then
28:         go to END;
29:     end if
30:     if z > 0.5 then
31:         z ⇐ 1 − z; AddInverter(ckt);
32:     end if
33:     z = z/0.5; AddAND(ckt, 0.5);
34: end if
35: END: return ckt, z;
```

Algorithm 11 builds the circuit from the output back to the inputs. The circuit is built up gate by gate when calling the function ReduceDigit(ckt, z), shown in Algorithm 12. Here the function AddInverter(ckt) attaches an inverter to the input of the circuit ckt and then changes the input of the circuit to the input of the inverter. The function AddAND(ckt, p) attaches a fanin-two AND gate to the input of the circuit and then changes the input of the circuit to one of the inputs of the AND gate. The other input of the AND gate is connected to a random input source of probability p. In Algorithm 12, Lines 3–5 correspond to Case 4 in the proof; Lines 6–9 correspond to Case 3 in the proof; Lines 10–19 correspond to Case 1 in the proof; Lines 20–34 correspond to Case 2 in the proof.

The synthesized circuit has a number of gates that is linear in the number of digits after the required value's decimal point, since at most three AND gates and three inverters are needed to generate a value with n digits after the decimal point from a value with $(n-1)$ digits after the decimal point.[1] The number of primary inputs of the synthesized circuit is at most $3n+1$.

Example 18.2 We show how to generate the probability value 0.757. Based on Algorithm 11, we can derive a sequence of operations that transform 0.757 to 0.7:

$$0.757 \xrightarrow{1-} 0.243 \xrightarrow{/0.4} 0.6075 \xrightarrow{1-} 0.3925 \xrightarrow{/0.5} 0.785$$
$$\xrightarrow{1-} 0.215 \xrightarrow{/0.5} 0.43,$$
$$0.43 \xrightarrow{/0.5} 0.86 \xrightarrow{1-} 0.14 \xrightarrow{/0.4} 0.35 \xrightarrow{/0.5} 0.7.$$

Since 0.7 can be realized as $0.7 = 1-(1-0.4)\times 0.5$, we obtain the circuit shown in Fig. 18.4. (Note that here we use a black dot to represent an inverter.) □

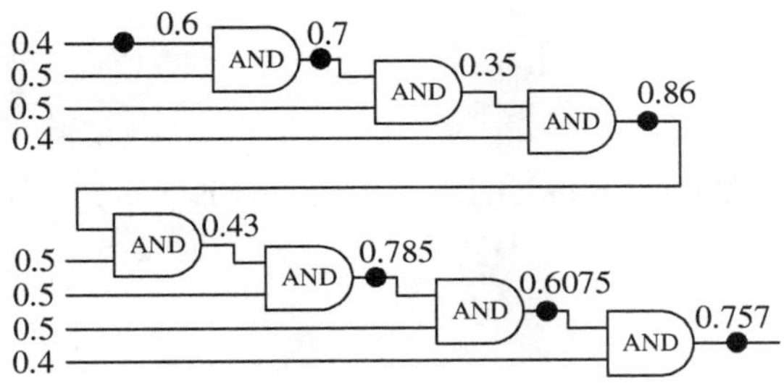

Fig. 18.4 A circuit taking input probabilities from the set $S = \{0.4, 0.5\}$ generating a decimal output probability of 0.757

18.3.2 Generating Decimal Probabilities from the Input Probability Set $S = \{0.5, 0.8\}$

Given a probability set $S = \{0.4, 0.5\}$, the algorithm in the previous section produces a circuit with at most $3n+1$ inputs to generate a decimal probability of n digits. If we use the set $S = \{0.5, 0.8\}$, then we can do better in terms of the number of inputs. With this set, we can synthesize a circuit with at most $2n$ inputs that generates a decimal probability of n digits. To prove this, we need the following lemma.

Lemma 18.1 *Given an integer $n \geq 2$, for any integer $0 \leq m \leq 10^n$, there exist integers $0 \leq a_i \leq 2^n\binom{n}{i}$, $i = 0, 1, \ldots, n$, such that $m = \sum_{i=0}^{n} a_i 4^i$.*

Proof Define $s_k = \sum_{i=0}^{k} 2^n \binom{n}{i} 4^i$. We first prove the following statement:

[1] In Case 3, z is transformed into $w = 1-2z$ where w is in Case 1(a). Thus, we actually need only three AND gates and one inverter for Case 3. For the other cases, it is not hard to see that we need at most three AND gates and three inverters.

Given $0 \leq k \leq n$, for any integer $0 \leq m \leq s_k$, there exist integers $0 \leq a_i \leq 2^n\binom{n}{i}$, $i = 0, 1, \ldots, k$, such that $m = \sum_{i=0}^{k} a_i 4^i$.

We prove the above statement by induction on k.

Base case: When $k = 0$, we have $s_0 = 2^n$. For any integer $0 \leq m \leq 2^n$, let $a_0 = m$. Then $0 \leq a_0 \leq 2^n\binom{n}{0}$. The statement is true for $k = 0$.

Inductive step: Assume the statement holds for $k - 1$ $(k \leq n)$. Consider the statement for k. There are two cases for $0 \leq m \leq s_k$.

1. $0 \leq m \leq 2^n\binom{n}{k}4^k$. Let $a_k = \left\lfloor \frac{m}{4^k} \right\rfloor$. Then,

$$0 \leq a_k \leq \frac{m}{4^k} \leq 2^n\binom{n}{k}$$

and

$$0 \leq m - a_k 4^k < 4^k \leq 2^n 4^{k-1} \leq \sum_{i=0}^{k-1} 2^n\binom{n}{i}4^i = s_{k-1}.$$

Based on the induction hypothesis, there exist integers $0 \leq a_i \leq 2^n\binom{n}{i}$, $i = 0, 1, \ldots, k-1$, such that

$$m - a_k 4^k = \sum_{i=0}^{k-1} a_i 4^i.$$

Therefore, $m = \sum_{i=0}^{k} a_i 4^i$, where $0 \leq a_i \leq 2^n\binom{n}{i}$, for $i = 0, 1, \ldots, k$.

2. $2^n\binom{n}{k}4^k < m \leq s_k$. Let $a_k = 2^n\binom{n}{k}$. Then,

$$0 < m - a_k 4^k \leq s_k - 2^n\binom{n}{k}4^k = s_{k-1}.$$

Based on the induction hypothesis, there exist integers $0 \leq a_i \leq 2^n\binom{n}{i}$, $i = 0, 1, \ldots, k-1$, such that

$$m - a_k 4^k = \sum_{i=0}^{k-1} a_i 4^i.$$

Therefore, $m = \sum_{i=0}^{k} a_i 4^i$, where $0 \leq a_i \leq 2^n\binom{n}{i}$, for $i = 0, 1, \ldots, k$.

Thus, the statement is true for all $0 \leq k \leq n$.

Note that when $k = n$,

$$s_k = \sum_{i=0}^{n} 2^n\binom{n}{i}4^i = 2^n(4+1)^n = 10^n.$$

Thus, for any integer $0 \le m \le 10^n = s_n$, there exist integers $0 \le a_i \le 2^n\binom{n}{i}$, $i = 0, 1, \ldots, n$, such that $m = \sum_{i=0}^{n} a_i 4^i$. □

With the above lemma, we can prove the following theorem.

Theorem 18.2 *For any decimal fraction of n ($n \ge 2$) digits, there exists a combinational circuit with $2n$ inputs generating that decimal probability with input probabilities taken from the set $S = \{0.5, 0.8\}$.*

Proof Consider combination logic with $2n$ inputs $x_1, x_2, \ldots, x_{2n}$ with input probabilities set as

$$P(x_i = 1) = \begin{cases} 0.8, & i = 1, \ldots, n, \\ 0.5, & i = n+1, \ldots, 2n. \end{cases}$$

For $n + 1 \le i \le 2n$, since $P(x_i = 1) = 0.5$, we have $P(x_i = 1) = P(x_i = 0) = 0.5$. Therefore, the probability of a certain input combination occurring only depends on the values of the first n inputs or, more precisely, only depends on the number of ones in the first n inputs. Thus, there are in total $2^n\binom{n}{i}$ $(0 \le i \le n)$ input combinations whose probability of occurring is $0.8^i \cdot 0.2^{n-i} \cdot 0.5^n$.

Suppose the given decimal fraction of n digits is $q = \frac{m}{10^n}$, where $0 \le m \le 10^n$ is an integer. Then, based on Lemma 18.1, there exist integers $0 \le a_i \le 2^n\binom{n}{i}$, $i = 0, 1, \ldots, n$, such that $m = \sum_{i=0}^{n} a_i 4^i$.

For each $0 \le i \le n$, since $0 \le a_i \le 2^n\binom{n}{i}$, we are able to choose a_i out of $2^n\binom{n}{i}$ input combinations whose probability of occurring is $0.8^i \cdot 0.2^{n-i} \cdot 0.5^n$; let the combinational logic evaluate to one for these a_i input combinations. Thus, the probability of the output being one is the sum of the probability of occurrence of all input combinations for which the logic evaluates to one, which is

$$\sum_{i=0}^{n} a_i 0.8^i \cdot 0.2^{n-i} \cdot 0.5^n = \sum_{i=0}^{n} a_i 4^i \cdot 0.1^n = \frac{m}{10^n} = q. \qquad \square$$

Remarks: Like Theorem 18.1, Theorem 18.2 implies a procedure for synthesizing combinational logic to generate a required decimal fraction. Although this procedure will synthesize a circuit with fewer inputs than that synthesized through Algorithm 11, the number of two-input logic gates in this circuit may be greater. Moreover, for this procedure, we must judiciously choose a_i out of $2^n\binom{n}{i}$ input combinations with probability $0.8^i \cdot 0.2^{n-i} \cdot 0.5^n$ of occurring as the minterms of the Boolean function, in order to minimize the gate count. In contrast, Algorithm 11 produces a circuit directly.

18.4 Sets with a Single Element that Can Generate Arbitrary Decimal Probabilities

In Section 18.3, we showed that there exist input probability sets of two elements that can be used to generate arbitrary decimal fractions. A stronger question is whether we can further reduce the size of the set down to one, i.e., whether there exists a real number $0 \le p \le 1$ such that any decimal fraction can be generated from p with combinational logic.

The first answer to this question is that there is no *rational* number p such that an arbitrary decimal fraction can be generated from that p through combinational logic. To prove this, we first need the following lemma.

Lemma 18.2 *If the probability $\frac{1}{2}$ can be generated from a rational probability p through combinational logic, then $p = \frac{1}{2}$.*

Proof Obviously, $0 < p < 1$. Thus, we can assume that

$$p = \frac{a}{b}, \tag{18.4}$$

where both a and b are positive integers, satisfying that $a < b$ and $(a, b) = 1$.

Moreover, we can assume that $a \ge b - a$. Otherwise, suppose that $a < b - a$. Since we can generate $\frac{1}{2}$ from p, we can also generate $\frac{1}{2}$ from $p^* = 1 - p$ by using an inverter to convert p^* into p. Note that $p^* = \frac{a^*}{b^*}$, where $a^* = b - a$ and $b^* = b$, satisfying that $a^* > b^* - a^*$. Thus, we can assume that $a \ge b - a$.

Suppose that the combinational logic generating $\frac{1}{2}$ from p has n inputs. Let l_k $(k = 0, 1, \ldots, n)$ be the number of input combinations that evaluate to one and have exactly k ones. Note that $0 \le l_k \le \binom{n}{k}$, for $k = 0, 1, \ldots, n$.

Since each input of the combinational logic has probability p of being 1, we have

$$\frac{1}{2} = \sum_{k=0}^{n} l_k (1 - p)^{n-k} p^k. \tag{18.5}$$

Let $c = b - a$. Based on (18.4), we can rewrite (18.5) as

$$b^n = 2 \sum_{k=0}^{n} l_k a^k c^{n-k}. \tag{18.6}$$

From (18.6), we can show that $a = 1$, which we prove by contradiction.

Suppose that $a > 1$. Since $0 \le l_0 \le \binom{n}{0} = 1$, l_0 is either 0 or 1. If $l_0 = 0$, then from (18.6), we have

$$b^n = 2\sum_{k=1}^{n} l_k a^k c^{n-k} = 2a \sum_{k=1}^{n} l_k a^{k-1} c^{n-k}.$$

Thus, $a|b^n$. Since $(a, b) = 1$, the only possibility is that $a = 1$ which is contradictory to our hypothesis that $a > 1$. Therefore, we have $l_0 = 1$. Together with binomial expansion

$$b^n = \sum_{k=0}^{n} \binom{n}{k} a^k c^{n-k},$$

we can rewrite (18.6) as

$$c^n + \sum_{k=1}^{n} \binom{n}{k} a^k c^{n-k} = 2c^n + 2\sum_{k=1}^{n} l_k a^k c^{n-k},$$

or

$$c^n = a \sum_{k=1}^{n} \left(\binom{n}{k} - 2l_k \right) a^{k-1} c^{n-k}. \tag{18.7}$$

Thus, $a|c^n$. Since $(a, b) = 1$ and $c = b - a$, we have $(a, c) = 1$. Thus, the only possibility is that $a = 1$, which is contradictory to our hypothesis that $a > 1$.

Therefore, we proved that $a = 1$. Together with the assumption that $b - a \leq a < b$, we get $b = 2$. Thus, p can only be $\frac{1}{2}$. □

Now, we can prove the original statement:

Theorem 18.3 *There is no rational number p such that an arbitrary decimal fraction can be generated from that p with combinational logic.*

Proof We prove the above statement by contradiction. Suppose that there exists a rational number p such that an arbitrary decimal fraction can be generated from it through combinational logic.

Since an arbitrary decimal fraction can be generated from p, $0.5 = \frac{1}{2}$ can be generated. Thus, based on Lemma 18.2, we have $p = \frac{1}{2}$.

Note that $0.2 = \frac{1}{5}$ is also a decimal number. Thus, there exists combinational logic which can generate the decimal fraction $\frac{1}{5}$ from $p = \frac{1}{2}$. Suppose that the combinational logic has n inputs. Let m_k $(k = 0, 1, \ldots, n)$ be the number of input combinations that evaluate to one and that have exactly k ones.

Since each input of the combinational logic has probability $p = \frac{1}{2}$ of being 1, we have

$$\frac{1}{5} = \sum_{k=0}^{n} m_k \left(1 - \frac{1}{2}\right)^{n-k} \left(\frac{1}{2}\right)^k,$$

or

$$2^n = 5 \sum_{k=0}^{n} m_k,$$

which is impossible since the right-hand side is a multiple of 5.

Therefore, we proved the statement in the theorem. □

Thus, based on Theorem 18.3, we have the conclusion that if such a p exists, it must be an irrational number.

On the one hand, we note that if such a value p exists, then 0.4 and 0.5 can be generated from it. On the other hand, if p can generate 0.4 and 0.5, then p can generate arbitrary decimal numbers, as was shown in Theorem 18.1. The following lemma shows that such a value p that could generate 0.4 and 0.5 does, in fact, exist.

Lemma 18.3 *The polynomial* $g_1(t) = 10t - 20t^2 + 20t^3 - 10t^4 - 1$ *has a real root* $0 < p < 0.5$. *This value* p *can generate both* 0.4 *and* 0.5 *through combinational logic.*

Proof First, note that $g_1(0) = -1 < 0$ and that $g_1(0.5) = 0.875 > 0$. Based on the continuity of the function $g_1(t)$, there exists a $0 < p < 0.5$ such that $g_1(p) = 0$. Let polynomial $g_2(t) = t - 2t^2 + 2t^3 - t^4$. Thus, $g_2(p) = 0.1$.

Note that the Boolean function

$$f_1(x_1, x_2, x_3, x_4, x_5) = (x_1 \vee x_2 \vee x_3 \vee x_4 \vee x_5) \wedge (\neg x_1 \vee \neg x_2 \vee \neg x_3 \vee \neg x_4 \vee \neg x_5)$$

has 30 minterms, $m_1, m_2, \ldots, m_{30}$. It is not hard to verify that with $P(x_i = 1) = p$ for $i = 1, 2, 3, 4, 5$, the output probability of f_1 is

$$\begin{aligned} p_1 &= 5(1-p)^4 p + 10(1-p)^3 p^2 + 10(1-p)^2 p^3 + 5(1-p)p^4 \\ &= 5g_2(p) = 0.5. \end{aligned}$$

Thus, the probability value 0.5 can be generated. The Boolean function

$$\begin{aligned} f_2(x_1, x_2, x_3, x_4, x_5) = {} & (x_1 \vee x_2 \vee x_3 \vee x_4) \wedge (x_1 \vee x_3 \vee \neg x_5) \\ & \wedge (\neg x_2 \vee x_3 \vee \neg x_5) \wedge (\neg x_1 \vee \neg x_2 \vee \neg x_4 \vee \neg x_5) \end{aligned}$$

has 24 minterms, $m_2, m_4, m_5, \ldots, m_8, m_{10}, m_{12}, m_{13}, \ldots, m_{24}, m_{26}, m_{28}, m_{29}, m_{30}$. It is not hard to verify that with $P(x_i = 1) = p$ for $i = 1, 2, 3, 4, 5$, the output probability of f_2 is

$$p_2 = 4(1-p)^4 p + 8(1-p)^3 p^2 + 8(1-p)^2 p^3 + 4(1-p)p^4$$
$$= 4g_2(p) = 0.4.$$

Thus, the probability value 0.4 can be generated. □

Based on Theorem 18.1 and Lemma 18.3, we have the following theorem.

Theorem 18.4 *With the set $S = \{p\}$, where p is the root of the polynomial $g_1(t) = 10t - 20t^2 + 20t^3 - 10t^4 - 1$ in the unit interval, we can generate arbitrary decimal fractions with combinational logic.*

18.5 Implementation

In this section, we will discuss algorithms to optimize circuits that generate decimal probabilities from the input probability set $S = \{0.4, 0.5\}$.

As shown in Example 18.2, the circuit synthesized by Algorithm 11 is in a linear style (i.e., each gate adds to the depth of the circuit). For practical purposes, we want circuits with shallower depth. We explore two kinds of optimizations to reduce the depth.

The first kind of optimization is at the logic level. The circuit synthesized by Algorithm 11 is composed of inverters and AND gates. We can reduce its depth by properly repositioning certain AND gates, as illustrated in Fig. 18.5.

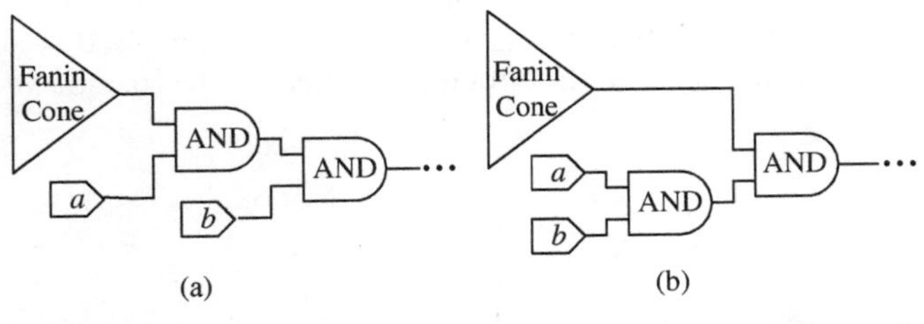

Fig. 18.5 An illustration of balancing to reduce the depth of the circuit. Here a and b are primary inputs. (**a**) The circuit before balancing. (**b**) The circuit after balancing

The second kind of optimization is at a higher level, based on the factorization of the decimal fraction. We use the following example to illustrate the basic idea.

Example 18.3 Suppose we want to generate the decimal fraction probability value 0.49.

Method based on Algorithm 11: We can derive the following transformation sequence:

$$0.49 \xrightarrow{/0.5} 0.98 \xrightarrow{1-} 0.02 \xrightarrow{/0.4} 0.05 \xrightarrow{/0.5} 0.1.$$

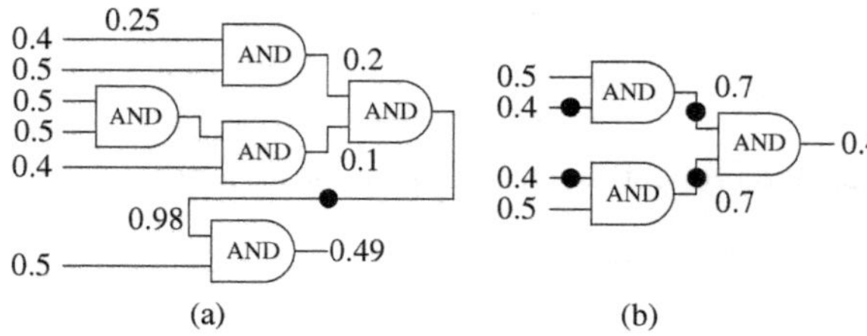

Fig. 18.6 Synthesizing combinational logic to generate probability 0.49. (**a**) The circuit synthesized through Algorithm 11. (**b**) The circuit synthesized based on fraction factorization

The synthesized circuit is shown in Fig. 18.6a. Notice that the circuit is balanced and it still has 5 AND gates and depth 4.[2]

Method based on factorization: Notice that $0.49 = 0.7 \times 0.7$. Thus, we can generate the probability 0.7 twice and feed these values into an AND gate. The synthesized circuit is shown in Fig. 18.6b. Compared to the circuit in Fig. 18.6a, both the number of AND gates and the depth of the circuit are reduced. □

Algorithm 13 shows the procedure that synthesizes the circuit based on the factorization of the decimal fraction. The factorization is actually carried out on the numerator. A crucial function is PairCmp(a_l, a_r, b_l, b_r), which compares the integer factor pair (a_l, a_r) with the pair (b_l, b_r) and returns a positive (negative) value if the pair (a_l, a_r) is better (worse) than the pair (b_l, b_r). Algorithm 14 shows how the function PairCmp(a_l, a_r, b_l, b_r) is implemented.

The quality of a factor pair (a_l, a_r) should reflect the quality of the circuit that generates the original probability based on that factorization. For this purpose, we define a function EstDepth(x) to estimate the depth of the circuit that generates the decimal fraction of a numerator x. If $1 \le x \le 9$, the corresponding fraction is $x/10$. EstDepth(x) is set as the depth of the circuit that generates the fraction $x/10$, which is

$$\text{EstDepth}(x) = \begin{cases} 0, & x = 4, 5, 6, \\ 1, & x = 2, 3, 7, 8, \\ 2, & x = 1, 9. \end{cases}$$

When $x \ge 10$, we use a simple heuristic to estimate the depth: we let EstDepth$(x) = \lceil \log_{10}(x) \rceil + 1$. The intuition behind this is that the depth of the circuit is a monotonically increasing function of the number of digits of x. The estimated depth of the circuit that generates the original fraction based on the factor pair (a_l, a_r) is

$$\max\{\text{EstDepth}(a_l), \text{EstDepth}(a_r)\} + 1. \tag{18.8}$$

The function PairCmp(a_l, a_r, b_l, b_r) essentially compares the quality of pair (a_l, a_r) and pair (b_l, b_r) based on (18.8). Further details are given in Algorithm 14.

---

[2] When counting depth, we ignore inverters.

Algorithm 13 ProbFactor(ckt, z)

1: {Given a partial circuit ckt and an arbitrary decimal fraction $0 \le z \le 1$.}
2: $n \Leftarrow$ GetDigits(z);
3: **if** $n \le 1$ **then**
4:     AddBaseCkt(ckt, z);
5:     Return ckt;
6: **end if**
7: $u \Leftarrow 10^n z$; $(u_l, u_r) \Leftarrow (1, u)$; {$u$ is the numerator of the fraction z}
8: **for** each factor pair (a, b) of integer u **do**
9:     **if** PairCmp$(u_l, u_r, a, b) < 0$ **then**
10:         $(u_l, u_r) \Leftarrow (a, b)$; {Choose the best factor pair for z}
11:     **end if**
12: **end for**
13: $w \Leftarrow 10^n - u$; $(w_l, w_r) \Leftarrow (1, w)$;
14: **for** each factor pair (a, b) of integer w **do**
15:     **if** PairCmp$(w_l, w_r, a, b) < 0$ **then**
16:         $(w_l, w_r) \Leftarrow (a, b)$; {Choose the best factor pair for $1 - z$}
17:     **end if**
18: **end for**
19: **if** PairCmp$(u_l, u_r, w_l, w_r) < 0$ **then**
20:     $(u_l, u_r) \Leftarrow (w_l, w_r)$; $z \Leftarrow w/10^n$;
21:     AddInverter(ckt);
22: **end if**
23: **if** IsTrivialPair(u_l, u_r) **then** {$u_l = 1$ or $u_r = u$}
24:     ReduceDigit(ckt, z); ProbFactor(ckt, z);
25:     Return ckt;
26: **end if**
27: $n_l \Leftarrow \lceil \log_{10}(u_l) \rceil$; $n_r \Leftarrow \lceil \log_{10}(u_r) \rceil$;
28: **if** $n_l + n_r > n$ **then** {Unable to factor z into two decimal fractions in the unit interval}
29:     ReduceDigit(ckt, z); ProbFactor(ckt, z);
30:     Return ckt;
31: **end if**
32: $z_l \Leftarrow u_l/10^{n_l}$; $z_r \Leftarrow u_r/10^{n_r}$;
33: ProbFactor(ckt_l, z_l); ProbFactor(ckt_r, z_r);
34: Add an AND gate with output as ckt and two inputs as ckt_l and ckt_r;
35: **if** $n_l + n_r < n$ **then**
36:     AddExtraLogic($ckt, n - n_l - n_r$);
37: **end if**
38: Return ckt;

In Algorithm 13, Lines 2–6 corresponds to the trivial fractions. If the fraction z is non-trivial, Lines 7–12 choose the best factor pair (u_l, u_r) of integer u, where u is the numerator of the fraction z. Lines 13–18 choose the best factor pair (w_l, w_r) of integer w, where w is the numerator of the fraction $1 - z$. Finally, the better factor pair of (u_l, u_r) and (w_l, w_r) is chosen. Here, we consider the factorization on both z and $1-z$, since in some cases the latter might be better than the former. An example is $z = 0.37$. Note that $1 - z = 0.63 = 0.7 \times 0.9$; this has a better factor pair than z itself.

After obtaining the best factor pair, we check whether we can utilize it. Lines 23–26 check whether the factor pair (u_l, u_r) is trivial. A factor pair is considered

Algorithm 14 PairCmp(a_l, a_r, b_l, b_r)

{Given two integer factor pairs (a_l, a_r) and (b_l, b_r)}
$c_l \Leftarrow$ EstDepth(a_l); $c_r \Leftarrow$ EstDepth(a_r);
$d_l \Leftarrow$ EstDepth(b_l); $d_r \Leftarrow$ EstDepth(b_r);
Order(c_l, c_r); {Order c_l and c_r, so that $c_l \leq c_r$}
Order(d_l, d_r); {Order d_l and d_r, so that $d_l \leq d_r$}
if $c_r < d_r$ **then** {The circuit w.r.t. the first pair has smaller depth}
    Return 1;
else if $c_r > d_r$ **then** {The circuit w.r.t. the first pair has larger depth}
    Return -1;
else
    if $c_l < d_l$ **then** {The circuit w.r.t. the first pair has fewer ANDs}
        Return 1;
    else if $c_l > d_l$ **then** {The circuit w.r.t. the first pair has more ANDs}
        Return -1;
    else
        Return 0;
    end if
end if

trivial if $u_l = 1$ or $u_r = 1$. If the best factor pair is trivial, we call the function ReduceDigit(ckt, z) (shown in Algorithm 12) to reduce the number of digits after the decimal point of z. Then we perform factorization on the new z.

If the best factor pair is non-trivial, Lines 27–31 continue to check whether the factor pair can be transformed into two decimal fractions in the unit interval. Let n_l be the number of digits of the integer u_l and n_r be the number of digits of the integer u_r. If $n_l + n_r > n$, where n is the number of digits after the decimal point of z, then it is impossible to utilize the factor pair (u_l, u_r) to factorize z. For example, consider $z = 0.143$. Although we could factorize $u = 143$ as 11×13, we could not utilize the factor pair (11, 13) for the factorization of 0.143. The reason is that either the factorization 0.11×1.3 or the factorization 1.1×0.13 contains a fraction larger than 1, which cannot be a probability value.

Finally, if it is possible to utilize the best factor pair, Lines 32–34 synthesize two circuits for fractions $u_l/10^{n_l}$ and $u_r/10^{n_r}$, respectively, and then combine these two circuits with an AND gate. Lines 35–37 check whether $n > n_l + n_r$. If this is the case, we have

$$z = u/10^n = u_l/10^{n_l} \cdot u_r/10^{n_r} \cdot 0.1^{n-n_l-n_r}.$$

We need to add an extra AND gate with one input probability $0.1^{n-n_l-n_r}$ and the other input probability $u_l/10^{n_l} \cdot u_r/10^{n_r}$. The extra logic is added through the function AddExtraLogic(ckt, m).

18.6 Empirical Validation

We empirically validate the effectiveness of the synthesis scheme that was presented in Section 18.5. For logic-level optimization, we use the "balance" command of the logic synthesis tool ABC [7], which we find very effective in reducing the depth of a tree-style circuit.[3]

Table 18.1 compares the quality of the circuits generated by three different schemes. The first scheme is called "Basic," which is based on Algorithm 11. It generates a linear-style circuit. The second scheme is called "Basic+Balance," which combines Algorithm 11 and the logic-level balancing algorithm. The third scheme is called "Factor+Balance," which combines Algorithm 13 and the logic-level balancing algorithm. We perform experiments on a set of target decimal probabilities that have n digits after the decimal point and average the results. The table shows the results for n ranging from 2 to 12. When $n \leq 5$, we synthesize circuits for all possible decimal fractions with n digits after the decimal point. When $n \geq 6$, we randomly choose 100000 decimal fractions with n digits after the decimal point as the synthesis targets. We show the average number of AND gates, average depth, and average CPU runtime in columns "#AND," "Depth," and "Runtime," respectively.

Table 18.1 A comparison of the basic synthesis scheme, the basic synthesis scheme with balancing, and the factorization-based synthesis scheme with balancing.

Number of digits	Basic		Basic+balance			Factor+balance				
	#AND	Depth	#AND	Depth	Runtime (ms)	#AND	Depth	Runtime (ms)	#AND Imprv. (%)	Depth Imprv. (%)
n			a_1	d_1		a_2	d_2		$100\frac{a_1 - a_2}{a_1}$	$100\frac{d_1 - d_2}{d_1}$
2	3.67	3.67	3.67	2.98	0.22	3.22	2.62	0.22	12.1	11.9
3	6.54	6.54	6.54	4.54	0.46	5.91	3.97	0.66	9.65	12.5
4	9.47	9.47	9.47	6.04	1.13	8.57	4.86	1.34	9.45	19.4
5	12.43	12.43	12.43	7.52	0.77	11.28	5.60	0.94	9.21	25.6
6	15.40	15.40	15.40	9.01	1.09	13.96	6.17	1.48	9.36	31.5
7	18.39	18.39	18.39	10.50	0.91	16.66	6.72	1.28	9.42	35.9
8	21.38	21.38	21.38	11.99	0.89	19.34	7.16	1.35	9.55	40.3
9	24.37	24.37	24.37	13.49	0.75	22.05	7.62	1.34	9.54	43.6
10	27.37	27.37	27.37	14.98	1.09	24.74	7.98	2.41	9.61	46.7
11	30.36	30.36	30.36	16.49	0.92	27.44	8.36	2.93	9.61	49.3
12	33.35	33.35	33.35	17.98	0.73	30.13	8.66	4.13	9.65	51.8

From Table 18.1, we can see that both the "Basic+Balance" and the "Factor+Balance" synthesis schemes have only millisecond-order CPU runtimes. Compared to the "Basic+Balance" scheme, the "Factor+Balance" scheme reduces by 10% the number of AND gates and by more than 10% the depth of the circuit for all

[3] We find that the other combinational synthesis commands of ABC such as "rewrite" do not affect the depth or the number of AND gates of a tree-style AND-inverter graph.

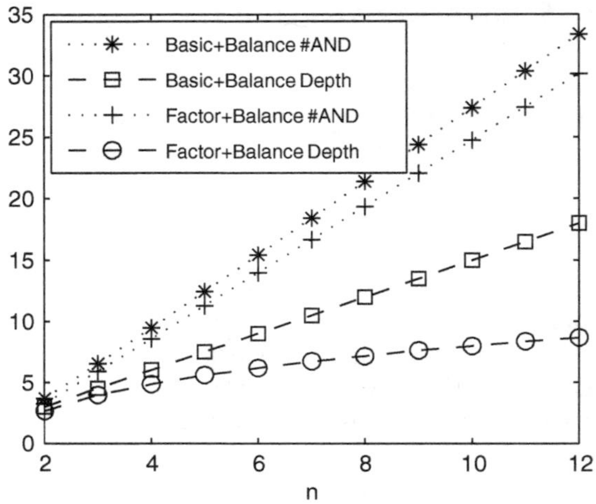

Fig. 18.7 Average number of AND gates and depth of the circuit versus n

n. The percentage of reduction on the depth increases with increasing n. For $n = 12$, the average depth of the circuit is reduced by more than 50%.

In Fig. 18.7, we plot the average number of AND gates and depth of the circuit versus n for both the "Basic+Balance" scheme and the "Factor+Balance" scheme. Clearly, the figure shows that the "Factor+Balance" scheme is superior to the "Basic+Balance" scheme. As shown in the figure, the average number of AND gates in the circuits synthesized by both the "Basic+Balance" scheme and the "Factor+Balance" scheme increases linearly with n. The average depth of the circuit synthesized by the "Basic+Balance" scheme also increases linearly with n. In contrast, the average depth of the circuit synthesized by the "Factor+Balance" scheme increases logarithmically with n.

18.7 Chapter Summary

In this chapter, we introduced the problem of synthesizing combinational logic to generate specified output probabilities from a given set of input probabilities. We focused on generating decimal output probabilities and aimed at finding a small set of input probabilities. We first showed that input probability sets consisting of two elements can be used to generate arbitrary decimal output probabilities. Next, as a mathematical result, we demonstrated that there exists a single input probability that can be used to generate any output decimal probability. That input probability cannot be rational; it must be an irrational root of a polynomial. We proposed algorithms to synthesize circuits that generate output decimal probabilities from the input probability set $S = \{0.4, 0.5\}$. To optimize the depth of the circuits, we proposed a method based on fraction factorization. We demonstrated the effectiveness of our algorithm with experimental results. The average depth of the circuits synthesized by our method is logarithmic in the number of digits of the output decimal probability.

References

1. Borkar, S., Karnik, T., De, V.: Design and reliability challenges in nanometer technologies. In: Design Automation Conference, p. 75. San Diego, CA, USA (2004)
2. Chakrapani, L., Korkmaz, P., Akgul, B., Palem, K.: Probabilistic system-on-a-chip architecture. ACM Transactions on Design Automation of Electronic Systems **12**(3), 1–28 (2007)
3. Cheemalavagu, S., Korkmaz, P., Palem, K., Akgul, B., Chakrapani, L.: A probabilistic CMOS switch and its realization by exploiting noise. In: IFIP International Conference on VLSI, pp. 535–541. Perth, Western Australia (2005)
4. Gill, A.: Synthesis of probability transformers. Journal of the Franklin Institute **274**(1), 1–19 (1962)
5. Gill, A.: On a weight distribution problem, with application to the design of stochastic generators. Journal of the ACM **10**(1), 110–121 (1963)
6. Karnik, T., Borkar, S., De, V.: Sub-90 nm technologies: Challenges and opportunities for CAD. In: International Conference on Computer-Aided Design, pp. 203–206. San Jose, CA, USA (2002)
7. Mishchenko, A., et al.: ABC: A system for sequential synthesis and verification (2007). http://www.eecs.berkeley.edu/~alanmi/abc/ (2010)
8. Nepal, K., Bahar, R., Mundy, J., Patterson, W., Zaslavsky, A.: Designing logic circuits for probabilistic computation in the presence of noise. In: Design Automation Conference, pp. 485–490. Anaheim, CA, USA (2005)
9. von Neumann, J.: Probabilistic logics and the synthesis of reliable organisms from unreliable components. In: C.E. Shannon, J. McCarthy (eds.) Automata Studies, pp. 43–98. Princeton University Press, Princeton, NJ (1956)
10. Qian, W., Riedel, M.D.: The synthesis of robust polynomial arithmetic with stochastic logic. In: Design Automation Conference, pp. 648–653. Anaheim, CA, USA (2008)
11. Wilhelm, D., Bruck, J.: Stochastic switching circuit synthesis. In: International Symposium on Information Theory, pp. 1388–1392. Toronto, ON, Canada (2008)

Chapter 19
Probabilistic Error Propagation in a Logic Circuit Using the Boolean Difference Calculus

Nasir Mohyuddin, Ehsan Pakbaznia, and Massoud Pedram

Abstract A gate-level probabilistic error propagation model is presented which takes as input the Boolean function of the gate, input signal probabilities, the error probability at the gate inputs, and the gate error probability and generates the error probability at the output of the gate. The presented model uses the Boolean difference calculus and can be efficiently applied to the problem of calculating the error probability at the primary outputs of a multilevel Boolean circuit with a time complexity which is linear in the number of gates in the circuit. This is done by starting from the primary inputs and moving toward the primary outputs by using a post-order – reverse Depth First Search (DFS) – traversal. Experimental results demonstrate the accuracy and efficiency of the proposed approach compared to the other known methods for error calculation in VLSI circuits.

19.1 Introduction

As CMOS hits nano-scale regime, device failure mechanisms such as cross talk, manufacturing variability, and soft error become significant design concerns. Being probabilistic by nature, these failure sources have pushed the CMOS technology toward stochastic CMOS [1]. For example, capacitive and inductive coupling between parallel adjacent wires in nano-scale CMOS integrated circuits (ICs) are the potential sources of cross talk between the wires. Cross talk can indeed cause flipping error on the *victim* signal [2]. In addition to the probabilistic CMOS, promising nanotechnology devices such as quantum dots are used in technologies such as quantum cellular automata (QCA). Most of these emerging technologies are inherently probabilistic. This has made reliability analysis an essential piece of circuit

---

N. Mohyuddin (✉)
Department of Electrical Engineering – Systems, University of Southern California, Los Angeles, CA 90089, USA
e-mail: mohyuddi@usc.edu

Based on Mohyuddin, N.; Pakbaznia, E.; Pedram, M.; "Probabilistic error propagation in logic circuits using the Boolean difference calculus," Computer Design, 2008. ICCD 2008. IEEE International Conference on pp. 7–13, 12–15 Oct. 2008 [2008] IEEE.

S.P. Khatri, K. Gulati (eds.), *Advanced Techniques in Logic Synthesis, Optimizations and Applications*, DOI 10.1007/978-1-4419-7518-8_19,

design. Reliability analysis will be even more significant in designing reliable circuits using unreliable components [3, 4].

Circuit reliability will thus be an important trade-off factor which has to be taken care of similar to traditional design trade-off factors such as performance, area, and power. To include the reliability into the design trade-off equations, there must exist a good measure for the circuit reliability, and there must exist fast and robust tools that, similar to timing analyzer and power estimator tools, are capable of estimating circuit reliability at different design levels. In [5] authors have proposed a probabilistic transfer matrix (PTM) method to calculate the output signal error probability for a circuit while [6] presents a method based on the probabilistic decision diagrams (PDDs) to perform this task.

In this chapter we first introduce a probabilistic gate-level error propagation model based on the concept of Boolean difference to propagate errors from inputs to output of a general gate. We then apply this model to account for the error propagation in a given circuit and finally estimate the error probability at the circuit outputs. Note that in the proposed model a gate's Boolean function is used to determine the error propagation in the gate. An error at an output of a gate is due to its input(s) and/or the gate itself being erroneous. The internal gate error in this work is modeled as an output flipping event. This means that, when a faulty gate makes an error, it flips (changes a "1" to a "0" and a "0" to a "1") its output value that it would have generated given the inputs, Von Neumann error model. In the rest of this chapter, we call our circuit error estimation technique the Boolean Difference-based Error Calculator, or BDEC for short, and we assume that a defective logic gate produces the wrong output value for every input combination. This is a more pessimistic defect model than the stuck-at-fault model.

Authors in [5] use a PTM matrix for each gate to represent the error propagation from the input(s) to the output(s) of a gate. They also define some operations such as matrix multiplication and tensor product to use the gate PTMs to generate and propagate error probability at different nodes in a circuit level by level. Despite of its accuracy in calculating signal error probability, PTM technique suffers from the extremely large number of computational-intensive tasks namely regular and tensor matrix products. This makes the PTM technique extremely memory intensive and very slow. In particular, for larger circuits, size of the PTM matrices grows too fast for the deeper nodes in circuit making PTM an inefficient or even infeasible technique of error rate calculation for a general circuit. References [8] and [9] developed a methodology based on probabilistic model checking (PMC) to evaluate the circuit reliability. The issue of excessive memory requirement of PMC when the circuit size is large was successfully addressed in [10]. However, the time complexity still remains a problem. In fact, the authors of [10] show that the run time for their space-efficient approach is even worse than that of the original approach.

Boolean difference calculus was introduced and used by [11] and [12] to analyze single faults. It was then extended by [13] and [14] to handle multiple fault situations; however, they only consider stuck-at-faults and they do not consider the case when the logic gates themselves can be erroneous and hence a gate-induced output error may nullify the effect of errors at the gate's input(s). Reference [6] has used Boolean difference to estimate the switching activity of Boolean circuits when multiple inputs switch, but error estimation using Boolean difference in the presence of multiple faults is more involved due to the fact that certain combinations of faults may cancel each other. In [15] authors use Bayesian networks to calculate the output error probabilities without considering the input signal probabilities.

The author in [6] uses probabilistic decision diagrams (PDD) to calculate the error probabilities at the outputs using probabilistic gates. While PDDs are much more efficient than PTM for average case, the worst-case complexity of both PTM- and PDD-based error calculators is exponential in the number of inputs in the circuit.

In contrast, we will show in Section 19.5 that BDEC calculates the circuit error probability much faster than PTM while achieving as accurate results as PTM's. We will show that BDEC requires a single pass over the circuit nodes using a post-order (reverse DFS) traversal to calculate the errors probabilities at the output of each gate as we move from the primary inputs to the primary outputs; hence, complexity is O (N) where N is the number of the gates in the circuit, and O (.) is the big O notation.

19.2 Error Propagation Using Boolean Difference Calculus

Some key concepts and notation that will be used in the remainder of this chapter are discussed next.

19.2.1 Partial Boolean Difference

The partial Boolean difference of function $f(x_1, x_2, \ldots, x_n)$ with respect to one variable or a subset of its variables [14] is defined as

$$\begin{aligned} \frac{\partial f}{\partial x_i} &= f_{x_i} \oplus f_{\bar{x}_i} \\ \frac{\partial f}{\partial \left(x_{i_1} x_{i_2} \ldots x_{i_k}\right)} &= f_{x_{i_1} x_{i_2} \ldots x_{i_k}} \oplus f_{\bar{x}_{i_1} \bar{x}_{i_2} \ldots \bar{x}_{i_k}} \end{aligned} \tag{19.1}$$

where $\oplus$ represents XOR operator and f_{x_i} is the co-factor of f with respect to x_i, i.e.,

$$\begin{aligned} f_{x_i} &= f(x_1, \ldots, x_{i-1}, x_i = 1, x_{i+1}, \ldots, x_n) \\ f_{\bar{x}_i} &= f(x_1, \ldots, x_{i-1}, x_i = 0, x_{i+1}, \ldots, x_n) \end{aligned} \tag{19.2}$$

Higher order co-factors of f can be defined similarly. The partial Boolean difference of f with respect to x_i expresses the condition (with respect to other variables) under which f is sensitive to a change in the input variable x_i. More precisely, if the logic values of $\{x_1, \ldots, x_{i-1}, x_{i+1}, \ldots, x_n\}$ are such that $\partial f / \partial x_i = 1$, then a change in the input value x_i, will change the output value of f. However, when $\partial f / \partial x_i = 0$, changing the logic value of x_i will not affect the output value of f.

It is worth mentioning that the *order-k partial Boolean difference* defined in (19.1) is different from the *kth Boolean difference* of function f as used in [13], which is denoted by $\partial^k f / \partial x_{i_1} \ldots \partial x_{i_k}$. For example, the second Boolean difference of function f with respect to x_i and x_j is defined as

$$\frac{\partial^2 f}{\partial x_i \partial x_j} = \frac{\partial}{\partial x_i}\left(\frac{\partial f}{\partial x_j}\right) = f_{x_i x_j} \oplus f_{\bar{x}_i x_j} \oplus f_{x_i \bar{x}_j} \oplus f_{\bar{x}_i \bar{x}_j} \tag{19.3}$$

Therefore, $\partial^2 f/\partial x_i \partial x_j \neq \partial f/\partial(x_i x_j)$.

19.2.2 Total Boolean Difference

Similar to the partial Boolean difference that shows the conditions under which a Boolean function is sensitive to change of any of its input variables, we can define *total Boolean difference* showing the condition under which the output of the Boolean function f is sensitive to the *simultaneous* changes in all the variables of a subset of input variables. For example, the total Boolean difference of function f with respect to $x_i x_j$ is defined as

$$\frac{\Delta f}{\Delta(x_i x_j)} = \frac{\partial f}{\partial(x_i x_j)}\left(x_i x_j + \bar{x}_i \bar{x}_j\right) + \frac{\partial f}{\partial(\bar{x}_i x_j)}\left(\bar{x}_i x_j + x_i \bar{x}_j\right) \tag{19.4}$$

where $\Delta f/\Delta(x_i x_j)$ describes the conditions under which the output of f is sensitive to a simultaneous change in x_i and x_j. That is, the value of f changes as a result of the simultaneous change. Some examples for simultaneous changes in x_i and x_j are transitioning from $x_i = x_j = 1$ to $x_i = x_j = 0$ and vice versa or from $x_i = 1, x_j = 0$ to $x_i = 0, x_j = 1$ and vice versa. However, transitions in the form of $x_i = x_j = 1$ to $x_i = 1, x_j = 0$ or $x_i = 1, x_j = 0$ to $x_i = 0, x_j = 0$ are not simultaneous changes. Note that $\partial f/\partial(x_i x_j)$ describes the conditions when a transition from $x_i = x_j = 1$ to $x_i = x_j = 0$ and vice versa changes the value of function f.

It can be shown that the total Boolean difference in (19.4) can be written in the form of

$$\frac{\Delta f}{\Delta\left(x_i x_j\right)} = \frac{\partial f}{\partial x_i} \oplus \frac{\partial f}{\partial x_j} \oplus \frac{\partial^2 f}{\partial x_i \partial x_j} \tag{19.5}$$

The total Boolean difference with respect to three variables is

$$\begin{aligned} \frac{\Delta f}{\Delta(x_1 x_2 x_3)} &= \frac{\partial f}{\partial(x_1 x_2 x_3)}\left(x_1 x_2 x_3 + \bar{x}_1 \bar{x}_2 \bar{x}_3\right) \\ &+ \frac{\partial f}{\partial(x_1 x_2 \bar{x}_3)}\left(x_1 x_2 \bar{x}_3 + \bar{x}_1 \bar{x}_2 x_3\right) \\ &+ \frac{\partial f}{\partial(x_1 \bar{x}_2 x_3)}\left(x_1 \bar{x}_2 x_3 + \bar{x}_1 x_2 \bar{x}_3\right) \\ &+ \frac{\partial f}{\partial(\bar{x}_1 x_2 x_3)}\left(\bar{x}_1 x_2 x_3 + x_1 \bar{x}_2 \bar{x}_3\right) \end{aligned} \tag{19.6}$$

It is straightforward to verify that

$$\frac{\Delta f}{\Delta(x_1x_2x_3)} = \frac{\partial f}{\partial x_1} \oplus \frac{\partial f}{\partial x_2} \oplus \frac{\partial f}{\partial x_3} \oplus \frac{\partial^2 f}{\partial x_1 \partial x_2} \oplus \frac{\partial^2 f}{\partial x_2 \partial x_3} \oplus \frac{\partial^2 f}{\partial x_1 \partial x_3} \oplus \frac{\partial^3 f}{\partial x_1 \partial x_2 \partial x_3} \tag{19.7}$$

In general total Boolean difference of a function f with respect to an n-variable subset of its inputs can be written as

$$\frac{\Delta f}{\Delta(x_{i_1}x_{i_2}\ \dots\ x_{i_n})} = \sum_{j=0}^{2^{n-1}-1} \left.\frac{\partial f}{\partial \overleftrightarrow{x}}\right|_{m_j} \left(m_j + m_{2^n-j-1}\right) \tag{19.8}$$

where m_j's are defined as

$$\begin{aligned} m_0 &= \bar{x}_{i_1}\bar{x}_{i_2}\ \dots\ \bar{x}_{i_{n-1}}\bar{x}_{i_n} \\ m_1 &= \bar{x}_{i_1}\bar{x}_{i_2}\ \dots\ \bar{x}_{i_{n-1}}x_{i_n} \\ &\vdots \\ m_{2^n-1} &= x_{i_1}x_{i_2}\ \dots\ x_{i_{n-1}}x_{i_n} \end{aligned} \tag{19.9}$$

and we have

$$\left.\frac{\partial f}{\partial \overleftrightarrow{x}}\right|_{m_j} = \frac{\partial f}{\partial\left(x_1^*x_2^*\ \dots\ x_{n-1}^*x_n^*\right)} \quad \text{where } m_j = x_1^*x_2^*\ \dots\ x_{n-1}^*x_n^* \tag{19.10}$$

19.2.3 Signal and Error Probabilities

Signal probability is defined as the probability for a signal value to be "1." That is

$$p_i = \Pr\{x_i = 1\} \tag{19.11}$$

Gate error probability is shown by ε_g and is defined as the probability that a gate generates an erroneous output, independent of its applied inputs. Such a gate is sometimes called $(1-\varepsilon_g)$-reliable gate. Signal error probability is defined as the probability of error on a signal line. If the signal line is the output of a gate, the error can be due to either error at the gate input(s) or the gate error itself. We denote the error probability on signal line x_i by ε_i.

We are interested in determining the circuit output error rates, given the circuit input error rates under the assumption that each gate in the circuit can fail independently with a probability of ε_g. In other words, we account for the general case of multiple simultaneous gate failures.

19.3 Proposed Error Propagation Model

In this section we propose our gate error model in the Boolean difference calculus notation. The gate error model is then used to calculate the error probability and reliability at outputs of a circuit.

19.3.1 Gate Error Model

Figure 19.1 shows a general logic gate realizing Boolean function f, with gate error probability of ε_{g}. The signal probabilities at the inputs, i.e., probabilities for input signals being 1, are $p_1, p_2, \ldots, p_n$ while the input error probabilities are $\varepsilon_1, \varepsilon_2, \ldots, \varepsilon_n$. The output error probability is ε_z.

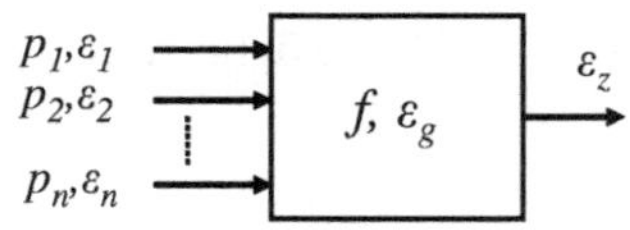

Fig. 19.1 Gate implementing function f

First consider the error probability equation for a buffer gate shown in Fig. 19.2. The error occurs at the output if (i) the input is erroneous and the gate is error free or (ii) the gate is erroneous and the input is error free. Therefore, assuming independent faults for the input and the gate, the output error probability for a buffer can be written as

$$\varepsilon_z = \varepsilon_{\mathrm{in}}(1 - \varepsilon_{\mathrm{g}}) + (1 - \varepsilon_{\mathrm{in}})\varepsilon_{\mathrm{g}} = \varepsilon_{\mathrm{g}} + (1 - 2\varepsilon_{\mathrm{g}})\varepsilon_{\mathrm{in}} \qquad (19.12)$$

where $\varepsilon_{\mathrm{in}}$ is the error probability at the input of the buffer. It can be seen from this equation that the output error probability for buffer is independent of the input signal probability. Note (19.12) can also be used to express the output error probability of an inverter gate.

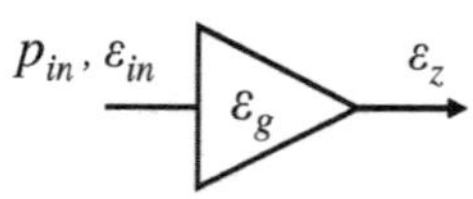

Fig. 19.2 A faulty buffer with erroneous input

We can model each faulty gate with erroneous inputs as an ideal (no fault) gate with the same functionality and the same inputs in series with a faulty buffer as shown in Fig. 19.3.

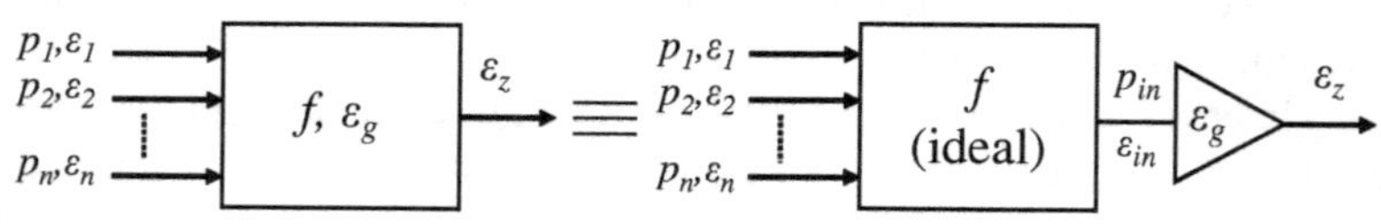

Fig. 19.3 The proposed model for a general faulty gate

Now consider a general 2-input gate. Using the fault model discussed above, we can write the output error probability considering all the cases of no error, single error, and double errors at the input and the error in the gate itself. We can write the general equation for the error probability at the output, ε_z, as

$$\varepsilon_z = \varepsilon_g + (1 - 2\varepsilon_g)\underbrace{\left(\begin{array}{l}\varepsilon_1(1-\varepsilon_2)\Pr\left\{\dfrac{\partial f}{\partial x_1}\right\} + (1-\varepsilon_1)\varepsilon_2 \Pr\left\{\dfrac{\partial f}{\partial x_2}\right\} \\ +\varepsilon_1\varepsilon_2 \Pr\left\{\dfrac{\Delta f}{\Delta(x_1x_2)}\right\}\end{array}\right)}_{\varepsilon_{\text{in}}} \tag{19.13}$$

where Pr{.} represents the signal probability function and returns the probability of its Boolean argument to be "1." The first and the second terms in ε_{in} account for the error at the output of the ideal gate due to single input errors at the first and the second inputs, respectively. Note error at each input of the ideal gate propagates to the output of this gate only if the other inputs are not masking it. The non-masking probability for each input error is taken into account by calculating the signal probability of the partial Boolean difference of the function f with respect to the corresponding input. The first two terms in ε_{in} only account for the cases when we have single input errors at the input of the ideal gate; however, error can also occur when both inputs are erroneous simultaneously. This is taken into account by multiplying the probability of having simultaneous errors at both inputs, i.e., $\varepsilon_1\varepsilon_2$, with the probability of this error to be propagated to the output of the ideal gate, i.e., the signal probability of the total Boolean difference of f with respect to x_1x_2.

For 2-input AND gate ($f = x_1x_2$) shown in Fig. 19.4 we have

$$\begin{aligned}
&\Pr\left\{\frac{\partial f}{\partial x_1}\right\} = \Pr\{x_2\} = p_2\,,\ \Pr\left\{\frac{\partial f}{\partial x_2}\right\} = \Pr\{x_1\} = p_1 \\
&\Pr\left\{\frac{\Delta f}{\Delta(x_1x_2)}\right\} = \Pr\{\bar{x}_1\bar{x}_2 + x_1x_2\} = (1-p_1)(1-p_2) + p_1p_2 \\
&\qquad = 1 - (p_1 + p_2) + 2p_1p_2
\end{aligned} \tag{19.14}$$

Plugging (19.14) into (19.13) and after some simplifications we have

$$\varepsilon_{\text{AND2}} = \varepsilon_g + \left(1 - 2\varepsilon_g\right)\left(\varepsilon_1 p_2 + \varepsilon_2 p_1 + \varepsilon_1\varepsilon_2\left(1 - 2\left(p_1 + p_2\right) + 2p_1p_2\right)\right) \tag{19.15}$$

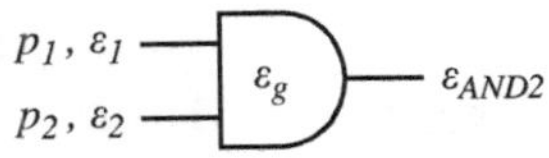

Fig. 19.4 A 2-input faulty AND gate with erroneous inputs

Similarly, the error probability for the case of 2-input OR can be calculated as

$$\varepsilon_{\text{OR2}} = \varepsilon_g + \left(1 - 2\varepsilon_g\right)\left(\varepsilon_1\left(1 - p_2\right) + \varepsilon_2\left(1 - p_1\right) + \varepsilon_1\varepsilon_2\left(2p_1p_2 - 1\right)\right) \tag{19.16}$$

And for 2-input XOR gate we have

$$\varepsilon_{\mathrm{XOR2}} = \varepsilon_{\mathrm{g}} + \left(1 - 2\varepsilon_{\mathrm{g}}\right)(\varepsilon_1 + \varepsilon_2 - 2\varepsilon_1\varepsilon_2) \tag{19.17}$$

It is interesting to note that the error probability at the output of the XOR gate is independent of the input signal probabilities. Generally, the 2-inpout XOR gate exhibits larger output error compared to 2-input OR and AND gates. This is expected since XOR gates show maximum sensitivity to input errors (XOR, like inversion, is an entropy-preserving function). The output error expression for a 3-input gate is

$$\varepsilon_z = \varepsilon_{\mathrm{g}} + (1 - 2\varepsilon_{\mathrm{g}}) \left(\begin{array}{l} \varepsilon_1(1 - \varepsilon_2 - \varepsilon_3 + \varepsilon_2\varepsilon_3)\,\Pr\left\{\dfrac{\partial f}{\partial x_1}\right\} \\ +\varepsilon_2(1 - \varepsilon_1 - \varepsilon_3 + \varepsilon_1\varepsilon_3)\,\Pr\left\{\dfrac{\partial f}{\partial x_2}\right\} \\ +\varepsilon_3(1 - \varepsilon_1 - \varepsilon_2 + \varepsilon_1\varepsilon_2)\,\Pr\left\{\dfrac{\partial f}{\partial x_3}\right\} \\ +\varepsilon_1\varepsilon_2(1 - \varepsilon_3)\,\Pr\left\{\dfrac{\Delta f}{\Delta(x_1x_2)}\right\} \\ +\varepsilon_2\varepsilon_3(1 - \varepsilon_1)\,\Pr\left\{\dfrac{\Delta f}{\Delta(x_2x_3)}\right\} \\ +\varepsilon_1\varepsilon_3(1 - \varepsilon_2)\,\Pr\left\{\dfrac{\Delta f}{\Delta(x_1x_3)}\right\} \\ +\varepsilon_1\varepsilon_2\varepsilon_3\,\Pr\left\{\dfrac{\Delta f}{\Delta(x_1x_2x_3)}\right\} \end{array} \right) \tag{19.18}$$

As an example of a 3-input gate, we can use (19.18) to calculate the probability of error for the case of 3-input AND gate. We can show that the output error probability can be calculated as

$$\varepsilon_{\mathrm{AND3}} = \varepsilon_{\mathrm{g}} + (1 - 2\varepsilon_{\mathrm{g}}) \left(\begin{array}{l} p_2p_3\varepsilon_1 + p_1p_3\varepsilon_2 + p_1p_2\varepsilon_3 \\ +p_3(1 - 2(p_2 + p_1) + 2p_1p_2)\varepsilon_1\varepsilon_2 \\ +p_1(1 - 2(p_2 + p_3) + 2p_2p_3)\varepsilon_2\varepsilon_3 \\ +p_2(1 - 2(p_1 + p_3) + 2p_1p_3)\varepsilon_1\varepsilon_3 \\ + \left(\begin{array}{l} 1 - 2(p_1 + p_2 + p_3) \\ +4(p_1p_2 + p_2p_3 + p_1p_3) - 6p_1p_2p_3 \end{array} \right) \varepsilon_1\varepsilon_2\varepsilon_3 \end{array} \right) \tag{19.19}$$

Now we give a general expression for a 4-input logic gate as

$$\varepsilon_z = \varepsilon_g + (1-2\varepsilon_g)\begin{pmatrix} \sum_i \varepsilon_i \left(1 - \sum_{j\neq i}\varepsilon_j + \sum_{(j,k)\neq i}\varepsilon_j\varepsilon_k - \sum_{(j,k,l)\neq i}\varepsilon_j\varepsilon_k\varepsilon_l\right)\Pr\left\{\frac{\partial f}{\partial x_i}\right\} \\ + \sum_{(i,j)}\varepsilon_i\varepsilon_j\left(1 - \sum_{k\neq i,j}\varepsilon_k + \sum_{(k,l)\neq i,j}\varepsilon_k\varepsilon_l\right)\Pr\left\{\frac{\Delta f}{\Delta(x_i x_j)}\right\} \\ + \sum_{(i,j,k)}\varepsilon_i\varepsilon_j\varepsilon_k\left(1 - \sum_{l\neq i,j,k}\varepsilon_l\right)\Pr\left\{\frac{\Delta f}{\Delta(x_i x_j x_k)}\right\} \\ + \varepsilon_1\varepsilon_2\varepsilon_3\varepsilon_4 \Pr\left\{\frac{\Delta f}{\Delta(x_1x_2x_3x_4)}\right\} \end{pmatrix} \tag{19.20}$$

The Boolean expression for a general k-input gate can be calculated in a similar manner.

19.3.2 Error Propagation in 2-to-1 Mux Using BDEC

We represent 2-to-1 Multiplexer (Mux) function as $f = as + b\overline{s}$. Using BDEC the output error probability in terms of the gate error probability, input signal probabilities, and input error probabilities is

$$\varepsilon_{\text{Mux2to1}} = \varepsilon_g + (1-2\varepsilon_g)\begin{pmatrix} \varepsilon_a(1-\varepsilon_b)(1-\varepsilon_s)\Pr\left(\frac{\partial f}{\partial a}\right) + \varepsilon_b(1-\varepsilon_a)(1-\varepsilon_s)\Pr\left(\frac{\partial f}{\partial b}\right) \\ +\varepsilon_s(1-\varepsilon_a)(1-\varepsilon_b)\Pr\left(\frac{\partial f}{\partial s}\right) + \varepsilon_a\varepsilon_b(1-\varepsilon_s)\Pr\left(\frac{\Delta f}{\Delta(ab)}\right) \\ +\varepsilon_a\varepsilon_s(1-\varepsilon_b)\Pr\left(\frac{\Delta f}{\Delta(as)}\right) + \varepsilon_b\varepsilon_s(1-\varepsilon_a)\Pr\left(\frac{\Delta f}{\Delta(bs)}\right) \\ +\varepsilon_a\varepsilon_b\varepsilon_s\Pr\left(\frac{\Delta f}{\Delta(abs)}\right) \end{pmatrix} \tag{19.21}$$

Now we stepwise show how to calculate various partial and total Boolean differences.

First we calculate all single variable partial Boolean differences as

$$\begin{aligned}\frac{\partial f}{\partial a} &= f_a \oplus f_{\overline{a}} \\ &= (s + b\overline{s}) \oplus (b\overline{s}) \\ &= \overline{(s+b\overline{s})}.(b\overline{s}) + (s+b\overline{s}).\overline{(b\overline{s})} \\ &= (\overline{s}.\overline{b\overline{s}}).(b\overline{s}) + (s.\overline{b\overline{s}} + b\overline{s}.\overline{b\overline{s}}) \\ &= (s.\overline{b\overline{s}}) \\ &= s.(\overline{b}+s) \\ &= s.\overline{b}+s \\ &= s.(1+\overline{b}) \\ &= s\end{aligned}$$

$$\begin{aligned}\frac{\partial f}{\partial b} &= f_b \oplus f_{\overline{b}} \\ &= (as+\overline{s}) \oplus (as) \\ &= \overline{(as+\overline{s})}.(as) + (as+\overline{s}).\overline{(as)} \\ &= (\overline{as}.s).(as) + (as.\overline{as} + \overline{s}.\overline{as}) \\ &= (\overline{s}.\overline{as}) \\ &= \overline{s}.(\overline{a}+\overline{s}) \\ &= \overline{as}+\overline{s} \\ &= \overline{s}.(\overline{a}+1) \\ &= \overline{s}\end{aligned}$$

$$\begin{aligned}\frac{\partial f}{\partial (s)} &= f_s \oplus f_{\overline{s}} \\ &= a \oplus b\end{aligned}$$

Then we calculate two-variable partial Boolean differences as

$$\frac{\partial f}{\partial (ab)} = f_{ab} \oplus f_{\overline{ab}} = 1 \oplus 0 = 1 \qquad \frac{\partial f}{\partial (\overline{a}b)} = f_{\overline{a}b} \oplus f_{a\overline{b}} = \overline{s} \oplus s = 1 \qquad \frac{\partial f}{\partial (as)} = f_{as} \oplus f_{\overline{as}} = 1 \oplus b = \overline{b}$$

$$\frac{\partial f}{\partial (a\overline{s})} = f_{a\overline{s}} \oplus f_{\overline{a}s} = b \oplus 0 = b \qquad \frac{\partial f}{\partial (bs)} = f_{bs} \oplus f_{\overline{bs}} = a \oplus 0 = a \qquad \frac{\partial f}{\partial (b\overline{s})} = f_{b\overline{s}} \oplus f_{\overline{b}s} = 1 \oplus a = \overline{a}$$

Finally we calculate three-variable partial Boolean differences as

$$\frac{\partial f}{\partial (abs)} = f_{abs} \oplus f_{\overline{abs}} = 1 \oplus 0 = 1 \qquad \frac{\partial f}{\partial (\overline{a}bs)} = f_{\overline{a}bs} \oplus f_{a\overline{bs}} = 0 \oplus 0 = 0 \qquad \frac{\partial f}{\partial (a\overline{b}s)} = f_{a\overline{b}s} \oplus f_{\overline{a}b\overline{s}} = 1 \oplus 1 = 0 \qquad \frac{\partial f}{\partial (ab\overline{s})} = f_{ab\overline{s}} \oplus f_{\overline{ab}s} = 1 \oplus 0 = 1$$

Next we calculate total Boolean differences as

$$\frac{\Delta f}{\Delta (as)} = \frac{\partial f}{\partial a} \oplus \frac{\partial f}{\partial s} \oplus \frac{\partial^2 f}{\partial a \partial s} = s \oplus (a \oplus b) \oplus 1 = s \oplus \overline{(a \oplus b)} = s \oplus (a \otimes b)$$

$$\frac{\Delta f}{\Delta (bs)} = \frac{\partial f}{\partial b} \oplus \frac{\partial f}{\partial s} \oplus \frac{\partial^2 f}{\partial b \partial s} = \overline{s} \oplus (a \oplus b) \oplus 1 = s \oplus (a \oplus b)$$

$$\frac{\Delta f}{\Delta (ab)} = \frac{\partial f}{\partial a} \oplus \frac{\partial f}{\partial b} \oplus \frac{\partial^2 f}{\partial a \partial b} = s \oplus \overline{s} \oplus 0 = 1$$

$$\frac{\Delta f}{\Delta (abs)} = \frac{\partial f}{\partial a} \oplus \frac{\partial f}{\partial b} \oplus \frac{\partial f}{\partial s} \oplus \frac{\partial^2 f}{\partial a \partial s} \oplus \frac{\partial^2 f}{\partial b \partial s} \oplus \frac{\partial^2 f}{\partial a \partial b} \oplus \frac{\partial^3 f}{\partial a \partial b \partial s}$$
$$= s \oplus \overline{s} \oplus (a \oplus b) \oplus 1 \oplus 1 \oplus 0 \oplus 0 = s \oplus \overline{s} \oplus (a \oplus b) = 1 \oplus (a \oplus b) = a \otimes b$$

Plugging these values in (19.21)

$$\varepsilon_{\text{Mux2to1}} = \varepsilon_{\text{g}} + (1 - 2\varepsilon_{\text{g}}) \begin{pmatrix} \varepsilon_a (1 - \varepsilon_b)(1 - \varepsilon_s)\, p_s + \varepsilon_b (1 - \varepsilon_a)(1 - \varepsilon_s)(1 - p_s) \\ + \varepsilon_s (1 - \varepsilon_a)(1 - \varepsilon_b) \Pr (a \oplus b) + \varepsilon_a \varepsilon_b (1 - \varepsilon_s) \\ + \varepsilon_a \varepsilon_s (1 - \varepsilon_b) \Pr (s \oplus (a \otimes b)) + \varepsilon_b \varepsilon_s (1 - \varepsilon_a) \Pr (s \oplus (a \oplus b)) \\ + \varepsilon_a \varepsilon_b \varepsilon_s \Pr (a \otimes b) \end{pmatrix} \quad (19.22)$$

19.3.3 Circuit Error Model

In this section we use the gate error model proposed in Section19.3.1 to calculate the error probability at the output of a given circuit. Given a multilevel logic circuit composed of logic gates, we start from the primary inputs and move toward the primary outputs by using a post-order (reverse DFS) traversal. For each gate, we calculate the output error probability using input signal probabilities, input error probabilities, and gate error probability and utilizing the error model proposed in Section 19.3.1. The signal probability for the output of each gate is also calculated based on the input signal probabilities and the gate function. The process of output error and signal probability calculation is continued until all the gates are processed. For each node z in the circuit, reliability is defined as

$$\chi_z = 1 - \varepsilon_z \tag{19.23}$$

After processing all the gates in the circuit and calculating error probabilities and reliabilities for all the circuit primary outputs, we can calculate the overall circuit reliability. Assuming that different primary outputs of the circuit are independent, the overall circuit reliability can be calculated as the product of all the primary outputs reliabilities, that is

$$\chi_{\text{circuit}} = \prod_i \chi_{\text{PO}_i} \tag{19.24}$$

The case of dependent primary outputs (which is obviously a more realistic scenario) requires calculation of spatial correlation coefficient as will be outlined further on in this chapter. The detailed treatment of spatial correlation coefficient calculation however falls outside the scope of this work.

This error propagation algorithm has a complexity of $O(2^k N)$ where k is the maximum number of inputs to any gate in the circuit (which is small and can be upper bounded a priori in order to give $O(N)$ complexity) and N is the number of gates in the circuit. This complexity should be contrasted to that of the PTM-based or the PDD-based approaches that have a worst-case complexity of $O(2^N)$. The trade-off is that our proposed approach based on post-order traversal of the circuit netlist and application of Boolean difference operator results in only *approximate* output error and signal probability values due to the effect of reconvergent fanout structures in the circuit, which create *spatial correlations* among input signals to a gate. This problem has been extensively addressed in the literature on improving the accuracy of signal probability calculators [16, 17]. Our future implementation of BDEC shall focus on utilizing similar techniques (including efficient calculation of spatial correlation coefficients) to improve the accuracy of proposed Boolean difference-based error calculation engine.

19.4 Practical Considerations

In this section we use the error models introduced in previous section to calculate the exact error probability expression at the output of a tree-structured circuit.

19.4.1 Output Error Expression

For the sake of elaboration, we choose a 4-input AND gate implemented as a balanced tree of 2-input AND gates as shown in Fig. 19.5. We can calculate the output error of this circuit by expressing the error at the output of each gate using (19.18).

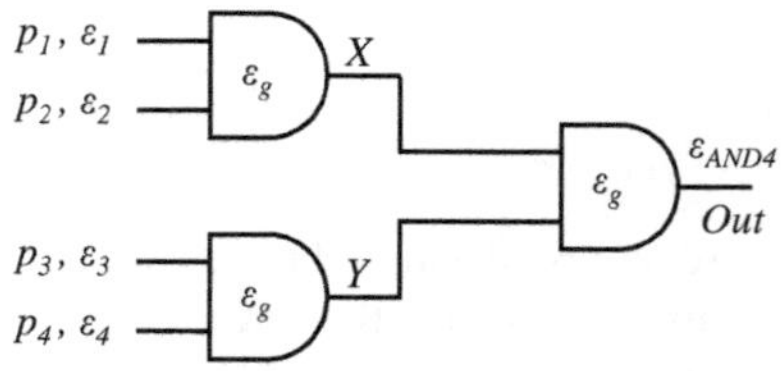

Fig. 19.5 Balanced tree implementation of 4-input AND gate

Equation (19.25) provides the exact output error probability of the circuit shown in Fig. 19.5 in terms of the input signal probabilities and input error probabilities where similar to [18] higher order exponents of the signal probabilities are reduced to first-order exponents.

$$\begin{aligned}
\varepsilon_{\text{AND4}} = {} & (p_2p_3p_4\varepsilon_1 + p_1p_3p_4\varepsilon_2 + p_1p_2p_4\varepsilon_3 + p_1p_2p_3\varepsilon_4) \\
& + p_3p_4\,(1 - 2\,(p_1 + p_2) + 2p_1p_2)\,\varepsilon_1\varepsilon_2 \\
& + p_2p_4\,(1 - 2\,(p_1 + p_3) + 2p_1p_3)\,\varepsilon_1\varepsilon_3 \\
& + p_2p_3\,(1 - 2\,(p_1 + p_4) + 2p_1p_4)\,\varepsilon_1\varepsilon_4 \\
& + p_1p_4\,(1 - 2\,(p_2 + p_3) + 2p_2p_3)\,\varepsilon_2\varepsilon_3 \\
& + p_1p_3\,(1 - 2\,(p_2 + p_4) + 2p_2p_4)\,\varepsilon_2\varepsilon_4 \\
& + p_1p_2\,(1 - 2\,(p_3 + p_4) + 2p_3p_4)\,\varepsilon_3\varepsilon_4 \\
& + p_4\,(1 - 2\,(p_1 + p_2 + p_3) + 4\,(p_1p_2 + p_1p_3 + p_2p_3) - 6p_1p_2p_3)\,\varepsilon_1\varepsilon_2\varepsilon_3 \\
& + p_2\,(1 - 2\,(p_1 + p_3 + p_4) + 4\,(p_1p_3 + p_1p_4 + p_3p_4) - 6p_1p_3p_4)\,\varepsilon_1\varepsilon_3\varepsilon_4 \\
& + p_3\,(1 - 2\,(p_1 + p_2 + p_4) + 4\,(p_1p_2 + p_1p_4 + p_2p_4) - 6p_1p_2p_4)\,\varepsilon_1\varepsilon_2\varepsilon_4 \\
& + p_1\,(1 - 2\,(p_2 + p_3 + p_4) + 4\,(p_2p_3 + p_2p_4 + p_3p_4) - 6p_2p_3p_4)\,\varepsilon_2\varepsilon_3\varepsilon_4 \\
& + \begin{pmatrix} 1 - 2\,(p_1 + p_2 + p_3 + p_4) \\ +4\,(p_1p_2 + p_1p_3 + p_1p_4 + p_2p_3 + p_2p_4 + p_3p_4) \\ -8\,(p_1p_2p_3 + p_1p_2p_4 + p_1p_3p_4 + p_2p_3p_4) \\ +14p_1p_2p_3p_4 \end{pmatrix} \varepsilon_1\varepsilon_2\varepsilon_3\varepsilon_4
\end{aligned} \tag{19.25}$$

In (19.25), without loss of generality, we assume $\varepsilon_g = 0$ in order to reduce the length of the expression.

Using symbolic notation along with higher order exponent suppression, the model presented in Section 19.3 can compute the exact output error probability in circuits with no reconvergent fanout. We will show in next section that by sacrificing little accuracy and using numerical values instead of symbolic notation, the computational complexity of our gate error model becomes linear in terms of the number of gates in the circuit.

19.4.2 Reconvergent Fanout

Figure 19.6 shows an example of a circuit with reconvergent fanout. It is clear from the figure that inputs to the final logic gate are not independent. Therefore, if the BDEC technique discussed in Section 19.3 is applied to this circuit, the calculated output error probability will not be accurate. In this section we describe a modification to the BDEC technique that improves the probability of error for the circuit in the presence of reconvergent fanout structures in the circuit.

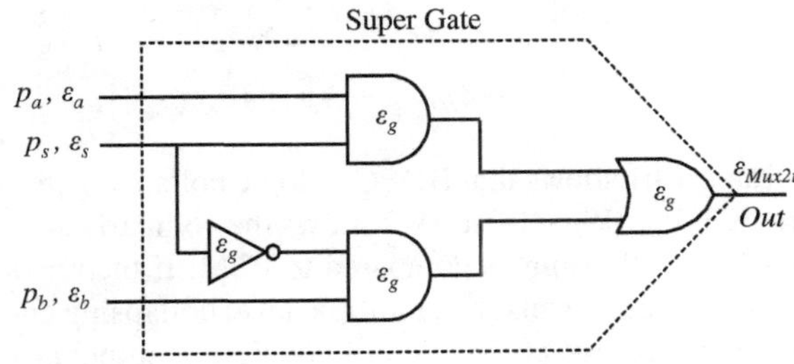

Fig. 19.6 Reconvergent fanout in a 2-to-1 multiplexer

Local reconvergent fanout such as the one depicted in Fig. 19.6 can be handled by collapsing levels of logic. For this example, we consider all the four gates in Fig. 19.6 as a single super gate and then apply the BDEC technique to this super gate. For the input to output error propagation in the original circuit, BDEC will ignore the internal structure of the super gate and only considers the actual function implemented by the super gate, 2-to-1 Mux in this case. The original implementation information can be taken into account by properly calculating the ε_g value for this new 3-input super gate.

The ε_g value for the collapsed gate is calculated using BDEC for the original circuit block before collapsing but assuming that the input error probabilities are zero. For example, for the circuit in Fig. 19.6 the error probability at the output of the top AND gate and the inverter using BDEC equations described in Section 19.3.1 and assuming input error probabilities to be 0 will be ε_g each. Similarly the error probability at the output of the bottom AND gate will be $\varepsilon_{AND2} = \varepsilon_g + (1 - 2\varepsilon_g)(\varepsilon_g p_b)$. Likewise we can calculate the expression for the error probability at the output of the OR gate which in this case will be the ε_g of the super gate. Equation (19.26) shows the final expression for the ε_g value of the collapsed gate; note that the ε_g value for the collapsed gate is also a function of the input signal probabilities. As discussed in Section 19.4.1, the error expression for the super gate ε_g has been obtained after suppressing the exponents of signal probabilities that are greater than "1" to "1." In contrast we do not suppress the exponents of ε_g values since this higher exponent may have correctly resulted from the fact that each gate in the circuit can fail with same error probability. On the other hand the signal probability higher exponent may have arisen from the fact that the error of some multiple fanout gate is propagated to a reconvergent fanout point through different paths in the circuit, and hence, these higher exponent must indeed be suppressed. So there is some inaccuracy in our proposed method. To be able to decide precisely whether or not the exponents of ε_g must be suppressed, we will have to use a unique symbol for each gate's error probability and propagate these unique symbols throughout the circuit while suppressing the higher exponents of each unique symbol. The results reported in Table 19.1 have been obtained using the estimation of super gate's ε_g from our implementation of BDEC in SIS [20], which does not include exponent suppression.

$$\begin{aligned}
\varepsilon_g^* = {} & 3\varepsilon_g - 5\varepsilon_g^2 + 2\varepsilon_g^3 \\
& + p_b p_s \varepsilon_g - p_a p_s \varepsilon_g - p_a p_b p_s \varepsilon_g \\
& - 3p_b \varepsilon_g^2 + 2p_a p_s \varepsilon_g^2 - 2p_b p_s \varepsilon_g^2 + 4p_a p_b p_s \varepsilon_g^2 \\
& + 8p_b \varepsilon_g^3 - 4p_a p_b p_s \varepsilon_g^3 - 4p_b \varepsilon_g^4
\end{aligned} \tag{19.26}$$

Table 19.1 shows that BDEC + logic collapsing produces accurate results for the circuit in Fig. 19.6. Table 19.2 shows the comparison of percent error for BDEC and BDEC + collapsing as compared to PTM. If the reconvergent fanout extends over multiple circuit levels then multiple level collapsing can be used but after few levels, the computational complexity of computing output error probability of a super gate with many inputs will become prohibitive and a trade-off between accuracy and complexity will have to be made.

Table 19.1 Output error probability with reconvergent fanout

S/N	Pa	Pb	Ps	ε_g	ε_a	ε_b	ε_s	BDEC	BDEC-CLP	PTM
1	0.5	0.5	0.5	0.05	0.05	0.05	0.05	0.1814	0.1835	0.1829
2	0.1	0.2	0.3	0.05	0.05	0.05	0.05	0.1689	0.1820	0.1852
3	0.5	0.6	0.7	0.05	0.05	0.05	0.05	0.1909	0.1827	0.1830
4	0.7	0.8	0.9	0.05	0.05	0.05	0.05	0.1858	0.1684	0.1668
5	0.5	0.5	0.5	0.001	0.001	0.001	0.001	0.0044	0.0044	0.0044
6	0.5	0.5	0.5	0.002	0.001	0.001	0.001	0.0072	0.0072	0.0072
7	0.5	0.5	0.5	0.01	0.01	0.01	0.01	0.0421	0.0422	0.0422
8	0.5	0.5	0.5	0.02	0.01	0.02	0.03	0.0842	0.0815	0.0814
9	0.5	0.5	0.5	0.08	0.08	0.08	0.08	0.2603	0.2645	0.2633
10	0.5	0.5	0.5	0.05	0.06	0.07	0.08	0.2027	0.2037	0.2032

In passing, we point out that the correlation coefficient method and partial collapse methods both tackle the same problem, that is, how to account for the correlations due to reconvergent fanout structures in VLSI circuits. The trade-off is that the correlation coefficient method has high complexity due to the requirement

Table 19.2 Percent error reduction in output error probability using BDEC +collapsing

S/N	Pa	Pb	Ps	ε_g	ε_a	ε_b	ε_s	BDEC error (%)	BDEC-CLP error (%)
1	0.5	0.5	0.5	0.05	0.05	0.05	0.05	0.84	0.33
2	0.1	0.2	0.3	0.05	0.05	0.05	0.05	8.78	1.75
3	0.5	0.6	0.7	0.05	0.05	0.05	0.05	4.34	0.15
4	0.7	0.8	0.9	0.05	0.05	0.05	0.05	11.39	0.94
5	0.5	0.5	0.5	0.001	0.001	0.001	0.001	0.02	0.01
6	0.5	0.5	0.5	0.002	0.001	0.001	0.001	0.01	0.01
7	0.5	0.5	0.5	0.01	0.01	0.01	0.01	0.21	0.08
8	0.5	0.5	0.5	0.02	0.01	0.02	0.03	3.48	0.12
9	0.5	0.5	0.5	0.08	0.08	0.08	0.08	1.13	0.46
10	0.5	0.5	0.5	0.05	0.06	0.07	0.08	0.25	0.24
11	–	–	–	–	–	–	**Average**	3.05	0.41

to calculate and propagate all correlation coefficients along with signal and error probabilities, whereas the partial collapse has high complexity due to the need to calculate output signals and error probabilities of super gates with a large number of inputs. In practice, the partial collapse of two or three levels of logic into each node (super gate) or the computation of only pairwise spatial correlations is adequate and provides high accuracy.

19.5 Simulation Results

In this section we present some simulation results for the proposed circuit reliability technique and we compare the results of our approach with those of PTM and PGM [19].

We implemented the proposed error calculator and algorithm (BDEC) in SIS [20]. SIS has been widely used by logic synthesis community for designing combinational and sequential logic circuits. We extended existing logic simulation in SIS with faulty circuit simulation based on Monte Carlo simulation technique. We attached a probability function with each node which flips the correct output of the node with a predefined error probability. We used this Monte Carlo simulation to form a reference to compare BDEC results for medium and large circuits.

We added a new BDEC module to the existing SIS package. While simulating a logic circuit, BDEC module models each gate as a probabilistic gate. We used the built-in co-factor function in SIS to develop partial Boolean difference and total Boolean difference functions that are used to propagate single and simultaneous multiple errors from the inputs to the output of the gate, respectively. We have also implemented level collapsing to overcome the inaccuracies introduced because of local reconvergent fanouts. Note that while collapsing levels of logic, we do not change the original logic network; instead, we simply recalculate and update the error and signal probability at the output of the nodes that have reconvergent fanout structures inside their corresponding super gate.

In the past, SIS has been used to apply various delay, area, and power-level optimizations to logic circuits. By incorporating BDEC module to SIS, we expect that researches will be able to use SIS to develop reliability-aware optimizations for logic circuits. For example, given a library of gates with different levels of reliability, design a circuit with given functionality that minimizes area, delay, and power overheads while meeting a given reliability constraint.

Regarding simulation results in this section, for simplicity, but without loss of generality, we assume all gates in a circuit have the same gate error probability ε_g. All primary inputs are assumed to be error free and spatiotemporally uncorrelated. Moreover, signal probability for all the inputs was set to 0.5. The gate error probability was set to 0.05. We thus present results that show how efficiently BDEC can calculate the output reliability for circuits with high primary input count. Running our MATLAB 7.1-based implementation of PTM on a computer system with 2 GB of RAM, we observed that typically for circuits with 16 or more inputs, PTM reported out of memory error. BDEC, however, does the calculations much faster and more efficient than PTM.

Table 19.3 shows the results for reliability calculation for some tree-structured circuits. For example, "8-Input XOR BT" (BT for Balanced Tree) refers to 8-input XOR function implemented using 2-input XOR gates in three levels of logic, whereas "8-Input XOR Chain" refers to the same function realized as a linear chain of seven 2-input XOR gates. We also show results for two 16-input circuits with balanced tree implementation of 2-input gates having layers of 2-input AND, OR, or XOR gates. First letter of gate name is used to show the gates used in each level. For example, AOXO means that the circuits consist of four levels of logic with AND, OR, XOR, and OR gates at the first, second, third, and fourth levels, respectively. Since the complexity of the PTM approach increases with the number of primary inputs exponentially, all the circuits in Table 19.3 are chosen to have relatively small number of primary inputs. Second and third columns of this table compare the execution times for PTM and BDEC, respectively, while the forth and the fifth columns compare the output reliability for the two approaches. It can be seen that our proposed BDEC technique achieves highly accurate reliability values, i.e., the reliability values are different than PTM ones by at most 0.1% for the circuits reported in Table 19.3. More importantly, Table 19.3 shows the difference between the scaling trend of the execution time in both PTM and BDEC techniques. In PTM, the execution time increases exponentially when we move from smaller circuits to larger circuits in Table 19.3, whereas in BDEC the change in the execution time when we move from smaller circuits to the larger ones in Table 19.3 is really small. For the two cases, 16-input XOR chain and 16-input AND chain, the system runs out of memory while executing PTM technique. This shows that execution of PTM technique for even relatively small circuits needs a huge amount of system memory.

Another important advantage of the proposed BDEC technique which can be observed from Table 19.3 is that the complexity of this technique mainly depends on the number of the gates in the circuit; however, the complexity of PTM technique depends on several other factors such as number of the inputs, width and depth of

Table 19.3 Circuit reliability for tree-structured circuits having relatively small number of PIs

		Execution time (ms)		Circuit reliability	
Benchmarks	**# of gates**	**PTM**	**BDEC**	**PTM**	**BDEC**
8-Input XOR BT	7	0.790	0.011	0.7391	0.7391
16-Input XOR BT	15	1664.5	0.017	0.6029	0.6029
16-Input XOR chain	15	Out of Memory	0.015	Out of Memory	0.6029
8-Input AND BT	7	0.794	0.010	0.9392	0.9382
16-Input AND BT	15	1752.2	0.017	0.9465	0.9462
16-Input AND chain	15	Out of Memory	0.016	Out of Memory	0.9091
16-input AOXO BT	15	1769.3	0.017	0.7622	0.7616
16-input OXAX BT	15	1593.1	0.017	0.7361	0.7361

the circuit, and number of the wire crossovers. In other words, efficiency (execution time and memory usage) of PTM depends not only on the number of the gates in the circuit but also on the circuit topology. This is a big disadvantage for PTM making it an infeasible solution for large and/or topologically complex circuits.

It is worth mentioning that although the complexity of Boolean difference equations increases exponentially with the number of the inputs of the function; this does not increase the complexity of the BDEC technique. The reason is the fact that using gates with more than few inputs, say 4, in the actual implementation of any Boolean function is not considered as a good design practice. This makes the complexity of calculating Boolean difference equations small. On the other hand for a fixed library of gates, all the Boolean difference equations can be calculated offline, so there is no computational overhead due to calculating the Boolean difference equations in BDEC.

Table 19.4 shows the results, execution time, and reliability calculation for some of synthesized tree-structured circuits with relatively larger number of inputs. Since the complexity of the PTM is really high for these circuits we only show the results for BDEC. Some of the circuits in Table 19.4 are the larger versions of the circuits reported in Table 19.3. We have also included 16-and 32-bit ripple carry adder (RCA) circuits. Results for two benchmark circuits, I1 and C18, are also included in this table.

Table 19.4 Circuit reliability for tree-structured circuits having relatively large number of PIs

Circuit	**# of gates**	**Execution time (ms)**	**Circuit reliability**
64-Input XOR (BT)	63	0.046	0.5007
64-Input XOR (Chain)	63	0.043	0.5007
64-Input AND (BT)	63	0.054	0.9475
64-Input AND (Chain)	63	0.051	0.9091
64-Input AOXAOXBT	63	0.054	0.6314
64-Input XAOXAOBT	63	0.053	0.9475
16-Bit RCA	80	0.115	0.0132
32-Bit RCA	160	0.216	0.0002
I1	46	0.054	0.3580
C18	6	0.013	0.8032

From the results of Tables 19.3 and 19.4 we note that circuits that use more XOR gates will incur smaller output reliability under a uniform gate failure probability. Furthermore, moving XOR gates closer to the primary outputs results in lower output reliability. Therefore, in order to have more reliable designs, we must have less number of XOR gates close to the primary outputs.

Table 19.5 compares the results for PTM, PGM [19], and BDEC for some more general circuits. Note FA1 and FA2 are two different implementations of full adder circuit. The former is XOR/AND implementation and the latter is NAND only implementation. Also Comp. is a 2-bit comparator circuit. We report the results for our implementation of PTM and BDEC; however, since we were not able to produce the results of PGM, we took the reported results in [19]. As it can be seen from this table, BDEC shows better accuracy as compared to PGM.

Table 19.5 Circuit reliability and efficiency of BDEC compared to PGM and PTM

	Execution time (ms)		**Circuit reliability ($\varepsilon_g = 0.05$)**			**% Error compared to PTM**	
Circuit	**BDEC**	**PTM**	**BDEC**	**PGM**	**PTM**	**BDEC**	**PGM**
2–4 Decoder	0.014	6.726	0.7410	0.7397	0.7479	0.92	1.10
FA1	0.013	2.399	0.7875	0.7898	0.8099	2.77	2.48
FA2	0.017	3.318	0.6326	0.5933	0.6533	3.17	9.18
C17	0.012	2.304	0.7636	0.7620	0.7839	2.59	2.79
Comp.	0.014	0.937	0.7511	0.7292	0.8264	9.11	11.76
Avg. Err.	–	–	–	–	–	3.71	5.46

Table 19.6 shows the results of running BDEC for somewhat larger benchmark circuits. In the last column, we report the results for some of the circuits that were analyzed in [5] to compare the run times of running PTM with that of BDEC. PTM results were reported for technology-independent benchmarks whereas BDEC results are for benchmark circuits mapped to a cell library in 65 nm CMOS technology. PTM results were generated using a system with 3 GHz Pentium 4 processor whereas BDEC results are generated from a system with 2.4 GHz dual-core processor. One can see that BDEC (which has very low memory usage) is orders of magnitude faster than PTM.

Table 19.7 shows how BDEC execution times linearly scale with the number of gates. As it was mentioned in the introduction of this chapter, the worst-case time complexity of previously proposed techniques such as PTM and PDD is exponential in terms of the number of the gates in the circuit.

Table 19.6 Run time comparison between BDEC and PTM for some large benchmark circuits

Benchmark	**# of gates**	**PIs**	**POs**	**BDEC exec time (s)**	**PTM exec times (s)**
C17	6	5	2	7.00E-06	0.313
Pcle	71	19	9	2.40E-05	4.300
z4ml	74	7	4	2.20E-05	0.840
Mux	106	21	1	2.80E-05	2.113
9symml	252	9	1	5.20E-05	696.211

Table 19.7 Circuit reliability for large benchmark circuits

Benchmark	# of gates	PIs	POs	BDEC exec time (μs)	BDEC reported reliability
Majority	22	5	1	9.0	0.6994
Decod	66	5	16	18.0	0.2120
Count	139	31	16	38.0	0.0707
frg1	143	28	3	48.0	0.6135
C880	442	60	26	96.0	0.0038
C3540	1549	50	22	358.0	0.0003
alu4	2492	12	8	577.0	0.0232
t481	4767	16	1	1710.0	0.8630

Table 19.8 shows how BDEC execution times and reliability calculations compared to those of Monte Carlo (MC) simulations. We could not run PTM for larger circuits because of out of bound memory requirements to store probability transfer matrices hence we resorted to MC simulations. In most of the cases we ran 10,000 iterations of MC simulations where each input changed with the probability of 0.5. In the case of higher input count we ran up to 1 M iterations to get more accurate results, but the execution times reported in fourth column of Table 19.8 are for 10,000 iterations in each case. Since overall circuit reliability for multi-output circuits tend to be very low, we also report BDEC calculated minimum output reliability for single output in the last column of Table 19.8.

Table 19.8 BDEC Circuit reliability compared to MC simulations for large benchmark circuits

Benchmark	# of gates	POs	MC exec time (s)	BDEC exec time (ms)	MC reported reliability	BDEC reported reliability	% Error	Min single output reliability
majority	22	1	0.25	2244	0.6616	0.6994	5.71	0.6994
decod	66	16	0.69	6234	0.2159	0.2120	1.81	0.8820
pcle	71	9	0.82	6899	0.2245	0.2270	1.11	0.8401
cordic	116	2	1.26	10093	0.5443	0.5299	2.65	0.7220
sct	143	15	1.54	13086	0.1310	0.130	0.76	0.7988
frg1	143	3	1.59	13864	0.5895	0.6135	4.07	0.7822
b9	147	21	1.64	14118	0.0271	0.0261	3.69	0.7223
lal	179	19	2.52	18001	0.0924	0.0990	7.14	0.8067
9symml	252	1	2.90	27225	0.7410	0.6189	16.48	0.6189
9sym	429	1	4.93	48039	0.7705	0.6398	16.96	0.6398
C5315	2516	123	33.34	267793	0.0000	0.0000	0.00	0.5822
Average	–	–	–	–	–	–	5.49	–

19.6 Extensions to BDEC

19.6.1 Soft Error Rate (SER) Estimation Using BDEC

As technology scales down, the node-level capacitance (which is a measure of the stored charge at the output of the node) and the supply voltage decrease; hence, soft error rates are increasing exponentially [21]. Soft errors in CMOS ICs are caused by a particle (Alpha, energetic neutron, etc.) striking a node which is holding some data value. Soft errors in general result in discharging of a node capacitance which in a combinational circuit means a "1" to "0" transition. This type of error is thus different from Von Neumann error discussed so far in this chapter. A soft error in SRAM can change the logic value stored in the SRAM and thus can be thought as a flipping error.

To use BDEC for soft error rate estimation of combinational logic circuits, we modify the BDEC equations developed in Section 19.3.1. We still use the Boolean difference calculus method to find out the conditions when an error on one or more inputs will affect the output of the gate. We also assume a sufficiently large latching window for a soft error so that such an error can in the worst case propagate to the primary output(s) of the target combinational circuit. In the following, we show the equations to calculate the soft error rate at the output of a buffer, a 2-input AND gate, and a 2-input XOR gate. Note $\varepsilon_{\text{g, soft}}$ in the following equations means the probability that a soft error at the output of the gate will cause the output to transition from logic "1" to logic "0."

To calculate the soft error rate expression at the output of a buffer, we note that soft error happens only when the input is "1" and either of the input or of the output is affected. That is

$$\varepsilon_{\text{buf, soft}} = p_{\text{in}} \left(\varepsilon_{\text{in, soft}} + \varepsilon_{\text{g, soft}} - \varepsilon_{\text{in, soft}} \varepsilon_{\text{g, soft}} \right) \tag{19.27}$$

where $\varepsilon_{\text{in, soft}}$ is the soft error rate at the input, and the term in the parentheses is the probability of error at the input or the output.

To calculate the soft error rate at the output of a 2-input AND gate, we pay attention to the truth table of this gate knowing that soft error can only make "1" to "0" changes. This leads us to the fact that the only time that the output value of a 2-input AND gate is affected by a soft error is when both inputs are "1" and an error occurs at any of the inputs or at the output. Therefore, the soft error rate at the output of a 2-input AND gate is written as

$$\varepsilon_{\text{AND2, soft}} = p_1 p_2 \begin{pmatrix} \varepsilon_{1,\text{ soft}} + \varepsilon_{2,\text{ soft}} + \varepsilon_{\text{g, soft}} \\ -\varepsilon_{1,\text{ soft}} \varepsilon_{2,\text{ soft}} - \varepsilon_{1,\text{ soft}} \varepsilon_{\text{g, soft}} - \varepsilon_{2,\text{ soft}} \varepsilon_{\text{g, soft}} \\ +\varepsilon_{1,\text{ soft}} \varepsilon_{2,\text{ soft}} \varepsilon_{\text{g, soft}} \end{pmatrix} \tag{19.28}$$

Similarly, the soft error rate at the output of a 2-input XOR gate can be calculated by looking into its truth table and realizing that the output value can be affected by a soft error when (i) exactly one input is "1" and one input is "0" and soft error changes the logic-1 input or the output or (ii) both inputs are "1" and soft error changes one and only one of these logic-1 inputs. Therefore, the soft error rate at the output of a 2-input XOR gate is calculated as

$$\begin{aligned}\varepsilon_{\text{XOR2, soft}} = {} & p_1 (1 - p_2) \left(\varepsilon_{1,\text{ soft}} + \varepsilon_{\text{g, soft}} - \varepsilon_{1,\text{ soft}}\varepsilon_{\text{g, soft}}\right) \\ & + (1 - p_1)\, p_2 \left(\varepsilon_{2,\text{ soft}} + \varepsilon_{g,soft} - \varepsilon_{2,\text{ soft}}\varepsilon_{\text{g, soft}}\right) \\ & + p_1 p_2 \left(\varepsilon_{1,\text{ soft}} \left(1 - \varepsilon_{2,\text{ soft}}\right) + \left(1 - \varepsilon_{1,\text{ soft}}\right) \varepsilon_{2,\text{ soft}}\right) \end{aligned} \quad (19.29)$$

Similarly we can derive error equations for other types of gate functions.

19.6.2 BDEC for Asymmetric Erroneous Transition Probabilities

BDEC for Von Neumann fault model assumed equal probability of error for a "0" to "1" and "1" to "0" erroneous transition. But this may not always be the case, for example, in dynamic and domino logic families, the only possible erroneous transition during the evaluate mode is from "1" to "0." In these situations, the solution is to independently calculate the overall circuit error probability using the low-to-high and high-to-low probability values and gate-level error rates. Both circuit error rates are then reported.

19.6.3 BDEC Applied to Emerging Nanotechnologies

A quantum-dot cellular automaton (QCA) [22] is a binary logic architecture which can miniaturize digital circuits to the molecular levels and operate at very low-power levels [23]. QCA devices encode and process binary information as charge configurations in arrays of coupled quantum dots, rather than current and voltage levels. One unique aspect of QCA is that both wires and gates are constructed from quantum dots. Each dot consists of a pair of electrons that can be configured in two different ways to represent a single bit of information. Hence in QCA both gates and wire are subject to bit-flip errors. QCAs have two main sources of error: (1) decay (decoherence)—when electrons that store information are lost to the environment and (2) switching error—when the electrons do not properly switch from one state to another due to background noise or voltage fluctuations [23]. BDEC uses Von Neumann (bit-flip) fault model, hence it is thus well suited to calculate errors in QCAs. In QCA wires/interconnects can also make bit-flip errors, hence BDEC must be extended to be used for QCAs. This extension in BDEC is straightforward and requires a simple replacement of each interconnect in the circuit with a probabilistically faulty buffer.

19.7 Conclusions

As technology scales down circuit reliability is becoming one of the main concerns in VLSI design. In nano-scale CMOS regime circuit reliability has to be considered in the early-design phases. This shows the need for fast reliability calculator tools that are accurate enough to estimate overall circuit reliability. The presented error/reliability calculator, BDEC, takes primary input signal and error probabilities and gate error probabilities and computes the reliability of the circuit. BDEC benefits from a linear-time complexity with number of the gates in the circuit. Compared to PTM which generates accurate reliability results, BDEC generates highly accurate results that are very close to PTM ones. We showed that the efficiency, execution time, and memory usage of BDEC is much better than those for PTM.

BDEC can find application in any combinatorial logic design where reliability is a major concern. Presently BDEC can be applied to combinatorial circuits only, sequential logic is not supported. BDEC can be easily enhanced to be applied to sequential logic. Current version of BDEC uses level collapsing to reduce the effect of reconvergent fanout. In future BDEC can be enhanced to use spatial correlations between the signals to further reduce the inaccuracies introduced because of reconvergent fanouts.

References

1. Krishnaswamy, S.: Design, analysis, and test of logic circuits under uncertainty. Dissertation, University of Michigan at Ann Arbor (2008)
2. Rabey, J.M., Chankrakasan, A., Nikolic, B.: Digital Integrated Circuits. Prentice Hall, pp. 445–490 (2003)
3. Hu, C. Silicon nanoelectronics for the 21st century. Nanotechnology **10**(2), 113–116 (1999)
4. Bahar, R.I., Lau, C., Hammerstrom, D., Marculescu, D., Harlow, J., Orailoglu, A., Joyner, W.H. Jr., Pedram, M.: Architectures for silicon nanoelectronics and beyond. Computer **40**(1), 25–33 (2007)
5. Krishnaswamy, S., Viamontes, G.F., Markov, I.L., Hayes, J.P.: Accurate reliability evaluation and enhancement via probabilistic transfer matrices. Proceedings of Design, Automation and Test in Europe (DATE), pp. 282–287 (2005)
6. Abdollahi, A. Probabilistic decision diagrams for exact probabilistic analysis. Proceedings of International Conference on Computer Aided Design (ICCAD) (2007)
7. Mehta, H., Borah, M., Owens, R.M., Irwin, M.J.: Accurate estimation of combinational circuit activity. Proceedings of the Design Automation Conference, (DAC), pp. 618–622 (1995)
8. Bhaduri, D., Shukla, S.: NANOPRISM: A tool for evaluating granularity versus reliability trade-offs in nano architectures. In: Proceedings of 14th ACM Great Lakes Symposium VLSI, pp. 109–112. (2004)
9. Norman G, Parker D, Kwiatkowska M, Shukla, S.: Evaluating the reliability of NAND multiplexing with PRISM. IEEE Transactions on Computer-Aided Design Integrated Circuits System **24**(10), 1629–1637 (2005)
10. Bhaduri, D., Shukla, S.K., Graham, P.S., Gokhale, M.B.: Reliability analysis of large circuits using scalable techniques and tools. IEEE Transactions on Circuits and Systems **54**(11), 2447–2460 (2007)
11. Sellers, F.F., Hsiao, M.Y., Bearnson, L.W.: Analyzing errors with the boolean difference. IEEE Transactions on Computers **17**(7), 676–683 (1968)
12. Akers, S.B. Jr. (1959) On the theory of Boolean functions. SIAM Journal on Applied Mathematics **7**, 487–498
13. Ku, C.T., Masson, G.M.: The Boolean difference and multiple fault analysis. IEEE Transactions on Computers **c-24**, 62–71 (1975)
14. Das, S.R., Srimani, P.K., Dutta, C.R.: On multiple fault analysis in combinational circuits by means of boolean difference. Proceedings of the IEEE **64**(9), 1447–1449 (1976)

15. Rejimon, T., Bhanja, S.: An accurate probabilistic model for error detection. Proceedings of 18th International Conference on VLSI Design, pp. 717–722. (2005)
16. Ercolani, S., Favalli, M., Damiani, M., Olivo, P., Ricco, B.: Testability measures in pseudo-random testing. IEEE Transactions on CAD **11**, 794–800 (1992)
17. Marculescu, R., Marculescu, D., Pedram, M.: Probabilistic modeling of dependencies during switching activity analysis. IEEE Transactions on Computer Aided Design **17**(2), 73–83 (1998)
18. Parker, K.P., McCluskey, E.J.: Probabilistic treatment of general combinational networks. IEEE Transactions on Computers **24**(6), 668–670 (1975)
19. Han, J., Gao, J.B., Jonker, P., Qi, Y., Fortes, J.A.B.: Toward hardware-redundant fault-tolerant logic for nanoelectronics. IEEE Transactions on Design and Test of Computers **22–24** 328–339 (2005)
20. Sentovich, E.M., Singh, K.J., Lavagno, L., Moon, C., Murgai, R., Saldanha, A., Savoj, H., Stephan, P.R., Brayton, R.K., SangiovanniVincentelli, A.: SIS: A system for sequential circuit synthesis," U.C. Berkeley, Technical Report, (1992)
21. Li, L., Degalahal, V., Vijaykrishnan, N., Kandemir, M., Irwin, M.J.: Soft error and energy consumption interactions: A data cache perspective. In: Proceedings of the international symposium on low power electronics and design ISPLD'04 (2004)
22. Lent, C.S., Tougaw, P.D., Porod, W., Bernstein, G.H.: Quantum cellular automata. Nanotechnology **4**, 49–57 (1993)
23. Rejimon, T., Bhanja, S.: Probabilistic error model for unreliable nano-logic gates. In: Proceedings NANO, 47–50 (2006)

Chapter 20
Digital Logic Using Non-DC Signals

Kalyana C. Bollapalli, Sunil P. Khatri, and Laszlo B. Kish

Abstract In this chapter, a new type of combinational logic circuit realization is presented. These logic circuits are based on non-DC representation of logic values. In this scheme, logic values can be represented by signals that are uncorrelated, for example, distinct values represented by independent stochastic processes (noise from independent sources). This provides a natural way of implementing multivalued logic. Signals driven over long distances could take advantage of this fact and can share interconnect lines. Alternately, sinusoidal signals can be used to represent logic values. Sinusoid signals of different frequencies are uncorrelated. This property of sinusoids can be used to identify a signal without ambiguity. This chapter presents a logic family that uses sinusoidal signals to represent logic 0 and logic 1 values. We present sinusoidal gates which exploit the anti-correlation of sinusoidal signals, as opposed to uncorrelated noise signals. This is achieved by employing a pair of sinusoid signals of the same frequency, but with a phase difference of 180°. Recent research in circuit design has made it possible to harvest sinusoidal signals of the same frequency and 180° phase difference from a single resonant clock ring, in a distributed manner. Another advantage of such a logic family is its immunity from external additive noise. The experiments in this chapter indicate that this paradigm, when used to implement binary valued logic, yields an improvement in switching (dynamic) power.

20.1 Introduction

With the recent slowing of Moore's law, unconventional ways to perform computation are being investigated. Signals that are not DC in nature are being used to represent logic values and their properties are being used for deterministic computation.

K.C. Bollapalli (✉)
NVIDIA Corporation, San Jose, CA, USA
e-mail: kbollapalli@nvidia.com

Based on Bollapalli, K.C.; Khatri, S.P.; Kish, L.B.; "Implementing digital logic with sinusoidal supplies," Design, Automation & Test in Europe Conference & Exhibition (DATE), 2010, pp. 315–318, 8–12 March 2010 © [2010] IEEE.

S.P. Khatri, K. Gulati (eds.), *Advanced Techniques in Logic Synthesis, Optimizations and Applications*, DOI 10.1007/978-1-4419-7518-8_20,

An example of a signal that is not DC in nature (and can be used for deterministic computation) is thermal noise generated by resistors [5, 6, 10]. Thermal noise generated by resistors are independent stochastic processes. The numerous problems that complicate the design of modern processors and ASICs motivates us to explore such logic schemes. These problems include process variations, thermal noise, and supply noise, to name a few. These issues introduce non-ideal waveforms in circuits that represent logic values using DC voltages. In the case of noise-based circuits, these problems introduce new noise on signals. These noise signals introduced being independent stochastic processes, they interfere non-destructively, thus not altering the properties of the output signals.

A new type of deterministic logic scheme was introduced in [5], which utilized the orthogonality of independent stochastic processes (noise processes). Generally, for arbitrary independent reference noise sources $V_i(t)(i = 1, \ldots, N)$:

$$\{V_i(t)V_j(t)\} = \delta_{i,j} \tag{20.1}$$

where $\delta_{i,j}$ is the Kronecker symbol (for $i = j$, $\delta_{i,j} = 1$, otherwise $\delta_{i,j} = 0$) and {} represents the correlation operator. Due to (20.1), the $V_i(t)$ processes can be represented by orthogonal unit vectors in a multidimensional space, thus we can use the term *logic basis vectors* for the reference noises and introduce the notion of an *N-dimensional logic space*, with N logic state vectors in it. Deterministic logic implies a logic framework where identification of logic values is independent of any notion of probability (as opposed to quantum computing).

By using this multidimensional space along with linear superposition, logic vectors and their superpositions can be defined, which results in a large number of different logic values, even with a relatively low number N of basis vectors. For example, when using binary superposition coefficients that have only on/off possibilities of the reference noises, the number of possible logic values is 2^{2^N}. It is important to emphasize that such a logic state vector is transmitted in a single wire [5], not on an ensemble of parallel wires. Thus, using a linear binary superposition of basis vectors, $X(t) = \Sigma_{i=1}^{N} a_i V_i(t)$, where the $X(t)$ represents a single number with N bit resolution, residing in a single wire. The coefficients a_i are weights for each of the reference noise sources $V_i(t)$. It was also shown that the noise-based logic scheme can replace classical digital gates, and the basic Boolean building elements (INVERTER, AND, OR, and XOR) were introduced in the noise-based scheme [5] in a deterministic way. The noise-based logic has several potential advantages, such as reduced error propagation and power dissipation, even though it may need larger numbers of circuit elements and binary computations may be slower.

Another powerful property of noise-based logic is that the product of two different orthogonal basis vectors (independent stochastic signals) is orthogonal to both the original noises. This property yields a logic hyperspace. If $i \neq k$ and $H_{i,k}(t) \equiv V_i(t)V_k(t)$ then for all $n = 1, \ldots, N$ can be used to construct

$$\{H_{i,k}(t)V_n(t)\} = 0 \tag{20.2}$$

In this manner, we can construct a product of all possible ${}^{N}C_i$ noise basis elements ($1 \leq i \leq N$), yielding a logic hyperspace with $2^N - 1$ elements. Hence a wire can carry up to 2^{2^N-1} logic values, yielding an extremely powerful multivalued logic scheme.

In this chapter we explore sinusoid signals to represent logic values. Sinusoid signals with different frequencies share many properties with noise from independent sources. Sinusoid signals with different frequencies do not correlate and are immune to additive noise. These properties ensure that only signals with the same frequency can correlate, and hence can be used to deterministically identify signals. This property of sinusoid signals could be exploited to implement different logic values using different frequencies and different phases. Let S_i^k represent the sinusoid of frequency i and phase k and let $<,>$ represents the correlation operator, then

$$\begin{aligned} \delta_{i,j}^{k,l} &:= < S_i^k, S_j^l > \\ i \neq j &\rightarrow \delta_{i,j}^{k,l} = 0 \\ i = j &\rightarrow \delta_{i,j}^{k,l} = \cos(k-l) \end{aligned} \tag{20.3}$$

Correlation between sinusoids is computed as the average of the product of the two sinusoids. In this chapter we use the above-mentioned ideas and present a way to realize binary valued logic gates that utilize sinusoidal supplies. We also explore design considerations that are induced due to the nature of the supplies and the methodology of computation. Logic gates using sinusoid supplies present us with some attractive advantages

- *Additive noise immunity*: Additive noise does not affect the orthogonality of logic values, as opposed to the traditional representation of logic values, where signal integrity degrades with additive noise.
- *Multivalued logic*: Sinusoids with different frequencies do not correlate. Multivalued logic can be realized by using sinusoidal signals of different frequencies, where different logic values can be represented by different frequencies or as a superimposition of multiple frequencies. Multivalued logic could be used to reduce the logic depth and thereby reduce circuit area.
- *Interconnect reduction*: The above-mentioned property of sinusoids (non-correlation of sinusoids of different frequencies) could also be used for multiple signals to share a single wire. This could be exploited by signals that have to be routed over large distances, thereby reducing wiring area.
- Using N sinusoids, a logic hyperspace with 2^{2^N-1} values can be constructed.

The contributions of this chapter are

- demonstrating that logic gates can be implemented with sinusoidal supply signals and
- exploring design considerations induced by the methodology of computation and quantifying the induced design parameters.

The remainder of this chapter is organized as follows. Section 20.2 discusses some previous work in this area. In Section 20.3 we describe our method to implement logic gates using sinusoidal supplies. Section 20.4 presents experimental results, while conclusions are drawn in Section 20.5.

20.2 Previous Work

There has been significant amount of research in conventional ways of representing logic. Recently, newer ways to represent logic values have been envisioned. In [11], authors have proposed using independent stochastic vectors for noise robust information processing. In their approach, an input signal is compared with a stored reference signal using a comparator. The number of rising edges in the output of the comparator is computed. The authors claim that as the similarity between signals increases as the number of rising edges counted would also increase. The authors present results averaged over a large number of simulations and show that similar signals, on an average, result in larger count values. Their work presents a way to correlate two signals, but their implementation requires complex hardware units (a comparator and a counter). Also, their method cannot deterministically identify a signal and therefore it belongs to the class of stochastic logic. In [5, 6], authors have proposed using noise for deterministic logic and digital computing. Independent noise sources are uncorrelated (orthogonal) stochastic processes. The author presents an approach to implement noise-based logic gates. In a binary logic system, there would be two reference noise signals (one signal per logic value) with which the input is correlated. A noise signal can be identified by correlating it with all the reference noise signals. A correlator could be realized as an analog multiplier followed by a time averaging unit like a low-pass filter. The result of the time averaging unit can then be used to drive the output of the gate with either logic 0 noise or logic 1 noise value or their superimposition. However, there are many open problems and questions with utilizing the noise-based logic. Calculations indicate a factor of 500 slowdown compared to the noise bandwidth (otherwise error rates are higher than traditional digital gates). Significant power reductions can only be expected with small noise amplitudes, which needs preamplifier stages in the logic gates (which results in extra power consumption). In [7], the authors build a multidimensional logic hyperspace with properties similar to quantum Hilbert space. They provide a noise-based deterministic string search algorithm which outperforms Grover's quantum search engine [7]. In this chapter, we seek an alternative way to implement logic with non-DC values. Among the features we desire are

- a representation in which the time for computing correlation is low and can be deterministically predicted and
- a representation that allows us to choose sources that are not very sensitive to load.

In the next section we describe our approach to realizing such a logic family, which uses some starting ideas from [6].

20.3 Our Approach

We describe our approach with an example. Figure 20.1 illustrates the block diagram of a basic gate. This gate consists of two mixers, two integrators (low-pass filters), and an output driver. The input signal (*in*) is mixed (multiplied) with reference sinusoidal signals (S_0 and S_1) which represent the logic 0 and logic 1 values. Equation (20.4) describes the result of mixing two sinusoidal signals:

$$< \sin(\omega_1 t + \delta_1), \sin(\omega_2 t + \delta_2) >= \frac{1}{2} \times (\cos((\omega_1 - \omega_2)t + (\delta_1 - \delta_2)) - \cos((\omega_1 + \omega_2)t + (\delta_1 + \delta_2))) \quad (20.4)$$

If $\omega_1 = \omega_2$, the low-frequency term of (20.4) $(\cos((\omega_1 - \omega_2)t + (\delta_1 - \delta_2)))$ would result in a constant. The output of the mixer is integrated using a low-pass filter (LPF). The LPF filters out the high-frequency component of (20.4). If the signals that are being mixed are in phase ($\delta_1 = \delta_2$) the result of the integrator is a positive constant. If the signals are out of phase ($\delta_1 = \delta_2 \pm \pi$) the result of integration is a negative constant. Thus if sinusoidal sources S_0 and S_1 are of the same frequency and are out of phase by 180°, they would correlate negatively. We now describe a way of practically realizing two signals of the same frequency and a 180° phase difference.

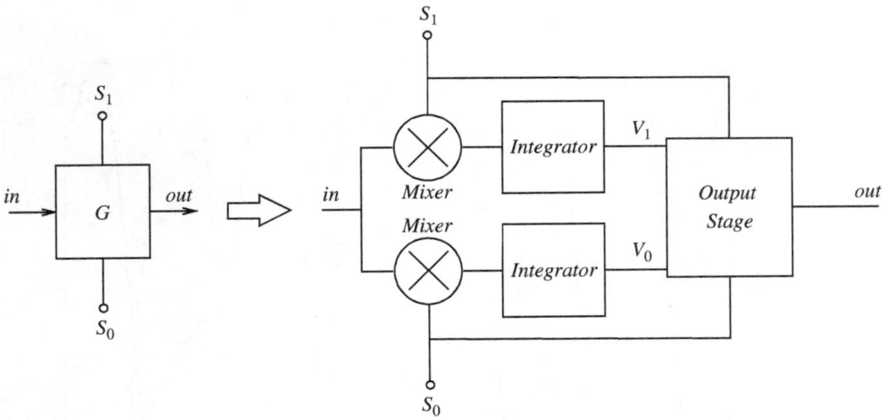

Fig. 20.1 Logic gate

20.3.1 Standing Wave Oscillator

Recent research in the field of low-power high-frequency clock generation has shown the feasibility of realizing a distributed LC oscillator on chip [3, 4, 9]. These oscillators use the parasitic inductance of on-chip interconnect, along with the par-

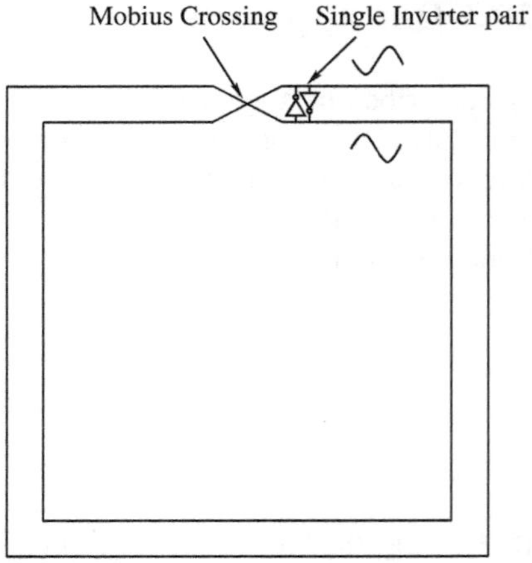

Fig. 20.2 Distributed resonating oscillator

asitic capacitance presented by interconnect as well as semiconductor devices to realize a high-frequency resonant oscillator. Figure 20.2 shows a schematic description of one such resonant oscillator. Such oscillators include a negative resistance structure to compensate for the resistive losses in the interconnect. In the example in Fig. 20.2, the cross-coupled inverter pair acts as the negative resistance element.

Figure 20.3 illustrates the two out-of-phase sinusoidal signals that can be extracted from the standing wave oscillator of Fig. 20.2

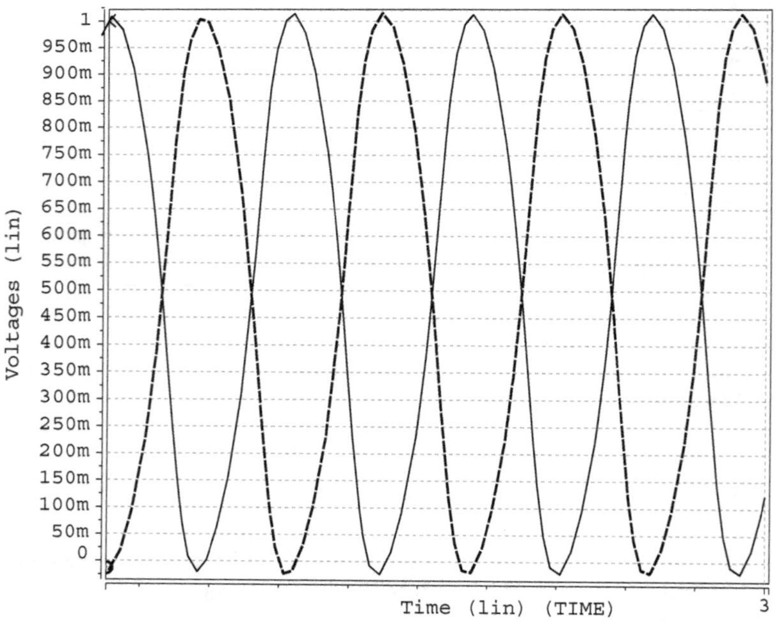

Fig. 20.3 Out-of-phase sinusoid sources

20.3.2 A Basic Gate

We now describe a basic inverter gate. To be able to use these gates in large integrated circuits, each of the components of the gate has to be implemented with as few devices as possible. Figure 20.4 illustrates the structure of a inverter in our implementation. The structure of any sinusoidal supply-based logic gate would be very similar to the gate in Figure 20.4. The input of the gate is a sinusoid signal. The input signal is correlated with all the reference sinusoidal sources (two sources in this instance). Based on the logic value identified on the wire, and based on the logic implemented in the gate, a corresponding sinusoidal signal is driven out. A combination of a multiplier and an integrator is used as a correlation operator. The output of the integrator is then used to enable pass gates which drive out the appropriate reference sinusoidal signal.

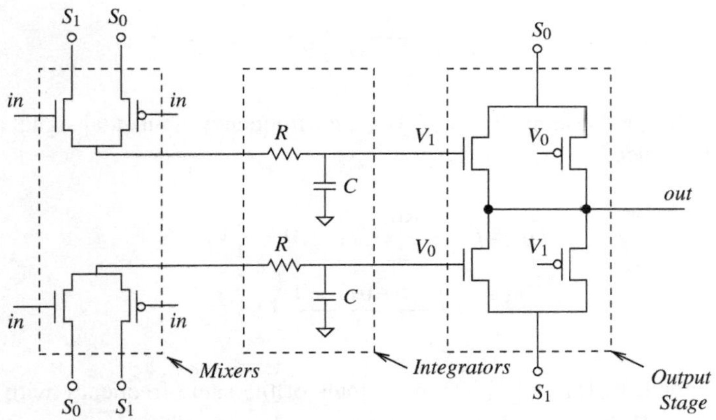

Fig. 20.4 Logic gate

20.3.2.1 Multiplier

To implement a multiplier with two devices, we bias an NMOS device and a PMOS device in the linear region. Equation (20.5) presents the formula for the drain current in an NMOS when the device is in linear region. The drain current is dependent on a product of gate voltage (V_{gs}) and drain voltage (V_{ds}). The drain current can thus be used to extract the result of multiplication:

$$I_{ds}(t) = \beta \left(V_{gs}(t) - V_{th} - \frac{V_{ds}(t)}{2} \right) V_{ds}(t) \tag{20.5}$$

$$I_{ds}(t) = \beta \left(V_{gs}(t) V_{ds}(t) - V_{th} V_{ds}(t) - \frac{V_{ds}(t)^2}{2} \right)$$

If this drain current is driven through a suitable low-pass filter to block the high-frequency components, we can extract the average drain current. Let $\overline{I}$ indicates the average value of $I(t)$:

$$\overline{I_{ds}} = \beta\left(\overline{V_{gs}V_{ds}} - \overline{V_{th}V_{ds}} - \frac{\overline{V_{ds}^2}}{2}\right)$$

Let $V_{gs}(t)$ and $V_{ds}(t)$ be signals of the same frequency and a phase difference of α. Without loss of generality, let $V_{gs}(t) = V_1 \sin(\omega t)$ and $V_{ds} = V_1 \sin(\omega t + \alpha)$. Then from (20.4),

$$\overline{V_{gs}V_{ds}} = \frac{\cos(\alpha)}{2} V_1^2$$

and

$$\overline{V_{ds}V_{ds}} = \frac{1}{2} V_1^2$$

Also since V_{th} is a constant, $V_{th}V_{ds}(t)$ is a high-frequency sinusoidal signal and thus $\overline{V_{th}V_{ds}} = 0$. Hence,

$$\overline{I_{ds}} = \beta\left(\frac{\cos(\alpha)}{2} V_1^2 - 0 - \frac{1}{4} V_1^2\right)$$

$$\overline{I_{ds}} = \left(\frac{2\cos(\alpha) - 1}{4}\right)\beta V_1^2 \tag{20.6}$$

If the signals $V_{gs}(t)$ and $V_{ds}(t)$ are signals of the same frequency with $\alpha = 0$, $\overline{I_{ds}}$ is positive. If they are of same frequency with $\alpha = 180°$, $\overline{I_{ds}}$ is negative. The above result shows that when signals correlate, the result is an average positive drain current, but when the signals do not correlate there is an average negative drain current.

A similar effect can be achieved with a PMOS device. The drain current of a PMOS transistor in linear region is given by

$$I_{sd}(t) = -\beta(V_{gs}(t) - V_{th} - \frac{V_{ds}(t)}{2})V_{ds}(t) \tag{20.7}$$

By the same argument as above, if V_{gs} and V_{ds} are sinusoidal signals of the same frequency, same amplitude, and differ in phase by α,

$$\overline{I_{sd}} = -\left(\frac{2\cos(\alpha) - 1}{4}\right)\beta V_1^2 = \left(\frac{1 - 2\cos(\alpha)}{4}\right)\beta V_1^2 \tag{20.8}$$

If the signals V_{gs} and V_{ds} are in phase ($\alpha = 0$), $\overline{I_{sd}}$ is negative. But if $\alpha = 180°$, $\overline{I_{sd}}$ is positive.

From (20.6) and (20.8) we can conclude that the drain current is positive when the NMOS device mixes signals that are in phase and is positive when PMOS mixes signals that are out of phase. Thus an NMOS device mixing the input with one reference sinusoidal signal and PMOS mixing the same input with the other reference sinusoidal signal forms a good *complementary* mixer. Figure 20.4 shows the circuit diagram of the above-described mixer. The total average drain current that can be driven by such a mixer is (from (20.6) and (20.8))

$$\overline{I} = \left(\frac{2\cos(\alpha) - 1}{4} + \frac{1 - 2\cos(\pi - \alpha)}{4}\right)\beta V_1^2$$

$$\overline{I} = \left(\frac{2\cos(\alpha) - 1}{4} + \frac{1 + 2\cos(\alpha)}{4}\right)\beta V_1^2$$

$$\overline{I} = \cos(\alpha)\beta V_1^2 \qquad (20.9)$$

20.3.2.2 Low-Pass Filter

The choice of the R and C values of the integrator (Fig. 20.4) poses an optimization problem. A small RC product ensures that the voltage on the capacitor reaches the final DC value early, but a large portion of the sinusoid will be present at output node. On the other hand a large value of RC product ensures that a smaller AC signal is visible on the capacitor but the DC voltage on the capacitor would take longer to reach its final DC value. We have

$$\frac{1}{\mathrm{RC}} = 3F$$

where F is the corner frequency of the low-pass filter. From the above equation we can compute the magnitude of input signal that is visible on the output of the low-pass filter as

$$\frac{\frac{1}{j\omega C}}{R + \frac{1}{j\omega C}} = \frac{1}{j\omega \mathrm{RC} + 1} = \frac{1}{\frac{j\omega}{3F} + 1} \qquad (20.10)$$

Based on the input frequency, a low-pass filter can be designed to minimize the peak to peak amplitude of input signal visible after the correlator. The resistance is realized using a poly-silicon wire, and the capacitance is realized as a gate capacitance.

20.3.2.3 Output Stage

Once the logic values on the inputs have been identified, they can be used to compute and drive out the correct reference signal on the gate output. To be able to achieve a full rail swing, it is necessary to have a complementary pass gate structure. Figure 20.4 highlights the pass gate structure of the output stage of a simple inverter. To enable correct functionality we appropriately connect the correlator outputs to the reference signals in the pass gate structure. The positive correlation outputs (V_1 in Fig. 20.4) drive the NMOS gates of the pass gate structure while the negative correlation outputs (V_0 in Fig. 20.4) drive the PMOS gates of the pass gate structure. To implement a buffer, the S_0 and S_1 connections in the pass gate structure (output stage) are reversed.

20.3.2.4 Complex Gates

In the same manner as described for an inverter, more complex gates can also be designed. Figure 20.5 illustrates the design of a sinusoidal supply-based NAND2 gate. Both the inputs of the NAND2 gate are correlated with both reference signals to identify the logic values on the input signals. Again, a complimentary pass gate structure (shown in Fig. 20.5) is used to drive out the correct reference signal.

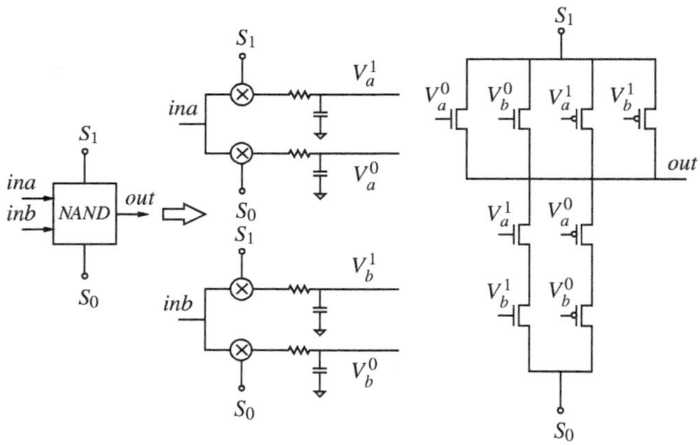

Fig. 20.5 NAND gate

20.3.3 Interconnects

In the sinusoidal supply-based logic style, information propagates around the circuit as high-frequency AC signals, and these signals are correlated. Hence it is important to estimate the effect of interconnect delays on the signals. All realistic circuits are plagued with parasitic capacitances and inductances. The effect of these parasitics on a signal can be modeled as a transformation of these signals by the transfer function of the parasitics. Identical signals could reach a gate by traversing different paths (i.e., different parasitics). These signals could therefore have different phases and different magnitudes when they are correlated. We quantify the error induced by such a phase shift. If the input signal gets phase shifted by α and is correlated with the reference signals, its correlation value reduces by a factor of $\cos(\alpha)$, from (20.9)

$$\frac{\overline{I}_\alpha}{\overline{I}_0} = \frac{\cos(\alpha)\beta V_1^2}{\cos(0)\beta V_1^2} = \cos(\alpha)$$

As the phase difference increases, the correlation between a signal and its like source decreases. To avoid this effect, the frequency of the reference sinusoidal signal should be chosen to be as low as possible. A fixed time shift of a signal presents itself as a smaller phase shift at lower frequencies and hence can tolerate a larger time shift without losing correlation. However, lower frequencies require a larger RC product in the low-pass filter of the gate. A larger RC product leads to a slower gate and also requires larger area to implement R and C. This presents us with an optimization problem which we solve empirically in the next section.

20.4 Experimental Results

In all our experiments we use a predictive technology model (PTM) 45 nm, metal gate, high-K dielectric, strained silicon model card [1]. All the sinusoid signals (unless otherwise stated) are 1 V peak to peak signals, with an average voltage of 0.5 V. This choice of average voltage is driven by the fact that the resonant oscillator that produces the sinusoidal supplies (Fig. 20.3) produces sinusoids whose average value is $\frac{V_{DD}}{2}$. The choice of the amplitude is technology dependent but the technique of sinusoidal supply-based circuit design is not restricted by the choice of peak to peak voltage.

We first quantify the power consumed by a circuit designed with sinusoidal supply-based gates. The total power consumed is the power consumed by the gates as well as the power consumed by the standing wave oscillator [3].

20.4.1 Sinusoid Generator

We modified the sinusoid generator described in Section 20.3.1 to consume as little power as possible. Results from [3] indicate that the power consumed is a few milliwatts. However, these power numbers are from experiments with very little capacitive load on the oscillator. However, when the oscillator is used to generate the sinusoidal supplies in our approach, a large number of sinusoidal supply-based gates are connected to it, and they load the oscillator to a greater extent. As the load on the oscillator increases, the frequency of oscillation decreases and the power consumption increases. Figure 20.6 plots the variation in power and frequency of the oscillator as the load on the oscillator (in terms of the number of gates connected to it) increases. The nature of the load presented by the gates on the oscillator is capacitive, and it arises from the diffusion and gate terminals of the transistors of the gates. The frequency of oscillation varies with capacitive load C on the oscillator, as $F = \frac{1}{\sqrt{LC}}$. On the other hand, the only source of power loss in the oscillator is the resistive loss. The power consumed due to resistive losses is I^2R, where I is the transient current, whose value is determined by the characteristic impedance of the transmission line used in the oscillator:

$$I = \frac{V}{Z_0}$$

The characteristic impedance of a transmission line is given by

$$Z_0 = \sqrt{\frac{L}{C}}$$

where L is the parasitic inductance in the transmission line and C is the parasitic capacitance. Thus the power consumed in the ring is

$$P = I^2R = \left(\frac{V}{Z_0}\right)^2 R$$
$$P = \frac{V^2C}{L}R = \frac{V^2CR}{L} \tag{20.11}$$

The power consumed in the oscillator is thus expected to grow linearly with the capacitive load on the oscillator. This is confirmed in Fig. 20.6, which plots the power of the oscillator from HSPICE [8] simulations. In this simulation the NMOS devices of the inverter pair had a width of 40 μm. The ring was implemented in metal 8 and its length was 80 μm. This trend of increasing power with load continues until the negative resistance structure can no longer compensate for the losses in the oscillator ring, which occurs at about 1800 gates, from Fig. 20.6.

For the oscillator shown in Fig. 20.2, the negative resistance is provided by the cross-coupled inverter pair. As discussed earlier, to be able sustain oscillations at larger loads, we require a larger cross-coupled inverter pair. A larger inverter pair also presents its own parasitic capacitance to the oscillator, thus offsetting the gains of using a larger inverter pair. Figure 20.7 plots the change in frequency of oscillation and power consumed in a oscillator, with varying inverter pair size. Oscillations do not set in until the inverter pair can sustain the resistive losses. Once oscillations start, increasing the inverter pair size increases power linearly while the frequency reduces as discussed earlier.

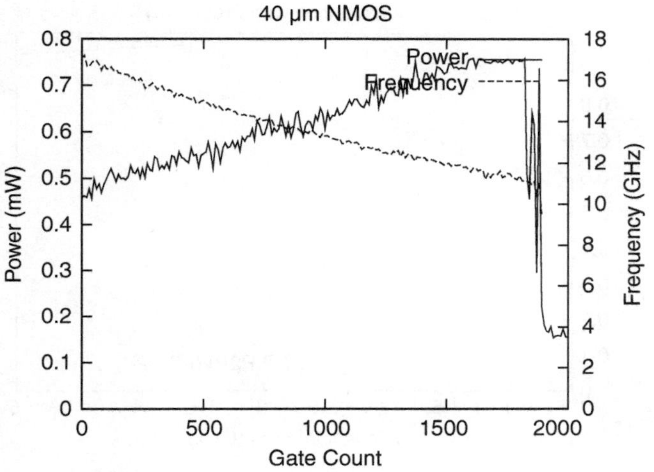

Fig. 20.6 Change in oscillator frequency and power with load

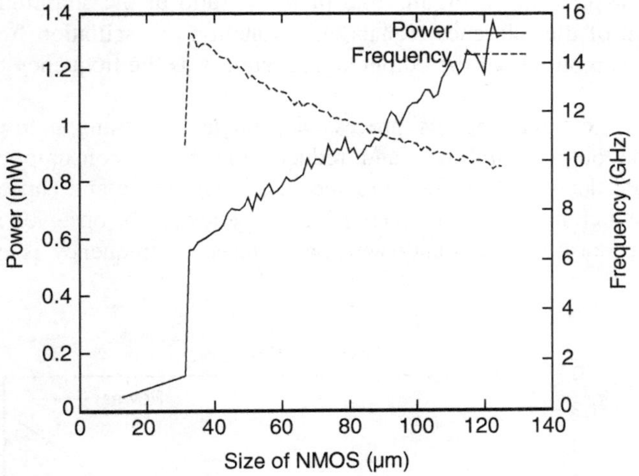

Fig. 20.7 Change in frequency and power with inverter pair size

20.4.2 Gate Optimization

As discussed in the Section 20.3.2.2 the design of the low-pass filter in a gate is an optimization problem. The optimal low-pass filter design is dependent on the frequency of operation. To find the best frequency of operation we compare the performance of a simple gate (buffer) with a fanout of four at various frequencies. The output stage of a gate also acts as a low-pass filter (the resistance of the output stage transistors along with the diffusion and gate capacitances of the fanout gates

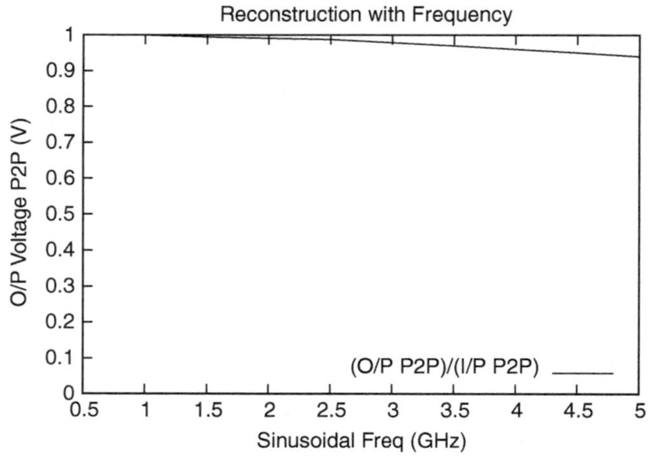

Fig. 20.8 Output amplitude with varying frequency

forms the low-pass filter). Figure 20.8 plots the ratio of the amplitude of output signal to that of the reference signals, as a function of oscillation frequency. As expected, the amplitude of the output signal reduces as the frequency of operation increases.

The choice of frequency is process dependent. Choosing a low frequency increases the output amplitude and reduces the power consumption but also increases gate delay and the value of the filter's resistance and capacitance. The increase in switching power with increasing frequency can be observed in Fig. 20.9, while the decrease in gate delay with an increase of frequency is portrayed in Fig. 20.10.

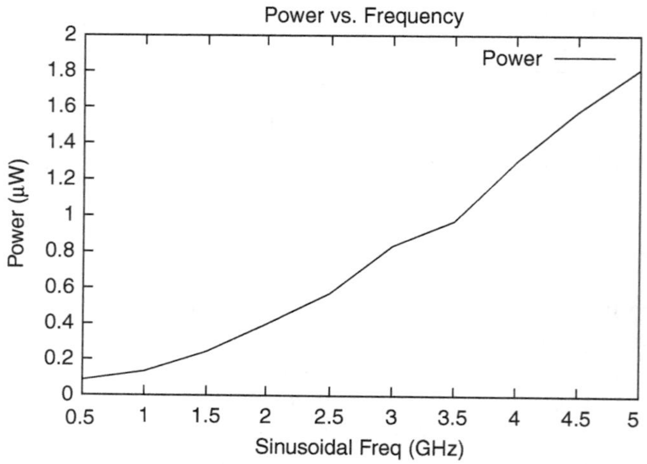

Fig. 20.9 Change in power with frequency

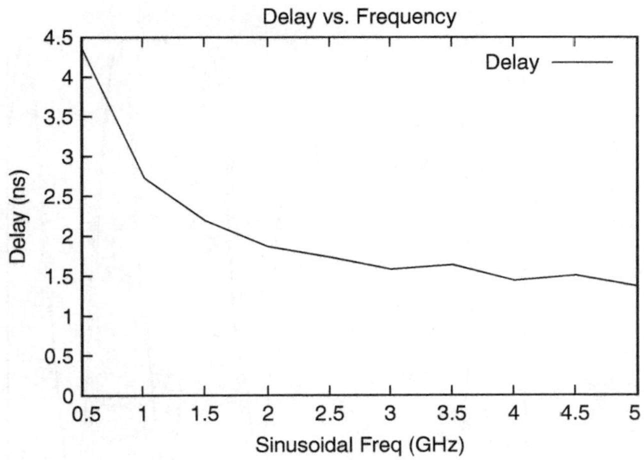

Fig. 20.10 Change in delay with frequency

The choice of frequency is also based on circuit performance requirements. For the rest of the discussion, we choose an operating frequency of 1 GHz. The next optimization explored is the choice of the R and C values in the low-pass filter of a gate. As discussed in Section 20.3.2.2 a large RC product increases the delay of computation, while a small RC product would cause the voltage of the capacitor to have a larger AC component. Mixing two 1 GHz sinusoids results in a DC component and a 2 GHz component. Thus the low-pass filter has to be designed to suppress the 2 GHz frequency signals. Realizing a large on-chip capacitance requires a large amount of active area. Thus area is minimized by realizing a larger resistance. A large resistance can be implemented in resistive poly-silicon, which offers a resistance of a few MΩ per square [2]. A 1 MΩ of resistance can therefore be practically realized on-chip with a small area footprint using resistive poly-silicon. The gate capacitance of the minimum size device in the 45 nm technology node is of the order of 0.1 fF [2]. Thus an RC product of about 1 ns can be achieved with very low area.

To be able to suppress a signal of frequency, 2 GHz we implement a low-pass filter with a corner frequency of $\frac{2}{3}$ GHz. The filter has a resistor of 1 MΩ. The capacitance C is realized as the sum of gate capacitance presented by output stage and an additional capacitor (in the form of gate capacitance), such that the RC product is 1.5 ns.

20.4.3 Gate Operation

We implemented and simulated (in HSPICE) simple gates based on the optimizations discussed. Simulation results of these gates are presented next. Figure 20.11 illustrates the operation of a simple buffer switching its output. The non-sinusoidal

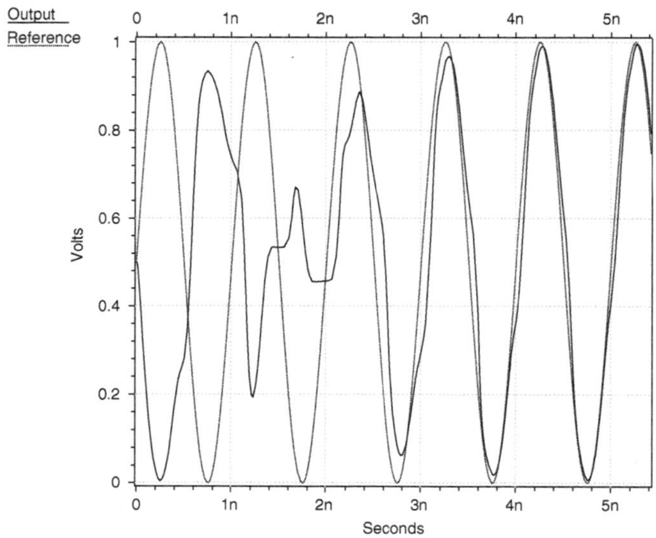

Fig. 20.11 Buffer operation

looking line represents the actual output of the buffer, while the sinusoidal line is the reference signal that the buffer is expected to drive out. The output signal takes a non-sinusoid form when the output of the correlator is switching due to a change in the input signal. The switching delay of the sinusoidal supply based inverter was about 2.55 ns.

Figure 20.12 presents the HSPICE simulation results from the operation of four NAND2 gates. Each of the NAND2 gates is driven by the four possible input combinations, respectively. Three of the four gates' outputs evaluate to the reference signal representing the logic value 1. The remaining gate evaluates to the reference signal representing the logic value 0, as shown in Fig. 20.12.

We now compare the power of sinusoid-based logic gates with traditional static CMOS gates. We present numbers from HSPICE simulations of a static CMOS inverter and a sinusoidal supply-based inverter, both implemented in a 45 nm PTM process. The static CMOS inverter is minimum sized, and sized to ensure equal rise and fall times, and is made of devices with same threshold voltage as the devices in the output stage of the sinusoid gate. The static CMOS inverter is operated with a 1 V supply. The power consumed by the static CMOS inverter over a 1 ns (1 GHz clock) duration (during which the inverter switches once) is 1.75 μW. The power consumed by a sinusoidal supply-based gate (with supply's oscillation frequency of 1 GHz) for 1 computation is 500 nW, which is lower than the static CMOS gate. At this operating point the sinusoidal supply-based inverter completes 1 computation in 2 ns (500 MHz of operation frequency) as opposed to 40 ps by a CMOS inverter. The leakage power of the static CMOS inverter at 1 V supply is 17 nW as opposed to 120 nW for a sinusoidal supply-based inverter. The high leakage power of sinusoidal supply-based inverter is due to the fact that the output stage switches from rail to

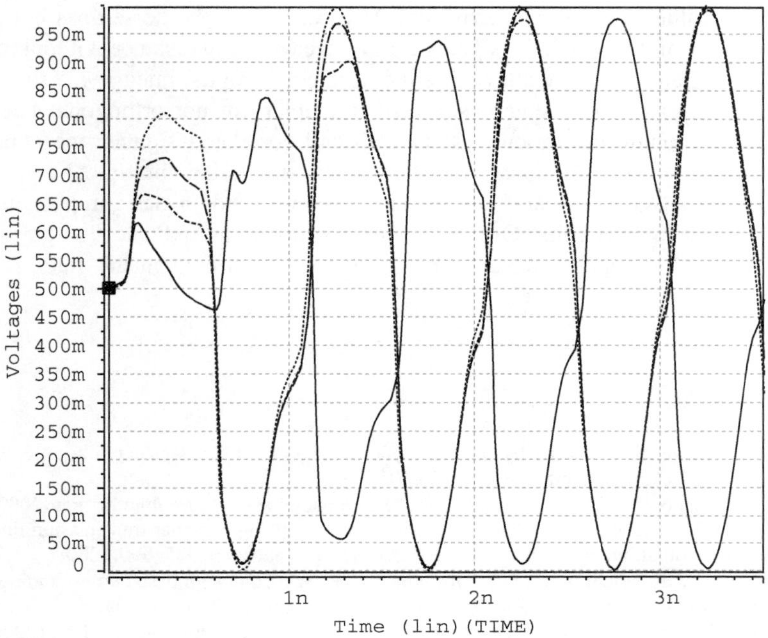

Fig. 20.12 NAND gate operation

rail even during steady state. Therefore the sinusoidal supply-based gates have a lower dynamic power (power per computation) and high leakage power, and are thus more suitable for implementing high activity circuits. Although the delay and the power-delay product (PDP) of the sinusoidal supply-based gates are larger than their CMOS counterparts (for binary valued gates), we expect that using multivalued sinusoidal supply-based gates would improve the PDP due to the possible reduction in logic depth. The bulk of the power consumed in the sinusoidal gate is in the output stage, due to the ON resistance of the passgates. If these passgates are implemented using relays, the power consumption of the sinusoidal gates can be dropped significantly. In this case, the dynamic and steady-state power of the sinusoidal gate are, respectively, $25X$ and $6X$ lower than the corresponding numbers for a static CMOS inverter. For our implementation, the oscillator power averaged over the gates powered by it is $0.05X$ of the gate power (during steady-state operation).

20.5 Conclusions

Sinusoidal supply-based logic could be a viable means to implement logic. Such circuits exhibit several advantages, being significantly immune to additive noise while elegantly implementing multivalued logic. Since sinusoids of any two frequencies are orthogonal, multiple sinusoidal signals can share the same wire and their logic values can still be recovered. Also, since additive noise does not affect the frequency of the signal, the signals could traverse a long distance without requiring to be buffered, thereby alleviating the problem of signal integrity. It should be noted that the two logic states used in this chapter are not orthogonal but anti-correlated. Though we present results with anti-correlated signals, the approach can be extended to orthogonal vector logic by employing multiple frequencies. We present design considerations in this chapter to aid in designing sinusoidal supply-based logic circuits. Our circuit-level simulations indicate that sinusoidal supply-based logic gates consume lower switching power than the static CMOS counterparts.

References

1. Zhao, W., Cao, Y.: Predictive technology model for nano-CMOS design exploration. J. Emerging Technol. Comput. System **3**(1), 1 (2007)
2. The International Technology Roadmap for Semiconductors. http://public.itrs.net/(2009)
3. Cordero, V.H., Khatri, S.P.: Clock distribution scheme using coplanar transmission lines. In: Design Automation, and Test in Europe (DATE) Conference, pp. 985–990. (2008)
4. Karkala, V., Bollapalli, K.C., Garg, R., Khatri, S.P.: A PLL design based on a standing wave resonant oscillator. In: International Conference on Computer Design (2009)
5. Kish, L.B.: Thermal noise driven computing. Applied Physics Letters **89**(144104) (2006)
6. Kish, L.B.: Noise-based logic: Binary, multi-valued, or fuzzy, with optional superposition of logic states, Physics Letters A **373**, 911–918 (2009)
7. Kish, L.B., Khatri, S., Sethuraman, S.: Noise-based logic hyperspace with the superposition of $2N$ states in a single wire. Physics Letters A **373**, 1928–1934 (2009). DOI: 10.1016/j.physleta.2009.03.059
8. Meta-Software, I.: HSPICE user's Manual. Meta-Software, Campbell, CA
9. O'Mahony, F., Yue, P., Horowitz, M., Wong, S.: Design of a 10 GHz clock distribution network using coupled standing-wave oscillators. In: DAC '03: Proceedings of the 40th Conference on Design automation, pp. 682–687. ACM (2003)
10. Peper, F.: Noise-Driven Computation and Communication. The Society of Instrument and Control Engineers (SICE), pp. 23–26. Japan, Catalog (2007)
11. Ueda, M., Ueda, M., Takagi, H., Sato, M., Yanagida, T., Yamashita, I., Setsune, K.: Biologically-inspired stochastic vector matching for noise-robust information processing. Physica **387**, 4475–4481 (2008)

Chapter 21
Improvements of Pausible Clocking Scheme for High-Throughput and High-Reliability GALS Systems Design

Xin Fan, Miloš Krstić, and Eckhard Grass

Abstract Pausible clocking-based globally asynchronous locally synchronous (GALS) systems design has been proven a promising approach to SoCs and NoCs. In this chapter, we study the throughput reduction and synchronization failures introduced by the widely used pausible clocking scheme and propose an optimized scheme for higher throughput and more reliable GALS system design. The local clock generator is improved to minimize the acknowledge latency, and a novel input port is applied to maximize the safe timing region for clock tree insertion. Simulation results using the IHP 0.13 μm standard CMOS process demonstrate that up to one-third increase in data throughput and an almost doubled safe timing region for clock tree insertion can be achieved in comparison to the traditional pausible clocking schemes. This work contributes to high-throughput and high-reliability GALS systems design.

21.1 Introduction

The ongoing technology scaling and device miniaturization enable novel and advanced applications. While the consumers can get more innovative and higher integrated products, today the chip designers are facing challenges. The requirements are continuously increasing in terms of more complexity, more performance, more mobility, less power, less cost, less time-to-market, etc. On the other hand, the new technologies are introducing even harder challenges for the design process including large process variability and problems with reliability and signal integrity. Consequently, application of the standard digital design flow based on the synchronous paradigm usually leads to problems to deliver the most complicated systems.

X. Fan (✉)
Innovations for High Performance Microelectronics, Frankfurt (Oder), Brandenburg, Germany
e-mail: fan@ihp-microelectronics.com

Based on Xin Fan; Krstic, M.; Grass, E.; "Analysis and optimization of pausible clocking based GALS design," Computer Design, 2009. ICCD 2009. IEEE International Conference on pp. 358–365, 4–7 Oct. 2009 [2009] IEEE.

S.P. Khatri, K. Gulati (eds.), *Advanced Techniques in Logic Synthesis, Optimizations and Applications*, DOI 10.1007/978-1-4419-7518-8_21,

One of the main issues is the system integration of complex high-performance digital design. It is not uncommon that the clock signal cannot even propagate from one side of the chip to the other, with acceptable clock skew, latency, and power consumption. It is hard to expect then the application of standard methods, architectures, and tools can give satisfactory results. Therefore, new design approaches appear, for example, in the direction of communication centric networks-on-chip (NoCs).

One alternative to the classical synchronous integration is the application of globally asynchronous locally synchronous (GALS) technology. This method, on one hand, reduces main system integration challenges and eases the timing closure by removing the global clock. On the other hand, the other design parameters are improved, including power consumption, noise profile, and in some cases also performance.

The most straightforward way to GALS systems is to insert synchronizer circuits between different clock domains [1]. Normally a synchronizer consists of two or more cascaded flops, and it introduces latency in the data transfer. Another approach to GALS systems is the use of asynchronous FIFOs [2, 3], and this results in overheads in both area and power. In recent years, an alternative method to GALS design, which is mainly based on pausible local clocks, has been developed [4–12]. Communication between asynchronous modules is achieved using a pair of request–acknowledge handshaking signals and the local clocks are paused and stretched, if necessary, to avoid metastability in data transfer. As a latest example, this scheme is applied in [13] to implement a dynamic voltage frequency scaling NoC. Figure 21.1 depicts a point-to-point GALS system based on this scheme and its handshaking waveforms. The design and evaluation of asynchronous wrappers are presented in detail in [14, 15].

21.2 Analysis of Pausible Clocking Scheme

21.2.1 Local Clock Generators

Generally speaking, the local clock generators used in pausible clocking-based GALS systems need to have the following three features: (a) create a clock signal with required clock period for each locally synchronous module, (b) stop the clock signals for transferring data safely across different clock domains, and (c) support concurrent transfer requests from multiple communication channels.

Figure 21.2 illustrates the typical structure of local clock generators applied in GALS design [8, 9, 13, 16]. A programmable delay line is employed to generate the clock signal *LClk*, and a Muller-C element is utilized to withhold the rising edge of the next clock pulse if any port request is acknowledged by the MUTEX elements. To support multi-port requests, an array of MUTEX elements is deployed in Fig. 21.2 to arbitrate between port requests *Reqx* and the request clock *RClk*.

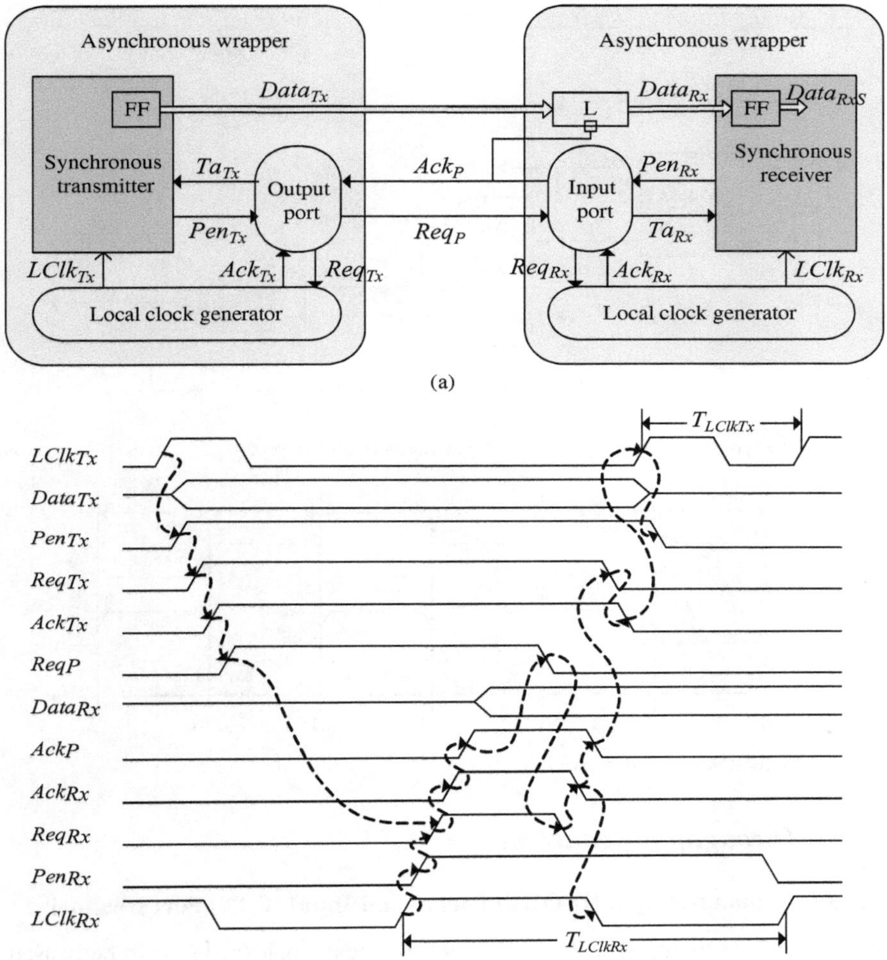

Fig. 21.1 A point-to-point GALS system (**a**) and its handshaking waveforms (**b**)

21.2.2 Clock Acknowledge Latency

We define the request–acknowledge window (RAW) in a local clock generator as the duration in each cycle of LCLK when port requests can be acknowledged. For the clock generator in Fig. 21.2, its RAW is the off-phase of RClk, as shown in Fig. 21.3. Considering the 50% duty cycle of LClk, the duration of RAW and the maximum acknowledge latency of the clock generator are deduced below:

$$t_{RAW} = t_{RCLK=0} \approx t_{LCLK=1} \approx T_{LCLK}/2 \qquad (21.1)$$

$$\max(Latency_{Ack}) = t_{RCLK=1} = T_{LCLK} - t_{RAW} \approx T_{LCLK}/2 \qquad (21.2)$$

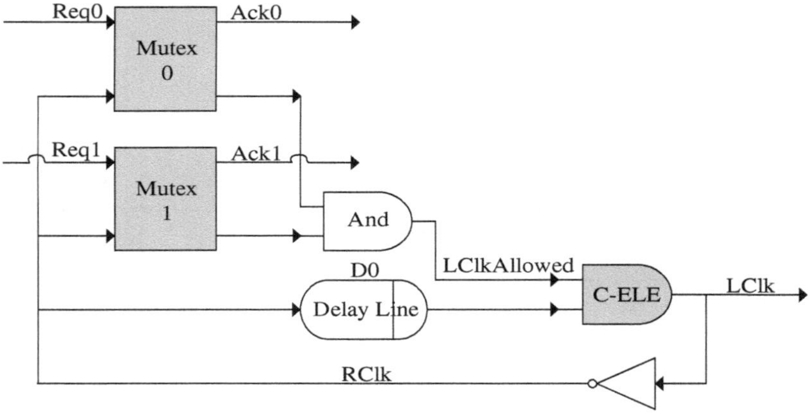

Fig. 21.2 The typical structure of local clock generators (with two ports)

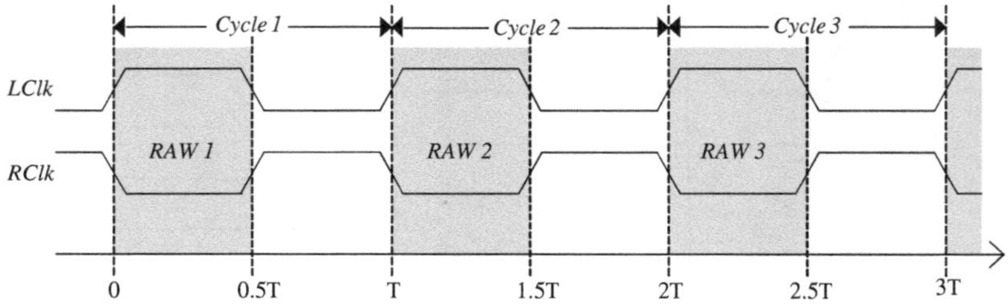

Fig. 21.3 Request acknowledged window

21.2.3 Throughput Reduction

21.2.3.1 Demand-Output (D-OUT) Port to Poll-Input (P-IN) Port Channel

For the receiver equipped with a P-IN port, the local clock $LClk_{Rx}$ will be paused after $Req_{Rx}+$ occurs [8]. Req_{Rx} will be asserted after a Req_P+ is detected, which is generated by the output port in the transmitter running at an independent clock. Therefore, without loss of generality, the arrival time of $\mathrm{Req_P}+$, and then the arrival time of $Req_{Rx}+$, can be modeled as a uniformly distributed random variable within a period of $LClk_{Rx}$. Since $t_{RAW} = T_{LCLKRx}/2$ in the above clock generator, there is 50% probability that a $Req_{Rx}+$ is extended to be acknowledged in the next *RAW*. Moreover, because the data is sampled in the receiver at the next rising edge of $LClk_{Rx}$, $Data_{RxS}$ will be delayed for one cycle of $LClk_{Rx}$.

For the transmitter equipped with a D-OUT port, its local clock $LClk_{Tx}$ is paused before Req_P is asserted by the output port, and $LClk_{Tx}$ will not be released until Req_P gets acknowledged [8]. Since Req_P will not be acknowledged by the input port until $Req_{Rx}+$ is acknowledged by the clock generator on the receiver side, there is maximum a $T_{LClkRx}/2$ latency in acknowledging Req_P as well. Consequently, the latency in the receiver is propagated into the transmitter. If the period of $LClk_{Rx}$ is much longer than that of $LClk_{Tx}$, this latency will result in a multi-cycle suspension in $LClk_{Tx}$. Since data is processed synchronously to $LClk_{Tx}$ in the transmitter, the suspension in $LClk_{Tx}$ eventually results in a delay in data transfer.

For instance, considering a D-OUT port to P-IN port communication channel, where the periods of $LClk_{Tx}$ and $LClk_{Rx}$ satisfy the following (21.3), and the requests in the transmitter, Req_{Tx}, are asserted every N cycles of $LClk_{Tx}$ as shown in (21.4):

$$(N-1) \cdot T_{LCLKTx} = \frac{3}{2} \cdot T_{LCLKRx} \tag{21.3}$$

$$T_{\mathrm{Req}Tx} = N \cdot T_{LCLKTx} \tag{21.4}$$

After Req_{Tx} is asserted in the transmitter, a Req_P+ will be generated by the output port controller. If a Req_P+ arrives in the receiver in the on-phase of $RClk_{Rx}$, it will not be acknowledged until $RClk_{Rx}$ turns low. Once $RClk_{Rx}-$ happens, which corresponds to $LClk_{Rx}+$ occurring, Req_P will be acknowledged and then $LClk_{Tx}$ will be released. As a result, the next $LClk_{Tx}+$ and $LClk_{Rx}+$ are automatically synchronized to occur at almost the same time. Then according to condition (21.3), all the following Req_P+ and $RClk_{Tx}+$ will occur simultaneously, and the extension in Req_P+ and the suspension in $LClk_{Tx}$ will appear periodically. As an example, Fig. 21.4 illustrates a waveform fragment in this case with $N = 7$.

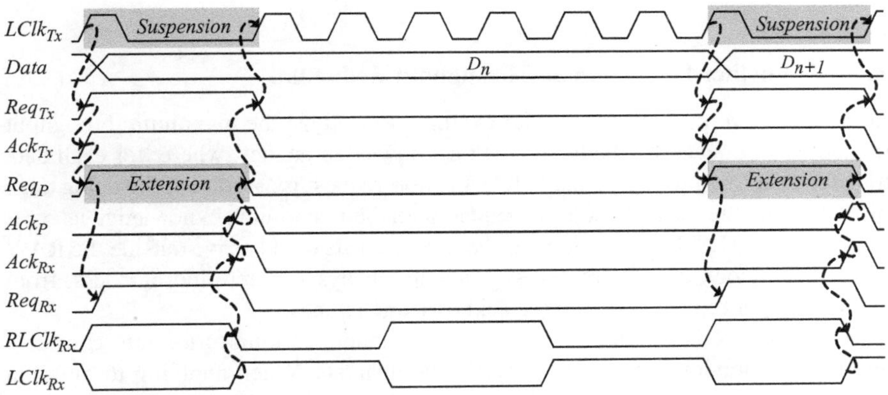

Fig. 21.4 Extension in $\mathrm{Req_P}$ and suspension in $\mathrm{LClk_{Tx}}$

Every time $LClk_{Tx}$ is suspended, its inactive phase will be stretched for a period of $(T_{LClkRx}/2-T_{LClkTx})$. The throughput reduction R_{Tx} due to the suspension of $LClk_{Tx}$ is deduced in (21.5), and with the increase in the value of N, we see the limit of R_{Tx} reaches one-third, as shown in (21.6):

$$R_{Tx} = (\frac{1}{2} \cdot T_{LCLKRx} - T_{LCLKTx})/T_{ReqTx} = \frac{N-4}{3 \cdot N} \tag{21.5}$$

$$\lim_{N \to +\infty} R_{Tx} = \lim_{N \to +\infty} \frac{N-4}{3 \cdot N} = \frac{1}{3} \tag{21.6}$$

21.2.3.2 Other Point-to-Point Channels

A similar analysis can be applied on the other three point-to-point communication channels. Table 21.1 presents the impacts of acknowledge latency on handshake signals and local clocks in four channels. Attentions need to be paid on D-OUT port to D-IN port channel. In this case, no matter whether there is data ready to be transferred or not, the clocks on both sides will be paused as soon as the ports get enabled. As a result, this channel is prone to unnecessarily very long suspensions in $LClk_{Rx}$ as well as $LClk_{Tx}$, and a significant throughput drop can be introduced.

Table 21.1 Impacts of acknowledge latency

Channel type	Extended signals	Suspended clock	Maximum suspension
D-OUT to P-IN	Req_P+	$LClk_{Tx}$	$T_{LClkRx}/2$
P-OUT to D-IN	Ack_P+	$LClk_{Rx}$	$T_{LClkTx}/2$
P-OUT to P-IN	Req_P+, Ack_P+	$LClk_{Rx}$	$T_{LClkTx}/2$
D-OUT to D-IN	NO	NO	0

21.2.3.3 Further Discussion on Throughput Reduction

In the above we present a particular example to analyze the maximum throughput reduction introduced by the local clock acknowledge latency, where, for each data transfer, according to (21.3) and (21.4), the port request is asserted at the rising edge of RCLK and the local clock is suspended with the maximum acknowledge latency. In practical GALS design, as long as the port requests could arrive outside the RAW of the local clock generators, the systems are always suffered, more or less, from the throughput reduction due to the acknowledge latency.

Another issue is we take only the single channel communication into consideration in the previous section to simplify the analysis. When applying to multiple channels concurrent communication systems, however, such latency may result in even more substantial throughput reduction. First, the acknowledge latency from clock generators could occur more frequently when port requests are asserted independently from different channels. Second, at the same time there could be a number of port requests extended and a number of local clocks suspended. Third, each local clock could be suspended for up to a half-cycle of the longest clock signal running in the GALS system, due to the propagation of the acknowledge latency across communication channels.

21.2.4 Synchronization Failures

21.2.4.1 $\Delta_{LClkRx} < T_{LClkRx}$

A benefit from GALS design is to simplify the globally distributed clock tree by a set of local clock trees. For the pausible clocking scheme, however, a crucial issue is to avoid synchronization failures caused by the local clock tree delays on receiver side [7, 12]. As the clock tree insertion delays are irrelevant to the handshake signals' propagation delays, $LClk_{RxDly}+$, can arrive at the sampling flip–flop FF simultaneously with loading data into the input port latch L. Then metastability occurs in FF. For the receiver clock tree delay satisfying $\Delta_{LClkRx} < T_{LClkRx}$, [12, 17] reveals that there are two timing regions in each cycle of $LClk_{Rx}$, as depicted in Fig. 21.5, where negligible synchronization failure probability can be expected.

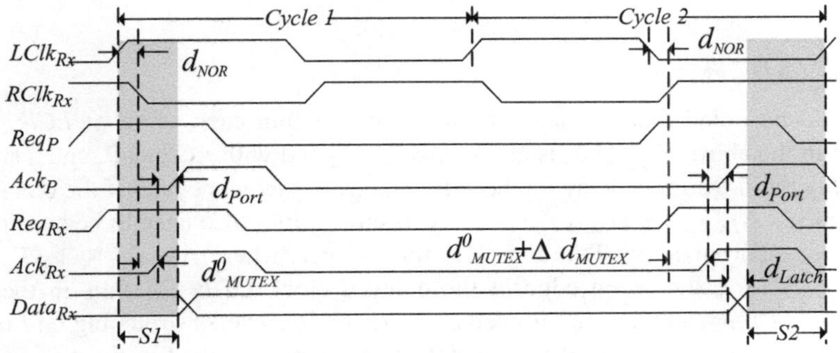

Fig. 21.5 Safe regions for $\Delta_{LClkRx} < T_{LClkRx}$

In Fig. 21.5, *Cycle* 1 illustrates the situation that the data is safely sampled by FF before L turns transparent. It contributes the safe timing region S1 of Δ_{LClkRx} as shown in (21.7), where $d^0{}_{MUTEX}$ and d_{Port} denote the MUTEX delay time from $Req_{Rx}+$ to $Ack_{Rx}+$ without metastability and the asynchronous port delay from $Ack_{Rx}+$ to Ack_P+

$$0 \le \Delta_{LCLKRx} < d_{NOR} + d^0_{MUTEX} + d_{Port} - t_{hold} \tag{21.7}$$

On the contrary, *Cycle* 2 represents another situation that data is safely sampled after it is latched in L. We draw the pessimistic case that $Req_{Rx}+$ happens concurrently with $RClk+$ and metastability occurs in the MUTEX. This leads to the safe timing region S2 shown in (21.8), where Δd_{MUTEX} denotes the additional delay of the MUTEX to resolve metastability, and d_{Latch} is the delays of L from asserting the gate enable Ack_P+ to data being stable

$$T_{LCLKRx}/2+(d_{NOR}+d^0_{MUTEX}+\Delta d_{MUTEX}+d_{Port}+d_{Latch})+t_{setup} \le \Delta_{LCLK} < T_{LCLKRx} \tag{21.8}$$

It can be seen in (21.8) that the width of safe timing region S2, W_{S2}, is actually dominated by the receiver clock period T_{LClkRx}. In below we analyze W_{S2} in two typical cases:

(a) If $T_{LCLKRx}/2 \approx d_{NOR}+d^0_{MUTEX}+\Delta d_{MUTEX}+d_{Port}+d_{Latch}+t_{setup}$ then $W_{S2} \approx 0$. It means that the MUTEX needs to consume half of a clock cycle to resolve metastability. As a result, only the safe region $S1$ is left for safely inserting clock tree in the receiver, which is rather narrow in width to be the propagation delay through a number of simple gates shown in (21.7).
(b) *If* $T_{LCLKRx}/2 \gg d_{NOR} + d^0_{MUTEX} + \Delta d_{MUTEX} + d_{Port} + d_{Latch} + t_{setup}$ then $W_{S2} \approx T_{LCLKRx}/2$. With the increase of T_{LClkRx}, W_{S2} is widened, but the hazard timing region is extended as well. With the increase in T_{LCLKRx}, the ratio of W_{S2} to T_{LCLKRx} reaches at most only 50%.

21.2.4.2 $\Delta_{LClkRx} \geq T_{LClkRx}$

It should be noted that the safe timing regions within each cycle of $LClk_{Rx}$, as shown in the above Fig. 21.5, is automatically aligned with $LClk_{Rx}+$, and a stretch in $LClk_{Rx}$ will lead to a delay on the safe regions of the next cycle. If the clock tree delay meets $\Delta_{LClkRx} \geq T_{LClkRx}$, this delay in safe regions also need to be considered for data synchronization. Take Fig. 21.6 for instance, where $\Delta_{LClkRx} \approx 1.8T_{LClkRx}$. During *Cycle 1*, the rising edge of the delayed clock $LClk_{RxDly}$ falls in the safe region of $LClk_{Rx}$, and data is sampled correctly by FF. Then a stretching on $LClk_{Rx}$ happens, and the safe regions in *Cycle 2* are delayed as well. But there is another $LClk_{RxDly}+$ scheduled in the clock tree, which arrives at FF without any delay. This eventually leads to the sampling conflict.

To analyze the potential stretching on $LClk_{Rx}$, the behavior of input port controllers used in the receiver has to be taken into account, as presented in below:

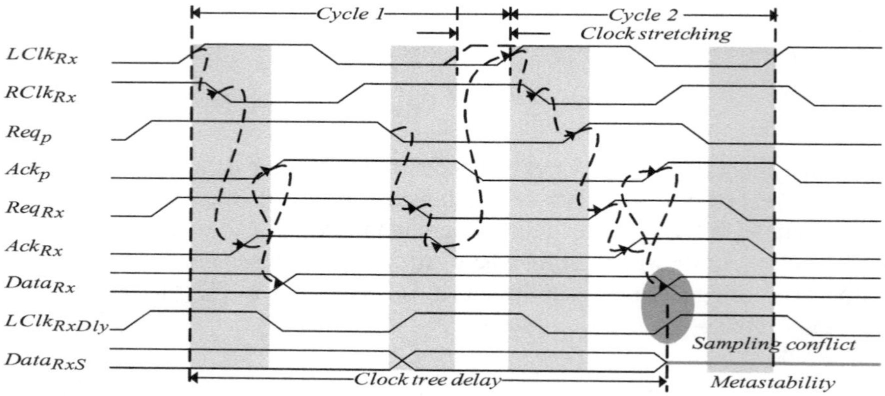

Fig. 21.6 An example of synchronization failures for $\Delta_{LCLKRx} > T_{LCLKRx}$

(a) P-IN port

In this situation, a maximum $T_{LClkTx}/2$ suspension on each cycle of $LClk_{Rx}$ could be introduced by the acknowledge latency as shown in Table 21.1. So the stretching on $LClk_{Rx}$, and then the delay of safe regions, is up to $(T_{LCLKTx}/2 - T_{LCLKRx})$. Since T_{LCLKTx} is independent from T_{LCLKRx}, this delay could be long enough to mismatch the safe regions of successive cycles of $LClk_{Rx}$, as illustrated in Fig. 21.7. Moreover, if $\Delta_{LCLKRx} > 2T_{LCLKRx}$, more than one cycle of $LClk_{Rx}$ could be stretched within the clock tree delay and an accumulated delay in safe regions should be considered.

(b) D-IN port

In this case, $LClk_{Rx}$ is paused and stretched by the input port controller until Req_P- is triggered by the output port controller. The stretching on $LClk_{Rx}$ as well as the delay in safe regions is unpredictable. That means no common safe region exists for the multi-cycle clock delay.

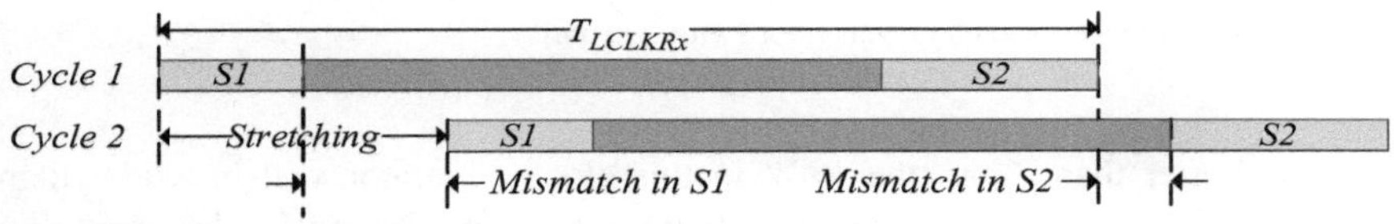

Fig. 21.7 Mismatch on safe timing regions due to clock stretching

Now we can conclude that, for the clock tree delay exceeding one clock period, clock stretching must be taken into account, and no matter what type of input port is utilized, the traditional scheme could hardly provide safe regions for clock tree insertion. In fact, in most of the reported pausible clocking systems, the local clock trees were deliberately distributed with $\Delta_{LClkRx} < T_{LClkRx}$ [14, 15, 18, 19].

21.3 Optimization of Pausible Clocking Scheme

21.3.1 Optimized Local Clock Generator

Figure 21.8 presents a simple solution to widen the *RAW* of local clock generators. Two delay lines, the programmable delay line $D0$ followed by the fixed one $D1$, which lengths are set in (21.9) and (21.10), are used to construct the ring oscillator. The request clock *RClk* is generated by an AND operation between *LClkB*, being the inverted signal of *LClk*, and L_0, the delayed version of *LClkB* after the programmable delay line $D0$.

$$d_{D0} = T_{LClK}/2 - (d_{D1} + d_{C\text{-}ELE} + d_{NOR}) \tag{21.9}$$

$$d_{D1} = d_{AND0} + d_{AND1} + (d^0_{MUTEX} + \Delta d_{MUTEX}) \tag{21.10}$$

The on-phase duration of *RClk* is the propagation delay through the path $MUTEX- > AND0- > C - ELEMENT- > INV- > AND1$. If such a

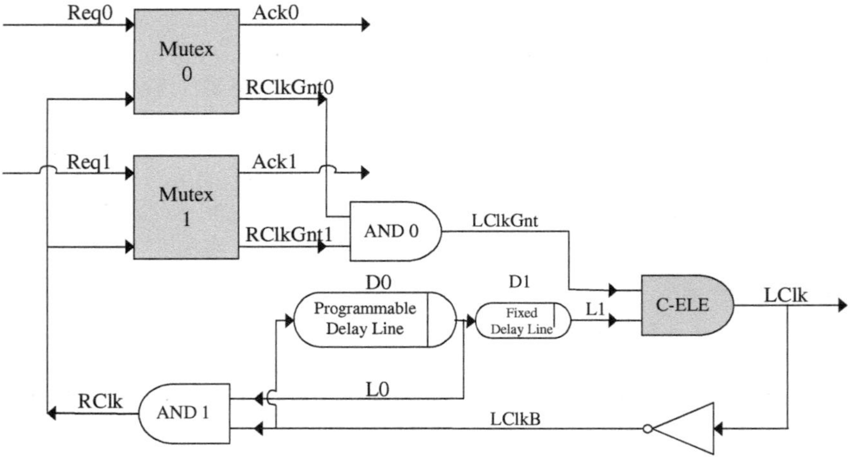

Fig. 21.8 An optimized local clock generator (with two ports)

delay is shorter than T_{LClk}, the *RAW* in this clock generator will be wider than that in Fig. 21.2. It is well known that there is no upper bound on the resolution time of the MUTEX elements [20]. A practical solution is to estimate the resolution time based on the mean time between failures (MTBF). From [1, 17], 40 FO4 inverter delays are sufficient for metastability resolution, i.e., for a MTBF of 10,000 years. Take the IHP 0.13 μm CMOS process, for instance, where d_{FO4} is 30 ps, and the resolution time of MUTEX in that technology can be estimated as $d_{MUTEX} = d^0_{MUTEX} + \Delta d_{MUTEX} \approx 40 d_{\mathrm{FO4}} \approx 1.2\,\mathrm{ns}$. Therefore, the length of $D1$ can be fixed at $50 d_{\mathrm{FO4}}$, including the delays of gates $AND0$ and $AND1$ according to (21.10). Based on the length of $D1$, the active phase of *RClk*, the duration of the *RAW*, as well as the acknowledge latency in the optimized clock generator are deduced in (21.11). It reveals that the duration of *RAW* is determined by the period of *LClk*. If $T_{LCLK} > 100 d_{\mathrm{FO4}}$, normally which represents the shortest clock cycle for standard cells based SoCs [17], the optimized clock generator provides a wider *RAW* than the traditional one:

$$\begin{aligned} &t_{RCLK=1} \approx d_{D1} + d_{C-ELE} + d_{NOR} \approx d_{D1} \\ &t_{RAW} = t_{RCLK=0} = T_{LCLK} - t_{RCLK=1} \approx T_{LCLK} - d_{D1} \,. \\ &\max(Latency_{Ack}) = t_{RCLK=1} \approx d_{D1} \end{aligned} \tag{21.11}$$

21.3.2 Optimized Input Port

21.3.2.1 Double Latching Mechanism

To widen the safe regions for the clock tree delay meeting $\Delta_{LCLKRx} < T_{LCLKRx}$, a double latching mechanism is proposed in Fig. 21.9. Since data latches $L1$ and $L2$ are enabled by the acknowledge signals of the MUTEX, there is only one latch

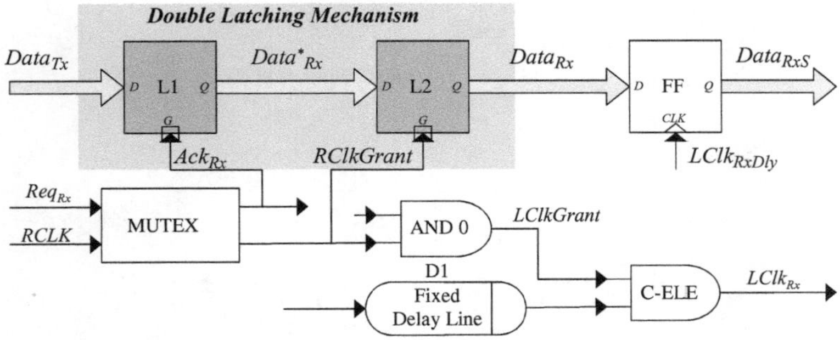

Fig. 21.9 Double latching mechanism

transparent at any time. Therefore, data is transferred by two mutually exclusive coupling latches in this scheme, instead of the single latch L in Fig. 21.1.

During the off-phase of *RClk*, *RClkGnt* remains low and $Data_{Rx}$ is locked in $L2$ stably. Any $LClk_{RxDly}+$ arriving at FF in the inactive phase of *RClk* samples $Data_{Rx}$ safely. If *RClk* turns high, *RClkGnt+* is triggered to load $Data^*_{Rx}$ into $L2$. Once $Req_{Rx}+$ occurs simultaneously with *RClk+*, *RClkGnt* will be asserted by the MUTEX in a random resolution time. Therefore, the safe timing region for clock tree distribution in this double latching mechanism is the off-phase period of *RClk*, which equals to its *RAW* duration as shown in (21.11).

Analyze the *RAW* duration in the following two typical cases:

(a) *If* $T_{LCLKRx} \approx 2d_{D1}$, *then* $W_S \approx d_{D1} \approx T_{LCLKRx}/2$. It occurs when d_{D0} is programmed to be 0 and only d_{D1} is valid. The optimized clock generator is then working as the traditional clock generator, while providing a half-cycle wide safe region.
(b) *If* $T_{LCLKRx} \gg d_{D1}$, *then* $W_S \approx T_{LCLKRx}$. With the increase of T_{LCLKRx}, the safe region is widened, but the hazard region is fixed to be d_{D1}. The percentage of W_S in T_{LCLKRx} will reach almost 100%, covering the entire clock period.

Comparing to the traditional structure analyzed in Section 21.2.4.1, this scheme provides a safe timing region with almost doubled width for clock tree insertion at any clock period, as demonstrated in Fig. 21.10.

21.3.2.2 Optimized Input Port Controller

An optimized signal transition graph of P-IN port controller, along with its logic synthesized with Petrify, is presented in Fig. 21.11 [21]. Once $LClk_{Rx}$ has been paused, which is indicated by $Ack_{Rx}+$ from the clock generator, the input port controller will assert both Ack_p and Ta_{Rx}, and then trigger $Req_{Rx}-$ to de-assert Ack_{Rx}, which signifies the release of $LClk_{Rx}$. The transitions are highlighted in Fig. 21.11a, and their delay time determines the on-phase period of Ack_{Rx} as well as the maximum stretching on $LClk_{Rx}$. Fig. 21.11b shows the longest delay path

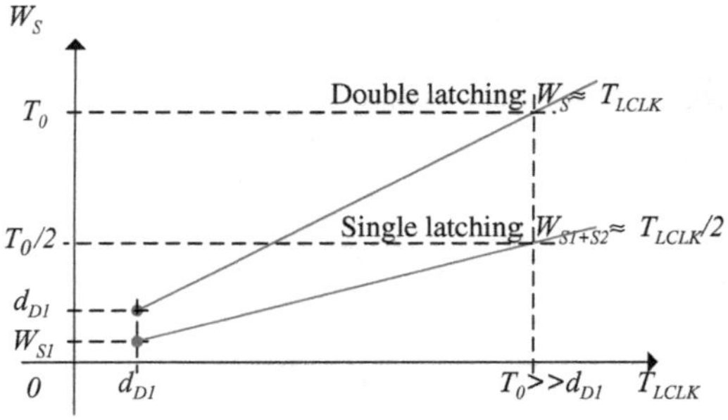

Fig. 21.10 Comparison of W_S in two kinds of latching mechanisms

from $Ack_{Rx}+$ to $Req_{Rx}-$, which consists of only 4 complex gates. So the stretching on $LClk_{Rx}$ introduced by the optimized controller is small and predictable, as shown below:

$$0 \leq \text{Stretching}_{LClkRx} \leq t_{AckRx=1} = d_{AckRx+\rightarrow ReqRx-} + d^0_{MUTEX} \tag{21.12}$$

We use n_{CT} to denote the maximum number of rising edges in the clock tree at one time. A common safe region among n_{CT} successive clock cycles is needed to insert clock tree. The location of the common safe region provided by the input port is shown in (21.13), and the width of safe regions for different n_{CT} is deduced in (21.14). A scenario for $n_{\text{CT}} = 3$ is depicted in Fig. 21.12 for example.

$$(n_{CT} - 1) \cdot (T_{LCLKRx} + t_{AckRx=1}) + t_{setup} < \Delta_{LCLKRx} < n_{CT} \cdot T_{LCLKRx} - d_{D1} - t_{hold} \tag{21.13}$$

$$W_S(n_{CT}) \approx (T_{LCLKRx} - d_{D1}) - (n_{CT} - 1) \cdot t_{AckRx=1} \tag{21.14}$$

21.4 Experimental Results

21.4.1 Input Wrapper Simulation

An asynchronous input wrapper consisting of the optimized clock generator and input port was designed at transistor level using the IHP 0.13 μm CMOS process. The delay slice as shown in [16], which delay is measured to be 0.13 ns in the experiment, is used to generate the delay lines. According to the analysis in Section 21.3.1, the delay of $D1$ is fixed to be 1.56 ns (12 delay slices), and the delay of $D0$ is programmed to be 0.52 ns (4 delay slices). Hence, the period of $LClk_{Rx}$ is 4.16 ns (240 MHz). In the input port controller, the on-phase duration of Ack_{Rx}, $t_{AckRx=1}$, which indicates the maximum clock stretching on $LClk_{Rx}$, is measured to be about 0.67 ns. Based on the above timing parameters, we derive the safe region widths from (21.14) for different clock tree depths n_{CT} as shown in Table 21.2. We

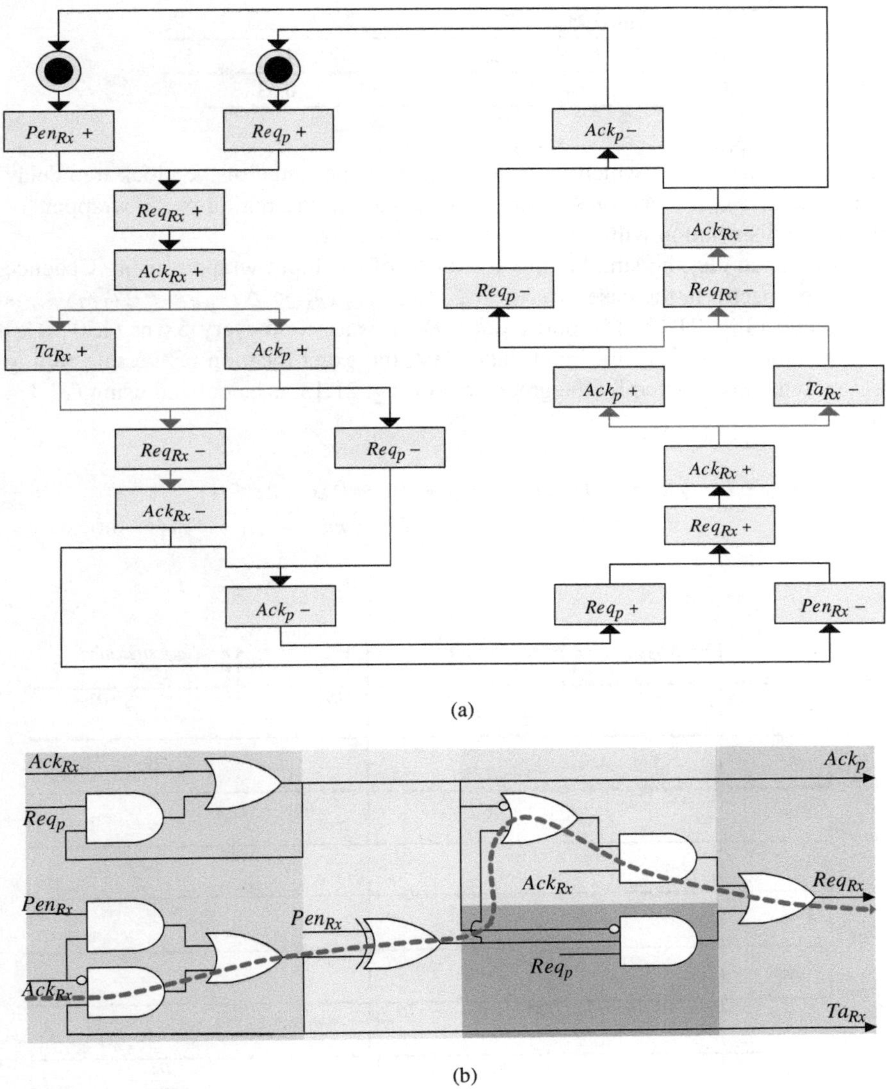

Fig. 21.11 An optimized P-IN port controller: (a) signal transition graph and (b) synthesized logic

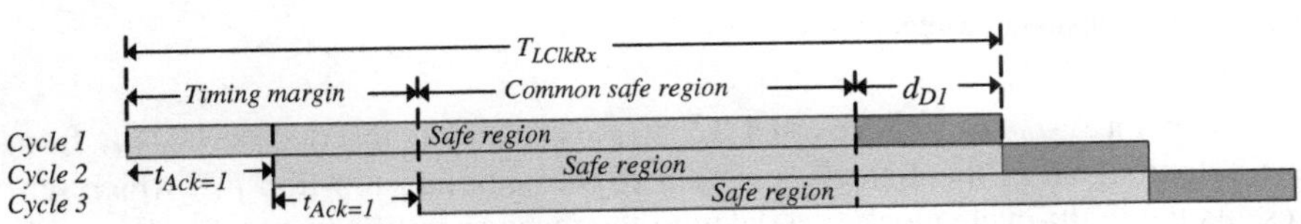

Fig. 21.12 An example of common safe region with $n_{\mathrm{CT}} = 3$

Table 21.2 n_{CT} and W_S ($T_{LClkRx} = 4.16$ ns)

n_{CT}	1	2	3	4	5
W_S (ns)	2.64	1.97	1.30	0.63	NA

see that, for $n_{CT} \leq 4$, which indicates a rather large range of the clock tree delay satisfying $\Delta_{LClkRx} < 4T_{LClkRx}$, there exists a safe timing region in the wrapper for clock tree distribution without synchronization failures.

As an example, the simulation waveforms of the input wrapper using Cadence Virtuoso Spectre in the case of $n_{CT} = 2$, i.e., $T_{LCLKRx} \leq \Delta_{LClkRx} < 2T_{LCLKRx}$, is presented in Fig. 21.13. The port request Req_P is asserted every 6.6 ns (150 MHz) in association with a 16-bit input data. First, the exact location of the safe timing region, which is covered by the green area in Fig. 21.13, is calculated using (21.13) in below:

$$4.95\,\text{ns} \approx T_{LCLKRx} + t_{AckRx=1} + t_{\text{setup}} = T_0 < \Delta_{LCLKRx} < T_1$$
$$= 2T_{LCLKRx} - d_{D1} - t_{\text{hold}} \approx 6.65\,\text{ns}$$

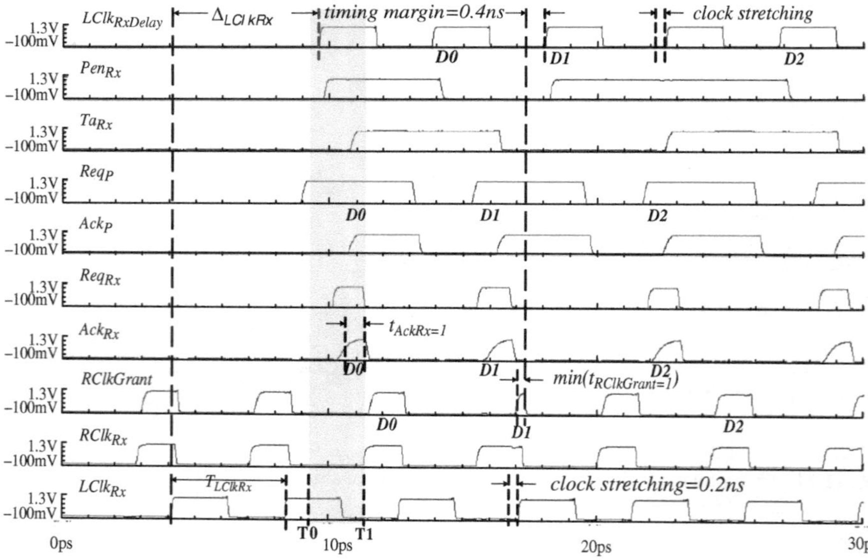

Fig. 21.13 Simulation waveforms with $n_{CT} = 2$

Second, a $\Delta_{LCLKRx} \approx 5.5\,\text{ns} \approx 1.32 \cdot T_{LCLKRx}$ delay is inserted on $LClk_{Rx}$ from the output signal of local clock generator to the sampling flip-flop FF, which represents the multi-cycle clock tree delay in the receiver falling inside the above safe region.

In Fig. 21.13 the handshaking signal waveforms when transferring three data items are illustrated. Each input data is first loaded in $L1$ when Ack_{Rx}=1, and then it is loaded in $L2$ when $RClkGnt = 1$, and finally it is sampled by FF at the next $LClk_{RxDly}+$. Even if $LClk_{Rx}$ is stretched accidentally and $RClkGnt+$ is delayed for $t_{AckRx=1}$, as shown in the transfer of data $D1$ in Fig. 21.13, there is a sufficient timing margin from loading data in $L2$ at $RClkGnt+$ to sampling data by FF at the next $LClk_{RxDly}+$. Therefore, safe data synchronization is achieved, and no additional latency in synchronization is required in the input wrapper.

21.4.2 Point-to-Point Communication

A point-to-point GALS system described in Section 21.2.3.1 is implemented at gate level to demonstrate the throughput increase from the optimized scheme. On the receiver side, the delay of the ring oscillator is fixed to be 12.48 ns, thus T_{LClkRx} is 24.96 ns, while on the transmitter side, the delay line is configured to generate a serial of clock periods T_{LClkTx}. At each value of T_{LClkTx}, the traditional scheme in Fig. 21.1 was first used to transfer 32 data items and then the proposed clock generator and input port were applied in the scheme running simulation for exactly the same duration. Table 21.3 presents the amount of data transfers accomplished using the optimized scheme and the percentage of improvement in data rate compared to the traditional scheme. It exhibits that the optimized scheme leads to much higher throughput and the increase becomes pronounced for the large value of N. As an example, Fig. 21.14 presents a waveform fragment for $N = 7$.

Table 21.3 Comparison in system throughput ($T_{\mathrm{LClkRx}} = 25$ ns)

T_{LClkTx} (ns)	9.36	6.25	5.72	4.16
N from Equation (21.3)	5	7	8	10
Throughput in Fig. 21.2	32	32	32	32
Throughput in Fig. 21.8	34	36	37	38
Improvement percentage	6.3%	12.5%	15.6%	18.8%
Improvement from Equation (21.5)	1/15	1/7	1/6	1/5

21.5 Conclusions

Pausible clocking-based GALS design has been widely applied in SoCs and NoCs design. In this chapter, we analyze the potential throughput reduction caused by the acknowledge latency of local clock generators and demonstrate the synchronization failures introduced by the uncertainty on clock stretching for multi-cycle clock tree insertion delays. Then the optimized pausible clocking scheme is proposed, where the local clock generator is first improved to minimize the acknowledge latency, and a novel input port is then proposed to maximize the safe region for clock tree delay. This work contributes to high throuhput and high reliability GALS systems design.

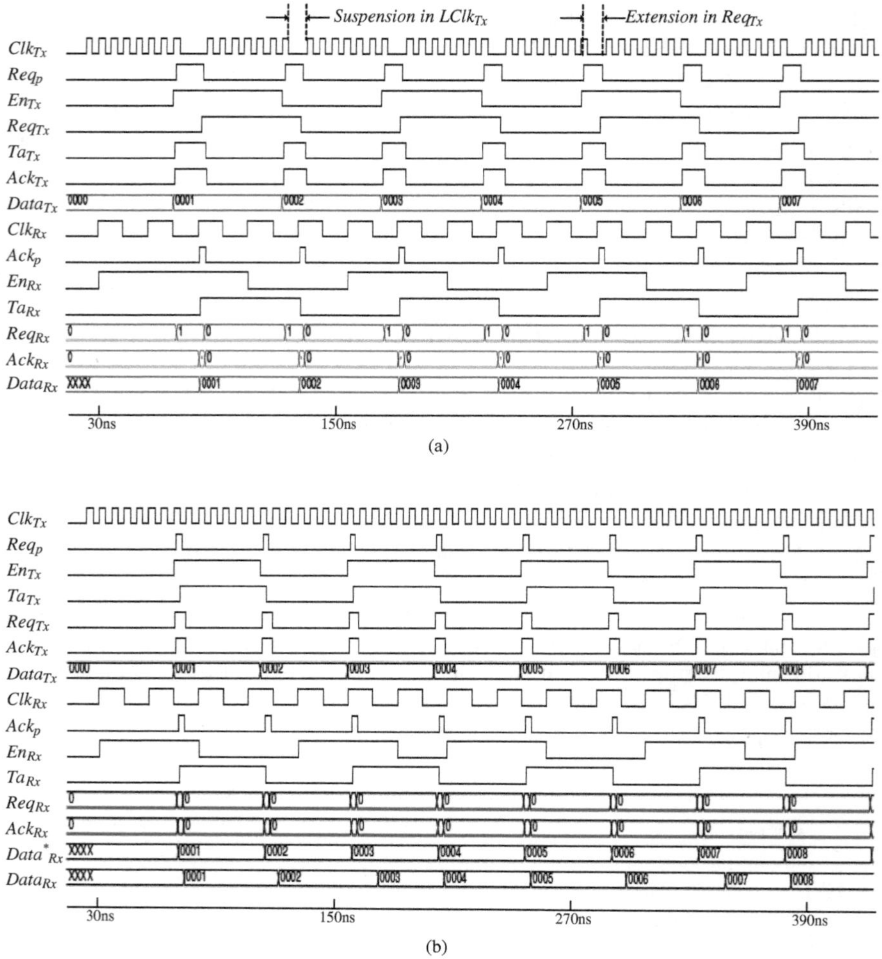

Fig. 21.14 Simulation waveforms (15–420 ns) (**a**) in traditional scheme and (**b**) in optimized scheme

Acknowledgments This work has been supported by the European Project GALAXY under grant reference number FP7-ICT-214364 (www.galaxy-project.org).

References

1. Ginosar, R.: Fourteen ways to fool your synchronizer. In: Proceedings of the International Symposium on Advanced Research in Asynchronous Circuits and Systems (ASYNC), pp. 89–96. Vancouver, BC, Canada (2003)
2. Greenstreet, Mark R.: Implementing a STARI chip. In: Proceedings of the International Conference on Computer Design (ICCD), pp. 38–43. Austin, TX, USA (1995)

3. Sutherland, I., Fairbanks, S. :GasP: a minimal FIFO control. In: Proceedings of the International Symposium on Advanced Research in Asynchronous Circuits and Systems (ASYNC) pp. 46–53, Salt Lake City, UT, USA (2001)
4. Chapiro, D.M.: Globally asynchronous locally synchronous systems. PhD thesis, Standford University, Stanford, CA (1984)
5. Yun, K.Y., Donohue, R.: Pausible clocking: A first step toward heterogeneous system. In: Proceedings of the International Conference on Computer Design (ICCD) pp. 118–123, Austin, TX, USA (1996)
6. Bormann, D.B., Cheung, P.: Asynchronous wrapper for heterogeneous systems. In Proceedings International Conference Computer Design (ICCD) pp. 307–314, Austin, TX, USA (1997)
7. Sjogren, A.E., Myers, C.J.: Interfacing synchronous and asynchronous modules within a high-speed pipeline. In Proceedings International Symposium Advanced Research in VLSI pp. 47–61, Ann Arbor, MI, USA (1997)
8. Muttersbach, J., Villiger, T., Fichtner, W.: Practical design of globally-asynchronous locally-synchronous systems. In: Proceedings of the International Symposium on Advanced Research in Asynchronous Circuits and Systems (ASYNC) pp. 52–59, Eilat, Israel (2000)
9. Moore, S., Taylor, G., Mullins, R., Robinson, P.: Point to point GALS interconnect. In Proceedings of the International Symposium on Advanced Research in Asynchronous Circuits and Systems (ASYNC) pp. 69–75, Manchester, UK (2002)
10. Kessels, J., Peeters, A., Wielage, P., Kim, S.-J.: Clock synchronization through handshaking. In: Proceedings of the International Symposium on Advanced Research in Asynchronous Circuits and Systems (ASYNC) pp. 59–68, Manchester, UK (2002)
11. Mekie, J., Chakraborty, S., Sharma, D.K.: Evaluation of pausible clocking for interfacing high speed IP cores in GALS framework. In: Proceedings of the International Conference on VLSI Design pp. 559–564, Mumbai, India (2004)
12. Dobkin, R., Ginosar, R., Sotirou, C.P.: Data synchronization issues in GALS SoCs. In: Proceedings of the International Symposium on Advanced Research in Asynchronous Circuits and Systems (ASYNC) pp. 170–179, Crete, Greece (2004)
13. Beigne, E., Clermidy, F., Miermont, S., Vivet, P.: Dynamic voltage and frequency scaling architecture for units integration within a GALS NoC. In: Proceedings of the International Symposium Networks-on-Chip (NoC) pp. 129–138, Newcastle upon Tyne, UK (2008)
14. Muttersbach, J.: Globally asynchronous locally synchronous architecture for VLSI systems. PhD thesis, ETH Zurich (2001)
15. Villiger, T., Kaslin, H., Gurkaynak, F. K., Oetiker, S., Fichter, W.: Self-time ring for globally-asynchronous locally-synchronous systems. In: Proceedings of the International Symposium on Advanced Research in Asynchronous Circuits and Systems (ASYNC) pp. 141–150, Vancouver, B.C., Canada (2003)
16. Moore, S. W., Taylor, G. S., Cunningham, P. A., Mullins, R. D., Robinson, P.: Self-calibrating clocks for globally asynchronous locally synchronous systems. In: Proceedings of the International Conference on Computer Design (ICCD) pp. 73–78, Austin, TX, USA (2000)
17. Rostislav (Reuven) Dobkin, Ran Ginosar, C.P. Sotiriou: High rate data synchronization in GALS SoCs. In: IEEE Transaction on VLSI Systems pp. 1063–1074, Volume: 14 Issue: 10 (2006)
18. Bormann, D.: GALS test chip on 130 nm process. In: Electronic Notes in Theoretical Computer Science, pp. 29–40, Volume 146, Issue 2, 2006
19. Gurkaynak, F. K.: GALS system design – side channel attach secure cryptographic accelerator. PhD thesis, ETH Zurich (2006)
20. Kinniment, D. J., Bystrov, A., Yakovlev, A.: Synchronization circuit performance. IEEE Journal of Solid State Circuits pp. 202–209, Volume 37, Issue 2 (2002)
21. Cortadella, J., Kishinevsky, M., Kondratyev, A., Lavagno, L., Yakovlev, A.: Petrify: A tool for manipulating concurrent specifications and synthesis of asynchronous controllers. IEICE Transaction on Information and Systems pp. 315–325, Volume: 80 (1997)

Subject Index

S.P. Khatri, K. Gulati (eds.), *Advanced Techniques in Logic Synthesis, Optimizations and Applications*, DOI 10.1007/978-1-4419-7518-8,

Kolibri 360 GmbH
Ferdinand-Jühlke-Straße 7
99095 Erfurt, Deutschland
produktsicherheit@kolibri360.de